Unconventional Oil and Gas Resources Handbook
Evaluation and Development

Unconventional Oil and Gas Resources Handbook
Evaluation and Development

Edited by

Y. Zee Ma
Schlumberger, Denver, CO, USA

Stephen A. Holditch
Texas A&M University, College Station, TX, USA

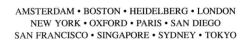

AMSTERDAM • BOSTON • HEIDELBERG • LONDON
NEW YORK • OXFORD • PARIS • SAN DIEGO
SAN FRANCISCO • SINGAPORE • SYDNEY • TOKYO

Gulf Professional Publishing is an imprint of Elsevier

Gulf Professional Publishing is an imprint of Elsevier
225 Wyman Street, Waltham, MA 02451, USA
The Boulevard, Langford Lane, Kidlington, Oxford, OX5 1GB, UK

Copyright © 2016 Elsevier Inc. All rights reserved.

No part of this publication may be reproduced or transmitted in any form or by any means, electronic or mechanical, including photocopying, recording, or any information storage and retrieval system, without permission in writing from the publisher. Details on how to seek permission, further information about the Publisher's permissions policies and our arrangements with organizations such as the Copyright Clearance Center and the Copyright Licensing Agency, can be found at our website: www.elsevier.com/permissions.

This book and the individual contributions contained in it are protected under copyright by the Publisher (other than as may be noted herein).

Notices
Knowledge and best practice in this field are constantly changing. As new research and experience broaden our understanding, changes in research methods, professional practices, or medical treatment may become necessary.

Practitioners and researchers must always rely on their own experience and knowledge in evaluating and using any information, methods, compounds, or experiments described herein. In using such information or methods they should be mindful of their own safety and the safety of others, including parties for whom they have a professional responsibility.

To the fullest extent of the law, neither the Publisher nor the authors, contributors, or editors, assume any liability for any injury and/or damage to persons or property as a matter of products liability, negligence or otherwise, or from any use or operation of any methods, products, instructions, or ideas contained in the material herein.

ISBN: 978-0-12-802238-2

Library of Congress Cataloging-in-Publication Data
A catalogue record for this book is available from the Library of Congress

British Library Cataloguing-in-Publication Data
A catalogue record for this book is available from the British Library

For information on all Gulf Professional Publishing publications
visit our website at http://store.elsevier.com/

Contents

List of Contributors vii
Preface xi

PART 1 GENERAL TOPICS

CHAPTER 1 Unconventional Resources from Exploration to Production 3
Y. Zee Ma

CHAPTER 2 World Recoverable Unconventional Gas Resources
Assessment 53
Zhenzhen Dong, Stephen A. Holditch, W. John Lee

CHAPTER 3 Geochemistry Applied to Evaluation of Unconventional
Resources 71
K.E. Peters, X. Xia, A.E. Pomerantz, O.C. Mullins

CHAPTER 4 Pore-Scale Characterization of Gas Flow Properties in Shale
by Digital Core Analysis 127
Jingsheng Ma

CHAPTER 5 Wireline Log Signatures of Organic Matter and Lithofacies
Classifications for Shale and Tight Carbonate
Reservoirs 151
*Y. Zee Ma, W.R. Moore, E. Gomez, B. Luneau, P. Kaufman, O. Gurpinar,
David Handwerger*

CHAPTER 6 The Role of Pore Proximity in Governing Fluid PVT Behavior
and Produced Fluids Composition in Liquids-Rich Shale
Reservoirs 173
Deepak Devegowda, Xinya Xiong, Faruk Civan, Richard Sigal

CHAPTER 7 Geomechanics for Unconventional Reservoirs 199
Shannon Higgins-Borchardt, J. Sitchler, Tom Bratton

CHAPTER 8 Hydraulic Fracture Treatment, Optimization, and Production
Modeling 215
Domingo Mata, Wentao Zhou, Y. Zee Ma, Veronica Gonzales

CHAPTER 9 The Application of Microseismic Monitoring in Unconventional
Reservoirs 243
*Yinghui Wu, X.P. Zhao, R.J. Zinno, H.Y. Wu, V.P. Vaidya,
Mei Yang, J.S. Qin*

CHAPTER 10	**Impact of Preexisting Natural Fractures on Hydraulic Fracture Simulation**	289
	Xiaowei Weng, Charles-Edouard Cohen, Olga Kresse	

PART 2 SPECIAL TOPICS

CHAPTER 11	**Effective Core Sampling for Improved Calibration of Logs and Seismic Data**	335
	David Handwerger, Y. Zee Ma, Tim Sodergren	
CHAPTER 12	**Integrated Hydraulic Fracture Design and Well Performance Analysis**	361
	Mei Yang, Aura Araque-Martinez, Chenji Wei, Guan Qin	
CHAPTER 13	**Impact of Geomechanical Properties on Completion in Developing Tight Reservoirs**	387
	S. Ganpule, K. Srinivasan, Y. Zee Ma, B. Luneau, T. Izykowski, E. Gomez, J. Sitchler	
CHAPTER 14	**Tight Gas Sandstone Reservoirs, Part 1: Overview and Lithofacies**	405
	Y. Zee Ma, W.R. Moore, E. Gomez, W.J. Clark, Y. Zhang	
CHAPTER 15	**Tight Gas Sandstone Reservoirs, Part 2: Petrophysical Analysis and Reservoir Modeling**	429
	W.R. Moore, Y. Zee Ma, I. Pirie, Y. Zhang	
CHAPTER 16	**Granite Wash Tight Gas Reservoir**	449
	Yunan Wei, John Xu	
CHAPTER 17	**Coalbed Methane Evaluation and Development: An Example from Qinshui Basin in China**	475
	Yong-shang Kang, Jian-ping Ye, Chun-lin Yuan, Y. Zee Ma, Yu-peng Li, Jun Han, Shou-ren Zhang, Qun Zhao, Jing Chen, Bing Zhang, De-lei Mao	
CHAPTER 18	**Monitoring and Predicting Steam Chamber Development in a Bitumen Field**	495
	Kelsey Schiltz, David Gray	
CHAPTER 19	**Glossary for Unconventional Oil and Gas Resource Evaluation and Development**	513
	Y. Zee Ma, David Sobernheim, Janz R. Garzon	
Index		527

List of Contributors

Aura Araque-Martinez
Weatherford International, Houston, TX, USA

Tom Bratton
Tom Bratton LLC, Denver, CO, USA

Jing Chen
College of Geosciences, China University of Petroleum, Beijing, China

Faruk Civan
Petroleum Engineering, University of Oklahoma, Norman, OK, USA

W.J. Clark
Schlumberger, Denver, CO, USA

Charles-Edouard Cohen
Production Operations Software Technology, Schlumberger, Rio de Janeiro, Brazil

Deepak Devegowda
Petroleum Engineering, University of Oklahoma, Norman, OK, USA

Zhenzhen Dong
Schlumberger, College Station, TX, USA

S. Ganpule
SPE Member, Denver, CO, USA

Janz R. Garzon
Schlumberger, Denver, CO, USA

E. Gomez
Schlumberger, Denver, CO, USA

Veronica Gonzales
Technology Integration Group TIG, Schlumberger, Denver, CO, USA

David Gray
Nexen Energy ULC, Calgary, AB, Canada

O. Gurpinar
Schlumberger, Denver, CO, USA

Jun Han
PetroChina Coalbed Methane Company Limited, Beijing, China

David Handwerger
Schlumberger, Salt Lake City, UT, USA

Shannon Higgins-Borchardt
Schlumberger, Denver, CO, USA

Stephen A. Holditch
Texas A&M University, College Station, TX, USA

T. Izykowski
Schlumberger, Denver, CO, USA

Yong-shang Kang
College of Geosciences, China University of Petroleum, Beijing, China; State Key Laboratory of Petroleum Resources and Prospecting, China University of Petroleum, Beijing, China

P. Kaufman
Schlumberger, Denver, CO, USA

Olga Kresse
Production Operations Software Technology, Schlumberger, Sugar Land, Texas, USA

W. John Lee
University of Houston, Houston, TX, USA

Yu-peng Li
EXPEC ARC, Saudi Aramco, Dhahran, Saudi Arabia

B. Luneau
Schlumberger, Denver, CO, USA

Jingsheng Ma
Institute of Petroleum Engineering, Heriot-Watt University, Edinburgh, UK

Y. Zee Ma
Schlumberger, Denver, CO, USA

De-lei Mao
PetroChina Coalbed Methane Company Limited, Beijing, China

Domingo Mata
Technology Integration Group TIG, Schlumberger, Denver, CO, USA

W.R. Moore
Schlumberger, Denver, CO, USA

O.C. Mullins
Schlumberger-Doll Research, Cambridge, MA, USA

K.E. Peters
Schlumberger, Mill Valley, CA, USA; Department of Geological & Environmental Sciences, Stanford University, Palo Alto, CA, USA

I. Pirie
Schlumberger, Denver, CO, USA

A.E. Pomerantz
Schlumberger-Doll Research, Cambridge, MA, USA

Guan Qin
University of Houston, Houston, TX, USA

J.S. Qin
Weatherford International, Houston, TX, USA

Kelsey Schiltz
Colorado School of Mines, Golden, CO, USA

Richard Sigal
Independent Consultant, Las Vegas, Nevada, USA

J. Sitchler
SPE Member, Denver, CO, USA

David Sobernheim
Schlumberger, Denver, CO, USA

Tim Sodergren
Alta Petrophysical LLC, Salt Lake City, UT, USA

K. Srinivasan
Schlumberger, Denver, CO, USA

V.P. Vaidya
Weatherford International, Houston, TX, USA

Chenji Wei
PetroChina Coalbed Methane Company Limited, Beijing, China

Yunan Wei
C&C Reservoirs Inc., Houston, TX, USA

Xiaowei Weng
Production Operations Software Technology, Schlumberger, Sugar Land, Texas, USA

H.Y. Wu
China University of Geosciences, Beijing, China

Yinghui Wu
China University of Geosciences, Beijing, China; Weatherford International, Houston, TX, USA

X. Xia
PEER Institute, Covina, CA, USA, Current address: ConocoPhillips, Houston, TX, USA

Xinya Xiong
Gas Technology Institute, Chicago, IL, USA

John Xu
C&C Reservoirs Inc., Houston, TX, USA

Mei Yang
Weatherford International, Houston, TX, USA

Jian-ping Ye
CNOOC China Limited, Unconventional Oil & Gas Branch, Beijing, China

Chun-lin Yuan
College of Geosciences, China University of Petroleum, Beijing, China

Bing Zhang
CNOOC China Limited, Unconventional Oil & Gas Branch, Beijing, China

Shou-ren Zhang
CNOOC China Limited, Unconventional Oil & Gas Branch, Beijing, China

Y. Zhang
University of Wyoming, Laramie, WY, USA

X.P. Zhao
ExGeo, Toronto, ON, Canada

Qun Zhao
Langfang Branch of Research Institute of Petroleum Exploration and Development, PetroChina, Hebei, China

Wentao Zhou
Production Product Champion, Schlumberger, Houston, TX, USA

R.J. Zinno
Weatherford International, Houston, TX, USA

Preface

While different definitions have been proposed for unconventional resources, we consider the subsurface hydrocarbon resources that are tight and must be developed using large hydraulic fracture treatments or methods to be unconventional reservoirs. These geological formations generally have very low permeability or high viscosity; they include tight gas sandstones, shale gas, coalbed methanes, shale oil, oil or tar sands, heavy oil, gas hydrates, and other low-permeability tight formations. On the other hand, conventional reservoirs are those that can be economically developed generally with vertical wellbores and without the use of massive stimulation treatments or the injection of heat.

Before the recent downturn in the oil industry, development of unconventional resources had a terrific run, especially in the North Americas. For example, the United States has been producing more oil than it had since the 1970s as a result of oil production from unconventional plays. While crude oil price went down dramatically in late 2014 and 2015, there may be a silver lining behind the recent price plunge. Historically, crude oil price has been very volatile, increasing and then decreasing suddenly, but the worldwide consumption of oil and gas keep increasing annually. The International Energy Agency has recently reported that the world oil demand is expanding at a fast pace and it was raising its estimate for demand growth in the long term. Also, the industry has historically been able to improve the efficiency of extracting hydrocarbons when facing the challenges of low prices. A downturn gives us some time to step back, review what has been done, and think about possible improvements and innovations.

Efficiency can be fostered by an integrated, multidisciplinary approach and innovative technologies for evaluation and development of unconventional resources. Large heterogeneity of unconventional formations and high extraction costs lead to substantial uncertainty and risk in developing these reservoirs. Multidisciplinary approaches in evaluating these resources are critical to successful development. This multidisciplinary approach is why we have assembled this handbook that covers a wide range of topics for developing unconventional resources, from exploration, to evaluation, to drilling, to completion and production. The topics in this book include theory, methodology, and case histories. We hope that these contents will help to improve the understanding, integrated evaluation, and effective development of unconventional resources.

Chapter 1 gives a general overview of how to evaluate and develop unconventional resources. It briefly presents a full development cycle of unconventional resources, includes exploration, evaluation, drilling, completion, and production. General methods, integrated workflows, and pitfalls used in these development stages are discussed.

Chapter 2 presents assessments of worldwide recoverable unconventional gas resources while characterizing the distribution of unconventional gas technically recoverable resources by integrating a Monte Carlo technique with an analytical reservoir simulator.

Chapter 3 deals with applications of geochemistry to unconventional resource evaluation. It overviews the current developments in the use of both organic and inorganic geochemistry to identify sweet spots in unconventional mudrock resource plays, and updates the understanding of kerogen structure. It also discusses strengths and weaknesses of current measurements, workflows for integrating geochemical and geomechanical measurements, and uses of three-dimensional basin and petroleum system models for characterizing unconventional plays.

Chapter 4 describes pore-scale characteristics of gas flow properties in shale by digital core analysis. The characterization of flow properties for shale gas reservoirs is important, but complex because of the tightness of pore space and the diverse chemical compositions of the matrix as well as nonDarcy flow. This chapter summarizes recent work on digital core analysis for flow characterization of shale gas.

Chapter 5 is an overview of wireline–log signatures of organic matter and presents lithofacies classifications for shale and tight-carbonate reservoirs. It presents several methods for lithofacies prediction using wireline logs. The classical petrophysical charts are integrated with statistical methods for lithofacies classification, through which wireline log signatures of organic shale are further highlighted.

Chapter 6 discusses the role of pore proximity in governing fluid Pressure-Volume-Temperature (PVT) behavior and produced fluids composition in liquids-rich shale reservoirs. It describes the shortcomings in conventional PVT models in predicting fluid behavior, and reviews the recent advances in describing the effects of pore proximity on storage and transport-related properties and the implications for well productivity and drainage area calculations. It also discusses the impact of pore proximity on fluid transport, well drainage areas, and long-term well performance.

Chapter 7 presents geomechanics for unconventional reservoirs. Geomechanical properties of the subsurface greatly influence both the drilling and hydraulic fracturing of a well in an unconventional reservoir. A mechanical earth model (MEM) is an estimate of the subsurface mechanical properties, rock strength, pore pressure, and stresses, and can be used to quantify the geomechanical behavior of the subsurface. These components in the MEM are described in this chapter, with an emphasis on applications to developing unconventional reservoirs.

Chapter 8 presents an overview on hydraulic fracture treatments and optimization, including fracture fluid selection, proppant selection, fracture design, estimating fracture properties, and completion strategies. It also discusses production modeling, and compares analytical and numerical modeling methods.

Chapter 9 first briefly presents the basic microseismic monitoring theory, data acquisition, and processing methods, and then focuses on the application of microseismic monitoring in unconventional reservoirs. Advances in the microseismic methodology have added value to existing datasets through interpretations of reservoir parameters. Microseismic data mitigates uncertainty during completions evaluation in unconventional reservoirs, which is critical as the economic challenges of development increase.

Chapter 10 discusses the impact of preexisting natural fractures on hydraulic fracture simulation. It first presents a hydraulic fracture network model that incorporates mechanical interaction between hydraulic fracture and natural fracture, and among hydraulic fractures. It then discusses fracture simulations for various configurations of natural fractures, the impact of the natural fractures on the hydraulic fracture network geometry, and the impact of the fracture geometry on proppant distribution in the fracture network.

Chapter 11 deals with an effective core sampling for improved calibration of logs and seismic data for unconventional reservoirs. Core data are commonly used to calibrate reservoir parameters with wireline logs. However, such a calibration is often suboptimal. This chapter discusses pitfalls in the calibration, and proposes guidelines for core sampling and adequate calibrations between various data in unconventional resource evaluation.

Chapter 12 discusses hydraulic fracture design and well performance analysis. The workflow presented in the chapter includes basic hydraulic fracture design, dynamic calibration of the design based on performance analysis, and completion optimization. The chapter gives an overview of fracture design, and discusses the well performance analysis in relation to the effective fracture geometry, and the various factors affecting well productivity. It also discusses optimal completion strategies and refracture treatments.

Chapter 13 discusses the impact of geomechanical properties on completion in developing tight reservoirs. Geomechanical properties generally vary across a field and have significant impact on the design of the completion. The chapter elaborates the impact of in situ stress on hydraulic fracture initiation, growth, connectivity, drainage, and well spacing. In the study, the MEM is integrated with hydraulic fracture and production modeling to optimize drainage strategy and reduce the development cost.

Chapters 14, 15, and 16 present case histories involving factors such as lithofacies, petrophysical analysis, reservoir modeling, and the development of tight gas sandstone reservoirs. Tight gas sandstones represent an important unconventional resource. Two schools of thought are discussed regarding the geologic control of tight gas sandstone reservoirs: continuous basin-centered gas accumulations and gas accumulation in low-permeability tight sandstones of a conventional trap. Differences and similarities between evaluating and developing tight-gas sandstone and shale reservoirs are also discussed.

Chapter 17 presents an overview of the evaluation and development of coalbed methane (CBM) reservoirs, with an example from Qinshui basin in Northern China. CBM is generated from methanogenic bacteria or thermal cracking of the coal. Most methane and other gases are retained in coal by adsorption, but free gas also occurs in fissures and pore systems. This chapter discusses CBM potential and producibility of coal bearing strata as these are strongly affected by the hydrogeological regime of formation waters and coal permeability.

Chapter 18 discusses the monitoring and prediction of steam chamber development in a bitumen field. Steam-assisted gravity drainage (SAGD) is an in situ thermal recovery method used to extract heavy oil and bitumen. The efficacy of SAGD depends on the presence of permeability heterogeneities. This study demonstrates how the integration of compressional and multicomponent seismic data using neural network is used to build a predictive model for steam chamber growth.

A glossary is presented, which contains the definitions of many terminologies commonly used in evaluating and developing unconventional resources.

To ensure the scientific standard of this book, the chapters were peer-reviewed along with the editorial review. We thank the reviewers for their diligent reviews as they have helped improve

Table 1 Reviewers of the Manuscripts	
Du, Mike C.	Psaila, David
Foley, Kelly	Prioul, Romain
Forrest, Gary	Royer, Jean-Jacques
Fuller, John	Sitchler, Jason
Hajizadeh, Yasin	Wei, Yunan
Han, Hongxue	Weng, Xiaowei
Handwerger, David	Yan, Qiyan
Higgins-Borchardt, Shannon	Zachariah, John
Liu, Shujie	Zhang, Xu
Marsden, Robert	Zhou, Jing
Moore, William R.	

the quality and depth of the manuscripts. Table 1 is a list of the technical reviewers. We also thank the authors who have worked diligently with the editors to produce this volume.

August 2015
Y. Zee Ma, Denver, Colorado, USA
Stephen A. Holditch, College Station, Texas, USA

PART 1

GENERAL TOPICS

CHAPTER 1

UNCONVENTIONAL RESOURCES FROM EXPLORATION TO PRODUCTION

Y. Zee Ma
Schlumberger, Denver, CO, USA

Evaluating and developing unconventional resources without using a multidisciplinary approach is analogous to the old tale about the blind men and the elephant. One grabs his long trunk, one touches his large ears, one pats his broad side, and each comes away with a totally different conclusion.

1.1 INTRODUCTION

Although some unconventional resources, such as heavy oil and oil sands, have been developed for some time, the large-scale production of oil and gas from deep unconventional resources, such as shale gas and shale oil, has occurred more recently. The success of the Barnett Shale in central Texas has led the way for a number of other successful shale plays in North America (Bowker, 2003; Parshall, 2008; Alexander et al., 2011; Ratner and Tiemann, 2014), including the Fayetteville, Haynesville, Marcellus, Woodford, Eagle Ford, Montney, Niobrara, Wolfcamp, and Bakken. While shale plays have been in the spotlight in the last decade or so, unconventional resources are much broader. It is worth systematically reviewing and classifying various hydrocarbon resources based on reservoir quality.

Conventional hydrocarbon resources typically accumulate in favorable structural or stratigraphic traps in which the formation is porous and permeable, but sealed by an impermeable layer that prevents hydrocarbon from escaping. These favorable subsurface structures possess migration pathways that link the source rocks to the reservoirs, and the formations have good reservoir quality and generally do not require a large-scale stimulation to produce hydrocarbon. Unconventional resources reside in tight formations, and are of lower reservoir quality and more difficult from which to extract hydrocarbons. On the other hand, unconventional resources are more abundant in the earth. The relative abundance of conventional and unconventional resources can be described by a triangle (Masters, 1979; Holditch, 2013), illustrated in Fig. 1.1. Unconventional resources represent a variety of geological formations, including tight gas sands, gas shales, heavy oil sands, coalbed methanes, oil shales, and gas hydrates. In fact, any hydrocarbon resources that are not conventional may be considered unconventional. For example, the tight basement formations that contain oil or gas can also be considered to be unconventional resources as they have lower permeability than the conventional reservoirs. Moreover, it is not always easy to classify a reservoir into a specific category. Some studies consider gas and oil shales, coalbed methane, and gas hydrates as unconventional while putting tight gas sandstones in the conventional category (Sondergeld et al., 2010). From the

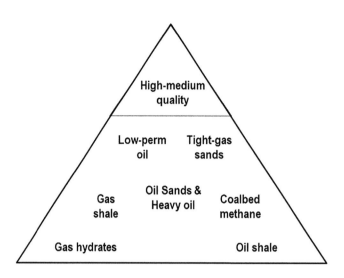

FIGURE 1.1

Resource triangle. *(Modified after Holditch. (2013).)* The high to medium-quality reservoirs in the upper region of the triangle are generally considered to be conventional resources; the other resources in the larger part of the triangle are considered to be unconventional reservoirs.

standpoint of reservoir characterization methodology, tight gas sandstone reservoirs (perhaps other tight formations as well) indeed share many characteristics of conventional reservoirs (Law and Spencer, 1989; Ma et al., 2011); they can be termed conventional unconventional resources or nonsource rock unconventional reservoirs, while shale reservoirs can be termed deep unconventional resources or source-rock unconventional resources. The same ambiguity can be applied to a specific reservoir; a shale formation may contain both oil and gas, or a hydrocarbon-bearing geological formation may contain a variety of lithofacies, both shale and other coarser-grained lithologies. Bakken oil-producing formations, for example, are mostly siltstones and other coarser-grained rocks that are sandwiched between the upper Bakken Shale and lower Bakken Shale (Theloy and Sonnenberg, 2013; LeFever et al., 1991).

The main characteristics of unconventional reservoirs include low-to-ultralow permeability and low-to-moderate porosity. As a result, hydrocarbon production from these reservoirs requires different extraction technologies than from conventional resources. An unconventional reservoir must be stimulated to produce hydrocarbons at an acceptable flow rate to recover commercial quantities of hydrocarbons. Permeability for unconventional reservoirs is mostly below 0.1 mD, and reservoirs with permeability above 0.1 mD are generally considered to be conventional. But this is not clear cut in practice because the permeability of a reservoir is never a constant value and the permeability heterogeneity of a reservoir, both conventional and unconventional, is generally high.

In most conventional reservoirs, shales are considered to be source rocks or seals as they have low permeability, but the Barnett Shale play has proved that shales can be important reservoirs. Shales are the most abundant sedimentary rock formations on Earth, but not all the shales are hydrocarbon plays, nor are all shale reservoirs made equal. There is no one unique geological or petrophysical parameter that can determine high hydrocarbon production, but two categories of variables are important: reservoir quality and completion quality (Miller et al., 2011; Cipolla et al., 2012).

Reservoir quality describes hydrocarbon potential, amount of hydrocarbon in place, and hydrocarbon deliverability of the rock formation. Completion quality describes stimulation potential or the ability to create and maintain fracture surface area. The important variables in reservoir quality include total organic carbon (TOC), thermal maturity, organic matter, mineralogical composition, lithology, effective porosity, fluid saturations, permeability, and formation pressure (Passey et al., 2010). Completion quality is highly dependent on geomechanical properties (Waters et al., 2011), and mineral composition of the formation, including rock fracturability, in situ stress regime, and the presence and characteristics of natural fractures (Weng et al., 2015).

Because of the tightness of unconventional formations, developing these resources is quite different from developing conventional reservoirs. Producing hydrocarbons from tight formations often requires not only drilling many more wells than developing a conventional reservoir, but also modern technologies, including horizontal drilling and hydraulic fracturing. The combination of horizontal drilling and multistage hydraulic fracturing has proven to be the key for economic production of hydrocarbons from many shale and other tight reservoirs. The important variables and parameters in these new technologies include horizontal well patterns, hydraulic fracturing design, stage count, and perforation clusters. Various well-spacing pilots are commonly used, and various hydraulic fracturing operation schemes are frequently tested for optimizing the well placement (Waters et al., 2009; Du et al., 2011; Warpinski et al., 2014).

As in other field development projects, uncertainty in developing unconventional resources is high because of the complex nature of subsurface formations. This is why more than ever, evaluating and developing unconventional resources requires an integrated, multidisciplinary approach. A multidisciplinary workflow can optimally integrate all the available information; geological, petrophysical, geophysical, and geomechanical variables can be incorporated into the workflow, enabling better characterization of reservoir properties, ranking critical parameters, optimizing production, and managing the uncertainty. An integrated workflow can be efficient in capturing the main characteristics of unconventional resources, and offers a quantitative means to effectively develop these fields. This requires study of many variables, including TOC, mineralogy, lithology, pore and pore throat geometry, anisotropy, natural fracture network, rock mechanical properties, in situ stress, structure and fault impact, and production interaction with the reservoir. Full characterization of these formations requires integrating cores, borehole images, well logs, seismic, and engineering data.

Without a truly integrated approach, prospecting and drilling in these fields are often somewhat subjective, with high uncertainty and risk. Three approaches can reduce the uncertainty in field development: acquiring adequate data, using better technologies, and using scientifically sound inference for interpreting various data and integrating different geoscience disciplines (Ma, 2011). The importance of data and technologies for reducing and managing uncertainty in field development is well known. Here we discuss the importance of inference in integrating various data for analyzing production data and reservoir properties.

Reservoir quality, completion quality, and completion effort for developing unconventional resources are multidimensional or composite variables that are made up of several other physical variables, such as kerogen, TOC, porosity, fluid saturation, and permeability for reservoir quality; stress, mineralogy, and mechanical properties for completion quality; and lateral length, proppant tonnage, and stage count for completion effort. Although these variables all impact production, a weak or no correlation, or even a reversed correlation, between production and such a variable, is frequently observed (Jochen et al., 2011; Gao and Du, 2012; Zhou et al., 2014). For example, lateral length or stage length and production rate may appear to have no, weak, or negative correlation (e.g., see Figures 12–17 in Miller et al., 2011). Similarly, one principle in selecting a hydraulic fracturing zone based on completion quality is to target the rocks that have a high Young's modulus and a low Poisson ratio, but what about rocks that exhibit simultaneously high Young's modulus and high Poisson ratio as in some cases, the Young's modulus is positively correlated with the Poisson ratio. This counterintuitive phenomenon, typically ignored in both evaluation and production data analyses, is related to a statistical phenomenon termed the Yule–Simpson's effect or Simpson's paradox (Ma et al., 2014). The crux of Simpson's paradox lies in interactions among many variables in a complex system (Simpson, 1951; Paris, 2012; Pearl, 2010; Li et al., 2013; Ma and Zhang, 2014); sensitivity analysis based on experiments often tests the effect of a variable while keeping all the other variables constant, but often, all the relevant variables change simultaneously and are correlated in real data. Without understanding the Yule–Simpson's effect in analyzing the relationships in a multivariate and nonlinear system, it is easy to make incorrect interpretations and decisions. An integrated approach is to understand the interaction of all the concerned variables in reservoir quality and completion quality so that the hydraulic fracturing treatment (HFT) is designed optimally for achieving the best economics.

Therefore, the evaluation of reservoir quality and completion quality must include evaluations of individual physical variables as many as possible. In practice, because of the limited availability of data, it is quite common to use a proxy to estimate reservoir quality, often based on correlation analysis. One of the pitfalls is to determine the overall reservoir quality using only a single or a limited number of physical variable(s). Some scientists call this type of reasoning atomistic inference, or seeing the forest from the trees (Gibbons, 2008; Lubinski and Humphreys, 1996). Although this type of inductive inference is common, and sometimes may be the only way of reasoning, geoscientists should attempt to mitigate the inferential bias. For example, does high kerogen content imply high reservoir quality in a shale formation? Although kerogen is an important component for evaluating reservoir quality for a shale reservoir, high kerogen content does not necessarily imply high reservoir quality. In fact, for a given TOC, high kerogen content implies a low maturity, and a small amount of oil or gas generated, as illustrated in Table 1.1. Conversely, low kerogen content may be a result of a significant amount of conversion of the original kerogen into oil or gas.

Table 1.1 Composition of Organic Carbon Content in Sedimentary Rock

Total Organic Carbon (TOC)		
EOM carbon	**Convertible carbon**	**Residual carbon**
Oil/gas	Kerogen	

EOM, extractable organic matter.
After Jarvie (1991).

Another pitfall is somewhat opposite to the atomistic inference—drawing inferences for individual reservoirs based on the aggregated statistics. For example, as multiple studies from different basins show a positive correlation between kerogen content and reservoir quality in unconventional formations, researchers may use the kerogen content as a proxy for reservoir quality. This may lead to an ecological inference bias (Robinson, 1950; Ma and Zhang, 2014). Although the kerogen content and amount of in-place hydrocarbon may be positively correlated using aggregated data, they are causally correlated inversely; as a matter of fact, when kerogen matures, the amount of kerogen in the formation decreases, and in-place hydrocarbon increases (Table 1.1).

In short, no single physical variable determines the quality of an unconventional reservoir, but many variables are important. Therefore, the evaluation of unconventional formations should use an integrated and multidisciplinary approach that includes studies of all the important parameters, whenever they are available, while considering their scales, heterogeneities, and relationships.

Unconventional resource development includes exploration, evaluation, drilling, completion, and production. Several geoscience disciplines are used in these processes, including petroleum system analysis, geochemistry, sequence stratigraphic interpretation, sedimentary analysis, seismic surveys, wireline log evaluation, core analysis, geomechanical study, petroleum and reservoir engineering, production data analysis, and many others. Table 1.2 presents important methods and parameters in evaluating and developing unconventional resources, especially those suitable for developing shale reservoirs. Development of some other unconventional resources, such as oil sands and tight gas sandstones, may use different evaluation methods, but most methods and parameters listed in the table can still be useful.

1.2 EXPLORATION AND EARLY APPRAISAL

Exploration strategies may be different for different unconventional resources; but for the most part, unconventional resources are found in basin-filling environments. In some cases, exploring a tight sandstone reservoir may be somewhat similar to exploring a conventional clastic reservoir, wherein source rock may or may not be proximal to the reservoir (Spencer, 1989; Law, 2002; Shanley, 2004). For shale reservoirs, although it is easy to find shale formations, identifying reservoir-quality shales can be difficult. Although shales are the most abundant sedimentary rocks on Earth, not all shales are hydrocarbon plays. Judging whether a given shale formation has a sufficient amount of hydrocarbon-bearing rocks requires detailed interpretations of geological, geophysical, geochemical, and engineering data, and integration of these disciplines. In particular, two fundamental analyses for exploration are petroleum system analysis and prospect assessment by integrating various data of different scales, including cores, logs, seismic data, and regional geological studies (Suarez-Rivera et al., 2013; Slatt et al., 2014).

1.2.1 PETROLEUM SYSTEM ANALYSIS AND MODELING

Petroleum system analysis can help understand the evolution of a sedimentary basin. Petroleum system modeling sheds light on the source and timing of hydrocarbon generation, migration routes, gas-in-place, maturation, and hydrocarbon type. These include kerogen quality and quantity,

Table 1.2 Methods and Important Parameters in Evaluating and Developing Shale and Other Tight Reservoirs

Development Stage	Exploration	Evaluation	Drilling and Completion	Production
Disciplines and methods	Petroleum system analysis,	Core, wireline logging,	Horizontal and vertical wells,	Production logging,
	Regional geology, sequence stratigraphy, core, logging	Seismic attributes, seismic inversion	Geosteering, well placement, hydraulic fracturing	Production data analysis and optimization
	Geochemistry	Geomechanics	LWD, MWD	Diagnosis
	Seismic survey	Integrated studies	Microseismic	Simulation
Issues and important parameters	Kerogen	Mineralogy, lithofacies	Drilling fluids	Calibrate reservoir and completion drivers to production,
	TOC	Porosity	Wellbore stability	Optimization
	Maturity	Fluid saturations	Well spacing	Refracturing
	Vitrinite reflectance	Matrix permeability,	Lateral length	Artificial lift
	Transformation ratio	Natural fractures, Fracture permeability,	Proppant type and quantity	Flowback management
	Adsorption	Pore pressure	Stage count	Water management,
	Expulsion/retention	Stress	Perforation clusters	Tracer
	Depositional environment	Anisotropy	Fracture complexity,	Leakoff
	Stacking patterns	Brittleness	Fracture network conductivity,	Feedback for evaluation,
	Reservoir continuity	Young's modulus	Fracture monitoring,	Feedback for future drilling and completion,
	Sweet spots	Poisson ratio	Water supply	Environmental considerations
	Seismic attributes		Environmental considerations	

LWD, logging while drilling; MWD, measurement while drilling

hyrocarbon generation, timing, migration, retention, transformation, thermal maturation, and TOC. Petroleum system analysis is important for both conventional and unconventional reservoirs to assess exploration risk before investment, but it is especially important for unconventional reservoirs found in a basin-filling environment.

1.2 EXPLORATION AND EARLY APPRAISAL

Important processes in petroleum system analysis include the following (Momper, 1979; Peters, 1986; Peters et al., 2015):

- Three main stages of evolution of sedimentary rocks include: (1) diagenesis in which no significant thermal transformation of organic matters occurs; (2) catagenesis in which organic matters are under thermal transformation and converted into hydrocarbon; and (3) metagenesis in which organic matters are under thermal transformation and the converted hydrocarbon may be overly cooked or destroyed.
- Organic-rich source rocks are exposed to heat during burial.
- Kerogen becomes partially converted to oil or gas.
- Converting kerogen into hydrocarbon often causes increased porosity and pressure within the source rock.
- When the pressure significantly exceeds geostatic pressure, the rock ruptures, oil and gas are expelled, the fractures close, and the source rock returns to near-pregeneration porosity and pressure.
- The above processes can be repeated in the formations as the source rock passes through the oil and gas-generation maturity.
- Oil expulsion is inefficient and a large proportion of the generated oil stays in the source rock, leading to shale reservoirs; the oil can be further converted first to condensate, then to wet gas and, ultimately, to dry gas and graphite if heat source is sufficient and temperature continuously rises.
- It is possible that the pressure of source rocks is higher than geostatic pressure, but not high enough for oil and gas to expel from the source rock, which causes overpressure in the source rock formation.

In shale gas and oil reservoirs, source rocks typically are also the reservoirs. Kerogen content is often an important variable to analyze reservoir quality. However, kerogen is only the basic material for generating hydrocarbons. Its conversion into hydrocarbon requires a long geological process for thermal maturation, under high temperature and high pressure. The amount of gas or oil generated is determined by the kerogen type and the heating rate, which can be estimated using kinetic modeling (Tissot and Welte, 1984). In some cases, hydrocarbon saturation is highly correlated to kerogen content, but not always, because low present-day kerogen content may either indicate low original kerogen during deposition, or partial conversion of high kerogen content that has thermally matured. These two cases can be distinguished with petroleum system modeling. Causally, the high kerogen content would mean that source rock is good; but it requires thermal maturation for the organic matters to enter the oil or gas window. Measuring kerogen, TOC, and maturity is necessary, and a number of methods have been proposed (Herron et al., 2014; Peters et al., 2015).

Kerogen type and its evolution into hydrocarbon can be described by a van Krevelen diagram using hydrogen and oxygen indices (Fig. 1.2(a)). Hydrocarbon generation takes place as kerogen is under thermal transformation in burial conditions (Fig. 1.2(b)). In the thermal transformation of kerogen to hydrocarbon, the rate of maturation depends on the type of kerogen, and heat and pressure related to burial (Fig. 1.2(b)). Migration of hydrocarbon typically occurs in conventional formations. But mature organic-rich shales that have not expelled all their oil are termed oil-bearing shales, such as the Bakken, Monterey, and Eagle Ford formations. The organic-rich shales that are either more thermally mature or of a different kerogen type contain gas instead of oil, such as the Barnett, Fayetteville, and Marcellus formations.

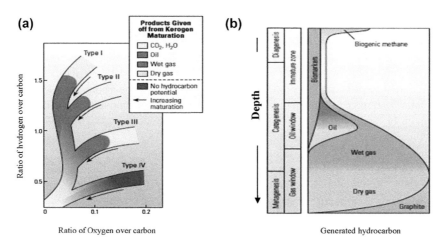

FIGURE 1.2

(a) van Krevelen diagram that describes kerogen types in relation to hydrogen and oxygen indices, and various types of hydrocarbon (oil, wet gas, dry gas), *(adapted from Allix et al. (2011))*. (b) Thermal transformation of kerogen into different types of hydrocarbon, *(adapted from Boyer et al. (2006))*.

Marine and lacustrine kerogens (Type I and II) tend to generate oil, but when the kerogen is overmature, gas is generated instead, especially for Type II (Passey et al., 2010; Peters et al., 2015). Type III kerogen, mostly from plant-generated organic materials, is more prone to gas or condensate generation. Type IV kerogen generally represent residual organic matters with low hydrogen index and little capacity for hydrocarbon generation.

Petroleum system modeling can aid the estimation of hydrocarbons in place. Kerogen contains carbon and other elements. As kerogen matures, carbon concentration increases. TOC is an important indicator for hydrocarbon resource base, and TOC from kerogen conversion can be estimated by the following equation (Tissot and Welte, 1984):

$$\text{TOC} = \frac{\text{Kerogen}}{Kvr} \times \frac{d_k}{d_b} \quad (1.1)$$

where TOC and kerogen are in weight percentage, d_k is the kerogen density, d_b is the bulk density, and Kvr is a maturity constant that can be represented by vitrinite reflectance. The latter depends on the type of kerogen and stage of maturation (diagenesis, catagenesis, and metagenesis; see, for example, Tissot and Welte, 1984). Another important parameter for the kerogen-hydrocarbon conversion is the transformation ratio (TR), which differs from vitrinite reflectance in terms of kinetic response (Peters et al., 2015).

1.2.2 RESOURCE PROSPECTING AND RANKING

Exploring for unconventional resources includes studies of regional geology and integrated analysis using core, log, seismic and geological data. In exploring shale reservoirs, quality screening is important since all the source rocks are not reservoir-quality plays (Evenick, 2013; Wang and Gale, 2009).

Table 1.3 Early Resource Play Assessment: Disciplines, Methods, and Parameters

Domain	Methods or Properties	Parameters and Descriptions
Charge	Source rock	Richness, thickness
	Maturity	Vitrinite reflectance
	Timing	Formation age, timing of maturation
	Seal and migration	Seal thickness, capillary entry pressure, migration pathway, and network
	Trap	Hydrocarbon trapping mechanism
	Preservation	Was the generated oil or gas preserved? Possible conversion of oil into gas? Migration into neighboring formations?
Reservoir properties	Depositional environment	Shoreface, channelized, bar-dominant, estuary, lacustrine, marine, and others
	Lithology, lithofacies	Mineralogical composition, lithofacies proportion, especially nonclay proportion
	Porosity	Average porosity, porosity by lithofacies, pore radius, nanopores, interparticle pores, intraparticle pores, organic-matter pores/kerogen-lined pores
	Fluid saturation, wettability	Water saturation, wettability type
	Permeability	Matrix permeability, fracture permeability, relative permeability, Klinkenberg permeability, nonDarcy flow, permeability under stress
Mechanical, physical properties	Pressure	Pore pressure, hydrostatic pressure, lithostatic pressure (overburden pressure), pressure gradient, normal pressure, overpressure, underpressure.
	Stress field	Regional stress, orientation, local stress, principal stresses, anisotropy, stress contrast
	Brittleness	Young's modulus, Poisson ratio, mineral composition
	Structural complexity	Formation bedding, faulting
	Natural fractures	Natural fracturing network, fracture permeability, DFN (discrete fracture network)
	Formation depth	Target zone depth from surface

Common source rock evaluation errors during prospect or play appraisals has been discussed elsewhere (Dembicki, 2009). One of the main objectives in exploration is the prospecting for hydrocarbon resources and their assessment. Important play elements include trap, charge access, reservoir storage and fluid properties, and other physical or mechanical properties. Table 1.3 presents methods and important parameters in early resource-play assessments. An integrated assessment process should integrate multidisciplinary data and present an overall play chance of success. It should enable transforming play fairway maps into a play chance map, and provide a basis for comparing geological risks, concessions, leads, and prospects. Understanding the limits of the elementary plays and how they contribute to the overall play chance enables ranking different resource opportunities and investment decisions. Figure 1.3 illustrates an example of combining elementary play properties into an overall assessment of play chance.

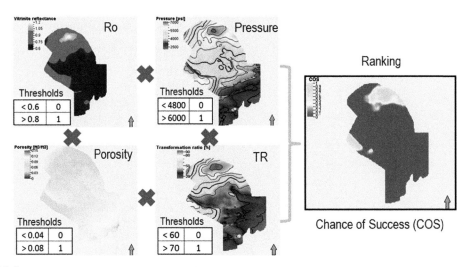

FIGURE 1.3

Example of play chance mapping by integrating multiple elementary play chance maps.

Adapted from Salter et al. (2014).

Seismic surveys are important tools for resource assessment; they are conducted to delineate the potential reservoir, and to interpret the structure and stratigraphic surfaces of the formations. Seismic analysis can help identify structural complexity, recognize analogous shale play systems, and determine possible sweet spots (Glaser et al., 2013). Seismic data can be used to identify faults and karsts (Du et al., 2011), helping assess the vertical connectivity between the hydrocarbon zone and the base formation. They can also be used for fault and stress analysis for a regional-scale interpretation. High-quality seismic data can be used to describe the reservoir properties through inversion, and attribute extraction and analysis.

1.3 EVALUATION

In developing both conventional and unconventional resources, the importance of evaluation lies in leveraging science and technology from geological, geophysical, petrophysical, and engineering studies to optimally produce the hydrocarbons from the subsurface. However, one misperception in developing unconventional resources is that shale reservoirs are "statistical" plays, meaning that the integrated studies based on various scientific disciplines are not useful to develop these types of reservoir, but the strategy should be to drill in uniformly or randomly selected locations. Schuenemeyer and Gautier (2014) compared the production costs between random drilling and smart drilling in the Bakken and Three Forks formations, and showed significant advantages of smart drilling based on the knowledge of the subsurface. In fact, the "statistical" play is simply a trial and error process, often leading to a large number of false positives (i.e., dry holes and mediocre producers; Ma, 2010), and false negatives (i.e., missing good prospects or prolific locations). In contrast, a comprehensive evaluation using a multidisciplinary approach can lead to fewer dry and mediocre wells (reduced false positives), and good locations are likely to be drilled (reduced false negatives), which increases the proportion of prolific wells and the overall productivity of the wells.

Evaluating unconventional resources is a complex process that involves the interpretation of seismic data, formation evaluation using wireline logs, core interpretation, abductive inference based on analogs, and integrated analyses using the available data and various geoscience disciplines. Important physical parameters or variables for unconventional reservoirs include sequence stratigraphy, fault system, kerogen content, TOC, mineralogy, lithology, pore structure and connectivity, natural fracture networks, rock mechanical properties, and in situ stress regime, among others. The main purpose of evaluation is to identify the sweet spots, and to assess the reservoir quality and completion quality. A good unconventional resource typically requires a certain level of reservoir quality and completion quality, as hydrocarbon production of an unconventional resource project depends on both (Suarez-Rivera et al., 2011). The questions to be answered in evaluation include:

- How much oil or gas is there?
- How producible are the resources?
- Is the reservoir easy to complete?
- What is the estimated ultimate recovery potential?

1.3.1 MINERALOGICAL COMPOSITION

As conventional resources reside in high-quality rocks, petrophysical investigations have focused on sandstone and carbonate facies, their porosity, fluid saturations, and permeability for the last century or so. In some unconventional formations, such as tight sandstones and carbonates, reservoirs are not the source rock; lithofacies interpretation and analysis based on cores and/or wireline logs often is an effective way for reservoir characterization and modeling (Ma et al., 2011). Although mineralogy plays an important role in understanding the lithofacies and rock types (Rushing et al., 2008) and the lithofacies may govern the fluid storage and flow, a detailed breakdown of mineral compositions is generally not necessary. In shale reservoirs, however, a description of their mineral composition is very important.

Shales can be mineralogically described using a ternary diagram with three dominant mineral components—silica, clay, and carbonates. Geochemical analysis, along with core and well log data, can help determine the position of a specific shale sample or reservoir on such a ternary plot. Figure 1.4(a) relates some of the well-known unconventional reservoirs to this mineralogical composition framework. These known reservoirs generally have less clay content than the average shale, which is calculated from all shales that include both organic and inorganic rocks. Notice also that these reservoirs have different mineral and lithological compositions, but they have been successful hydrocarbon projects, implying that mineralogy is not the only determining variable for reservoir quality. Conversely, other formations may have similar lithological compositions to some of these reservoirs, but they may or may not be good reservoirs depending on other variables. Moreover, for a given shale reservoir, mineralogical composition may vary widely as a function of vertical and lateral locations, as shown by the Barnett and Eagle Ford plays (Fig. 1.4(b)). These variations simply reflect the mineralogical heterogeneity of the formations.

An accurate mineralogical model is highly important for formation evaluation of shale resources. Individual shale reservoirs and even different locations in the same reservoir can vary

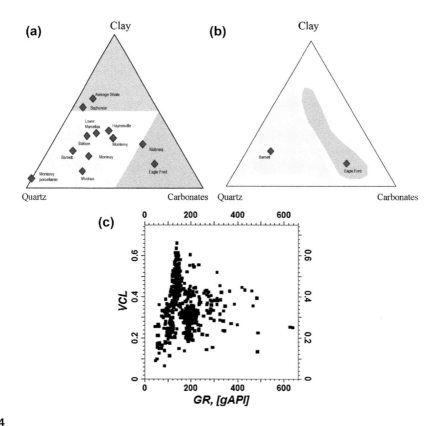

FIGURE 1.4

(a) Ternary plot of mineral compositions and relating several known shale reservoirs to this mineralogy-compositional triangle. *(Adapted from Grau et al. (2010))*. Notice that some of these reservoirs were projected on the plot based on limited data, and may not be accurate. For example, new data have shown that the Bakken is more calcareous (Theloy and Sonnenberg, 2013). (b) Heterogeneity in mineral composition in two shale-gas reservoirs. *(Adapted from Passey et al. (2010))*. Notice that the means on the ternary plot for these reservoirs in (a) and (b) are not necessarily in the middle of the areas because of the asymmetric frequency distributions. (c) Vclay and GR relationship from Marcellus Shale formations. The two variables have a weak, negative correlation, as opposed to a strong, positive correlation in a conventional reservoir (Ma et al., 2014).

considerably in mineralogy (Fig. 1.4). Understanding the mineralogical mix of shale formations is critical to analyze both reservoir quality and completion quality. Good organic shales have clay content that is significantly lower than that of conventional shales, and they have less clay-bound water. Most well-known productive shale reservoirs are highly siliceous, but a few known shale reservoirs have high calcareous content (e.g., Eagle Ford and Niobrara). When a shale reservoir has a relatively high clay content, such as the Bazhenov, it generally needs to have other favorable factors, such as a more intensive fracture network, to compensate for the negative effect of the high clay content.

One commonly used, continuously described, composition in formation evaluation is the volume of clay (Vclay). In conventional formations, Vclay is often estimated from the gamma ray (GR), but in shale reservoirs, GR is often weakly correlated to, and sometimes even negatively correlated to Vclay, which is often related to a statistical phenomenon called Simpson's paradox (Ma et al., 2014, 2015), as shown in Fig. 1.4(c). This often indicates a significant presence of organic matter in the shale, especially for shales deposited under marine conditions (Passey et al., 2010). A weak correlation could also reflect an anomaly of GR caused by a high concentration of radioactive minerals that are not necessarily associated with clay in tight gas sand reservoirs (Ma et al., 2011, 2014). Thus, estimating Vclay independently from GR is extremely important in evaluating shale reservoirs, and sometimes tight sandstone reservoirs as well. Geochemical logging provides basic mineralogical compositions of the rock, such as clay, quartz, carbonate, and pyrite, and can be an important tool for evaluating shale reservoirs for this reason. Moreover, for an accurate evaluation, clay should be further subdivided into its components, such as illite, chlorite, and smectite (smectite can causes expansion, leading to fracture closures). Similarly, carbonates should be subdivided into calcite, dolomite, and phosphate so that their compositions can be used to assess rock quality, such as to calculate the rock brittleness (Wang and Gale, 2009).

Quantification of TOC is very important to the evaluation of organic shales as TOC represents the weight percentage of carbon in the rock. TOC can be evaluated as part of evaluating the mineral composition or independent from it using various methods (Passey et al., 1990; Charsky and Herron, 2013; Gonzalez et al., 2013; Joshi et al., 2015). All these methods have some advantages and limitations; it is often useful to use several methods to reduce the uncertainty in the evaluation. One advantage of evaluating TOC as part of mineral composition is the so-called closure, implying all the components add to one when they are represented by fractions (Aitchison, 1986; Pawlowsky-Glahn and Buccianti, 2011).

1.3.2 PORES

Pore characterization is one of the most basic tasks in formation evaluation for both conventional and unconventional plays, because pore represents the storage capacity for hydrocarbon resources. In conventional formations or nonsource rock unconventional reservoirs, pore space mainly includes two types of pores: interparticle pores and intraparticle pores. In organic-rich shales, organic-matter pores or pores associated with kerogen often play an important role in the pore network (Loucks et al., 2012; Saraji and Piri, 2014), and as a result, porosity is often highly correlated to TOC (Alqahtani and Tutuncu, 2014; Lu et al., 2015). An organic-rich shale is typically composed of clay and nonclay mineral matrix, and contains pore space between these mineral components (Ambrose et al., 2012).

Rocks with high TOC often have high porosity because the transformation from kerogen to hydrocarbons often leads to an increase in pore volume. When TOC is high and significant amounts of kerogen transforms to hydrocarbons, more pores are generated and they tend to be interconnected, which explains why TOC is often highly correlated to permeability as well. Intra and interparticle pores may decrease as maturity increases, and they may be water-filled pores. Organic-matter pores increase as maturity increases; they should be mostly hydrocarbon-filled pores.

The porosity is often measured using a volume displacement or gas-filling method after crushing the rock to a specific particle size (Luffel et al., 1992; Bustin et al., 2008; Passey et al., 2010; Bust et al., 2011). Total porosity represents the pore space in the mineral matrix and kerogen (if present), including the space occupied by hydrocarbon, mobile and capillary-bound water, and clay-bound water.

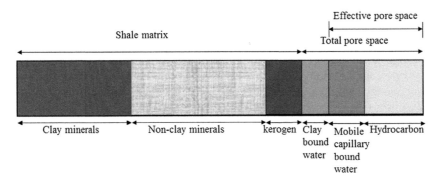

FIGURE 1.5

Illustration of total and effective pore space, and other components in an organic shale rock (synthesized after several publications: Eslinger and Pevear, 1988; Eslinger and Everett, 2012, and others).

Effective porosity does not include clay-bound water (Fig. 1.5). Determining porosity in shale reservoirs is complicated by very small pores and the presence of smectite in low-maturity clays that contain interlayer water. Passey et al. (2010) noticed that there were significant differences in quantifying porosity in shales among various labs because of different definitions of porosity and measuring conditions.

Porosity measured in the absence of reservoir stress is generally different from porosity in reservoir conditions. This has a significant impact on estimating hydrocarbon saturation and in-place resources (Passey et al., 2010).

1.3.3 HYDROCARBON SATURATION AND TYPES

Pore space in rock contains oil, gas, or water; evaluating fluid saturations is an integral part of analyzing reservoir quality. In tight gas sandstone reservoirs, water saturation is often negatively correlated to porosity (Ma et al., 2011). In shale gas reservoirs, gas-in-place includes adsorbed and free gas. In general, the free gas saturation should be positively correlated to effective porosity as it represents gas-filled porosity. The adsorbed gas is a function of pressure, TOC content, and isotherm (Boyer et al., 2006). Direct quantification of adsorbed gas is difficult; an isotherm measured from core samples can be used. Adsorption isotherm measurements enable the evaluation of the maximum adsorption capacity of gas by organic matter as a function of pressure (Boyer et al., 2006). The Langmuir isotherm model is one of the earliest models for adsorption, in which a monolayer and flat surface of pores are assumed, among other assumptions (Ma, 2015). Figure 1.6 shows the Langmuir isotherm function that enables computation of the adsorbed gas based on the pressure. However, the assumptions in the Langmuir isotherm model may not be always realistic for adsorption in shale gas reservoirs. Other heterogeneous multilayer adsorption models have been proposed (Ma, 2015).

1.3.4 PERMEABILITY

Permeability is very important to the producibility of hydrocarbon from the subsurface formation for both conventional and unconventional reservoirs (see e.g., Frantz et al., 2005). In tight gas sandstone

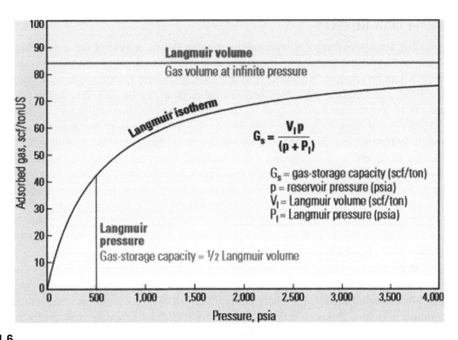

FIGURE 1.6

Langmuir isotherm or adsorbed gas as a function of pressure.

Adapted from Boyer et al. (2006).

reservoirs, permeability is highly dependent on lithofacies and generally is highly correlated to effective porosity (Ma et al., 2011). In shale reservoirs, both the kerogen type and maturity have an impact on the permeability and deliverability of the source rock. In thermally mature shale reservoirs, porosity and permeability have been likely enhanced by the kerogen conversion into oil or gas. The permeability to gas in kerogen-lined pores should be much greater than pores within the inorganic matrix. The presence of connected, kerogen-lined pores is the key to producing hydrocarbon from the shale. These systems are permeable to gas and should not imbibe water from hydraulic stimulation. Maturation of kerogen causes shrinkage and develops a connected hydrocarbon–wet pore system. This is why a good correlation between permeability and the amount of TOC is commonly observed. Natural and drilling-induced fractures are important to the production of gas as they enhance hydraulic fractures. It is also important to understand the concept of wettability and pore connectivity. Separating porosity and permeability created through the diagenesis and catagenesis of kerogen from matrix porosity and permeability can help better characterize the pore systems in organic shales. Kerogen and kerogen-related porosity are oil-prone wetting, and facilitate the flow.

Fracture permeability is also important for developing tight reservoirs. Natural fracture systems can be modeled using a discrete fracture network (DFN) and the reservoirs may be modeled by a single porosity and dual permeability method or a dual porosity and dual permeability method (Du et al., 2010, 2011). Natural fractures can be reactivated or enhanced further by hydraulic fracturing treatment (Weng et al., 2015).

1.3.5 COMPLETION QUALITY

Completion quality for unconventional reservoirs describes the rock quality for completing an oil or gas well, notably for hydraulic fracturing. It is concerned with selecting the best zone in the formation in which the rock can be adequately fractured hydraulically and the fractures can stay open after the pressure is reduced. For the most part, completion quality is driven by rock mechanics and mineralogical composition; geomechanical analysis is an integral part of interpreting completion quality (Cipolla et al., 2012; Chong et al., 2010; Warpinski et al., 2013). Some of the important variables in completion include fracture containment, facture conductivity, fluid sensitivity, ability of fractures to retain the surface area, and rock mechanics for stimulated reservoir volume (SRV) (Zoback, 2007; Mayerhofer et al., 2008). Hydraulic stress gradients impact the amount of pressure to open and close a fracture, and in situ stress and anisotropy impact the fracture geometry, orientation, and containment complexity.

The type and amount of clay minerals are important variables for completion quality as they are related to the brittleness of the rock (Jarvie et al., 2007; Wang and Gale, 2009; Slatt et al., 2014). Stress is high where the clay content is high, especially if significant quantities of smectite are present. In general, quartz-rich rocks tend to be brittle and are more conducive for completion; clay-rich rocks are more ductile as they generally have a high Poisson ratio, a low Young's modulus, and a high fracture closure stress. For example, the Barnett shale tends to be brittle as it is rich on silica and has a high Young's modulus and low Poisson ratio (King, 2010, 2014). Carbonate-rich rocks tend to be moderately brittle with limestone being less brittle than dolostone (Wang and Gale, 2009). Brittleness index can be defined as (Jarvie et al., 2007; Wang and Gale, 2009):

$$BI = \frac{Q}{Q + C + Cl} \quad (1.2)$$

or more accurately

$$BI = \frac{Q + Dol}{Q + Dol + Lm + Cl + TOC} \quad (1.3)$$

where BI is brittleness index, and all other variables are the rock components: Q is quartz, C is carbonate, Cl is clay, Dol is dolomite, Lm is limestone, and TOC is total organic carbon.

It is also possible to use an average of a normalized Young's modulus and Poisson ratio to represent the BI (Rickman et al., 2008). However, as noted earlier, an occurrence of Yule–Simpson's phenomenon will likely invalidate this method because Young's modulus and Poisson ratio may be positively correlated in aggregated data, but negatively correlated in disaggregated data (for causal/physical analysis, see e.g., Vernik et al., 2012).

A slight variation in clay content and clay type can also create significant vertical segmentation of the organic shale, and segmentation for the completion. Apart from clay content, there are other properties of the geologic formation that impact the fracturing. For example, beddings in the formation are natural weaknesses that impact the hydraulic fracture height containment (Chuprakov and Prioul, 2015).

The in situ stress dramatically impacts the fracturing geometry, including hydraulic fracture propagation and containment (Han et al., 2014). A high minimum horizontal stress contrast leads to

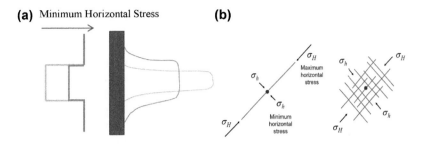

FIGURE 1.7

Impact of stress and mechanical properties on hydraulic fracture geometry. (a) Illustration of the effect of minimum horizontal stress on fracture geometry. Higher stress contrast yields greater fracture lengths and containment, while lower stress contrast yields shorter fractures. (b) Impact of stress anisotropy on fracture growth *(adapted from Sayers and Le Calvez (2010))*; while high stress anisotropy tends to create narrower fracture fairways, and low stress anisotropy creates wide fracture fairways.

a deeper fracture (Fig. 1.7(a)); similarly, a high Young's modulus or low Poisson ratio tends to give a deep fracture. Moreover, the anisotropy of the in situ stress also controls the fracture growth, including the geometry and size of the fractures (Fig. 1.7(b)).

1.3.6 PRESSURE

Hydrocarbon production is highly impacted by reservoir pressure. Overpressure can be an important production drive mechanism for tight formations (Meckel and Thomasson, 2008). Overpressure is theorized to be the result of hydrostatic isolation of a volume, thus limiting equalization of the pressure gradient of the reservoir with that of the surrounding rocks. Isolation can be caused by lithological anisotropy, low porosity and permeability, and a restricted fracture or fault network. Other events that can generate abnormal pressures include changes in fluid type, kerogen maturity, and tectonic compression. When the source rock generates a large amount of gas in tight formations, gas cannot escape easily and cause overpressure in the system.

Pore pressure prediction is important for drilling in tight formations because an accurate prediction of pore pressure can improve drilling efficiency, more easily deal with borehole adverse conditions, and avoid blowouts (Zhang and Wieseneck, 2011). Pore pressures are difficult to measure and estimate, especially for shale reservoirs because of the low porosity and permeability. In some cases, they can be modeled using mud log and seismic velocity data or estimated using rock physics (Bowers, 1995; Sayers, 2006). Pore pressure field in unconventional petroleum systems can also be modeled.

1.3.7 INTEGRATED EVALUATION

The importance of integration for developing unconventional resources was briefly discussed earlier. Here several specific aspects of integration in evaluating unconventional resources are discussed, including the composite or aggregated variable problem (especially, ecological inference bias or

Simpson's paradox; Robinson, 1950; Ma, 2009), nonlinearity, spurious correlation, and weak or reversed correlation in analyzing and integrating various data.

For most reservoir evaluation projects, logs are the most readily available data, and core data are much sparser, but can provide insights about the important drivers for hydrocarbon production. Porosity, fluid saturation, and permeability derived from core are used to calibrate log responses. Core descriptions can be used to validate petrophysical, geomechanical, and facies properties (Slatt et al., 2008). Pressure tests and well production analysis should also be integrated with logs, seismic data, and reservoir descriptions. Integration of petrophysical log data, completions, and seismic data is valuable for selection of drilling locations and production optimization (Chong et al., 2010; Du et al., 2011; Glaser et al., 2013; Hryb et al., 2014).

Variables that describe reservoir quality discussed earlier are generally correlated, but the magnitudes of correlation may be quite variable. In analyzing and quantifying the relationships among kerogen, TOC, porosity, fluid saturation and permeability, in situ stress, clay content, and geomechanical properties, there are many pitfalls and it is easy to draw an inaccurate inference. These include the use of correlation as causation, incorrect appreciation of the relationships among three or more variables, and ignoring the correlation. For example, because of lack of data, one often draws inference of a reservoir property from its relationships with other properties. A related statistical paradox is that people generally think that when variables A and B are correlated positively, and variables B and C are correlated positively, then variables A and C must be correlated positively. But this is not necessarily true (Langford et al., 2001). In a system affected by multiple variables, such as reservoir quality, all the influencing variables and their relationships must be analyzed. The relationship between two variables may be affected by other interfering variables, which may lead to a spurious correlation or other forms of Simpson's paradox related to a third variable's effect (Ma et al., 2014). As an example, the resistivity and maturity of organic matter in shales are generally correlated positively, but in some Niobrara formations, the relationship between maturity and resistivity can be reversed to be negatively correlated because of the wettability change in conjunction with development of petroleum-expulsion fractures or other mechanism in transition from wet gas to dry gas (Al Duhailan and Cumella, 2014; Newsham et al., 2002).

It is useful to distinguish completion quality from reservoir quality in general, but it is also very important to analyze their correlation as they may or may not be highly correlated. Sometimes, geomechanical parameters are indicative of not only completion quality, but also reservoir quality. Britt and Schoeffler (2009) showed an example of separating prospective shales from nonprospective shales using static and dynamic Young's modulus correlation, yet Young's modulus is often used as a criterion for completion quality. In other cases, reservoir and completion qualities are not significantly or even inversely correlated, how to couple them is highly important for completion optimization.

Table 1.4 lists some of the most important parameters in evaluating shale reservoirs. A commonly used cutoff of TOC for a source rock is 2% in weight. Effective porosity should be more than 4% for shale gas reservoirs. Obviously, the higher the better; but it is uncommon for gas shales to have effective porosity more than 10%. Marcellus shales have relatively high porosity compared to most gas shales, but they also have higher clay content. Water saturation (Sw) should be less than 45%, and the lower the better. Higher Sw may be related to more water-wet pores within the inorganic

Table 1.4 Important Parameters in Evaluating Shale Reservoirs

Parameter	Critical or Desired Values	Data Sources
TOC	>2% (Weight)	Leco TOC, Rock-Eval
Thermal maturity	Oil window: $0.5 < Ro < 1.3$ Gas window: $1.3 < Ro < 2.6$	Vitrinite reflectance, Rock-Eval
Mineralogy	Clay <40%, quartz or carbonate >40%	X-ray diffraction, Spectroscopy, log-based
Average porosity	>4%	Core, logs
Average water saturation	<45%	Core, Capillary pressure, log-based
Average permeability	>100 nanoDarcy	Mercury injection capillary pressure, nuclear magnetic resonance, gas expansion
Oil or gas-in-place	Gas: free and adsorbed gas > 100 Bcf/section	Log-based, integrated evaluation
Natural fracture	Moderate to dense, and contained in the target zone	Seismic, image log
Wettability	Oil-prone wetting of kerogen	Special core analysis
Formation lateral continuity	Continuous	Sequence stratigraphy with core, logs, and regional data
Hydrocarbon type	Oil or thermogenic gas	Geochemistry, Rock-Eval
Pressure	Overpressure is preferable	Log-based, seismic
Reservoir temperature	>230 °F	Drill stem test
Stress	<2000 psia net lateral stress	Logs, image log, seismic
Young's modulus	>3.0 MM psia	Acoustic logs, cores
Poisson ratio	<0.25	Acoustic logs, cores

Expanded and synthesized from a number of shale reservoir studies (Boyer et al., 2006; Sondergeld et al., 2010).

minerals—pores that may not contain producible hydrocarbon. Analysis of both reservoir quality and completion quality is important for successfully developing an unconventional reservoir.

It is possible to assign a favorability score for each of the reservoir and completion–quality parameters and then compare the reservoir to analog reservoirs. In practice, only a small number of these parameters may be evaluated because of limited data. These reservoir and completion–quality parameters cannot be linearly combined; it is not straightforward to derive a composite score of quality. For example, a reservoir with 10% porosity and 1% weight TOC may not be as good as a reservoir with 5% porosity and 2% weight TOC. Radar plots provide a useful way to visually analyze qualities and compare them with analog reservoirs. Figure 1.8 shows an example of ranking reservoir and completion qualities based on the availability of data using radar plots. As more quality parameters become available, the ranking becomes more complex, but analysis of more parameters should be helpful to make a more informed decision. The comparison with a well-known analog reservoir using a radar plot can also be useful to identify strengths and weaknesses, and subsequently helps the completion design (Fig. 1.8(d)).

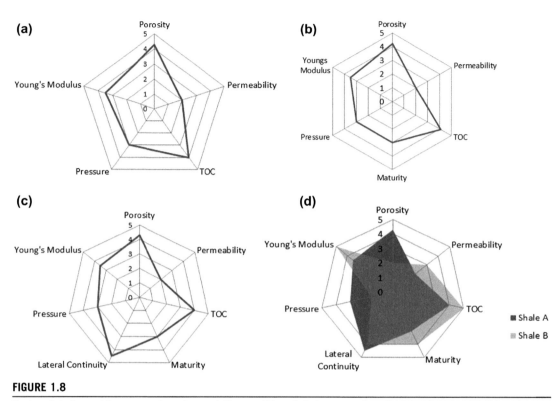

FIGURE 1.8

Radar plots for ranking reservoir and completion qualities of shale reservoirs with 6-level scores (0–5 from low to high, decimals can be used as well). (a) 5-parameter ranking; (b) 6-parameter ranking; (c) 7-parameter ranking; (d) 7-parameter ranking; and analog comparison.

1.4 DRILLING

As a result of developing unconventional resources, an increasing number of horizontal wells have been drilled. In the US, for example, more than 70% of the rigs were drilling horizontal wells in 2013 versus only 30% in 2005 (RigData, 2014). This growth comes mainly from shale plays and is related to hydraulic fracture stimulation because horizontal wells generally have more contact with the target formation, which is especially important for increasing production from source rock formations (Warpinski et al., 2008). By increasing the lateral length in a horizontal well, the well can stay in the targeted geological zone and thus increasing its contact with the formation (Fig. 1.9).

Well spacing regulations, reduced environmental footprint, and increased reservoir contact are some of the main reasons why extended laterals are becoming more common. Not only multiple wells can be drilled from a single pad, the directional S-shaped wells can be drilled using specially designed rigs with a skid system, which allows substantial reduction of footprint and optimization of surface

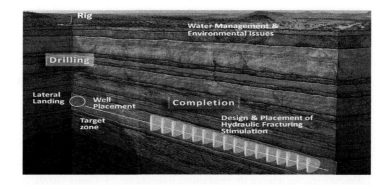

FIGURE 1.9

Illustration of hydraulic fracturing from a horizontal well drilled into a subsurface source-rock zone.

Adapted from Allix et al. (2011).

facilities (Pilisi et al., 2010). Drilling technology, including interactive three-dimensional (3D) well path design and rotary steerable systems, helps drill longer laterals with greater accuracy. Longer laterals have generally improved the production performance, but longer laterals are more costly; optimal lateral length and number of stages can be found based on the optimization of net present value (NPV) incorporating the production and costs (details are discussed in the next section). After a number of years of production in several known reservoirs, such as the Barnett, Fayetteville, and Haynesville formations, an optimal lateral length along with stage count can be found using the production history with various completion designs. In some cases, the lateral length is pushed beyond the limits of coiled tubing.

Other issues related to drilling include challenge in multiwell pads, staying in the target geological zones or optimal well placement (Kok et al., 2010), wellbore stability (Higgins-Borchardt et al., 2015), drilling fluid selection, lost circulation kicks and blowouts (Pilisi et al., 2010; Zhang and Wieseneck, 2011). Subsurface formations are heterogeneous and shale reservoir quality can be highly variable in a relatively narrow vertical or horizontal section. Horizontal or directional wells can be challenging because the geological layers are not always horizontal and often difficult to track; many formations are distorted, displaced, or undulated, making it difficult for the lateral to stay in the zone. Real-time measurements, such as logging while drilling (LWD) and measurement while drilling (MWD), can help depict the subsurface formation (Baihly et al., 2010). A more accurate image of the subsurface formation can be depicted by coupling LWD with the 3D seismic surveys, allowing the placement of the laterals in the correct geological layers. An accurate mapping of the subsurface is like a global positioning system for well placement as it enables optimal landing and steering the lateral in the right zone.

Drilling direction is important as breakdown will happen more frequently if a horizontal well is drilled in the direction of maximum horizontal stress. High-pressure, high-temperature reservoirs can be problematic for drilling. Advanced drilling fluids provide borehole strengthening and limit lost circulation while adhering to environmental regulations. The right fluid can also help drill faster with ease. Optimal drilling fluids vary from one play to another. Other important drilling considerations for tight reservoirs are listed in Table 1.5.

Table 1.5 Important Considerations in Drilling and Completion of Tight Reservoirs

Variables and Parameters	Specifications, Relevance, and Effects	References
Well spacing	Impacts the production; spacing should be determined together with other variables, such as reservoir quality and fracture design.	Malayalam et al. (2014); Warpinski et al. (2014); Yu and Sepehrnoori (2014)
Lateral length	Impacts the contact area between wellbore and formation, highly influential to production.	Zhou et al. (2014); Malayalam et al. (2014)
Stage count	Directly impacts the contact area between wellbore and formation, highly influential to production.	Aviles et al. (2013); Zhou et al. (2014); Malayalam et al. (2014)
Fracture spacing	Fracture spacing impacts the production; it should be optimized with several variables, including reservoir quality, completion quality, proppant effect, and mechanical interaction.	Cheng (2009); Roussel et al. (2012); Morrill and Miskimins (2012); Jin et al. (2013); Wu and Olson (2013); Sanaei and Jamili (2014)
Cased and cemented versus open hole	Cased and cemented completion enables isolation of fracture stages, and development of complex fractures. Openhole completion has a cost advantage.	IIseng (2005); Nelson and Huff (2009); Darbe and Ravi (2010); Daneshy (2011); Ermila et al. (2013); Pavlock et al. (2014).
Tubular selection	Need to ensure pumping slickwater at a high rate and ensure HFT with an adequate number of stages.	Palisch et al. (2008); Teodoriu (2012).
Selection of fracturing point	Division of the wellbore into equally spaced zones is a common way to select fracture initiation points, but it may be suboptimal because of the heterogeneity of the formation. Mud logging for gas composition; LWD, MWD for formation stratigraphy; sonic logs for stress variation.	Waters et al. (2009); Baihly et al. (2010); Tollefsen et al. (2010); Ketter et al. (2006)
Simultaneous versus sequential fracturing	Sequential fracturing may include acid stage, pad stage, prop sequence stage, and flushing stage; simultaneous fracturing requires close monitoring of equipment, and multiple large pads.	Mutalik and Gilbson (2008); Waters et al. (2009); Olsen and Wu (2012); Wu and Olson (2013); Viswanathan et al. (2014)
Fracturing fluids	Fluids are used to create fractures in the formation and to carry proppant into fractures to keep them from closing up. Slickwater is the most common one; other types include cross-linked gel, gas-assisted, and hybrid fluids. Choice depends on whether it meets proppant placement need, and increases fracture contact area.	Palisch et al. (2008); Britt et al. (2006); Britt and Schoeffler (2009); Vincent (2011); Montgomery (2013); McKenna (2014)
Additives	Common additives include friction reducer, oxygen scavenger and scale inhibitor. Special additives may include enhanced oil recovery chemicals for facilitating conductive pathways, recovering water, and minimizing damage from water blocking.	Kaufman et al. (2008); Holditch 1999
Hydraulic fracture interaction with natural fractures	Natural fractures include open and closed fractures; both of them impact HFT, fracture geometry and complexity, and SRV.	Gale et al. (2007); Du et al. (2011); Aimene et al. (2014); Huang et al. (2014); Busetti et al. (2014); Smart et al. (2014); Weng et al. (2015)

Table 1.5 Important Considerations in Drilling and Completion of Tight Reservoirs—cont'd		
Variables and Parameters	**Specifications, Relevance, and Effects**	**References**
Perforation	Placement efficiency; perforation clusters/fracture stage, interval length, distance between clusters, shot density, charge type.	Wutherich and Walker (2012); Kraemer et al. (2014)
Proppant	Type and amount of proppant impact the flow capacity in the fractures; it could prevent flowback. The common proppant is sand with various specifications. Small mesh bauxite and ceramic can also be used. Proppant design has a significant impact on fracture conductivity, allowing open channels in the proppant pack improves the fracture conductivity.	Cipolla et al. (2009); Olsen et al. (2009); Gillard et al. (2010); d'Huteau et al. (2011); Gao and Du (2012); McKenna (2014); Greff et al. (2014); Mata et al. (2015)
Drilling fluids	Oil-based, water-based	McDonald (2012); Guo et al. (2012); Stephens et al. (2013); Young and Friedheim (2013)

1.5 COMPLETION AND STIMULATION

Tight reservoirs commonly require hydraulic fracturing to be productive; for shale reservoirs, the combination of horizontal well and HFT is often the key for commercial production. HFT is a process of creating or restoring fractures in a geological formation using fluids and proppants to stimulate production from oil and gas wells. The completion of hydraulically fractured wells involves many processes and variables, including completion type (cased and cemented or open hole), stage design, lateral landing, hydraulic fracture geometry, fracturing fluid, proppant type/size/schedule, hydraulic fracture monitoring, optimizing the hydraulic fracture treatment, stimulation of reservoir volume, and production analysis (Mayerhofer et al., 2008; Chong et al., 2010; King, 2010, 2014; Gao and Du, 2012; Agrawal et al., 2012; Manchanda and Sharma, 2014). A good completion should include generating, maintaining, and monitoring of fracture networks, and optimally placing the perforation clusters. This requires understanding the geological and geomechanical properties, and natural fractures of the formation, heterogeneity in the reservoir, and optimized stimulation.

An accurate mechanical earth model (MEM) is important for developing shale reservoirs. The MEM should include the distribution of mechanical properties and stress field of the subsurface formations that include not only the targeted reservoir zone, but also the layers above and below it because the stresses of the adjacent bounding formations impact the fracture height and length. The modeling process integrates data from logging, core, drilling, and completions, and facilitates the understanding of the mechanical properties and the stress state throughout the stratigraphic column. A good MEM enables the understanding of the distribution and possible reactivation of natural fractures during the hydraulic stimulation, and a reasonable prediction of hydraulic fracture geometry (orientation, length, height growth, and aperture).

The completion design in developing unconventional reservoirs should be based on a combination of geomechanical and reservoir properties. It is important to distinguish completion effort from completion quality. While completion quality describes the properties of the rocks that make completion straightforward or complicated, completion effort describes the various completion methods and tools used in the completion. The combination of completion quality and completion effort is the completion efficiency.

Completion effectiveness can be improved when hydraulic fracturing and perforating operations take into account the anisotropies in stress and other reservoir parameters. Evenly fracturing all the clusters in a heterogeneous zone may not be optimal as the lateral penetrates a heterogeneous reservoir. Sequenced fracturing in combination with an effective fluid can increase the effective fracture length (Kraemer et al., 2014). Typically, without knowledge of heterogeneity of the formation, a uniform design for perforation clusters is used, but detailed knowledge of the lateral heterogeneity can help place the perforation clusters in optimal locations, and enhances the hydrocarbon production of perforated clusters (d'Huteau et al., 2011).

Other important variables in completion design include fracture spacing, cased and cemented versus open hole, tubular selection, selection of fracturing point, simultaneous versus sequential fracturing, fracture initiation points, proppants, fracturing fluids, additives, hydraulic fracture interaction with natural fractures, and perforation strategies (King, 2010, 2014). These variables impact the fracture complexity, SRV, and effectiveness of a fracturing treatment. Table 1.5 lists important considerations in well completion of a tight formation. Three topics are discussed below.

1.5.1 FRACTURE GEOMETRY AND COMPLEXITY

Some of the most important considerations in fracture design include: (1) the fracture azimuth and orientation are mainly determined by the state of in situ stress; (2) rocks with lower Young's modulus tend to have greater hydraulic fracture widths; (3) stress contrast is critical for fracture containment and length; (4) planar hydraulic fractures occur more likely without presence of natural fractures; (5) transverse hydraulic fractures will likely occur if a horizontal well is drilled in the direction of minimum horizontal stress; (6) effective fracture length is limited by the ability of the fracturing fluid to transport the proppant over long distances; and (7) high injection rates generally improve proppant transport and fracture length.

Rock properties impact vertical height and horizontal shape of fractures, including width, length, and fracture complexity index (Daniels et al., 2007; Cipolla et al., 2008). Hydraulic fracture height is governed mainly by the in situ stresses (Han et al., 2014), and can be also affected by the fracturing-fluid viscosity and injection rate. Once a propagating fracture reaches a horizon where the stress contrast is greater than the net pressure in the fracture, it will stop. There is a strong relationship between net pressure and fracture width. Generally high injection rate increase fracture width. Fracture width is important as narrow fracture width can result in early screenouts. A gradual increasing net pressure after the initial growth period often indicates a vertical fracture extending in length. Effective fracture length is often limited by the ability of the fracturing fluid to transport proppant over long distances. Fracture-height growth and fracture width affect propped-fracture half-length for a given treatment size. Generally high injection rate improves proppant transport and increasing the effective fracture length.

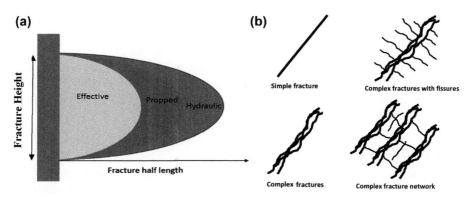

FIGURE 1.10

(a) Illustration of a hydraulic fracture geometry. (b) Illustration of different levels of fracture complexity (some of them can be natural fractures or reactivated natural fractures, synthesized from several publications, Fisher et al., 2002; Cipolla et al., 2008).

Fractures can be simple or complex (Fig. 1.10(b)), and the stress anisotropy and the nature of a natural fracture system are the key drivers to hydraulic fracture complexity (Sayers and Le Calvez, 2010; Weng et al., 2015). The interaction between the hydraulic fractures and the preexisting natural fractures is an important consideration in hydraulic fracture design. Moreover, when the horizontal maximum and minimum stresses have similar magnitude, fractures tend to grow in various directions and fracture network may be complex. Fracture complexity has a strong impact on the production (Warpinski et al., 2008, 2013; Waters et al., 2011; Mata et al., 2014), but notice that although fracture complexity increases the reservoir contact, it may pose challenge in creating a durable proppant pack with sufficient hydraulic continuity (Vincent, 2013).

1.5.2 FRACTURING FLUID AND PROPPANT

Fracturing fluids are additives or chemicals that are used to treat the subsurface formation in order to stimulate the flow of oil or gas. The commonly used types of fracturing fluids include water-based fluids (including slickwater, linear gels and cross-linked fluids), oil-based fluids, foamed fluids and viscoelastic surfactants (polymer-free fluids).

Fracturing fluid should have stable viscosity during pumping, and needs to be broken at the end of fracturing job. They should have a certain level of viscosity for proppant transport and controlling the fracture net pressure. The viscosity of the fracturing fluid also impacts the fracture geometry. Moreover, mineralogy impacts the hydraulic fracturing fluid selection. For example, the formations that have more than 50% clay are difficult for completion and the fluid selection is more delicate.

The use of proppants in HFT is to support the opening of the fractures so that the reservoir and the wellbore are connected for hydrocarbon flow. Naturally, the type and amount of proppant impact the flow capacity of the fractures and production (Gao and Du, 2012; McKenna, 2014; Greff et al., 2014). The proppant used in most shale reservoirs are sands with various specifications, including similarity in grain size, roundness in shape, and crush resistance. Small mesh bauxite and ceramic are also used. The main principle of using and selecting proppant for hydraulic fracturing is to prop the fractures while not blocking the flow between the reservoir and the wellbore. Important considerations in selecting proppants should include closure stress, conductivity, and the bottom-hole flowing pressure.

Proppant distribution in the fracture network is highly important for the effectiveness of hydraulic fractures (Cipolla, 2009). When the proppant is concentrated in a primary planar fracture, the effective fracture length is limited even when the hydraulic fracture may be large. When fracture growth is complex, the average proppant concentration tends to be low and the proppant may not materially impact well performance.

1.5.3 FRACTURING STAGES

Longer laterals and more stages have generally increased the production in shale plays (Fig. 1.11(a)), which is why the stage count has gone up for many shale plays. For example, in Marcellus and Eagle Ford, the average fracturing stage count was less than 10 in 2008, and was increased to more than 16 in 2012; in Bakken, it was around 14 in 2008, and was increased to more than 27 in 2012 (RigData, 2014). However, because the complexity and cost of completion increase as the stage count increases, a large stage count does not always increase the economic value (Fig. 1.11(b)). In some plays, such as the Barnett, Fayetteville, and Haynesville, the optimal stage count has been well studied. In other plays, operators are still trying to test longer laterals and increased stage count.

1.6 PRODUCTION

One main concern in hydrocarbon production from tight formations is sustainability. The wells may begin producing at very high rates, then they fall off very sharply, and finally they level off at a much lower rate. On the other hand, when wells are drilled in high reservoir-quality zones, shale wells can last a very long time when the hydrocarbon-bearing deposition is relatively continuous. Maximization of the recovery factor for each well is important for economic success of the overall project. This

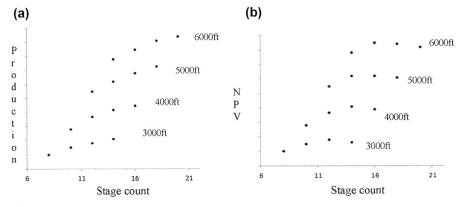

FIGURE 1.11

(a) Illustration of the impact of lateral length and stage count on production. (b) Illustration of the impact of lateral length and stage count on NPV while incorporating the completion cost.

Adapted from Malayalam et al. (2014).

requires production data analysis, diagnosis, and optimization. Production surveillance and forecasting can help optimally produce hydrocarbons and enhance the ultimate recovery.

Because of the lack of detailed knowledge of the subsurface formation, the most common practice is to drill and complete a well in the same way along the lateral, but the heterogeneity in the rock properties often leads to highly uneven production from different perforation clusters. It is suboptimal when wellbores are treated as if the rocks are the same all along the lateral. The production from the wells can be evaluated, and changes can be made to improve the future wells. Production logging can be conducted to identify where the gas or oil is coming from; it may show that only some of the perforation clusters contribute to the production, while other perforation clusters produce little (Baihly et al., 2010). Sometimes more than 40% of the perforations do not contribute to production (Miller et al., 2011). By comparing the producing clusters with the rock strength log, it is possible to identify the areas of low stress and high brittleness in the formation, and locating the fracture positions and analyzing their geometries. Some fracturing techniques can sequentially isolate fractures at the wellbore to ensure every cluster is fractured and contributes to the well performance. Comprehensive production workflows reduce the time to analyze the data, help identify problems, and optimize production.

Important processes in production include analysis of production drivers, calibration of drivers to actual production, refracturing, artificial lift, flowback management, and water management, among others.

1.6.1 PRODUCTION DRIVERS

Production drivers include variables in reservoir quality, completion quality and effort; these variables are related to charge access, structural complexity, natural fractures, pore space, fluid properties, pressure, geomechanics, and well placement. Table 1.6 gives a list of common production drivers and their relevance to production. In practice, calibrating the reservoir quality and completion quality drivers to production is critical to understand and enhance the production.

1.6.2 CALIBRATING THE DRIVERS TO PRODUCTION

Production from shale reservoirs is highly variable as a function of the location in a field, and there are many causes for the very heterogeneous productions (Baihly et al., 2010; Miller et al., 2011). Reservoir quality, completion quality, and completion effort all impact the production. To understand the impact of these variables, it is important to analyze individual physical variables, such as kerogen, TOC, porosity, fluid saturation, and permeability for reservoir quality; stress, mineralogy, and mechanical properties for completion quality; and lateral length, proppant tonnage, and stage count for completion effort. Causally, all these variables impact production, a weak or no correlation, or even a reversed correlation, between production and such variables is frequently observed (Jochen et al., 2011; Gao and Du, 2012; Gao and Gao, 2013; Zhou et al., 2014). For example, lateral length or stage length and production rate may appear to have no, weak, or negative correlation (Miller et al., 2011).

An example of negative correlation between oil production and stage length from a shale oil reservoir is shown in Fig. 1.12(a). Without insightful analysis, one may conclude that the longer the lateral length, the worse the production! In fact, this phenomenon is a manifestation of Simpson's paradox in complex systems with interaction of multiple variables (Pearl, 2010, 2014; Ma and Zhang, 2014).

Table 1.6 Production Drivers for Well Performance

Domain/Category	Parameters and Variables	Relevance
Petroleum system and charge access	Kerogen types and content, TOC, maturity (thermal maturation), fluid properties, pore pressure	TOC is a measure of organic richness or capacity for hydrocarbon generation; maturity is a measure of the conversion of kerogen into hydrocarbon; kerogen type impacts the fluid type and quality
Structure, stratigraphy, diagenesis, depth	Stacking patterns, formation (lateral and vertical) continuity, net thickness, faults, natural fractures, karsts	Lateral borehole staying in the best zone ensures more contact among wellbore, fractures, and the productive zone
Static reservoir quality	Mineralogy, porosity, saturation,	Storage, oil or gas-in-place
Flow-related reservoir quality	Matrix permeability, fracture permeability	Fluid mobility and deliverability from formation to wells
Completion quality	Stress field, mechanical properties (Young's modulus, Poisson ratio, brittleness), mineralogy	Hydraulic fracturing containment, fracture complexity, wellbore stability
Pressures, temperature, and fluid properties	Pore pressure, overburden pressure, temperature, wettability	Impact on drilling and hydrocarbon productivity, cementing
Drilling/completion effort and effectiveness	Well placement, geosteering, LWD, hydraulic fracture geometry, all the parameters in Table 1.5	Stay in zone, maximum contact of well and fractures to the productive zone, enhanced SRV

The lateral length per stage that is negatively correlated to production is caused by the third variable effect—for a given total lateral length, a longer lateral length per stage implies a smaller stage count; stage count is generally positively, highly correlated to production, as discussed in the previous section (also see Zhou et al., 2014), but exceptions occur (Miller et al., 2011; Gao and Gao, 2013) because of the effect of other variables.

When a weak or reversed correlation between production and "reservoir quality" is observed, people are perplexed with the fact that "the better the reservoir quality, the lower the hydrocarbon production!" Two possible causes for this occurrence include an inaccurate definition of reservoir quality and suboptimal completion. For example, when the reservoir quality is only partially represented by the static component, such as bulk volume of hydrocarbon (i.e., porosity times hydrocarbon saturation), the dynamic component, such as permeability, may be more highly correlated to production, and leads to a negative correlation between the static reservoir quality and the production. This occurs frequently in a single porosity and dual permeability system. Figure 1.12(b) shows such an example, wherein the correlation between the static reservoir quality and one-year oil production is −0.231. The negative correlation is caused by the dual permeability that includes the matrix and fracture components. The dual permeability system can be interpreted as two different reference classes (Ma, 2010; Ma et al., 2011); when the matrix and fracture static reservoir qualities are separately analyzed, the correlation between production and the static reservoir quality is high for both the systems, 0.748 for matrix reservoir quality, and 0.992 for fracture reservoir quality.

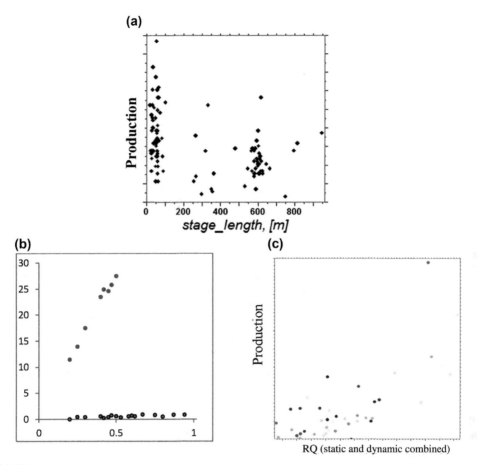

FIGURE 1.12

(a) Negative correlation, −0.438, between the production and stage length in a tight oil reservoir. (b) Cross plot between production (y-axis) and static reservoir quality (x-axis) in a dual-permeability system. Red (gray in print versions) dots represent fracture properties. (c) Cross plot between production (y-axis) and reservoir quality (x-axis). Color represents different wells.

It is also possible to generate a composite reservoir quality by combining the static and dynamic ones. Then the composite reservoir quality will be highly, positively correlated with the production (Fig. 1.12(c)). Conversely, Yule–Simpson's effect (reversal or not) occurs in a dual permeability system when the fracture permeability is not explicitly incorporated in the reservoir quality.

A reversal of correlation between reservoir quality and production can be also caused by a suboptimal completion. When the completion efficiency is the same for all the wells, production should have a positive correlation with reservoir quality. When completion efficiency is not the same for all the wells, the correlation between the production and reservoir quality is reduced or reversed. For example, a reversal of correlation between production and reservoir quality may occur when the

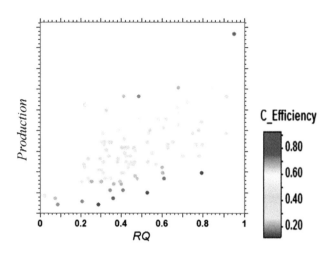

FIGURE 1.13

Cross plot between the production and reservoir quality (combined static and dynamic qualities) overlaid with completion efficiency (C_Efficiency). Notice that some wells have reservoir quality (RQ) greater than 0.8, but their productions are significantly lower than some other wells with RQ less than 0.5. The former wells are underperformers and should be candidates for recompletion.

completion efficiency is stronger for a moderate reservoir-quality zone while the completion efficiency is lower for a high reservoir-quality zone.

Therefore, it is possible to identify ineffective completions and optimize them. Figure 1.13 shows a cross plot between the production and reservoir quality overlaid with the completion efficiency. For very low reservoir-quality rocks, production is low regardless of the completion efficiency, but for moderate to high reservoir-quality rocks, completion efficiency determines the production. Completion efficiency includes two components: completion quality (CQ) and completion effort. It is very important to distinguish them because the completion efficiency is impacted by both the quality of rock for completion (i.e., CQ) and the intervention or effort (i.e., HFT).

In summary, it is very important to understand Simpson's paradox in analyzing the relationships between a potential driver and the production by considering third variables' effects. Discerning Simpson's paradox can help screen and rank the importance of drivers and identify inefficient completion.

1.6.3 MICROSEISMIC MONITORING

Microseismicity is a low-magnitude earthquake with frequencies in the order of 0.1–10 kHz, and it is a geophysical remote-sensing technique that enables the detection of hydraulic fracturing activities and imaging hydraulic fracturing stimulations. As discussed earlier, the effectiveness of a hydraulic fracturing treatment depends on a number of factors, including reservoir quality and completion quality. A variety of heterogeneous properties in the formation make the impact of the HFT highly variable (Baihly et al., 2010; Miller et al., 2011). Microseismic monitoring enables mapping the fracture propagation and effectiveness of the HFT (Le Calvez et al., 2006; Daniels et al., 2007;

Warpinski, 2009; Maxwell et al., 2011; Warpinski et al., 2013). The interpretation of microseismic patterns generally include fracture length, fracture height, fracture azimuth, fracture network complexity, fracture locations, interaction between hydraulic and natural fractures and stimulated rock volume (Xu et al., 2009; Maxwell and Cipolla, 2011; Cipolla et al., 2011). Other applications of microseismic monitoring include real-time fracture analysis and control, refracturing and diversion, fracture treatment design, completion optimization, and well placement.

Because microseismic events are small shear slippages induced by changes in stress and pressure while hydraulic fractures are rather tensile openings, microseismic interpretation can be vague (Warpinski, 2009; Maxwell and Cipolla, 2011). Determining the event location and fracture geometry thus has significant uncertainty.

1.6.4 REFRACTURING TREATMENT

Refracturing can be sometimes economically more advantageous than drilling and fracturing new wells. Main reasons for refracturing include (Moore and Ramakrishnan, 2006; Vincent, 2010): (1) initial fracturing was ineffective in developing fracture complexity for high production rates; and (2) small fractures and microfractures were not well propped and production rates decreased dramatically after a high initial rate. The problem with refracturing treatment is that not all the wells with refracturing will be productive enough. Some wells may have produced the majority of their extractable resources in the initial HFT, while other wells may still have a lot to produce from the relevant reservoir formations. Understanding the relationships among reservoir quality, completion efficiency, and production is a good way to identify the refracturing candidate wells. For example, if a high RQ well underperforms many of moderate RQ wells, then it may be because of the initial ineffective completion and the well should be a candidate for refracturing. In Fig. 1.13, some wells have RQ index greater than 0.8, but they underperform some other wells with an RQ index less than 0.5, and thus they should be considered for refracturing. One caveat for selecting a well for refracturing treatment is to select the worst-performance wells. Refracturing a bad well could make the well better if the RQ of the formation around the well is good, but it will be waste of money if the RQ is poor. Furthermore, even if the RQ is good, but CQ is poor, refracturing will be challenging.

Refracturing can alter the stresses of the reservoir, create new fractures, reopen existing fractures, reorient fractures along a different azimuth, enlarge existing fractures, increase fracture conductivity by other means, add perforations or diversion to contact untapped reservoir rocks, reactivate natural fractures or rearrange an existing proppant pack and/or repressurize an existing well before offset drilling and completions (Jacobs, 2015). It can restore well productivity to near-original or even higher rates of production, extending the productive life of the well.

1.6.5 ARTIFICIAL LIFT

In many tight reservoirs, the early production rate is very high because of the hydraulic fracturing stimulation, but it falls off sharply after a relatively short period. Artificial lift may be employed to lift the trapped hydrocarbon and mitigate the rapid declining production rate (Lane and Chokshi, 2014). It involves fracturing fluid recovery and choke management. One of the challenges for horizontal drilling of tight formations is to drain hydrocarbons from long, deep lateral wells. In a broad sense, artificial lifting should be part of production optimization and integrated planning for field development. Performance monitoring and surveillance of the wells will help design different lift systems.

1.6.6 TRACER, LEAKOFF, AND FLOWBACK

Chemicals and radioactive isotope tracers are useful for both stimulation of reservoir and tracing fluid flowback (Woodroof et al., 2003; Sullivan et al., 2004; Curry et al., 2010; Munoz et al., 2009). Radioactive tracers are used to mark fracturing entry points, and chemical tracers are used for tracing fluid return efficiency for different stages, and recording water volumes, salinity, production rates, and pressures (King, 2010).

Leakoff into surrounding porous formations and flowback in the reverse direction of fracturing fluid must be carefully dealt with in hydraulic fracturing. The leakoff can be modeled by Carter's analytical model or numerical simulation (Copeland and Lin, 2014). Flowback can be modeled to estimate reservoir parameters, such as effective fracture half-length, pore volume, percentage of injected fluid left in the formation, and to forecast hydrocarbon recovery (Ezulike and Dehghanpour, 2014). The amount of fracturing fluid recovery in the flowback depends on HFT design and fluid type (Sullivan et al., 2004; Crafton and Gunderson, 2007).

Tight sandstones generally have a higher recovery rate than shale reservoirs as shales often have smaller and more complex fracture geometry (King, 2010). Disposal of flowback fluids is very important to the protection of both surface and ground water as the flowback brine may contain radioactive elements, organic compounds and metals (Lane and Peterson, 2014; Balashov et al., 2015). The majority of flowback fluids are disposed of in underground injection wells; but advances in flowback fluid treatment technology offer the possibility of using flowback fluids for other uses, such as reuse for other HFTs and irrigation (Gupta and Hlidek, 2010). Balashov et al. (2015) presented a model that described the flowback chemistry change for better long-term planning for brine disposal, and understanding of the interaction between hydraulic fractures and shale matrix.

1.6.7 HEAVY OIL AND PRODUCTION FROM OIL SANDS

As shown in Fig. 1.1, oil sands and heavy oil are part of unconventional portfolio. Heavy oil and oil from bitumen represent a large worldwide hydrocarbon resource (Alboudwarej et al., 2006). Oil is considered to be heavy when its API gravity is lower than 22.3 according to the U.S. Department of Energy's standards; bitumen has API lower than 10 (Chopra et al., 2010; Schiltz and Gray, 2015). Bitumen in oil sands is highly viscous, and its gas-free viscosity is greater than 10,000 cP. Source-rock bitumen was generated during the early maturation of kerogen; crude bitumen represents degraded remnants of oil after exposure to water and bacteria. Its nuclear log response is similar to oil, and it typically exhibits very high resistivity.

Bitumen and heavy oil deposits occur in many parts of the world, but most of these resources are found in Canada, Venezuela and the United States. Production of heavy oil typically requires sophisticated recovery technologies (Alboudwarej et al., 2006), such as cold production techniques augmented by solvent-injection, vapor-assisted petroleum extraction, cyclic-steam stimulation, steamflooding, and steam-assisted gravity drainage. For effective steam stimulation, it is important to accurately identify the area of bitumen presence through lithofacies analysis and modeling as shale layers limit growth of steam chamber. Another challenge is to place the producer wells based on the delineation of bitumen zones and water zones because the water zones may absorb the steam heat, preventing it from heating the bitumen.

1.6.8 WATER MANAGEMENT

Water management in developing unconventional resources is a very important issue for both the hydrocarbon exploitation from tight formations (Lutz et al., 2013; Li et al., 2015; Kang et al., 2015) and impact on the social use of water (Taylor, 2014). It includes acquiring water, mixing water with chemicals and additives, injecting the mixed water fluids into the wells, flowback of fluids from the wells, and wastewater treatment. Provision and recycling of fracturing water and management, treatment, and disposal of produced water create a significant cost burden. Cost-effective techniques for managing oilfield water require a comprehensive understanding of reservoir characteristics, production volumes, hydrogeology, engineering design, and environmental considerations (Lane and Peterson, 2014). The fracking process uses millions of gallons of high-pressure water mixed with sand and chemicals to break apart the rocks for stimulating oil and gas. Techniques have been proposed to enable frac-fluid recycling and water conservation (Gupta and Hlidek, 2010). Treating produced water includes removing chemicals and other suspended solids. Minimizing the use of chemicals is advantageous for flowback-water reuse.

1.7 UNCONVENTIONAL GLOBALIZATION

Some unconventional resources, such as oil sands, heavy oil, and tight gas sandstones, have been developed for quite some time (Dusseault, 2001; Law et al., 1986; Law and Spencer, 1989). But the commercial success of the Barnett Shale was only a decade or so ago, which has led the way for a number of developments of other shale plays in the US, from the Fayetteville to Haynesville, Woodford to Eagle Ford, Marcellus to Niobrara, and Bossier to Bakken, among others (Fig. 1.14). In the last few years, the lessons learned in the United States shale plays have being applied in global shale reservoirs. Shale gas reservoirs are even being evaluated in many Mideast countries, where conventional reservoirs are among the world's most prolific producers. The lessons learned in the Barnett Shale and other shale plays in the United States are often transferrable to other worldwide shale plays, or at least can be used as starting points for the learning curve in new areas. A globalization of unconventional resources development appears to be happening, although a number of challenges exist, including the volatility of crude oil price and natural gas price.

Based on the historic trends of supply and demand in the United States, the impact of the production from shale and tight sandstone gas reservoirs is dramatic. The United States Department of Energy forecasts an increasing supply from the shale gas reservoirs for the foreseeable future (EIA, 2013). In other parts of the world, these trends may be duplicated as new areas are opened to unconventional resources.

1.7.1 GLOBAL UNCONVENTIONAL RESOURCES

Globally, unconventional resources are vast but ill-defined. One systematic evaluation to identify the distribution of worldwide unconventional resources was performed by Rogner (1997). In 2000, the United States Geological Survey estimated that the remaining recoverable gas from conventional reservoirs totaled about 12,000 Tcf, with about 3000 Tcf of natural gas already produced. The total estimated gas-in-place from all reservoirs globally is 50,000 Tcf.

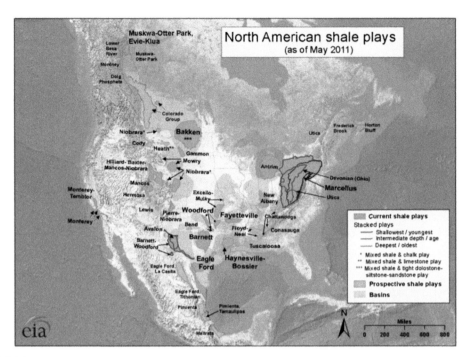

FIGURE 1.14

Shale plays in North America.

Source: EIA (2013).

Several studies have been released in the last a few years, including a report by Advanced Resources International (ARI, 2011; 2013), and reports by Energy Information Administration (EIA) (Kuuskraa et al., 2011, 2013). EIA (2013) updated its prior assessment of shale gas resources, and made an initial assessment of shale oil resources by analyzing 137 shale formations in 41 countries outside the United States.

According to EIA (2013), the top 10 countries with the most recoverable resources for oil and natural gas are listed in Table 1.7. Argentina has the second highest estimated shale gas reserves in the world, at 802 Tcf. It is ranked the fourth highest in the world at 27 billion barrel (bbl) of shale oil reserves. Russia has is 75 billion bbl of shale oil reserves in addition to its large conventional crude oil reserves. Some believe that the Bazhenov in western Siberia itself could hold hundreds of billions of barrels of oil. China has the world's largest estimated shale gas reserves at 1115 Tcf (trillion cubic feet) and the world's third largest shale oil reserves at 32 billion bbl.

1.7.2 NATIONAL SECURITY MATTER AND ENVIRONMENTAL CONCERNS

Different countries regard shale reservoirs, hydraulic fracturing, and other unconventional extraction technologies very differently. Many countries are attracted by the economics of potential new resources, and see it as a matter of national security (Kerr, 2010; NPC, 2011; Maugeri, 2013; Zuckerman, 2013),

Table 1.7 Top 10 Countries with Technically Recoverable Shale Gas and Shale Oil Resources

	Country	Shale Gas (Tcf)	Country	Shale Oil (billion bbl)
1	China	1115	Russia	75
2	Argentina	802	US	58 (48)[a]
3	Algeria	707	China	32
4	US	665 (1161)[a]	Argentina	27
5	Canada	573	Libya	26
6	Mexico	545	Venezuela	13
7	Australia	437	Mexico	13
8	South Africa	390	Pakistan	9
9	Russia	285	Canada	9
10	Brazil	245	Indonesia	8
	World total	7299 (7795)		345 (335)

[a]There are two different estimates for United States shale gas and shale oil reserves.
Source: EIA (2013).

but some countries are more concerned with environmental impact of developing unconventional resources (Wilber, 2012; Prud'homme, 2014; AP, 2014). In 2013, the United Kingdom's government announced tax breaks to kick-start shale drilling, arguing that "if we do not back" fracking technology, "we will miss a massive opportunity to help families with their bills and make our country more competitive. Without it, we could lose ground in the tough global race" (Daily Telegraph, 2013). On the other hand, France, Bulgaria, and a few other countries have banned hydraulic fracturing because of concerns about potential environmental damages.

How we treat hydraulic fracturing impacts the policies regarding the development of shale and other tight reservoirs. United States oil and gas production from unconventional resources has increased dramatically, reducing its reliance on imports from foreign countries. The United States total production of natural gas has increased from the middle of the last decade to 2012 by approximately 30%, despite the reduced production from onshore and offshore conventional sources (Fig. 1.15). As a result, natural gas price is much less expensive in the United States than in Asia and Europe. This has helped the United States to recover its economy much more quickly after the 2008 financial crisis than Europe and other parts of the world (Maugeri, 2013; Prud'homme, 2014). The EIA (2013, 2014) has projected further increasing natural gas production from both tight gas sandstone and shale gas reservoirs (Fig. 1.15).

China has an ambitious shale gas production target of between 60 and 100 billion m^3 by 2020, as outlined in its 12th Five Year Plan. China is the world's largest energy consumer, and is exploring the possibility of shale gas not only because of the environmental benefits of using natural gas as opposed to coal, but also because China holds the world's largest reserve of natural gas in shales. The government released a detailed plan in 2013 that aims to rebalance its energy mix so that coal, responsible for heavy pollution, accounts for less than 65% of the overall energy supply by 2017. The domestic shale development could reduce reliance on gas imports, allowing China to grow its economy more sustainably and less dependent on foreign imports.

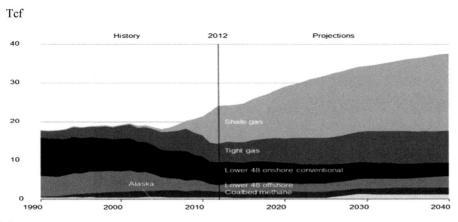

FIGURE 1.15

United States natural gas production by source and projection to 2040.

Source: EIA (2014).

1.7.3 CHALLENGES IN GLOBAL DEVELOPMENT OF UNCONVENTIONAL RESOURCES

1.7.3.1 Uncertainty and Risk in Resource Estimates

Estimates of world recoverable unconventional resources vary widely because of many technical difficulties, including lack of detailed data, limited understanding of subsurface formations, and difficult assessment of possible technological breakthroughs for enhancing recovery.

An estimated 470 Tcf is buried in various European shale formations (Zuckerman, 2013; EIA, 2013). Poland initially had an estimate of 187 Tcf of shale reserves, and has been eager to develop these shale resources. Unfortunately, this estimate was cut by 90% in 2012 because of the deep formations and difficulty of extraction.

Similarly, EIA downgraded the oil reserves of the Monterey Formation dramatically from an initial estimate of 13.7 billion bbl (EIA, 2011) to about 0.6 billion bbl in 2014 (Sieminski et al., 2014; CCST, 2014). The original EIA estimate is based on a series of highly skewed assumptions.

1.7.3.2 Technologies for Developing Difficult Unconventional Resources

While the estimates of worldwide unconventional resources are quite large, it is important that technologies are developed to produce some of the difficult subsurface resources (deeper and/or tighter) economically.

While there have been a number of reports outlining the economic and environmental benefits that shale gas could present for China, the possibility of actually developing shale gas in a large scale in China has on several occasions been thrust into doubt. China's spending in shale gas is in some cases four times as much as what the United States spends to develop its fields. Depth and tight formations are some of the sticking points to shale development in China.

1.7.3.3 Water Stress

Developing unconventional resources uses a large amount of fresh water. The oil and natural gas reserves may exist, and could be technically recoverable with horizontal drilling and hydraulic fracturing, but when there is a water shortage, the resources cannot be developed easily. In the 20 countries with the largest shale gas and tight oil resources, the World Resources Institute (WRI) analyzed the level of water stress across every play in each country. Globally, around 38% of shale gas reserves are located in areas dealing with high to extremely high water stress. Drilling and hydraulic fracturing requires up to 25 MM L of water per well, meaning shale resources are difficult to develop where fresh water is in short supply. Shale plays in 40% of those countries face high water stress or arid conditions, including China, Algeria, Mexico, South Africa, Libya, Pakistan, Egypt, and India (Table 1.8). For example, China's more challenging shale and other tight rock formations require more water to fracture each well, but China's water system is already strained. China has about 20% of the world's population, but only 7% of its water resources; its water supply per capita is only one-third of the global average, with wide regional disparity. Some argue that water is the biggest threat to the sustainability of China's economic growth ahead of energy. More than half of its shale gas reserves are located in areas suffering from high to extremely high water stress. To date, much of the drilling has taken place in the Sichuan basin in central China, an area of medium to high water stress. The Tarim

Table 1.8 Water Stress Level for 20 Countries with the Largest Technically Recoverable Shale Gas and Tight Oil Resources

Rank	Largest Shale Gas		Largest Tight Oil	
	Country	Water stress over shale plays	Country	Water stress over shale plays
1	China	High	Russia	Low
2	Argentina	Low to medium	United States	Medium to high
3	Algeria	Arid; low water to use	China	High
4	Canada	Low to medium	Argentina	Low to medium
5	United States	Medium to high	Libya	Arid; low water to use
6	Mexico	High	Australia	Low
7	Australia	Low	Venezuela	Low
8	South Africa	High	Mexico	High
9	Russia	Low	Pakistan	Extremely high
10	Brazil	Low	Canada	Low to medium
11	Venezuela	Low	Indonesia	Low
12	Poland	Low to medium	Columbia	Low
13	France	Low to medium	Algeria	Arid; low water use
14	Ukraine	Low to medium	Brazil	Low
15	Libya	Arid; low water to use	Turkey	Medium to high
16	Pakistan	Extremely high	Egypt	Arid; low water to use
17	Egypt	Arid; low water use	India	High
18	India	High	Paraguay	Medium to high
19	Paraguay	Medium to high	Mongolia	Extremely high
20	Colombia	Low	Poland	Low

Source: WRI (Reig et al., 2014). Note: Some ranks for estimates of resources from WRI are different than in Table 1.7.

basin is another geologically promising shale gas region, but it suffers from severe water shortages. South Africa faces similar water constraints, with the largest known shale play located in a region suffering from extreme water stress. Similarly, most of Mexico's known shale gas resources are located in the northern water-stressed deserts.

Many countries around the world are deciding whether or not to develop their shale and other tight gas and oil resources, partly because of the concern for water shortage (Reig et al., 2014). Potential business risks associated with freshwater availability to hydraulic fracturing should be assessed before developing an unconventional reservoir, although techniques that enable frack-fluid recycling and water conservation can mitigate the water shortage problem (Gupta and Hlidek, 2010; Lane and Peterson, 2014).

1.7.3.4 Environmental Considerations
Hydraulic fracturing used to be conducted away from residential areas. Many wells are now drilled near or in residential areas. This requires a great deal of effort to minimize the impact of drilling and production operations. Understandably, some residents worry about the environmental impacts of fracking despite being much cleaner than coal (DOE, 2009). There are concerns about possible groundwater contamination. General or specific risks and impacts to surface and ground water have been studied for some development areas (King, 2012; Vengosh et al., 2014; Lane and Peterson, 2014).

The concern around groundwater contamination primarily centers on whether the created fractures are contained within the target formation so that they do not contact underground sources of drinking water. A report from a federal study on hydraulic fracturing found no evidence that chemicals or brine water from the gas drilling process moved upward to contaminate drinking water at a site in western Pennsylvania (AP, 2014). The study was conducted with independent monitoring of a drilling site during the fracking process for 18 months. Researchers found that the chemical-laced fluids used to free gas stayed about 5000 ft below drinking water supplies. Tracer fluids, seismic monitoring, and other tests were used or performed to look for problems about how fracking affects adjacent rock structures. The report found that the hydraulic fractures could extend up to 1900 ft from the base of the well, likely caused by existing fault lines in the Marcellus Shale or other formations above it. More studies will be conducted in the future.

1.8 CONCLUSIONS
Evaluation and development of unconventional resources are more complex than for conventional resources. We must use an integrated approach, from evaluating TOC and thermal maturity of the petroleum system to evaluating petrophysical variables (porosity, fluid saturation, and permeability) and geomechanical properties. Drilling and completion design and production optimization must be based on reservoir quality and completion quality in order to improve the economics of developing unconventional resources. Many pitfalls exist in evaluating unconventional resources and in optimizing production of these resources. Discerning the composite variable effect, spurious correlation, and spurious weak or reversed correlation can help optimize the completion design for enhancing production.

As the world economy grows, developing unconventional resources is becoming more and more important, albeit with ups and downs. The production of commercial quantities of natural gas and oil from shales was uncommon at the beginning of the twenty-first century, but improved technology has enabled large-scale production of oil and gas from tight formations. However, there are still many challenges on the road of producing energy sources from unconventional resources. Better use of existing technologies and developing new technologies can help exploit these resources to benefit the society and minimize the environmental impact at the same time.

LIST OF ABBREVIATIONS

3D Three-dimensional
BBOE Billions of barrels of oil equivalent
COS Chance of success
CQ Completion quality
DFT Discrete fracture network
EOM Extractable organic matter
EUR Estimated ultimate recovery
GR Gamma ray
HFT Hydraulic fracturing treatment
LWD Logging while drilling
MEM Mechanical earth model
MWD Measurement while drilling
NPV Net present value
RQ Reservoir quality
SRV Stimulated reservoir volume
Sw Water saturation
TOC Total organic carbon
USGS US Geological Survey
WRI World Resources Institute

UNITS

bbl barrel
Bcf billion cubic feet
billion m^3 billion cubic meters
ft feet
G$_s$ gas storage capacity
L liter
mD millidarcy
MM million
p pressure
P$_l$ Langmuir pressure
Psi pounds per square inch
psia pounds per square inch absolute
Tcf trillion cubic feet
Vclay volume of clay
V$_l$ Langmuir volume

ACKNOWLEDGMENTS

The authors thank Schlumberger Ltd for permission to publish this work, and several reviewers for commenting on the manuscript.

REFERENCES

Agrawal, A., Wei, Y., Holditch, S., 2012. A Technical and Economic Study of Completion Techniques in Five Emerging US Gas Shales: A Woodford Shale Example. Society of Petroleum Engineers. http://dx.doi.org/10.2118/135396-PA.

Aimene, Y.E., Nairn, J.A., Ouenes, A., August 25–27, 2014. Predicting Microseismicity from Geomechanical Modeling of Multiple Hydraulic Fractures Interacting with Natural Fractures—Application to the Marcellus and Eagle Ford. Paper URTeC 1923762, presented at the URTeC, Denver, Colorado, USA.

Aitchison, J., 1986. The Statistical Analysis of Compositional Data. Chapman & Hall, London.

Alboudwarej, H., et al., 2006. Highlighting Heavy Oil. Oilfield Review, Summer 2006.

Al Duhailan, M.A., Cumella, S., August 25–27, 2014. Niobrara Maturity Goes up, Resistivity Goes down; What's Going on? Paper URTeC 1922820, presented at the URTeC, Denver, Colorado, USA.

Alexander, T., et al., 2011. Shale gas revolution. Oilfield Review 23 (3), 40–57.

Allix, P., et al., 2011. Coaxing oil from shale. Oilfield Review 22 (4), 4–16.

Alqahtani, A., Tutuncu, A.N., August 25–27, 2014. Quantification of Total Organic Carbon Content in Shale Source Rocks: An Eagle Ford Case Study. Paper URTeC 1921783, presented at the URTeC, Denver, Colorado, USA.

Ambrose, R.J., Hartman, R.C., Diaz-Campos, M., Akkutlu, I.Y., Sonfergeild, C.H., 2012. Shale gas-in-place calculations Part I: New pore-scale considerations. SPE Journal 17 (1).

AP, 2014. Landmark Fracking Study Finds No Water Pollution. Associated Press web: http://bigstory.ap.org/article/landmark-fracking-study-finds-no-water-pollution (last accessed 02.10.14.).

ARI, 2011, 2013. Advanced Resources International web: http://www.adv-res.com/ (last accessed 01.12.14.).

Aviles, I., Baihly, J., Liu, G.H., 2013. Multistage stimulation in liquid-rich unconventional formations. Oilfield Review 25 (2), 26–33.

Baihly, et al., 2010. Unlocking the Shale Mystery: How Lateral Measurements and Well Placement Impact Completions and Resultant Production. Paper SPE 138427 presented at the SPE Annual Technical Conference and Exhibition.

Balashov, V.N., Engelder, T., Gu, X., Fantle, M.S., Brantley, S.L., 2015. A model describing flowback chemistry changes with time after Marcellus Shale hydraulic fracturing. AAPG Bulletin 99 (1), 143–154.

Bowers, G., 1995. Pore pressure estimation from velocity data: accounting for overpressure mechanism besides undercompaction. SPE Drilling and Completion 10 (2), 89–95.

Bowker, K.A., 2003. Recent developments of the Barnett shale play, Fort Worth Basin. West Texas Geological Society Billetin 42 (6), 4–11.

Boyer, C., et al., 2006. Producing gas from its source. Oilfield Review 1 (3), 36–49.

Britt, L.K., et al., 2006. Waterfracs: We Do Need Proppant after All. Paper SPE 102227 presented at the SPE Annual Technical Conference and Exhibition, 24–27 September, San Antonio, Texas, USA.

Britt, L.K., Schoeffler, J., October 4–7, 2009. The Geomechanics of a Shale Play. Paper SPE 125525 presented at the SPE Annual Technical Conference and Exhibition, New Orleans, LA, USA.

Busetti, S., Jiao, W., Reches, Z., 2014. Geomechanics of hydraulic fracturing microseismicity: Part 1. Shear, hybrid, and tensile events. AAPG Bulletin 98 (11), 2439–2457.

Bust, V.K., Majid, A.A., Oletu, J.U., Worthington, P.F., 2011. The Petrophysics of Shale Gas Reservoirs: Technical Challenges and Pragmatic Solutions. IPTC 14631, IPTC held in Bangkok Thailand, February 7–9, 2012.

REFERENCES

Bustin, R.M., et al., 2008. Impact of Shale Properties on Pore Structure and Storage Characteristics. SPE 119892, presented at the SPE Shale Gas Production Conference, Ft. Worth, TX, USA, 16–18 November.

CCST, 2014. Advanced Well Stimulation Technologies in California, California Council on Science and Technology Report. Lawrence Berkeley National Laboratory Pacific Institute, p. 32.

Charsky, A., Herron, S., 2013. Accurate, Direct Total Organic Carbon (TOC) Log from a New Advanced Geochemical Spectroscopy Tool: Comparison with Conventional Approaches for TOC Estimation. AAPG Search & Discovery. #41162.

Cheng, Y., October 4–7, 2009. Boundary Element Analysis of the Stress Distribution around Multiple Fractures. Paper SPE 125769 presented at the SPE Annual Technical Conference and Exhibition, New Orleans, LA, USA.

Chong, K.K., Grieser, B., Jaripatke, O., Passman, A., 2010. A Completion Roadmap to Shale-Play Development. SPE 130369, presented at the CPS/SPE International Oil & Gas Conference and Exhibition in China, Beijing, China, 8–10 June.

Chopra, S., Lines, L., Schmitt, D.R., Batzle, M., 2010. Heavy-oil reservoirs: their characterization and production. In: Heavy Oils: Reservoir Characterization and Production Monitoring, Society of Exploration Geophysicists, Geophysical Developments, vol. 13, pp. 1–69.

Chuprakov, D.A., Prioul, R., 2015. Hydraulic Fracture Height Containment by Weak Horizontal Interfaces. Society of Petroleum Engineers. http://dx.doi.org/10.2118/173337-MS.

Cipolla, C., et al., February 7–9, 2012. Appraising Unconventional Resource Plays: Separating Reservoir Quality from Completion Quality. IPTC 14677, International Petroleum Technology Conference, Bangkok, Thailand.

Cipolla, C.L., Fitzpatrick, T., Williams, M.J., Ganguly, U.K., 2011. Seismic-to-Simulation for Unconventional Reservoir Development. Society of Petroleum Engineers. http://dx.doi.org/10.2118/146876-MS.

Cipolla, C.L., Lolon, E., Dzubin, B.A., October 4–7, 2009. Evaluating Stimulation Effectiveness in Unconventional Gas Reservoirs. Paper SPE 124843 presented at the SPE Annual Technical Conference and Exhibition, New Orleans, LA, USA.

Cipolla, C.L., September 2009. Modeling Production and Evaluating Fracture Performance in Unconventional Gas Reservoir. Paper SPE 118536, Distinguished Author Series, Journal of Petroleum Technology.

Cipolla, C.L., Warpinski, N.R., Mayerhofer, M.J., October 20–22, 2008. Hydraulic Fracture Complexity: Diagnosis, Remediation, and Exploitation. Paper SPE 115771 presented at the SPE Asia Pacific Oil and Gas Conference and Exhibition, Perth, Australia.

Copeland, D.M., Lin, A., August 25–27, 2014. A Unified Leakoff and Flowback Model for Fractured Reservoirs. Paper URTeC 1918356 presented at the URTeC, Denver, Colorado, USA.

Crafton, J.W., Gunderson, D., November 11–14, 2007. Stimulation Flowback Management: Keeping a Good Completion Good. Paper SPE 110851 presented at the SPE Annual Technical Conference and Exhibition, Anaheim CA, USA.

Curry, M., et al., 20–22 September 2010. Less Sand May Not Be Enough. Paper SPE 131783 presented at the SPE Annual Technical Conference and Exhibition, Florence Italy.

Daily Telegraph, 2013. http://www.telegraph.co.uk (last accessed 10.10.14.)

Daneshy, A.A., 2011. Hydraulic Fracturing of Horizontal Wells: Issues and Insights. Society of Petroleum Engineers. http://dx.doi.org/10.2118/140134-MS.

Daniels, J., et al., October 12–14, 2007. Contacting More of the Barnett Shale through an Integration of Real-Time Microseismic Monitoring, Petrophysics, and Hydraulic Fracture Design. Paper SPE 110562 presented at the SPE ATCE Conference, Anaheim, California, USA.

Darbe, R.P., Ravi, K., 2010. Cement Considerations for Tight Gas Completions. Paper SPE 132086 presented at the SPE Deep Gas Conference and Exhibition, 24–26 January, Manama, Bahrain.

Dembicki, H., 2009. Three common source rock evaluation errors made by geologists during Prospect or play appraisals. AAPG Bulletin 93 (3), 341–356.

D'Huteau, E., et al., 2011. Open-channel fracturing—a fast track to production. Oilfield Review 23 (3), 4–17.

DOE, 2009. Modern Shale Gas Development in the United States: A Primer. Groundwater Protection Council and All Consulting. DOE Report, DE-FG26–04NT15455.

Du, C., et al., June 8–10, 2010. Modeling Hydraulic Fracturing Induced Fracture Networks in Shale Gas Reservoirs as a Dual Porosity System. Paper SPE 132180 presented at the CPS/SPE International Oil & Gas Conference and Exhibition, Beijing, China.

Du, C., et al., 2011. An integrated modeling workflow for shale gas reservoirs. In: Ma, Y.Z., LaPointe, P. (Eds.), Uncertainty Analysis and Reservoir Modeling, AAPG Memoir 96.

Dusseault, M.B., 2001. Comparing venezuelan and Canadian heavy oil and tar sands. In: Proceedings of Petroleum Society's Canadian International Petroleum Conference 2001-061.

EIA, 2011, 2013, 2014. Technically Recoverable Shale Oil and Shale Gas Resources: An Assessment of 137 Shale Formations in 41 Countries outside the United States. US Energy Information Administration, Department of Energy (2013), web: www.eia.gov (last accessed 06.09.14.).

Ermila, M., Eusters, A., Mokhtari, M., 2013. Improving Cement Placement in Horizontal Wells of Unconventional Reservoirs Using Magneto-Rheological Fluids. Paper SPE 168904 presented at the Unconventional Resources Technology Conference, 12–14 August, Denver, Colorado, USA.

Eslinger, E., Pevear, D., 1988. Clay minerals for petroleum Geologists and engineers. SEPM Short Course 22.

Eslinger, E., Everett, R.V., 2012. Petrophysics in gas shales. In: Breyer, J.A. (Ed.), Shale Reservoirs-Giant Resources for the 21st Century: AAPG Memoir 97, pp. 419–451.

Evenick, J.C., August 2013. Not All Source Rocks Are Source Rock Plays: Screening Quality from Quantity. Paper URTeC 1581891 presented at the URTeC, Denver, Colorado, USA.

Ezulike, O., Dehghanpour, H., August 25–27, 2014. A Workflow for Flowback Data Analysis—creating Value out of Chaos. Paper URTeC 1922047 presented at the URTeC, Denver, Colorado, USA.

Fisher, M.K., et al., September 29–October 2, 2002. Integrating Fracture Mapping Technology to Optimize Stimulations in the Barnett Shale. Paper SPE 77441 presented at the SPE ATCE Conference, San Antonio, Texas, USA.

Frantz, J.H., et al., October 9–12, 2005. Evaluating Barnett Shale Production Performance Using an Integrated Approach. Paper SPE 96917 presented at the SPE ATCE Conference, Dallas, Texas, USA.

Gale, J., Reed, R., Holder, J., 2007. Natural fractures in the barnett shale and their importance for hydraulic fracture treatments. AAPG Bulletin 91 (4), 603–622.

Gao, C., Du, C., October 2012. Evaluating the Impact of Fracture Proppant Tonnage on Well Performances in Eagle Ford Play Using the Data of Last 3–4 Years. Paper SPE 160655 presented at the SPE Annual Technical Conference and Exhibition, San Antonio TX, USA.

Gao, C., Gao, H., September 30–October 2, 2013. Evaluating Early-Time Eagle Ford Well Performance Using Multivariate Adaptive Regression Splines (MARS). Paper SPE 166462 presented at the SPE Annual Technical Conference and Exhibition, New Orleans, LA, USA.

Gibbons, L., 2008. Nature + Nurture > 100%. The American Journal of Clinical Nutrition 87 (6), 1968.

Gillard, M.R., Medvedev, O.O., Hosein, P.R., Medvedev, A., Peñacorada, F., d'Huteau, E., 2010. A New Approach to Generating Fracture Conductivity. Society of Petroleum Engineers. http://dx.doi.org/10.2118/135034-MS.

Glaser, et al., 2013. Seeking the sweet spot: reservoir and completion quality in organic shales. Oilfield Review 25 (4), 16–29.

Gonzalez, J., Lewis, R., Hemingway, J., Grau, J., Rylander, E., Pirie, I., August 2013. Determination of Formation Organic Carbon Content Using a New Neutron-induced Gamma Ray Spectroscopy Service that Directly Measures Carbon. Paper URTeC 1576810 presented at the URTeC, Denver, Colorado, USA.

Grau, J., et al., October 18–22, 2010. Organic carbon content of the Green river oil shale from nuclear Spectroscopy logs. In: Proceedings of 30th Oil Shale Symposium. Colorado School of Mines, Golden, Colorado, USA.

Greff, K., Greenbauer, S., Huebinger, K., Goldfaden, B., August 2014. The Long-term Economic Value of Curable Resin-coated Proppant Tail-in to Prevent Flowback and Reduce Workover Cost. Paper URTeC 1922860, presented at the URTeC, Denver, Colorado, USA, 25-27.

REFERENCES

Guo, Q., Ji, L., Rajabov, V., Friedheim, J., October 22–24, 2012. Marcellus and Haynesville Drilling Data: Analysis and Lessons Learned. Paper SPE 158894, presented at the SPE Asia Pacific Oil and Gas Conference and Exhibition held in Peth, Australia.

Gupta, D.V.S., Hlidek, B.T., February 2010. Frac-fluid recycling and water conservation. SPE Production and Operations 65–69. http://dx.doi.org/10.2118/119478-PA.

Han, H., Higgins-Borchardt, S., Mata, D., Gonzales, V., 2014. In-situ and Induced Stresses in the Development of Unconventional Resources. SPE 171627.

Herron, et al., August 25–27, 2014. Clay Typing, Mineralogy, Kerogen Content and Kerogen Characterization from DRIFTS Analysis of Cuttings or Core. Paper URTeC 1922653 presented at the URTeC, Denver, Colorado, USA.

Higgins-Borchardt, S., Sitchler, J., Bratton, T., 2015. Geomechanics for unconventional reservoirs. In: Ma, Y.Z., Holditch, S., Royer, J.J. (Eds.), Unconventional Resource Handbook: Evaluation and Development. Elsevier.

Holditch, S.A., December 1999. Factors affecting water blocking and gas flow from a hydraulic fractured gas well. Journel of Petroleum Technology.

Holditch, S.A., 2013. Unconventional oil and gas resource development—Let's do it right. Journal of Unconventional Oil and Gas Resources 1–2, 2–8.

Hryb, D., et al., 2014. Unlocking the True Potential of the Vaca Muertta Shale via an Integrated Completion Optimization Approach. SPE. Paper 170580.

Huang, J., et al., August 25–27, 2014. Natural-Hydraulic Fracture Interaction. Paper URTeC 1921503, presented at the URTeC, Denver, Colorado, USA.

Ilseng, J.R., et al., 2005. Should Horizontal Sections Be Cemented and How to Maximize Value? Paper SPE 94288 presented at the SPE Production Operations Symposium, 16–19 April, Oklahoma City, Oklahoma, USA.

Jarvie, D.M., 1991. In: Merrill, R.K. (Ed.), Total Organic Carbon (TOC) Analysis. AAPG Treatise of Petroleum Geology, pp. 113–118.

Jarvie, D.M., Hill, R.J., Ruble, T.E., Pollastro, R.M., 2007. Unconventional shale-gas systems: the Mississippian Barnett Shale of North-Central Texas as one model for thermogenic shale-gas assessment. AAPG Bulletin 91, 475–499.

Jin, C.J., Sierra, L., Mayerhofer, M., August 2013. A Production Optimization Approach to Completion and Fracture Spacing Optimization for Unconventional Shale Oil Exploitation. Paper URTeC 1581809 presented at the URTeC, Denver, Colorado, USA.

Jochen, V., et al., 2011. Production Data Analysis: Unraveling Rock Properties and Completion Parameters. Paper SPE 147535 presented at Calgary, AB, Canada.

Jacob, T., 2015. Changing the equation: refracturing shale oil wells. JPT 2015 (4), 40–44.

Joshi, G.K., et al., January 26–28, 2015. Direct TOC Quantification in unconventional Kerogen-rich Shale Resource Play from Elemental Spectroscopy Measurements: A Case Study from North Kuwait. Paper SPE-172975, presented at SPE Middle East Unconventional Resources Conference and Exhibition.

Kang, et al., 2015. Coalbed methane development—an example from Qinshui basin, China. In: Ma, Y.Z., Holditch, S., Royer, J.J. (Eds.), Unconventional Resource Handbook: Evaluation and Development. Elsevier.

Kaufman, P.B., Penny, G.S., Paktinat, J., January 1, 2008. Critical Evaluation of Additives Used in Shale Slickwater Fracs. Society of Petroleum Engineers. http://dx.doi.org/10.2118/119900-MS.

Kerr, R.A., 2010. Natural gas from shale bursts onto the Scene. Science 328, 1624–1626.

Ketter, A.A., et al., September 24–27, 2006. A Field Study Optimizing Completion Strategies for Fracture Initiation in Barnett Shale Horizontal Wells. Paper SPE 103232 presented at the SPE ATCE Conference, San Antonio, Texas, USA.

King, G.E., September 19–22, 2010. Thirty Years of Gas Shale Fracturing. Paper SPE 133456 presented at the SPE Annual Technique and Exhibition, Florence, Italy.

King, G.E., February 6–8, 2012. Hydraulic fracturing 101: What Every Representative, Environmentalist, Regulator, Reporter, Investor, University Researcher, Neighbor, and Engineer Should Know about Estimating Frac Risk and Improving Performance in Unconventional Gas and Oil Wells. Paper SPE 152596 presented at the SPE Hydraulic Fracturing Technology Conference held in The Woodlands, TX USA.

King, G.E., 2014. 60 Years of Multi-Fractured Vertical, Deviated and Horizontal Wells: What Have We Learned? Paper SPE 17095 presented at the SPE Annual Technical Conference and Exhibition, 27–29 October, Amsterdam, The Netherlands.

Kok, J., et al., Nov. 2010. The Significance of Accurate Well Placement in Shale Gas Plays. Paper SPE 138438 Presented at the SPE Tight Gas Completions Conference, San Antonio, TX, USA 2–3.

Kraemer, C., et al., April 1–3, 2014. A Novel Completion Method for Sequenced Fracturing in the Eagle Ford Shale. Paper SPE 169010 presented at the SPE Unconventional Resources Conference, Held in The Woodlands, TX, USA.

Kuuskraa, V., Stevens, S., Van Leeuwen, T., Moodhe, K., 2011. World Shale Gas Resources: An Initial Assessment of 14 Regions outside the United States. Advanced Resources International, Inc. Prepared for US Energy Information Administration, Washington, DC.

Kuuskraa, V., Stevens, S., Van Leeuwen, T., Moodhe, K., 2013. World Shale Gas Resources: An Initial Assessment of 14 Regions outside the United States. Advanced Resources International, Inc. Prepared for US Energy Information Administration, Washington, DC.

Lane, A., Peterson, R., August 25–27, 2014. Evaluation Tool for Wastewater Treatment Technologies for Shale Gas Operations in Ohio. Paper URTeC 1922494, presented at the URTeC, Denver, Colorado, USA.

Lane, W., Chokshi, R., August 25–27, 2014. Considerations for Optimizing Artificial Lift in Unconventionals. Paper URTeC 1921823, presented at the URTeC, Denver, Colorado, USA.

Langford, E., Schwertman, N., Owens, M., 2001. Is the property of being positively correlated transitive? American Statistician 55 (4), 322–325.

Law, B.E., Spencer, C.W., 1989. Geology of Tight Gas Reservoirs in Pinedale Anticline Area, Wyoming, and Multiwell Experiment Site, Colorado. US Geologic Survey Bulletin 1886.

Law, B.E., 2002. Basin-centered gas systems. AAPG Bulletin 86 (11), 1891–1919.

Law, et al., 1986. Geologic characterization of low permeability gas reservoirs in selected wells, Greater Green River Basin, Wyoming, Colorado, and Utah. Geology of Tight Gas Reservoirs, AAPG Studies in Geology 24, 253–269.

Le Calvez, J.H., et al., October 11–13, 2006. Using Induced Microseismicity to Monitor Hydraulic Fracture Treatment: A Tool to Improve Completion Techniques and Reservoir Management. Paper SPE 104570 presented at the SPE Eastern Regional Meeting, Canton, Ohio, USA.

LeFever, J.A., Martiniuk, C.D., Dancsok, D.F.R., Mahnic, P.A., 1991. Petroleum potential of the middle member, Bakken Formation, Williston Basin, Saskatchewan Geological Society. Special Publication 6 (11), 74–94.

Li, Y., Tang, D., Xu, H., Elsworth, D., Meng, Y., 2015. Gelogical and hydrological controls on water coproduced with coalbed methane in Liulin, eastern Ordos basin, China. AAPG Bulletin 99 (2), 207–229.

Li, Y., et al., 2013. Experimental investigation of quantum Simpson's paradox. Physics Review A 88 (1), 015804.

Loucks, R.G., et al., 2012. Spectrum of pore types and networks in Mudrocks and a Descriptive Classification for matrix-related Mudrock pores. AAPG Bulletin 96 (6), 1071–1098.

Lu, J., Ruppel, S.C., Rowe, H.D., 2015. Organic matter pores and oil generation in the Tuscaloosa marine shale. AAPG Bulletin 99 (2), 333–357.

Lubinski, D., Humphreys, L.G., 1996. Seeing the forest from the trees: when predicting the behavior or status of groups, correlate means. Psychology, Public Policy, and Law 2, 363–376.

Luffel, D.L., Guidry, F.K., Curtis, J.B., 1992. Evaluation of Devonian Shale with New Core and Log Analysis Methods. SPE, 21297, JPT, 1192–1197.

Lutz, B.D., Lewis, A.N., Doyle, M.W., 2013. Generation, transport, and disposal of wastewater associated with Marcellus hale gas development. Water Resources Research 49 (2), 647–656.

REFERENCES

Ma, J., 2015. Pore-scale characterization of gas flow properties of shales by digital core analysis. In: Ma, Y.Z., Holditch, S., Royer, J.J. (Eds.), Unconventional Resource Handbook: Evaluation and Development. Elsevier.

Ma, Y.Z., et al., April 2014. Identifying hydrocarbon zones in unconventional formations by Discerning Simpson's paradox. Paper SPE 169496 presented at the SPE Western and Rocky Regional Conference.

Ma, Y.Z., et al., 2015. Lithofacies and Rock Type Classifications using wireline logs for shale and tight-carbonate reservoirs. In: Ma, Y.Z., Holditch, S., Royer, J.J. (Eds.), Unconventional Resource Handbook: Evaluation and Development. Elsevier.

Ma, Y.Z., et al., 2011. Integrated reservoir modeling of a Pinedale tight-gas reservoir in the greater Green river Basin, Wyoming. In: Ma, Y.Z., LaPointe, P. (Eds.), Uncertainty Analysis and Reservoir Modeling, AAPG Memoir 96, pp. 89–106.

Ma, Y.Z., 2010. Error types in reservoir characterization and management. Journal of Petroleum Science and Engineering 72, 290–301.

Ma, Y.Z., 2011. Uncertainty analysis in reservoir characterization and management. In: Ma, Y.Z., LaPointe, P. (Eds.), Uncertainty Analysis and Reservoir Modeling, AAPG Memoir 96.

Ma, Y.Z., 2009. Simpson's paradox in natural resource evaluation. Mathematical Geosciences 41 (2), 193–213.

Ma, Y.Z., Zhang, Y., 2014. Resolution of Happiness-Income paradox. Social Indicators Research 119 (2), 705–721. http://dx.doi.org/10.1007/s11205-013-0502-9.

Malayalam, A., et al., August 25–27, 2014. Multi-disciplinary Integration for Lateral Length, Staging and Well Spacing Optimization in Unconventional Reservoirs. Paper URTeC 1922270, presented at the URTeC, Denver, Colorado, USA.

Manchanda, R., Sharma, M.M., 2014. Impact of Completion Design on Fracture Complexity in Horizontal Shale Wells. Society of Petroleum Engineers. http://dx.doi.org/10.2118/159899-PA.

Masters, J.A., 1979. Deep basin gas trap, Western Canada. AAPG Bull 63 (2), 152.

Mata, D., et al., 1-3 April, 2014. Modeling the Influence of Pressure Depletion in Fracture Propagation and Quantifying the Impact of Asymmetric Fracture Wings in Ultimate Recovery. Paper SPE 169003 presented at the SPE Unconventional Resources Conference, The Woodlands, Texas, USA.

Mata, D., Zhou, W., Ma, Y.Z., Gonzalez, V., 2015. Hydraulic Fracture Treatment, Optimization and Production Modeling. In: Ma, Y.Z., Holditch, S. (Eds.), Unconventional Resource Handbook: Evaluation and Development. Elsevier, 2015.

Maugeri, L., 2013. The shale oil Boom: a US phenomenon, discussion Paper 2013-05 and report of Geopolitics of energy project. Harvard Kennedy School 66.

Maxwell, S.C., Pope, T.L., Cipolla, C.L., Mack, M.G., Trimbitasu, L., Norton, M., Leonard, J.A., 2011. Understanding Hydraulic Fracture Variability through Integrating Microseismicity and Seismic Reservoir Characterization. Society of Petroleum Engineers. http://dx.doi.org/10.2118/144207-MS.

Maxwell, S.C., Cipolla, C.L., 2011. What Does Microseismicity Tell Us about Hydraulic Fracturing? Society of Petroleum Engineers. http://dx.doi.org/10.2118/146932-MS.

Mayerhofer, M.J., et al., November 16–18, 2008. What is Stimulated Reservoir Volume? Paper SPE 119890 presented at the SPE Shale Gas Production Conference, Fort Worth, Texas, USA.

McDonald, M.J., January 1, 2012. A Novel Potassium Silicate for Use in Drilling Fluids Targeting Unconventional Hydrocarbons. Society of Petroleum Engineers. http://dx.doi.org/10.2118/162180-MS.

McKenna, J.P., August 25–27, 2014. Where Did the Proppant Go? Paper URTeC 1922843 presented at the URTeC, Denver, Colorado, USA.

Meckel, L.D., Thomasson, M.R., 2008. Pervasive tight-gas sandstone reservoirs: an overview. In: Cumella, S.P., Shanley, K.W., Camp, W.K. (Eds.), Understanding, Exploring, and Developing Tight-gas Sands, AAPG Hedberg Series 3, pp. 13–27 (Tulsa, OK).

Miller, C., Waters, G., Rylander, E., June 14–16, 2011. Evaluation of Production Log Data from Horizontal wells Drilled in Organic Shales. Paper SPE 144326 presented at the SPE Americas Unconventional Conference, The Woodlands, Texas, USA.

Momper, J.A., 1979. Generation of abnormal pressures through organic matter transformations. AAPG Bulletin 63 (8), 1424.

Montgomery, C., 2013. Fracturing fluids. In: Bunger, A.P., McLennan, J., Jeffrey, R. (Eds.), Effective and Sustainable Hydraulic Fracturing. INTECH, pp. 3–24.

Moore, L.P., Ramakrishnan, H., 2006. Restimulation: Candidate Selection Methodologies and Treatment Optimization. Society of Petroleum Engineers. http://dx.doi.org/10.2118/102681-MS.

Morrill, J.C., Miskimins, J.L., 2012. Optimizing Hydraulic Fracture Spacing in Unconventional Shales. http://dx.doi.org/10.2118/152595-MS. Paper SPE 152595.

Munoz, A.V., Asadi, M., Woodroof, R.A., Morals, R., May 31–June 3, 2009. Long-term Post-Frac Performance Analysis Using Chemical Frac-Tracer. Paper SPE 121380 presented at the SPE Latin American and Caribbean Petroleum Engineering Conference, Cartagena, Colombia.

Mutalik, P.N., Gilbson, R.W., 2008. Case History of Sequential and Simultaneous Fracturing of the Barnett Shale in Parker County. Paper SPE 116124 presented at the SPE Annual Technical Conference and Exhibition, 21–24 September, Denver, Colorado, USA.

Nelson, S.G., Huff, C.D., April 4–8, 2009. Horizontal Woodford Shale completion cementing practices in the Arkoma Basin, Southeast Oklahoma: A case history. Paper SPE 12074, presented at the SPE Production and Operations Symposium, Oklahoma City, OK, USA.

Newsham, K.E., Rushing, J.A., Chaouche, A., Bennion, D.B., 2002. Laboratory and Field Observations of an Apparent Sub Capillary-Equilibrium Water Saturation Distribution in a Tight Gas Sand Reservoir. SPE paper 75710, presented at the SPE Gas Technology Symposium, 5–8 April, Calgary, Alberta, Canada.

NPC, September 2011. Prudent Development: Realizing the Potential of North America's Abundant Natural Gas and Oil Resources. A Report of National Petroleum Council, Washington D.C, p. 68.

Olsen, J.E., Wu, K.W., 2012. Sequential versus Simultaneous Multizone Fracturing in Horizontal Wells: Insights From a Non-Planar, Multifrac Numerical Model. Paper SPE 152602 presented at the SPE Hydraulic Fracturing Technology Conference, 6–8 February, The Woodlands, Texas, USA.

Olsen, T.N., Bratton, T.R., Thiercelin, M.J., 2009. Quantifying Proppant Transport for Complex Fractures in Unconventional Reservoirs. http://dx.doi.org/10.2118/119300-MS. SPE paper 119300.

Palisch, T.T., Vincent, M.C., Handren, P.J., 2008. Slickwater Fracturing: Food for Thought. Paper SPE 115766 presented at the SPE Annual Technical Conference and Exhibition, 21–24 September, Denver, Colorado, USA.

Paris, M.G., 2012. Two quantum Simpson's Paradoxes. Journal of Physics A 45, 132001.

Parshall, J., 2008. Barnett shale showcases tight-gas development. Journal of Petroleum Technology 48–55.

Passey, et al., 1990. A practical model for organic Richness from porosity and resistivity logs. AAPG Bulletin 74 (12).

Passey, Q.R., et al., June 8–10, 2010. From Oil-Prone Source Rock to Gas-Producing Shale Reservoir—Geologic and Petrophysical Characterization of Unconventional Shale-Gas Reservoirs. Paper SPE 131350 presented at the CPS/SPE International Oil and Gas Conference and Exhibition, Beijing, China.

Pavlock, C., Bratcher, J., Leotaud, L., Tennison, B., 2014. Unconventional Reservoirs: Proper Planning and New Theories Meet the Challenges of Horizontal Cementing. Paper SPE 167749 presented at the SPE/EAGE European Unconventional Resources Conference and Exhibition, 25–27 February, Vienna, Austria.

Pawlowsky-Glahn, V., Buccianti, A. (Eds.), 2011. Compositional Data Analysis: Theory and Applications. Wiley, p. 400p.

Pearl, J., 2014. Comment: understanding Simpson's paradox. The American Statistician 68 (1), 8–13.

Pearl, J., 2010. Causality: Models, Reasoning and Inference, second ed. Cambridge University Press.

Peters, K.E., 1986. Guidelines for evaluating petroleum source rock using Programmed Pyrolysis. AAPG Bulletin 70 (3), 318–329.

REFERENCES

Peters, K.E., Xia, X., Pomerantz, D., Mullins, O., 2015. Geochemistry applied to evaluation of unconventional resources. In: Ma, Y.Z., Holditch, S., Royer, J.J. (Eds.), Unconventional Resource Handbook: Evaluation and Development. Elsevier.

Pilisi, N., Wei, Y., Holditch, S.A., 2010. Selecting Drilling Technologies and Methods for Tight Gas Sand Reservoirs. Society of Petroleum Engineers. http://dx.doi.org/10.2118/128191-MS. Paper SPE 128191, presented at the 2010 IADC/SPE Drilling Conference in New Orleans, Louisiana, USA, 2–4 February.

Prud'homme, A., 2014. Hydrofracking: What Everyone Needs to Know. Oxford University Press, New York p. 184.

Ratner, M., Tiemann, M., 2014. An overview of unconventional oil and natural gas. Congress Research Service 7-5700, R43148.

Reig, P., Luo, T., Proctor, J.N., 2014. Global Shale Gas Development: Water Availability and Business Risks. World Resources Institute Report (last accessed 09.10.14.). www.WRI.org.

Rickman, R., Mullen, M., Petre, E., Grieser, B., Kundert, D., September 21–24, 2008. A Practical Use of Shale Petrophysics for Stimulation Design Optimization. Paper SPE 115258 presented at the SPE Annual Technical Conference, Denver, CO, USA.

RigData, 2014. http://www.rigdata.com/index.aspx (last accessed 28.11.14.).

Robinson, W., 1950. Ecological correlation and behaviors of individuals. American Sociological Review 15 (3), 351–357. http://dx.doi.org/10.2307/2087176.

Rogner, H.H., 1997. An assessment of world hydrocarbon resources. Annual Review of Energy and the Environment 22, 217–262.

Roussel, N.P., Manchanda, R., Sharma, M.M., 2012. Implications of Fracturing Pressure Data Recorded during a Horizontal Completion on Stage Spacing Design. SPE paper 152631 presented at the SPE Hydraulic Fracturing Technology Conference, 6–8 February, The Woodlands, Texas, USA.

Rushing, J.A., Newsham, K.E., Blasingame, T.A., 2008. Rock Typing—Keys to Understanding Productivity in Tight Gas Sands. Paper SPE 114164 presented at the SPE.

Salter, R., Meisenhelder, J., Bryant, I., Wagner, C., August 25–27, 2014. An Exploration Workflow to Improve Success Rate in Prospecting Unconventional Emerging Plays. Paper URTeC 193507, presented at the URTeC, Denver, Colorado, USA.

Sanaei, A., Jamili, A., August 25–27, 2014. Optimum Fracture Spacing in the Eagle Ford Gas Condensate Window. Paper URTeC 1922964 presented at the URTeC, Denver, Colorado, USA.

Saraji, S., Piri, M., August 25–27, 2014. High-Resolution Three-dimensional Resolution Three-dimensional Characterization of Pore Networks in Shale Reservoir Rocks. URTeC paper 1870621, presented at the URTeC, Denver, Colorado, USA.

Sayers, C., 2006. An introduction to velocity-based pore-pressure estimation. The Leading Edge 25 (12), 1496–1500.

Sayers, C.M., Le Calvez, J., 2010. Characterization of microseismic data in gas shales using the radius of gyration tensor, 80th SEG Annual Meeting. Expanded Abstracts 29 (1), 2080–2084.

Schiltz, K., Gray, D., et al., 2015. Monitoring and predicting steam chamber development in a bitumen field: a seismic study of the Long Lake SAGD project. In: Ma, Y.Z., Holditch, S., Royer, J.-J. (Eds.), Unconventional Resource Handbook. Elsevier.

Schuenemeyer, J., Gautier, D., August 25–27, 2014. Probabilistic Resource Costs of Continuous Oil Resources in the Bakken and Three Forks Formations, North Dakota and Montana. Paper URTeC 1929983, presented at the URTeC, Denver, Colorado, USA.

Shanley, K.W., 2004. Fluvial reservoir description for a Giant low-permeability gas field, Jonah field, Green River Basin, Wyoming. In: Jonah Field: Case Study of a Tight-Gas Fluvial Reservoir, AAPG Studies in Geology 52, pp. 159–182.

Sieminski, A., June 16, 2014. In: US Oil and Natural Gas Outlook," IAEE International Conference, New York web link: http://www.eia.gov/pressroom/presentations/sieminski_06162014.pdf.

Simpson, E.H., 1951. The interpretation of interaction in Contingency tables. Journal of the Royal Statistical Society, Series B 13, 238–241.

Slatt, R.M., Singh, P., Philp, R.P., Marfurt, K.J., Abousleiman, Y., O'Brien, N.R., 2008. Workflow for stratigraphic characterization of unconventional gas shales. In: SPE 119891, SPE Sahel Gas Production Conference, Fort Worth, TX, USA, 16–18 November.

Slatt, R., et al., August 25–27, 2014. Sequence Stratigraphy, Geomechanics, Microseismicity, and Geochemistry Relationships in Unconventional Resource Shales. Paper URTeC 1934195, presented at the URTeC, Denver, Colorado, USA.

Smart, K.J., Ofoegbu, G.I., Morris, A.P., McGinnis, R.N., Ferrill, D.A., 2014. Geomechanical modeling of hydraulic fracturing: why mechanical stratigraphy, stress state, and pre-existing structure matter. AAPG Bulletin 98 (11), 2237–2361.

Sondergeld, et al., 2010. Petrophysical Considerations in Evaluating and Producing Shale Gas Resources. SPE 131768, presented at the 2010 SPE Unconventional Gas Conference, Pittsburgh, PA, USA, 23–25 February.

Spencer, C.W., 1989. Review of characteristics of low-permeability gas reservoirs in western United States. AAPG Bulletin 73, 613–629.

Stephens, M., He, W., Freeman, M., Sartor, G., August 12, 2013. Drilling Fluids: Tackling Drilling, Production, Wellbore Stability, and Formation Evaluation Issues in Unconventional Resource Development. Society of Petroleum Engineers. http://dx.doi.org/10.1190/URTEC2013-105.

Suarez-Rivera, R., Deenadayalu, C., Chertov, M., Hartanto, R.N., Gathogo, P., Kunjir, R., 2011. Improving Horizontal Completions on Heterogeneous Tight-shales. Society of Petroleum Engineers. http://dx.doi.org/10.2118/146998-MS.

Suarez-Rivera, R., et al., 2013. Development of a heterogeneous earth model in unconventional reservoirs, for early assessment of reservoir potential. ARMA 13-667.

Sullivan, R., Woodroof, R., Steinberger-Glaser, A., Fielder, R., Asadi, M., 2004. Optimizing Fracturing Fluid Cleanup in the Bossier Sand Using Chemical Frac Tracers and Aggressive Gel Breaker Deployment. Society of Petroleum Engineers. http://dx.doi.org/10.2118/90030-MS.

Taylor, T.L., August 25–27, 2014. Demonstrating Social Responsibility in Water Management. Paper URTeC 1922591, presented at the URTeC, Denver, Colorado, USA.

Teodoriu, C., 2012. Selection Criteria for Tubular Connection used for Shale and Tight Gas Applications. Paper SPE 153110 presented at the SPE/EAGE European Unconventional Resources Conference and Exhibition, 20–22 March, Vienna, Austria.

Theloy, C., Sonnenberg, S.A., August 2013. Integrating Geology and engineering: implications for Production in the Bakken Play, Williston Basin. Paper URTeC 1596247 presented at the Unconventional Resources Technology Conference, Denver, CO, USA.

Tissot, B.P., Welte, D.H., 1984. Petroleum Formation and Occurrence. Springer-Verlag, Berlin, Germany.

Tollefsen, E.M., et al., 2010. Unlocking the Secrets for Viable and Sustainable Shale Gas Development. Paper SPE 139007 presented at the SPE Eastern Regional Meeting, 13–15 October, Morgantown, West Virginia, USA.

Vengosh, A., et al., 2014. A critical Review of the risks to water resources from unconventional shale Gas development and hydraulic fracturing in the United States. Environmental Science and Technology 48, 8334–8343.

Vernik, L., Chi, S., Khadeeva, Y., 2012. In: Rock Physics of Organic Shale and Its Applications: Society of Exploration Geophysicists (SEG) Annual Meeting, 4–9 November. Las Vegas, SEG-2012-0184.

REFERENCES

Vincent, M.C., 2013. Five things you didn't want to know about hydraulic fractures. In: Bunger, A.P., McLennan, J., Jeffrey, R. (Eds.), Effective and Sustainable Hydraulic Fracturing. INTECH, pp. 81–93.

Vincent, M.C., October 30–November 2, 2011. Optimizing Transverse Fractures in Liquid-Rich Formations. Paper SPE 146376, SPE Annual Technical Conference and Exhibition, Denver, CO.

Vincent, M.C., 2010. Restimulation of Unconventional Reservoirs: When Are Refracs Beneficial? Society of Petroleum Engineers. http://dx.doi.org/10.2118/136757-MS. Paper SPE 136757.

Viswanathan, A., Watkins, H.H., Reese, J., Corman, A., Sinosic, B.V., 2014. Sequenced Fracture Treatment Diversion Enhances Horizontal Well Completions in the Eagle Ford Shale. Society of Petroleum Engineers. http://dx.doi.org/10.2118/171660-MS.

Wang, F.P., Gale, J.F., 2009. Screening Criteria for Shale-Gas systems. Gulf Coast Association of Geological Society Transactions 59, 779–793.

Warpinski, N., 2009. Microseismic Monitoring: Inside and Out. Society of Petroleum Engineers. http://dx.doi.org/10.2118/118537-JPT. November issue 80–85.

Warpinski, N.R., Mayerhofer, M.J., Agarwal, K., Du, J., 2013. Hydraulic-fracture geomechanics and microseismic-source mechanism. Society of Petroleum Engineers Journal 766–780.

Warpinski, N.R., Mayerhofer, M.J., Davis, E.J., Holley, E.H., August 25–27, 2014. Integrating Fracture Diagnostics for Improved Microseismic Interpretation and Stimulation Modeling. Paper URTeC 1917906, presented at the URTeC, Denver, Colorado, USA.

Warpinski, N.R., et al., February 10–12, 2008. Stimulating Unconventional Reservoirs: Maximizing Network Growth while Optimizing Fracture Conductivity. Paper SPE 114173 presented at the SPE Unconventional Reservoirs Conference, Keystone, Colorado, USA.

Waters, G.A., Lewis, R.E., Bentley, D.C., October 30–November 2, 2011. The Effect of Mechanical Properties Anisotropy in the Generation of Hydraulic Fractures in Organic Shale. Paper SPE 146776 Presented at the SPE ATCE, Denver, Colorado, USA.

Waters, G., et al., January 19–21, 2009. Simultaneous Hydraulic Fracturing of Adjacent Horizontal Wells in the Woodford Shale. Paper SPE 119635 presented at the SPE Hydraulic Fracturing Technology Conference, The Woodlands, Texas, USA.

Weng, X., Cohen, C., Kresse, O., 2015. Impact of pre-existing natural fractures on hydraulic fracture stimulation. In: Ma, Y.Z., Holditch, S., Royer, J.-J. (Eds.), Unconventional Resource Handbook. Elsevier.

Wilber, T., 2012. Under the Surface. Cornell University Press, Ithaca, New York and London, England, 272 p.

Woodroof, R.A., Asadi, M., Warren, M.N., 2003. Monitoring Fracturing Fluid Flowback and Optimizing Fluid Cleanup Using Frack Tracers. Paper SPE 82221, presented at the SPE European Formation Damage Conference, The Hague, The Netherlands, 13–14 May.

Wu, K., Olson, J.E., November 1, 2013. Investigation of the Impact of Fracture Spacing and Fluid Properties for Interfering Simultaneously or Sequentially Generated Hydraulic Fractures. Society of Petroleum Engineers. http://dx.doi.org/10.2118/163821-PA.

Wutherich, K.D., Walker, K.J., 2012. Designing Completions in Horizontal Shale Gas Wells—Perforation Strategies. SPE 155485.

Xu, W., Le Calvez, J., Thiercelin, M., June 15–17, 2009. Characterization of Hydraulically Induced Fracture Network Using Treatment and Microseismic Data in a Tight-Gas Formation: A Geomechanical Approach. Paper SPE 125237 presented at the SPE Tight Gas Completions Conference, San Antonio, Texas, USA.

Young, S., Friedheim, J., March 20, 2013. In: Environmentally Friendly Drilling Fluids for Unconventional Shale. Offshore Mediterranean Conference.

Yu, W., Sepehrnoori, K., 25-27 August, 2014. Optimization of Well Spacing for Bakken Tight Oil Reservoirs. Paper URTeC 1922108, presented at the URTeC, Denver, Colorado, USA.

Zhang, J., Wieseneck, J., October 30–November 2, 2011. Challenges and Surprises of Abnormal Pore Pressure in Shale Gas Formations. Paper SPE 145964, Presented at the SPE Annual Technical Conference and Exhibition, Denver Colorado, USA.

Zhou, Q., Dilmore, R., Kleit, A., Wang, J.Y., August 25–27, 2014. Evaluating Gas Production Performance in Marcellus Using Data Mining Technologies. Paper URTeC 1920211, presented at the URTeC, Denver, Colorado, USA.

Zoback, M.D., 2007. Reservoir Geomechanics, sixth ed. Cambridge University Press.

Zuckerman, G., 2013. The Frackers. Penguin, New York, USA, p. 404.

CHAPTER 2

WORLD RECOVERABLE UNCONVENTIONAL GAS RESOURCES ASSESSMENT

Zhenzhen Dong[1], Stephen A. Holditch[2], W. John Lee[3]

Schlumberger, College Station, TX, USA[1]; Texas A&M University, College Station, TX, USA[2]; University of Houston, Houston, TX, USA[3]

2.1 INTRODUCTION

As the world reserves of liquid hydrocarbons from conventional reservoirs peak and begin to decline, natural gas will play an increasingly important energy supply role. However, as the use of natural gas increases, additional supplies will be needed. To obtain additional natural gas supplies, the industry can develop unconventional gas resources that are often overlooked in the search for conventional hydrocarbons. Higher natural gas prices and significant technological advances have led to a dramatic increase in production of unconventional gas resources in the United States, and that trend is expected to continue unabated and to expand worldwide.

Three natural gas sources—coal bed methane (CBM), tight gas, and shale gas—comprise today's unconventional gas. Methane hydrate reservoirs, a future candidate, are still decades away from being a potential energy source, mainly due to their location and market conditions. Gas hydrates are found in the Arctic and in deep water, neither of which have any pipeline capacity available for taking the gas to market.

CBM is methane that is mostly adsorbed to the surface of the coal, although there can be some free gas in the coal also. CBM is considered an unconventional natural gas resource because it does not rely on "conventional" trapping mechanisms, such as a fault or anticline, or stratigraphic traps. Instead, CBM is "adsorbed" or attached to the molecular structure of the coals—an efficient storage mechanism as coals can contain as much as seven times the amount of gas typically stored in a conventional natural gas reservoirs.

Tight gas is the term commonly used to refer to low-permeability reservoirs that produce mainly dry natural gas. Many of the low-permeability reservoirs developed in the past are sandstone, but significant quantities of gas also are produced from low-permeability carbonates. In this chapter, production of gas from tight sandstones is the predominate theme. The best definition of a tight gas reservoir is "a reservoir that cannot be produced at economic flow rates nor recover economic volumes of natural gas unless the well is stimulated by a large hydraulic fracture treatment, by a horizontal wellbore, or by use of multilateral wellbores" (Holditch, 2006).

Shale gas refers to natural gas (mainly methane) in fine-grained, organic-rich rocks (gas shales). When talking about shale gas, the word shale does not refer to a specific type of rock. Instead, it describes rocks with more fine-grained particles (smaller than sand) than coarse-grained particles, such as shale (fissile) and mudstone (nonfissile), siltstone, fine-grained sandstone interlaminated with shale or mudstone, and carbonate rocks. Gas is stored in shales in three ways: (1) adsorbed gas is gas attached to organic matter or to clays; (2) free gas is gas held within the tiny spaces in the rock (pores, porosity or microporosity) or in spaces created by the rock cracking (fractures or microfractures); and (3) solution gas is gas held within other liquids, such as bitumen and oil. Gas shales are almost always source rocks that have not released all of their generated hydrocarbons. In fact, source rocks that are "tight" or "inefficient" at expelling hydrocarbons may be the best prospects for shale gas potential.

2.1.1 PETROLEUM RESOURCES MANAGEMENT SYSTEM (PRMS)

The terms "resources" and "reserves" have been used in the past and continue to be used to represent various categories of mineral and/or hydrocarbon deposits. In March 2007, the Society of Petroleum Engineers, the American Association of Petroleum Geologists, the World Petroleum Council, and the Society of Petroleum Evaluation Engineers jointly published the PRMS to provide an international standard for classification of oil and gas reserves and resources (Fig. 2.1(a)). Technically and economically recoverable resources are not formally defined in the system.

2.1.2 ENERGY INFORMATION ADMINISTRATION (EIA) CLASSIFICATION SYSTEM

According to the EIA (2010), technically recoverable resources are the subset of the total resource base that is recoverable with existing technology. The term "resources" represents the total quantity of hydrocarbons that are estimated, at a particular time, to be contained in: (1) known accumulations and

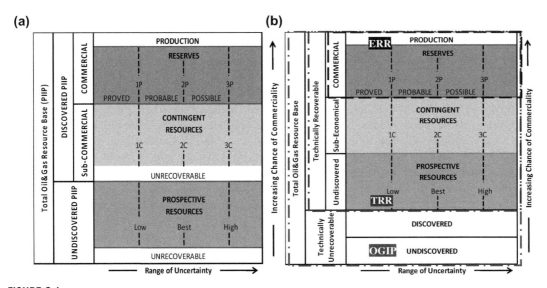

FIGURE 2.1

Flow chart and generalized division of resources and reserves categories (Dong et al., 2013b). (a) Resource Classification of PRMS and (b) EIA definitions mapped to PRMS categories.

(2) accumulations that have yet to be discovered (prospective resources). Economically recoverable resources are those resources that are technically recoverable and can be developed and marketed profitably. It is important to note that economically unrecoverable resources may, at some time in the future, become recoverable, as soon as the technology to produce them becomes less expensive or the characteristics of the market are such that companies can ensure a fair return on their investment by extracting the resources. Dong et al. (2013b) also considered TRR to be the resources that can be produced within a 25-year time period.

Dong et al. (2013b) rearranged categories of PRMS and showed how the estimates of technically and economically recoverable resources are classified (Fig. 2.1(b)). Commercial resources, which include cumulative production and reserves, are economically recoverable resources. Technically recoverable resources are the subset of the total resource base that includes commercial resources, contingent resources and prospective resources. Estimated ultimate recovery (EUR) is not a resources category, but a term that refers to the quantities of petroleum which are estimated to be potentially recoverable from an accumulation, including those quantities that have already been produced.

2.1.3 BASIN TYPES AND GLOBAL DISTRIBUTION OF BASINS

To determine whether the distribution of North American basin types is representative of the distribution of basin types in the rest of the world, Dong et al. (2012) compared the 26 North American basins evaluated in this study with 151 global basins in which giant oil and gas fields are located (Mann et al., 2001; Fig. 2.2). There is a similar distribution of basin types between the North American and global basins that have giant fields. For example, foreland basins account for 53% of global basins and 44% of North American basins (Fig. 2.2). It was acknowledged that this approach is not as robust as an integrated basin-by-basin assessment of basin type, reservoirs, source rocks, and resources, but such a detailed evaluation was impractical for a global study.

2.1.4 MONTE CARLO PROBABILISTIC APPROACH

Uncertainty exists in geologic and engineering data and, consequently, in the results of calculations made with these data. Probabilistic approaches are required to provide an assessment of uncertainty in

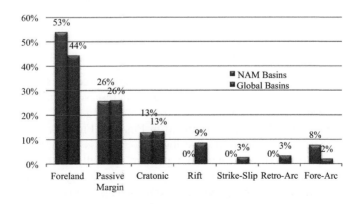

FIGURE 2.2

Comparison of basin types between North American and global basins.

resources estimates. Reservoir simulation coupled with stochastic methods (e.g., Monte Carlo) provides an excellent means to predict production profiles for a wide variety of reservoir characteristics and producing conditions. The uncertainty is assessed by generating a large number of simulations, sampling from distributions of uncertain geologic, engineering, and other important parameters. Uncertainty has been a subject of study for some time in CBM reservoirs. For example, Oudinot et al. (2005) coupled Monte Carlo simulation with a fractured reservoir simulator, COMET3, to assess the EUR in coalbed methane reservoirs.

2.2 METHODOLOGY

Dong et al. (2012) first established the probability distributions of unconventional original gas-in-place (OGIP) for seven world regions. Using the concept of the resource triangle, they proposed that one can estimate unconventional gas in place by knowing the volumes of oil and gas that exist in the conventional reservoirs. Following are the assumptions they made to assess the distribution of unconventional gas in place.

1. CBM OGIP is proportional to original coal in place,

$$\text{CBM OGIP} = A \times \text{Original Coal in place} \tag{2.1}$$

The values of "A" represent the distribution of the average gas content in coal seams.

2. The value of tight gas OGIP is proportional to the value of conventional OGIP,

$$\text{Tight Gas OGIP} = B \times \text{Conventional OGIP} \tag{2.2}$$

The distribution "B" was estimated from the distribution of tight gas and conventional original gas in place in North America.

3. The sum of coal bed OGIP and shale OGIP is proportional to the sum of tight gas OGIP and conventional hydrocarbons in place,

$$\text{CBM OGIP} + \text{Shale OGIP} = C \times (\text{Tight Gas OGIP} + \text{Conventional OOIP} + \text{Conventional OGIP}) \tag{2.3}$$

The values of "C" make up the distribution of the ratio, which was estimated from the distribution of original conventional hydrocarbon and unconventional gas resources in North America. A possible limitation of this approach is that it omits potential contributions of carbonate source rocks.

Dong et al. (2013b) then developed an unconventional gas resource assessment system (UGRAS), which integrates the Monte Carlo simulation with an analytical reservoir simulator, PMTx 2.0 (2012), to generate probability distributions of recovery factors (RFs) (Dong et al., 2013b). The workflow of the probabilistic reservoir model UGRAS is outlined in Fig. 2.3. First, an input file is created and uncertain parameters are assigned probability distributions. There is no limitation to the number of parameters that can be varied. The distribution types are typically normal, uniform, triangular, exponential, or lognormal. The reservoir model and probability density functions for uncertain parameters are varied until a reasonable match between simulated and actual cumulative distribution for several years of cumulative production is obtained. Finally, the resulting probability density functions

2.3 GLOBAL UNCONVENTIONAL GAS ORIGINAL GAS-IN-PLACE ASSESSMENT

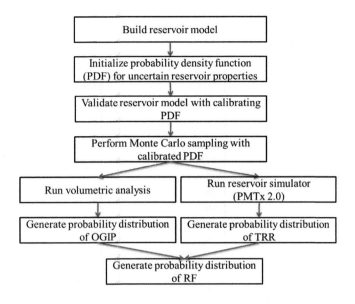

FIGURE 2.3

UGRAS flow chart (Dong et al., 2013a).

(after calibration) are sampled using Monte Carlo for volumetric analysis and flow simulation (PMTx 2.0) to generate the probability distributions of OGIP, TRR, and RF.

To derive a representative distribution of RFs for unconventional gas reservoirs, a distribution of OGIP and TRR must be generated for gas wells in the primary active producing basins of a country or a region. Thus, Dong et al. (2013a, 2014, 2015) used UGRAS to evaluate the most active unconventional gas reservoirs in the United States. Finally, they extended the distribution of RFs gained from our analyses of unconventional gas reservoirs in the United States to estimate technically recoverable unconventional gas resources for the seven world regions.

2.3 GLOBAL UNCONVENTIONAL GAS ORIGINAL GAS-IN-PLACE ASSESSMENT

Using the concept of the resource triangle, Dong et al. (2012) started assessing global unconventional gas OGIP with CBM, followed by tight gas and shale gas.

2.3.1 CBM OGIP

Rogner (1997) estimated 9000 Tcf of CBM gas in place worldwide (Table 2.1). The deterministic assessment was established by using the distribution of coal resources and estimated values for coal bed gas content. Based on Kuuskraa's study (1992), Rogner reported that the worldwide coal bed gas resources range from 2980 to 9260 Tcf. However, only the top 12 countries with coal resources were included in the assessment. While Rogner's initial work focused on these 12 major coal-bearing areas,

Table 2.1 Regional Level Assessments of Global OGIP in CBM Reservoirs, in Tcf

Region	Rogner (1997)	Dong et al. (2012) P50
Commonwealth of Independent States (CIS)	3957	859
North America (NAM)	3017	1629
Australasia (AAO)	1724	1348
Europe (EUP)	274	176
Latin America (LAM)	39	13
Africa (AFR)	39	18
Middle East (MET)	0	9
World	9051	4046

many other countries, such as Spain, Hungary, and France, have smaller but significant coal reserves and by extension, coal bed gas resources.

Dong et al. (2012) improved global CBM OGIP assessment by including more countries and generated the probabilistic distribution of CBM OGIP for seven world regions (Fig. 2.4). The largest CBM resource bases lie in North America, Australasia, and the CIS. However, much of the world's CBM recovery potential remains untapped. CBM will not be produced in some countries due to a lack of incentive to fully exploit the resource base, particularly in parts of the CIS where conventional natural gas is abundant.

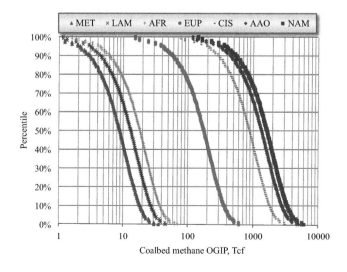

FIGURE 2.4

Probability distribution of coal bed methane OGIP for seven world regions (Dong et al., 2012).

2.3.2 TIGHT GAS OGIP

The first notable estimate of global tight gas OGIP was reported by Rogner (1997). He estimated 7400 Tcf OGIP, worldwide, in tight gas formations of seven world regions (Table 2.2). Since tight gas reservoirs are present in every petroleum province, Rogner allocated the regional OGIP in tight gas formations by weighting Kuuskraa and Meyer's (1980) estimated global tight gas volume of nearly 190 gigatonne of oil equivalent (Gtoe) with the regional distribution of conventional gas. However, Rogner's estimates did not quantify the considerable uncertainty in the size of tight gas regional OGIP.

Tight gas OGIP in the North America has increased significantly since the 1990s. Dong et al. (2012) reviewed tight gas OGIP assessments available in the published literature for 14 North American basins. They added the minimum and maximum reported tight gas OGIP estimates for each basin and obtained an overall range of 8748 to 13,105 Tcf in place for the 14 North American basins. The reported range greatly exceeds Rogner's (1997) estimate of 1317 Tcf for the total North American tight gas OGIP Rogner (1997). If the increase in North America holds similarly for other plays around the world, Rogner's global tight gas OGIP estimate will prove to be conservative. Thus, Dong et al. (2012) presented the probabilistic solution to establish the distributions of tight gas OGIP for the seven world regions (Fig. 2.5). Except for the Middle East and the CIS, the largest tight OGIP is in North America (Dong et al., 2012).

2.3.3 SHALE GAS ORIGINAL GAS-IN-PLACE

Rogner (1997) estimated shale-gas OGIP to be 16,000 Tcf for seven groupings of world countries (Table 2.3). However, Rogner's (1997) world estimate is most likely conservative, given the discoveries of significant shale gas worldwide, such as the Eagle Ford shale in the United States and the Mikulov shale in Austria. Actually, a basin-by-basin assessment of shale gas resources in five regions containing 32 countries, conducted by the EIA (2011), indicates that shale-gas OGIP (25,840 Tcf) is larger than estimated by Rogner in 1997 (16,112 Tcf), even accounting for the fact that Russia and the Middle East were not included in the EIA's study (but were included in Rogner's assessment)

Table 2.2 Regional Level Assessments of Global OGIP in Tight Gas Reservoirs, in Tcf

Region	Rogner (1997)	Dong et al. (2012) (P50)
Australasia (AAO)	1802	6253
North America (NAM)	1371	10,784
Latin America (LAM)	1293	3366
Commonwealth of Independent States (CIS)	901	28,604
Middle East (MET)	823	15,447
Africa (AFR)	784	4000
Europe (EUP)	431	3525
World	7405	72,182

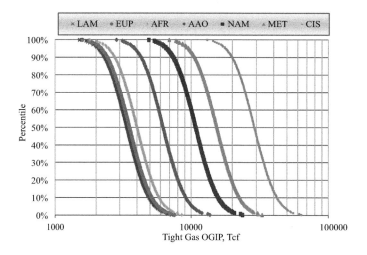

FIGURE 2.5

Probability distribution of tight gas OGIP for the seven world regions (Dong et al., 2012).

Table 2.3 Regional Level Assessments of Global OGIP in Shale Gas Reservoirs, in Tcf

Region	Rogner (1997)	EIA (2011)	Dong et al. (2012) P50
Australasia (AAO)	6151	7042	2690
North America (NAM)[a]	3840	5314	5905
Middle East (MET)	2547	N/A	15,416
Latin America (LAM)	2116	6935	3742
Commonwealth of Independent States (CIS)	627	N/A	15,880
Europe (EUP)	549	2587	2194
Africa (AFR)	274	3962	3882
World	16,103	25,840	50,220

N/A, not available.
[a]Includes United States and Canada.

(Table 2.3). However, neither Rogner's nor the EIA's estimates quantified the considerable uncertainty in shale gas OGIP.

Dong et al. (2012) presented a probabilistic solution and established the probability distributions of shale-gas OGIP for the seven world regions originally used by Rogner (Fig. 2.6). Except for the Middle East and the CIS, the largest differences between EIA and Dong et al. (2012) estimates are the shale-gas OGIP assessments for Australasia and Latin America.

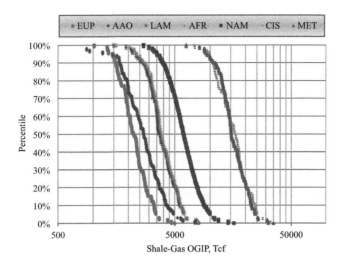

FIGURE 2.6

Probability distributions of shale-gas OGIP for seven world regions (Dong et al., 2012).

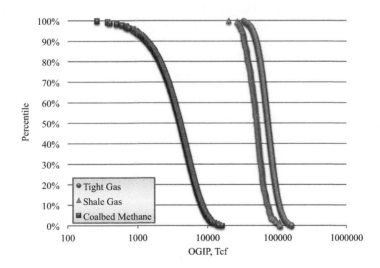

FIGURE 2.7

Probability distribution of world unconventional gas OGIP (Dong et al. 2013, 2014, 2015).

Dong et al. (2013, 2014, 2015) added the seven probability distributions of unconventional gas OGIP up with a correlation coefficient of zero and established the distribution of world unconventional gas OGIP with Monte Carlo simulation, respectively (Fig. 2.7). Global CBM resource endowment ranged from 1300 (P90) to 8000 (P10) Tcf worldwide, with a P50 value of 4000 Tcf (Fig. 2.7). Global tight OGIP was estimated to be between 48,000 (P90) and 105,000 (P10) Tcf, with a P50 value of 72,000 Tcf (Fig. 2.7). Global shale OGIP was estimated to be between 34,000 (P90) and 73,000 (P10) Tcf, with a P50 value of 50,000 Tcf (Fig. 2.7).

2.4 TECHNICALLY RECOVERABLE RESOURCES RECOVERY FACTOR

There are many publications in which one can find information on technically recoverable unconventional gas resources in limited geographic areas. However, little is known publicly about values of TRR for unconventional reservoirs on a global scale. To derive a representative distribution of RFs for unconventional gas reservoirs, Dong et al. (2013, 2014, 2015) developed the data sets, methodology, and tools to determine the values of technically recoverable unconventional gas resources in the primary active producing basins in the United States. A detailed simulation study of all the unconventional gas formations in the United States basins was not performed mainly due to lack of data. However, one or more small scale simulations performed on the active unconventional gas formations which have majority producing gas wells, substantial available public production data, and reservoir descriptions in the literature were considered representative of the basin as a whole.

2.4.1 COAL BED METHANE RECOVERY FACTORS

As of 2011, the top two producing CBM basins were the San Juan and Powder River basin which have the most producing well data available. Dong et al. (2014) generated general 25-year RFs of CBM for the Fruitland Formation in the San Juan Basin and the Big George coal in the Powder River Basin, respectively. Then, they derived a representative probability distribution of RFs for United States CBMformations from these two coals. The best fitting function for this distribution follows a log–logistic distribution, with a P50 value of 38% (Fig. 2.8).

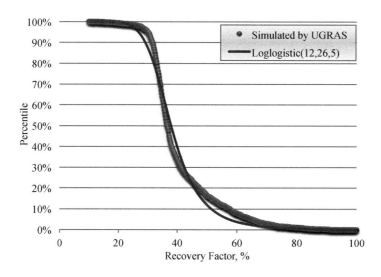

FIGURE 2.8

25-year recovery factor of United States coal bed methane formation.

2.4.2 TIGHT GAS RECOVERY FACTORS

As of 2011, the top two producing tight gas basins were Greater Green River and East Texas basins. The Lance formation in the Greater Green River basin has been extensively developed with vertical wells and hydraulic fracture treatments, especially, after 2000. In the East Texas basin, the Cotton Valley and Travis Peak tight gas formation have the majority of producing wells. Thus, Dong et al. (2013b) applied UGRAS to generate a general 25-year RFs for technically recoverable tight gas for the Lance formation in the Greater Green River basin, Cotton Valley, and Travis Peak formations in the East Texas basin. They then put all the realizations of 25-year RFs of the three study tight gas formations together with equal weight and generated a probability distribution of RFs for United States tight gas formations. The probability distribution of RFs is very high, with a range from 70% (P90) to 85% (P10) (Fig. 2.9) because the induced hydraulic fractures penetrate most of these three tight gas formations.

2.4.3 SHALE GAS RECOVERY FACTORS

Dong et al. (2015) applied the workflow of UGRAS to assess the distribution of OGIP, TRR, and RF for the five key shale gas plays in the United States, including the Barnett, the gas window of the Eagle Ford, the Marcellus, the Fayetteville, and the Haynesville. They put all the realizations of probability distributions of RFs from the five shale gas plays in the United States together with equal weight to derive a probability distribution of RFs for technically recoverable shale gas formations in the United States. The best fitting function for the representative distribution of RFs is a general Beta distribution with a mean value of 25% (Fig. 2.10).

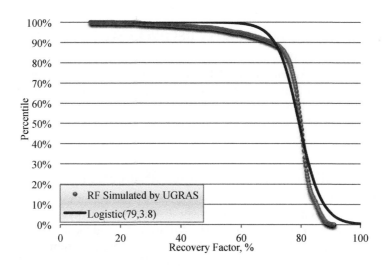

FIGURE 2.9

25-year recovery factor of United States tight gas formations.

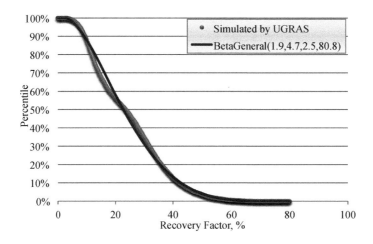

FIGURE 2.10

25-year recovery factor of United States shale gas plays.

2.5 GLOBAL RECOVERABLE UNCONVENTIONAL GAS RESOURCE EVALUATION

Dong et al. (2013, 2014, 2015) assumed that the distribution of RFs for technically recoverable unconventional gas globally are the same as in the United States, since there is similar distribution of basin types in the global basins and North American (Dong et al., 2012).

2.5.1 COAL BED METHANE TECHNICALLY RECOVERABLE RESOURCES

The distributions of CBM OGIP for the seven global regions (Fig. 2.4; Dong et al., 2012) and the probabilistic distribution of 25-year RFs for CBM reservoirs in the United States were established (Fig. 2.8; Dong et al., 2013a). They multiplied the distributions of CBM OGIP by the same distribution of RFs to estimate the technically recoverable resources from CBM reservoirs for the seven world regions (Fig. 2.11). Table 2.4 listed the P90, P50, and P10 of CBM TRR for the seven world regions.

2.5.2 TIGHT GAS TECHNICALLY RECOVERABLE RESOURCES

The distributions of tight OGIP for each of seven world regions were determined (Fig. 2.5; Dong et al., 2012). Dong et al. (2013b) established the probabilistic distributions of 25-year technically RFs for the tight gas in the United States (Fig. 2.9). Dong et al. (2013b) multiplied the distributions of tight OGIP by the distribution of RF to calculate the technically recoverable resource from tight gas reservoirs for the seven world regions (Fig. 2.12). Table 2.5 lists the P90, P50, and P10 tight gas TRRs estimated for the seven world regions by this study. Results of this work indicate the existence of significant global resources of technically recoverable tight gas. Large volumes of tight gas are likely to technically recoverable from the CIS and the Middle East.

2.5 UNCONVENTIONAL GAS RESOURCE EVALUATION

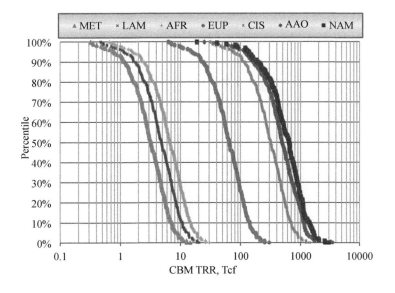

FIGURE 2.11
TRR from CBM reservoirs for seven world regions.

Table 2.4 Assessment Results for CBM TRR Worldwide, in Tcf

Region	P90	P50	P10
North America (NAM)	186	642	1382
Australasia (AAO)	159	488	1129
Commonwealth of Independent States (CIS)	109	318	698
Europe	22	65	143
Africa (AFR)	2	7	15
Latin America (LAM)	2	5	11
Middle East (MET)	1	3	7

2.5.3 SHALE GAS TRR

The distributions of shale OGIP for each of seven world regions were determined (Fig. 2.6; Dong et al., 2012). The representative probability distribution of 25-year TTR RF from shale gas plays in the United States was determined (Fig. 2.10). It was assumed that shale gas TRR RFs are the same globally as in the United States as there is similar distribution of basin types in the global basins that have giant fields and North American basins (Dong et al., 2012). Dong et al. (2015) then multiplied the distributions of shale OGIP by the distribution of 25-year RFs to assess technically recoverable resources from shale gas reservoirs for the seven world regions (Fig. 2.13). Significant technically

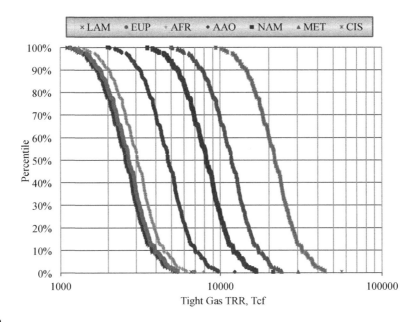

FIGURE 2.12

TRR of tight gas reservoirs for seven world regions.

Table 2.5 Assessment Results of Tight Gas TRR Worldwide, in Tcf			
Region	**P90**	**P50**	**P10**
Commonwealth of Independent (CIS)	14,504	21,745	31,919
Middle Ease (MET)	7832	11,743	17,237
North America (NAM)	5468	8198	12,034
Australasia (AAO)	3171	4754	6978
Africa (AFR)	2028	3041	4464
Europe (EUP)	1788	2680	3934
Latin America (LAM)	1707	2559	3756

recoverable shale gas resources exist in Middle East and CIS regions. Table 2.6 lists the P90, P50, and P10 shale gas TRRs estimated for the seven world regions.

Dong et al. (2013, 2014, 2015) assumed that the seven probability distributions of CBM, shale gas, and tight gas TRR are independent of each other. Thus, they added the seven probability distributions with a correlation coefficient of zero and generated a distribution of world CBM, shale gas, and tight gas TRR with Monte Carlo sampling, respectively (Fig. 2.14). According to these studies, global CBM TRR ranges from 500 (P90) to 3000 (P10) Tcf. Global tight gas TRR ranges from 37,000 (P90) to

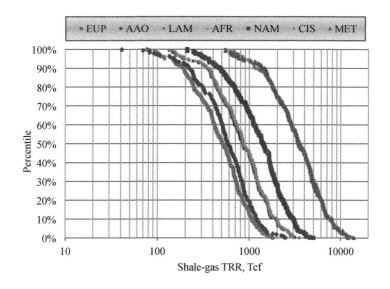

FIGURE 2.13

TRR from shale gas reservoirs for the seven world regions.

Table 2.6 Assessment Results of Shale Gas TRR Worldwide, in Tcf			
Region	P90	P50	P10
Middle Ease (MET)	1354	3415	7974
CIS	1136	3520	7541
North America (NAM)	466	1395	2975
Africa (AFR)	341	862	1991
Latin America (LAM)	342	836	1921
Australasia (AAO)	218	582	1184
Europe (EUP)	188	504	1068

80,000 (P10) Tcf. And global shale gas TRR ranges from 4000 (P90) to 24,000 (P10) Tcf (Fig. 2.14). Table 2.7 lists the P90, P50, and P10 of world CBM, shale gas and tight gas TRRs.

2.6 DISCUSSION

Dong et al.'s (2012, 2013, 2014, 2015) resource assessments are high-level assessments. Although they estimate resources for entire formations, they do not model reservoir and well properties on a well-by-well basis. Instead, they model each formation as a whole, using probability distributions that encompass the variability in reservoir properties across the formation as well as the uncertainty in these properties. Even though they calibrated the Eagle Ford and Marcellus dry gas forecasts against actual production data, there is uncertainty in these forecasts.

FIGURE 2.14

World unconventional gas TRR.

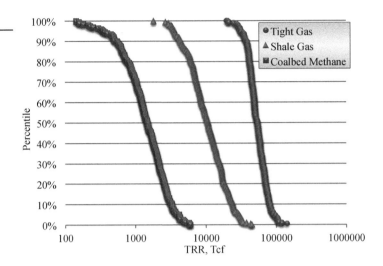

Table 2.7 World Unconventional Gas TRR, in Tcf

	P90	P50	P10
CBM	513	1499	3292
Tight gas	37,491	54,424	79,939
Shale gas	4413	10,723	24,397

The technology and tools described in this chapter can be useful in assessing probabilistic distribution of original gas in place and technically recoverable resources in CBM formations. However, it is important to acknowledge the uncertainties inherent in the data presented in this chapter. The input parameters used to generate production forecasts for the two coals were obtained from the literature and well data. The parameter values and forecasts were reviewed by operators and reserves evaluators in these formations to verify their reasonableness.

It was assumed a 25-year well life for calculation of TRR. It was also acknowledged the uncertainty in the production forecasts generated by the probabilistic analytical simulator. The probabilistic forecasts for the reservoirs were calibrated against actual cumulative production data, although this of course does not guarantee the accuracy of 25-year forecasts.

2.7 CONCLUSION

Using published assessments of 26 North American basins, published global assessments, and resource-triangle-based methodology, Dong et al. (2012) developed a global estimate of unconventional gas in place. With the help of the workflow of UGRAS, Dong et al. (2013b, 2014, 2015) estimated the distribution of technically recoverable resources, and reached the following conclusions.

1. Estimated global unconventional OGIP ranges from 83,300 (P10) to 184,200 (P90) Tcf. The P50 of our global unconventional OGIP assessments (125,700 Tcf) is four times greater than Rogner's (1997) estimate of 32,600 Tcf.
2. Global CBM in place is estimated to be 1300 (P10) to 8100 (P90) Tcf. Global CBM TRR is estimated to be 500 (P90) to 3000 (P10) Tcf, with a P50 value of 1600 Tcf. North America holds the largest amount of CBM in place and technically recoverable CBM resources.
3. The volume of global tight gas OGIP ranges from 49,000 (P10) to 105,000 (P90) Tcf. The volume of global tight gas TRR ranges from 37,000 (P90) to 80,000 (P10) Tcf, with a P50 value of 54,000 Tcf. The CIS region has the largest technically recoverable tight gas resources.
4. The amount of shale-gas OGIP worldwide is 34,000 (P90) to 73,000 (P10) Tcf, with TRR of 4000 (P90) to 24,000 (P10) Tcf. Significant technically recoverable shale gas resources exist in the CIS region and Middle East.
5. The technical RF of the two key CBM formations in the United States follows a log-logistic distribution, with a P50 value of 38%.
6. The RFs of tight gas in the United States follow an extreme value minimum distribution and range from 59% (P90) to 84% (P10), with a P50 value of 79%.
7. The probability distribution of technical RF from the five key shale gas plays in the United States follows a general Beta distribution ranging from 8% to 38%, with a mean value of 25%.

NOMENCLATURE

A Area, acres (1 acre = 43,560 sq. ft)
B_{gi} Gas formation volumetric factor, cf/scf
G_c Initial gas content, scf/lb
h Net pay, ft
p_i Discovery reservoir pressure, psi
p_L Langmuir pressure, pressure at $0.5 \times V_L$, psi
S_{wi} Water saturation, volume fraction of porosity filled with interstitial water
T Formation temperature, °R (460°F + °F at formation depth)
v_L Langmuir volume, gas volume at initial pressure, scf/ton
z_i Compressibility factor at p_i and T, dimensionless
ρ_c Bulk density, lb/cf
Φ_f Fracture porosity, fraction of rock volume available in fracture to store fluid
P90 Value for which there is at least a 90% probability that the value will equal or exceed the estimate, indicated by the 90th percentile on a cumulative probability distribution plot. Similarly for P50 and P10.
Betageneral ($\alpha 1, \alpha 2$, min, max) Beta distribution with defined minimum, maximum and shale parameters $\alpha 1$ and $\alpha 2$.
Gamma (α, β) Gamma distribution with shape parameter α and scale parameter β.
GEV (μ, σ, ξ) Generalized extreme value distribution with mean μ, standard deviation σ and shape parameter ξ.
InvGauss (μ, λ) Inverse Gaussian distribution with mean μ and shape parameter λ.
Logistic (α, β) Logistic distribution with location parameter α and scale parameter β.
Log–logistic (γ, β, α) Log–logistic distribution with location parameter γ, scale parameter β and shape parameter α.
Lognormal (μ, σ) Lognormal distribution with specified mean and standard deviation.

Pearson5 (α, β) Pearson type V (or inverse gamma) distribution with shape parameter α and scale parameter β.
Triangular (min, most likely, max) Triangular distribution with defined minimum, most likely and maximum value.
Uniform (min, max) Uniform distribution between minimum and maximum

REFERENCES

Dong, Z., Holditch, S.A., McVay, D.A., et al., 2015. Probabilistic assessment of world recoverable shale-gas resources. SPE Economics & Management 7 (2). SPE-167768-PA.

Dong, Z., Holditch, S.A., Ayers, W.B., et al., 2014. Probabilistic estimate of global coalbed methane recoverable resource. SPE Economics & Mangement 7 (2). SPE-169006-PA.

Dong, Z., Holditch, S.A., Ayers, W.B., 2013a. Probabilistic evaluation of global recoverable tight gas resources. SPE Economics & Mangement 7 (3). SPE-169006-PA.

Dong, Z., Holditch, S.A., McVay, D.A., 2013b. Resource evaluation for shale gas reservoirs. SPE Economics & Management 5 (1), 5–16. SPE-152066-PA.

Dong, Z., Holditch, S.A., McVay, D.A., et al., 2012. Global unconventional gas resource assessment. SPE Economics & Management 4 (4), 222–234. SPE-148365-PA.

EIA, 2010. The Natural Gas Resource Base. http://naturalgas.org/overview/ng_resource_base.asp.

EIA, 2011. World Shale Gas Resources: An Initial Assessment of 14 Regions outside the United States. Energy Infomration Administration, Washington, DC.

Holditch, S.A., 2006. Tight gas sands. SPE Journal of Petroleum Technology 58 (6), 86–93. SPE-103356-MS.

Kuuskraa, V.A., Meyers, R.F., 1980. Review of world resources of unconventional gas. In: Paper Present at IIASA Conference on Conventional and Unconventional World Natural Gas Resources, Laxenburg, Austria, 06/30/1980.

Kuuskraa, V.A., 1992. Hunt for quality basins goes abroat. Oil and Gas Journal 90 (40), 49–54.

Mann, P., Gahagan, L., Gordon, M.B., 2001. Tectonic setting of the world's giant oil and gas fields. AAPG Memoirs 78, 15–105.

Oudinot, A.Y., Koperna, G.J., Reeves, S.R., 2005. Development of a probabilistic forecasting and history matching model for coalbed methane reservoirs. In: Paper Presented at 2005 International Coalbed Methane Symposium, Tuscaloosa, Alabama, 05/05/2005.

PMTx 2.0, 2012. http://www.phoenix-sw.com/.

Rogner, H.H., 1997. An assessment of world hydrocarbon resource. Annual Review of Energy and the Environment 22, 217–262.

CHAPTER 3

GEOCHEMISTRY APPLIED TO EVALUATION OF UNCONVENTIONAL RESOURCES

K.E. Peters[1,2], X. Xia[3], A.E. Pomerantz[4], O.C. Mullins[4]

Schlumberger, Mill Valley, CA, USA[1]; Department of Geological & Environmental Sciences, Stanford University, Palo Alto, CA, USA[2]; PEER Institute, Covina, CA, USA, Current address: ConocoPhillips, Houston, TX, USA[3]; Schlumberger-Doll Research, Cambridge, MA, USA[4]

3.1 INTRODUCTION

The purpose of this chapter is to: (1) summarize current developments in the use of both organic and inorganic geochemistry to identify sweet spots in unconventional mudrock resource plays, and (2) to update the status of our understanding of kerogen structure. Kerogen is the insoluble particulate organic matter in sedimentary rock that includes residues of lipids and biopolymers, as well as reconstituted organic components (Tegelaar et al., 1989). For the purpose of this chapter, we define a sweet spot as a volume of rock with enhanced porosity, permeability, fluid properties, water saturation, and/or rock stress that is likely to produce more petroleum than surrounding rock (e.g., Shanley et al., 2004; Cander, 2012). Sweet spots can be identified both in map view and vertical profile.

3.1.1 SUBSURFACE EVOLUTION OF ORGANIC MATTER

Organic matter evolves in sediments after burial (Fig. 3.1). Biogenic methane originates by microbial activity on organic matter at shallow depth and temperatures less than ~80 °C. Cracking of kerogen at greater depths yields thermogenic petroleum, which might include methane, "wet" hydrocarbon gases (ethane, propane, butanes, and pentanes), condensate, and crude oil. Deep, dry gas can originate directly from the kerogen or by secondary cracking of trapped oil. Generation of oil, condensate, and hydrocarbon gas by cracking of kerogen does not continue indefinitely with depth, but ends when the kerogen is severely depleted in hydrogen. Some biomarkers (molecular fossils) survive diagenesis and much of catagenesis prior to complete destruction during late catagenesis and metagenesis (Peters et al., 2005). The depth scale in Fig. 3.1 differs depending on various factors, such as the geothermal gradient and the kinetics of petroleum generation for the specific kerogen.

Effective source rocks result from deposition and partial preservation of organic matter in fine-grained sediments, followed by thermal alteration, usually due to burial (McKenzie, 1978). Three stages in the evolution of sedimentary organic matter (Fig. 3.1) include: (1) diagenesis or transformations that occur prior to significant thermal alteration; (2) catagenesis or thermal transformation

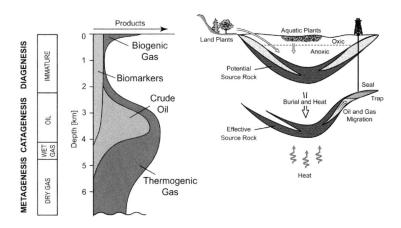

FIGURE 3.1

Schematic evolution of organic matter and volatile products with depth (left) parallels burial and preservation of potential (thermally immature) source rock that can later become effective (mature) source rock for conventional and unconventional petroleum accumulations (right). Preservation of organic matter is favored by anoxic conditions during diagenesis. Burial maturation of potential source rock can result in effective source rock for migrated petroleum. Unconventional petroleum remains in the source rock, which thus also becomes reservoir rock.

From Peters et al. (2005). Reprinted with permission by ChevronTexaco Exploration and Production Technology Company, a division of Chevron USA Inc.

of kerogen to petroleum at ~50 to 200 °C; and (3) metagenesis or thermal destruction of petroleum at >200 °C, but prior to greenschist metamorphism.

Catagenesis includes the generation of petroleum from thermally reactive kerogen (e.g., Mackenzie and Quigley, 1988). Much of the early-generated petroleum that originates from kerogen in oil-prone source rock is liquid that contains dissolved compounds in the gas phase at surface conditions. Above 150 °C, residual hydrogen-poor kerogen generates thermogenic hydrocarbon gas, which may initially contain dissolved condensate, but becomes "drier" or enriched in methane with further heating. The deadline for conversion of oil to gas varies due to complex factors, such as seal integrity, kinetics of cracking specific compounds, burial history, and the presence or absence of thermochemical sulfate reduction. For example, complete transformation of oil to gas in the Jurassic Smackover Formation in the Gulf of Mexico occurs by ~200 °C and vitrinite reflectance (R_o) of 2.0% (Peters et al., 2005). However, Tian et al. (2006) analyzed residual hydrocarbon gas generated by pyrolysis of Triassic oil from the Tarim Basin and concluded that complete cracking to methane requires $R_o > 2.4\%$.

The transformation ratio (TR) describes the extent of conversion of kerogen to petroleum in organic-rich source rock. R_o is another common organic maturity indicator based on the percentage of white light reflected from vitrinite phytoclasts selected from a dried and polished mixture of kerogen and epoxy. Phytoclasts are small particles of organic matter in the kerogen (Bostick, 1979). Vitrinite is a maceral that originates from higher plants. R_o and TR differ in kinetic response and therefore are not interchangeable. Measured kinetic response for vitrinite cracking begins at activation energy (E_a) of

~36 kcal/mol, whereas typical hydrocarbon generation kinetics begin ~10 kcal/mol later (Waples and Marzi, 1998). The rate of thermal decomposition of kerogen to petroleum differs for different source rocks (Tegelaar and Noble, 1994; Peters et al., 2015).

3.1.2 CONVENTIONAL VERSUS UNCONVENTIONAL RESOURCES

Current nomenclature for unconventional source rocks is poorly defined and in many cases is confusing. The terms mudrock, mudstone, and shale are commonly used interchangeably, despite different formal definitions. Strictly, mudrocks consist of >50% mud-size particles and are a class of fine-grained siliciclastic sedimentary rocks that include siltsone, claystone, mudstone, slate, and shale. Unlike mudstone, shale is finely bedded material that splits readily into thin layers. Strictly, shale oil is oil produced by retorting (pyrolysis) of oil shale, which is thermally immature source rock. We will use the term shale oil (also called "tight oil") to indicate oil produced from mudstone-rich rock units. In accordance with common usage, we will use the term "shale" to loosely encompass many different types of fine-grained sedimentary rocks. By this convention, "shale gas" consists of self-sourced thermogenic or biogenic hydrocarbon gases trapped within fine-grained, low permeability, organic-rich mudrock. In this chapter, we will focus on unconventional mudrock resources.

Conventional and unconventional resources are fundamentally different (Table 3.1). We define unconventional resources as rock or sediment units that require stimulation to produce retained petroleum due to the combined effects of petroleum viscosity and matrix permeability. By our definition, gas hydrates and gas shales are unconventional resources because very low permeability impedes the escape of hydrocarbon gases, such as methane, without stimulation. Likewise, oil sand is an unconventional resource because the high viscosity of biodegraded oil does not allow production from otherwise high-permeability sandstone without stimulation. Exploration for conventional resources focuses on petroleum that migrated from source rocks to reservoirs and traps and the timing of petroleum generation–migration–accumulation relative to trap formation is critical. Significant conventional petroleum accumulations can only occur when the trap exists before generation and migration.

3.1.3 EMPIRICAL MEASURES OF SWEET SPOTS

Current geochemical methods to identify sweet spots are largely empirical, i.e., most assume that certain characteristics of productive unconventional rock units are useful to predict sweet spots in analogous rock units. For example, Jarvie (2012a) list various characteristics of the main producing areas within the top 10 shale gas resource systems, including the Marcellus, Haynesville, Bossier,

Table 3.1 Key differences between conventional and unconventional petroleum plays

Conventional	Unconventional
Petroleum migrates to the reservoir rock	Petroleum remains in the source (also reservoir) rock
Oil-to-gas cracking is usually unimportant	Oil-to-gas cracking is important
Reservoir storage is within intergrain porosity	Cracking of kerogen creates organoporosity
Free gas is important	Proportions of free and adsorbed gas are important

Barnett, Fayetteville, Muskwa, Woodford, Eagle Ford, Utica, and Montney formations. In the main producing area, the Eagle Ford Formation has net thickness of 150–300 ft in the depth range 4000–10,000 ft, ~1000 nanodarcy permeability, 6–14% porosity, 60% carbonate content, ~1.2% R_o, present-day and original hydrogen indices of 80 and 411 mg hydrocarbon/g total organic carbon (TOC) 79% transformation ratio, present-day, and original TOC of 2.76 and 4.24 weight percent (wt %), and ethane stable carbon isotope "rollover" (discussed later).

Exploration for unconventional resources targets petroleum retained by the source rock or migrated a short distance within the source rock interval, in which the extent of maturation and geomechanical properties of the rock are critical for successful production. This chapter discusses the geochemical measurements used to characterize unconventional resources. Geomechanical measurements are discussed in more detail elsewhere. However, a short discussion follows on factors that contribute to the brittle character of some unconventional targets, which facilitates hydraulic fracturing to release trapped oil and gas.

The tendency of rock to fracture can be described in various ways (Altindag and Guney, 2010). Many interpreters assume that brittleness is proportional to Young's modulus and/or Poisson's ratio (Rickman et al., 2008), although some argue that computing brittleness from elastic properties is not physically meaningful (Vernik et al., 2012). Nevertheless, one brittleness index (BI) is as follows:

$$BI = \sigma_c/\sigma_d \tag{3.1}$$

where σ_c = compressive strength and σ_d = tensile strength (Aubertin et al., 1994; Ribacchi, 2000). Higher values of BI correspond to more brittle rock, although yield points vary due to nonlinear elasticity.

The above version of BI requires laboratory measurement of compressive and tensile strength. Jarvie et al. (2007) and Wang and Gale (2009) proposed more practical versions of BI based on the mineral compositions of rocks. The amounts of more brittle minerals (quartz or dolomite) in the numerator are divided by the sum of all mineral components in the denominator:

$$BI_{Jarvie} = Quartz/(Quartz + Calcite + Clay) \tag{3.2}$$

$$BI_{Wang\ \&\ Gale} = (Quartz + Dolomite)/(Quartz + Dolomite + Calcite + Clay + TOC) \tag{3.3}$$

Triangular mineralogy plots (Fig. 3.2) are another qualitative means to assess brittleness. Many mudrocks may be too enriched in ductile clays (<2 μm) for effective hydraulic fracturing, whereas source rocks that are rich in comparatively brittle calcite and dolomite (e.g., Eagle Ford Formation) or silica (Muskwa Formation) have more favorable geomechanical properties for unconventional production. Caution is recommended when considering clay content as an indicator of ductility, because the ductility of clay-rich rocks likely varies with mineralogy. For example, it is known that smectitic mudstones generally undergo chemical compaction at shallower depths than kaolinitic mudstones (Bjørlykke, 1998). Further study of the ductility of various clay minerals is needed.

Source rock is a key element in unconventional shale plays because it provides charge, reservoir, trap, and seal. In most shallow conventional reservoirs, secondary oil-to-gas cracking is unimportant. However, oil that is retained in deeply buried conventional reservoirs or in source rocks can undergo further cracking to hydrocarbon gas. Reduced porosity during burial is an important consideration for conventional plays, which contain free gas and oil in intergrain pore space. However, organoporosity increases during the progressive conversion of kerogen to solid bitumen, oil, and gas in unconventional

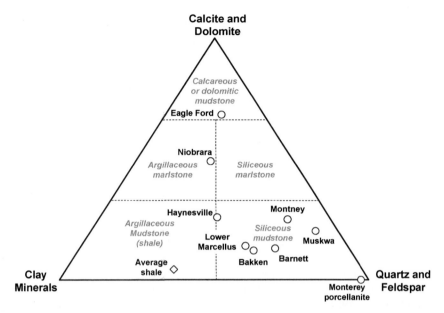

FIGURE 3.2

Mudrock ("shale") mineral composition can be used to indicate susceptibility to hydraulic fracturing; worldwide average mudrock composition (diamond) and some examples of source rocks.

Modified from Allix et al. (2011).

plays (Passey et al., 2010; Dahl et al., 2012). The creation of new porosity by kerogen cracking, results in pores that become filled with some of the generated hydrophobic hydrocarbons. Therefore, unconventional resources comprise free hydrocarbons in pore space and fractures as well as adsorbed within the kerogen or on rock surfaces. Organic-rich source rock that contains 10 wt% TOC may have >30 volume% of kerogen, which represents substantial pore space available for hydrocarbon gas and oil.

Many unconventional source rocks have both low porosity and permeability. Mudrock reservoirs contain silt (4–62.5 µm) and colloid-size (1–1005 nm) particles, but usually are dominated by clay-size particles (1–4 µm). Mudstone permeability can vary by three orders of magnitude at a given porosity (Dewhurst et al., 1999). In typical "tight" gas shale, permeability drops to a few nanodarcies (10^{-6} mD) in which connected pore throats may be only a few methane molecules in width. By comparison, concrete typically shows permeability in the range 0.1–1 mD and many conventional oil reservoir rocks have permeability in the range 100–10,000 mD. Tight shale gas reservoirs are typically low net-to-gross intervals (total pay footage divided by total thickness <10%) dominated by low permeability (<10 nD) rocks having little reservoir potential without interbedded units having higher porosity and permeability.

Unconventional gas and tight oil shales generally have at least 2 wt% TOC present-day, which may have been much higher prior to thermal maturation. Some of the porosity in these unconventional targets occurs in pores that developed in the kerogen or pyrobitumen as it underwent cracking to oil and gas. The storage capacity of many tight organic-rich shales, such as the Barnett Shale (Texas) and Haynesville Shale (Texas, Arkansas, Louisiana), is largely in organoporosity. Many tight shales show a correlation between petroleum saturation and TOC because much of the petroleum is trapped in pores that form in the kerogen during cracking (Passey et al., 2010; Dahl et al., 2012). Clay minerals can also supplement the sorption capacity of shales (Gasparik et al., 2012).

Organic-rich shales can be divided into the following general types (Jarvie, 2012b): (1) tight, i.e., no open fractures, (2) fractured, and (3) hybrid, i.e., having juxtaposed ductile organic-rich and brittle organic-lean lithofacies. The juxtaposed fine and coarse-grained units in hybrid shales facilitate short-range migration of petroleum generated in the fine-grained organic-rich units into the lean units. The result is a lack of correlation between petroleum saturation and TOC in hybrid shales. Hybrid shales, such as the Bakken Formation (Williston Basin), are some of the best targets for tight oil or gas production for two main reasons: (1) matrix and fracture porosity are generally more important in hybrid systems due to the intervening organic-lean units, and (2) expulsion and short-range migration from the organic-rich to the organic-lean units is favored (e.g., migration from the organic-rich upper and lower shale members of the Bakken Formation into the organic-lean and dolomitic middle member).

Migration fractionation causes retention of asphaltenes and resins in fine-grained, organic-rich units and preferential movement of the mobile saturate and aromatic hydrocarbons to organic-lean units (Tissot and Welte, 1984; Kelemen et al., 2006), which increases their productivity. In the Bakken Formation, the organic-rich upper and lower members are mudstones with extremely low porosity and permeability. These units are generally not productive, even when hydraulically fractured. However, the middle member of the Bakken Formation is organic-lean and brittle dolomite (Fig. 3.3), which represents an excellent unconventional production target at the proper level of thermal maturity (Pitman et al., 2001; Sarg, 2012).

3.2 DISCUSSION
3.2.1 ORGANIC GEOCHEMICAL AND PETROPHYSICAL CHARACTERIZATION

Rapid screening tools, such as Rock-Eval pyrolysis, TOC, and vitrinite reflectance (e.g., Peters and Cassa, 1994), can be used to: (1) map the areal distribution of effective source rock and thereby predict the migration paths of generated petroleum toward conventional accumulations, and (2) identify the vertical distribution of sweet spots within each source rock interval to better define unconventional resources. An effective source rock has generated and/or is currently generating and expelling petroleum. The same geochemical screening technology can be used to identify unconventional targets, but it must be supplemented by tools that predict geomechanical properties of the rock. Brittle lithologies, which contain significant amounts of carbonate or silicate minerals (Fig. 3.2), more readily release petroleum when stimulated by hydraulic fracturing compared with ductile lithologies. As discussed above, not all clay minerals have equivalent ductility.

3.2.1.1 Rock-Eval Pyrolysis
Rock-Eval pyrolysis rapidly determines the quality and thermal maturity of organic matter in rock samples (Espitalié et al., 1977; Tissot and Welte, 1984; Peters, 1986). As will be discussed, Rock-Eval 6 can also be used to determine TOC. Concentrations of free and adsorbed hydrocarbons (HC) released

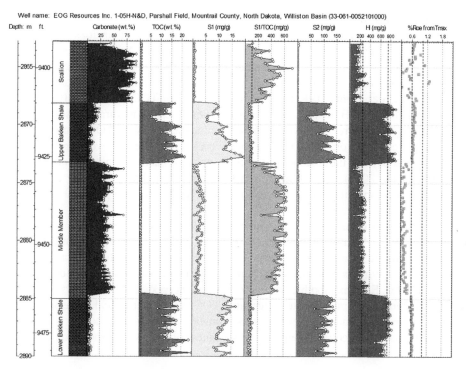

FIGURE 3.3

Organic geochemical log for a well from North Dakota shows TOC, Rock-Eval pyrolysis parameters (see text), and carbonate content. Intervals having oil saturation indices (OSI > 100 mg hydrocarbons/g rock) may produce unconventional petroleum provided they have sufficient brittle character (e.g., carbonate) to fracture during hydraulic stimulation. The Bakken Middle Member in this well is highly productive.

From Jarvie (2012b); American Association of Petroleum Geologists (AAPG)© 2012 reprinted by permission of the AAPG whose permission is required for further use.

by programmed heating of rock cuttings (~100 mg) in a stream of helium at 300 °C for 5 min are recorded as the area under the first peak on the pyrogram (S1; mg HC/g rock). The second peak on the pyrogram is composed of pyrolytic hydrocarbons generated by thermal breakdown of kerogen as the sample is heated from 300 to 600 °C (S2) (mg HC/g rock) at 25 °C/min (Fig. 3.3). CO_2 generated by kerogen degradation is retained during the heating from 300 to 390 °C and is analyzed as the third peak on the pyrogram (S3) (mg CO_2/g rock). Rock-Eval also records the temperature of maximum S2 hydrocarbon yield, T_{max}, which is a measure of maturity. The S2 and S3 peak areas, when calibrated and normalized to percent TOC, yield hydrogen index (HI), and oxygen index expressed as mg HC/g TOC and mg CO_2/g TOC, respectively. The production index is defined as $S1/(S1 + S2)$. PI is identical to TR, except in cases where some S1 has escaped from the rock.

We recommend Rock-Eval pyrolysis and TOC measurements approximately every 10 m (32.8 ft) in each well (Peters, 1986) for assessment of conventional targets. However, closer sample spacing (e.g., 1 m or less) may be required to properly identify sweet spots in unconventional rock units. Pitfalls in Rock-Eval interpretations are discussed in Peters (1986) and Peters and Cassa (1994).

3.2.1.2 Total Organic Carbon

TOC is a measure of the quantity of organic matter in rock samples that does not address organic matter quality (e.g., oil-prone, gas-prone, inert) or thermal maturity. Assuming that organic matter is 83 wt% carbon, total organic matter can be calculated by multiplying TOC by 1.2. Contrary to some earlier publications, there is no significant difference in the minimum TOC for carbonate versus noncarbonate source rock (Jones, 1987). Excellent quantities of TOC for thermally immature source rock are generally >4 wt%. However, TOC decreases with thermal maturation, so reasonable quantities of TOC for highly mature, unconventional source rock are >2 wt% TOC. Various methods have been proposed to reconstruct original TOC for source rocks that have achieved high thermal maturity (e.g., Peters et al., 2005) and these values are critical input for basin and petroleum system modeling (BPSM) (Peters et al., 2006). For BPSM, gross source rock thickness must be corrected for well deviation, structural complexity, and nonsource units that lack petroleum potential (Peters and Cassa, 1994).

Most sedimentary rock samples contain organic and carbonate carbon, which must be differentiated to determine TOC. Many methods are available to quantify organic carbon (Table 3.2); most geochemical methods are based on combustion followed by detection of the generated carbon dioxide using thermal conductivity, coulometry, or spectroscopy. These and other methods are discussed in Gonzalez et al. (2013). Each method has strengths and weaknesses. Table 3.2 shows some examples of two different categories of methods to determine TOC: (1) geochemical measurements on rock samples, and (2) direct and indirect well log measurements using wireline tools.

The following discussion covers strengths and weakness of some of the more common organic and inorganic geochemical and methods to determine TOC.

3.2.1.3 Geochemical Methods for TOC

The most common geochemical method to measure TOC involves acidification of the sample (usually ground cuttings) with 6 N HCl in a filtering crucible (filter acidification; Table 3.2) to remove carbonate minerals, removal of the filtrate by washing/aspiration, drying at $\sim 55\,°C$, and combustion in an elemental analyzer or Leco furnace with metallic oxide accelerator at $\sim 1000\,°C$. The CO_2 generated by combustion is analyzed using an infrared or thermal conductivity detector. Filter acidification–combustion is fast, but the results can be inaccurate for organic-lean, carbonate-rich or low-maturity rock samples. Thermally immature organic matter is susceptible to acid hydrolysis and loss of functional groups containing carbon during filtering. Peters and Simoneit (1982) compared TOC results for thermally immature rock samples based on filter acidification and a modified method that used nonfiltering crucibles in which hydrolyzate was retained. The results show that more than 10% of the TOC was lost as hydrolyzate using the filter acidification method.

Another geochemical method involves splitting the sample into two aliquots, where TOC is determined as the difference between total carbon (TC) from combustion of one aliquot minus carbonate carbon (C_{carb}) from coulometric measurement of CO_2 released upon acid treatment of the second aliquot (Table 3.2). This indirect TOC method is usually applied to organic-lean, carbonate-rich rocks and it is more time-consuming and requires more sample ($\sim 1-2$ g ground rock) than the direct method.

Early versions of the Rock-Eval pyroanalyzer (e.g., Rock-Eval 2) determined TOC as the sum of carbon in the pyrolyzate with that obtained by oxidizing the residual organic carbon at 600 °C. For small samples (~ 100 mg), this method provides reliable TOC that is comparable to that from the filter

Table 3.2 Some Geochemical and Wireline Methods to Determine Total Organic Carbon (TOC) in Boreholes

TOC Method	Approach	Typical Limitation	References
Geochemical (rock samples)			
Filter acidification	6 N HCl, wash, combust	Some organic carbon can be lost by hydrolysis	Peters and Simoneit (1982)
Nonfilter acidification	6 N HCl, combust	Oven corrosion	Peters and Simoneit (1982) and Wimberley (1969)
Total minus coulometric[a]	$TOC = TC - C_{carb}$	Slow, large (split) sample	Engleman et al. (1985)
Rock-Eval[b]	$TOC = PC + RC$	Incomplete combustion for models prior to v. 6	Lafargue et al. (1998)
Laser-induced pyrolysis	Volatilize residual petroleum potential	Requires lab core analysis; no response for inert carbon	Elias et al. (2013)
DRIFTS[c]	Spectroscopy of C–H bonds	Oil-based mud must be removed; kerogen type and maturity affect absorption	Herron et al. (2014)
Wireline (well logs)			
Gross gamma-ray (γ)	Available for most well logging runs	Responds mainly to U, not kerogen; depends on many factors, e.g., Eh/pH	Schmoker (1981)
Spectral γ-ray	TOC to total gamma calibration	Local calibration; uranium minerals interfere, e.g., phosphates	Fertl and Chilingar (1988)
Bulk density[d]	$TOC = (154.497/\rho) - 57.261$	Assumes inorganic density = 2.69 g/cm^3, underestimates TOC in clay and carbonate-rich rocks	Schmoker and Hester (1983)
Δlog R[e]	Log R = R/R$_{baseline}$ − $\chi(\rho - \rho_{baseline})$	Maturity sensitive; assumes similar properties for baseline and organic-rich units; clays interfere	Passey et al. (1990, 2010) and Issler et al. (2002)
Pulsed neutron-spectral γ-ray, LithoScanner[f]	$TOC = TC - TIC$	Requires separate capture and inelastic spectroscopy measurements; borehole and formation corrections for inorganic carbon	Pemper et al. (2009), Radtke et al. (2012), Charsky and Herron (2013), Gonzalez et al. (2013) and Aboud et al. (2014)

Eh, redox potential.
[a]TOC, total organic carbon; TC, total carbon; C_{carb}, carbonate carbon.
[b]PC, pyrolyzable carbon; RC, residual carbon (850 °C combustion).
[c]DRIFTS, diffuse reflectance infrared Fourier transform spectroscopy.
[d]ρ, bulk density, g/cm^3.
[e]$\rho_{baseline}$, bulk density in baseline organic-lean zone, g/cm^3; χ, scaling factor calculated after setting the baseline for the two curves in the organic-lean zone. The Δlog R separation of the two curves is scaled to maturity of the formation to determine TOC in wt%; R, resistivity, ohm/m; $R_{baseline}$, resistivity in the baseline organic-lean zone; $TOC = (\Delta log\ R) \times 10(2.297 - 0.1688 LOM)$ where LOM, level of organic metamorphism (Hood et al., 1975).
[f]$TIC = 0.12(calcite) + 0.13(dolomite)$.

acidification or coulometric methods. However, mature samples having vitrinite reflectance more than ~1% yield inaccurate TOC because of incomplete combustion at 600 °C. Rock-Eval 6 pyrolysis and oxidation reaches 850 °C, which yields more reliable TOC data, especially for highly mature rock samples (Lafargue et al., 1998).

Laser-induced pyrolysis (LIPS; Elias et al., 2013) generates high-resolution organic carbon logs (i.e., ~10,000 measurements on a 100 m core) and can be used to calibrate other well log data, such as γ-ray logs. Each LIPS measurement consists of two parts: (1) an initial low-power laser treatment removes volatile contaminants from the surface of the core, and (2) a second high-energy laser treatment on the same target area yields a signal generated by pyrolysis of the sample. These two laser treatments leave a small crater (1–3 mm diameter) on the surface of the core. The LIPS photoionization detector is insensitive to CO or CO_2 from decomposition of carbonates and thus measures only compounds containing carbon and hydrogen. However, LIPS does not measure nonpyrolyzable (inert) organic matter and therefore reflects only the residual petroleum generative potential and not TOC. Thus, LIPS organic carbon content for highly mature samples or samples that contain significant inertinite can be less than conventional TOC determined on the same sample.

Diffuse reflectance infrared Fourier transform spectroscopy (DRIFTS) is a rapid method to analyze core and cuttings samples for TOC, kerogen composition, and mineralogy (Table 3.2). DRIFTS is perhaps the only measurement of chemical composition robust enough to be employed in real time at the well site. Infrared (IR) spectra in general are discussed later in the chapter. IR spectroscopy measures the wavelength dependence of photon absorption in the infrared region. The wavelength of the absorption is characteristic of the molecular vibrational frequency, which corresponds to particular chemical groups, and the intensity of the absorption is proportional to the abundance of the chemical group. Both organic and inorganic materials respond at IR wavelengths and can be measured by DRIFTS.

DRIFTS measures the absorption of diffusely reflected IR photons. The technique has been used to measure qualitative and quantitative mineralogy and TOC in soils and sediments (Janik and Skjemstad, 1995; Vogel et al., 2008; Rosen et al., 2011). Unlike IR spectroscopy of kerogen, measurements of intact shale must handle overlapping absorption features of organic materials and minerals. The most readily interpreted organic features in DRIFTS analysis of whole shale are aliphatic and aromatic C—H stretches near 2900 cm^{-1}. Ratios such as CH_2/CH_3 and aliphatic C—H_x/aromatic C—H can be measured accurately by DRIFTS of whole shale. These peak ratios can be used to assign maturity, and the total signal in this region, corrected with a scaling factor that is determined from these peak ratios, can be used to measure TOC. Quantifying TOC from the strength of these adsorptions has historically been challenging for shale because the kerogen signal overlaps any oil-based mud signal that contaminates cuttings and because the relationship between TOC and the strength of absorption near 2900 cm^{-1} depends on kerogen type (Appendix) and maturity (Durand and Espitalié, 1976; Painter et al., 1981; Fuller et al., 1982; Ganz and Kalkreuth, 1991; Lin and Ritz, 1993; Ibarra et al., 1994, 1996; Riboulleau et al., 2000; Lis et al., 2005). Recently, those challenges have been addressed by development of a rapid method to remove oil-based mud and a correlation between the relative intensities of the different C—H stretching absorptions and TOC. The correlation allows DRIFTS to be used to estimate maturity, and the number and accuracy of minerals quantified by DRIFTS was improved (Herron et al., 2014). DRIFTS can now be used for simultaneous measurement of TOC, maturity, and mineralogy of core or cuttings. Analyses can be conducted in the laboratory or at the well site in order to provide data as the drill bit progresses in horizontal wells. Failure to adequately clean oil-based mud and drilling additives from cuttings results in inaccurate DRIFTS measurements.

3.2.1.4 Indirect Wireline TOC

Unlike TOC based on organic geochemical measurements of discrete rock samples, direct and indirect inorganic geochemical or wireline methods offer continuous TOC versus depth, limited only by the vertical resolution of the tool. Wireline logs can be used to indirectly estimate TOC based on empirical observations (generally assuming a constant inorganic matrix) or the TOC can be measured directly. Examples of empirical wireline logs for TOC include: (1) gross or spectral γ-ray logs, (2) bulk density logs, and (3) Δlog R (Table 3.2). We recommend that all TOC values inferred from well logs be calibrated to values measured using laboratory-based organic geochemistry.

For gross or spectral γ-ray logs (Table 3.2), one assumes that the amount of TOC is related to uranium (U) in the formation. Typical vertical resolution for gross or spectral γ-ray measurements is ~ 0.6–1.0 m (~ 2–3 ft). This approach has limited applicability because U content depends on many factors, such as the original U content in the formation water, deposition rate, and formation chemistry, including redox potential (Eh) and hydrogen ion concentration (pH). Use of U as a proxy for TOC in organic-rich mudstones requires careful calibration to determine the stratigraphic and regional limits of the correlation (Lüning and Kolonic, 2003).

For bulk density logs (Table 3.2), it is assumed that any change in bulk density is related solely to TOC. This commonly works because the grain density of kerogen is very low compared to the inorganic matrix, which does not change dramatically through the section of interest. However, this approach can fail when the inorganic grain density varies, e.g., due to changes in pyrite content.

ΔLog R is another empirical method in which the offset between the porosity and the resistivity logs is equated to TOC (Table 3.2). Sonic, density, or neutron logs having vertical resolution of ~ 0.6–1.0 m (~ 2–3 ft) can be used for porosity. ΔLog R represents this offset (adjusted for units of measurement), which correlates to TOC after one assumes a level of organic metamorphism (LOM; Hood et al., 1975). As with bulk density, this model assumes that any changes in the logs are based solely on kerogen and associated porosity. Any changes in the inorganic matrix surrounding the wellbore can lead to errors in TOC estimation.

3.2.1.5 Direct Wireline TOC

Direct wireline TOC measurements are generally more reliable than indirect methods. Examples of direct wireline TOC include: (1) carbon subtraction, (2) nuclear magnetic resonance (NMR)–geochemical, and (3) pulsed neutron–spectral γ-ray (Gonzalez et al., 2013). Total carbon includes both organic and inorganic carbon and can be measured using an inorganic geochemical log. The inorganic contribution is estimated by using Ca, Mg, and other elements to determine the amount of carbonate in the rock. For the carbon subtraction method, TOC is defined as the difference between total and inorganic carbon. This TOC comprises all forms of organic carbon, including kerogen, bitumen, oil, and gas. Unlike migrated oil, bitumen is indigenous and can be extracted from fine-grained rocks using common organic solvents.

For NMR–geochemical estimates of TOC (Gonzalez et al., 2013), the fluid-filled pore volume is determined using NMR logging where the proper hydrogen contributions for pore volumes are input. The NMR–geochemical method is commonly used for tight oil because the ratio of hydrogen to pore volume is well defined for oil (gas is more variable). The low-density volume (pore volume and kerogen) is estimated using the inorganic grain density from an inorganic geochemical logging tool with the log bulk density. The difference between this volume and the NMR volume is assumed to be the volume of kerogen. Depending on viscosity, bitumen may or may not be included in the kerogen volume. The estimate of kerogen volume (V_k) is as follows:

$$V_k = [(\rho_{mg} - \rho_l)/(\rho_{mg} - \rho_k)] - [\varphi_{NMR}(\rho_{mg} - \rho_f)/HI_f(\rho_{mg} - \rho_k)] \qquad (3.4)$$

where $\rho_l = \rho_{mg}(1 - V_f - V_k) + \rho_f V_f + \rho_k V_k$ and $\varphi_{NMR} = V_f HI_f$

V_f = volume pore fluid
ρ_{mg} = matrix density from geochemical log (not corrected for TOC)
ρ_k = kerogen density
ρ_f = pore fluid density
φ_{NMR} = total NMR porosity and
HI_f = fluid hydrogen index.

The conversion of kerogen volume to TOC is given by Tissot and Welte (1978):

$$TOC\ (wt\%) = (V_k/K_{vr})(\rho_k/\rho_b) \qquad (3.5)$$

where K_{vr} = kerogen conversion factor (i.e., 1.2–1.4) and ρ_b = bulk density.

Pulsed neutron–spectral γ-ray methods (Table 3.2) simultaneously measure carbon and the major elements that form carbonate minerals, particularly calcium and magnesium. TOC can be determined as the difference between total carbon and inorganic carbon at open-hole logging speeds. This method differs considerably from the current practice of estimating TOC from conventional log measurements based on empirical approaches. Two common techniques are the Schmoker density-log method and the Δlog R method. Other common methods to estimate TOC include uranium or γ-ray logs, although they require local calibration. The Schmoker density-log method uses the bulk density log and assumes that the change in density of the formation is due to presence or absence of low-density organic matter (~ 1.0 g/cm^3). The Δlog R method is based on separation of the sonic or density and resistivity curves and the level of thermal maturity to determine TOC.

3.2.2 ORGANIC GEOCHEMICAL LOGS AND ANCILLARY TOOLS

Organic geochemical logs can be used to identify sweet spots in unconventional plays because they show large numbers of rapid and inexpensive analyses at consistent vertical spacing (e.g., every 10 m or 32 ft) through the wellbore. The organic geochemical log in Fig. 3.3 includes TOC, Rock-Eval pyrolysis, and carbonate content for closely-spaced samples from a well in the Williston Basin, North Dakota. The oil saturation index (OSI = 100 × S1/TOC) is high, but unreliable for the Scallion Formation because both oil saturation (Rock-Eval S1, mg hydrocarbon/g rock) and TOC are very low. Although S1 is high in the upper and lower Bakken members, the OSI is low. These units do not produce petroleum, even when hydraulically stimulated. However, the middle member of the Bakken Formation is productive because, although S1 is lower than in the Upper and Lower Members, OSI is high (>100 mg HC/g TOC). T_{max} (converted to vitrinite reflectance, R_o; Eq. (3.6)) is suppressed in the Bakken Middle Member due to interference by heavy ends of oil in the S2 peak. In summary, the sweet spot consists of the middle member of the Bakken Formation because it combines favorable geochemical (high OSI) and geomechanical properties (high brittle carbonate content). High OSI does not always correspond to productive intervals, particularly when they are clay-rich and lack significant quantities of brittle components, such as calcite, dolomite, quartz, or feldspar (Fig. 3.2).

The following relationship can be used with caution to convert Rock-Eval pyrolysis T_{max} to R_o.

$$R_o(\text{calculated}) = (0.0180)(T_{max}) - 7.16 \tag{3.6}$$

Equation (3.6) is based a collection of shale samples that contain low-sulfur Type II and Type III kerogen (Jarvie et al., 2001). It works reasonably well for many Type II and Type III kerogens, but not for Type I kerogens.

3.2.2.1 Van Krevelen Diagrams

Van Krevelen diagrams characterize source rock organic matter on a plot of atomic O/C versus atomic H/C from elemental analysis, while modified van Krevelen diagrams use the oxygen versus hydrogen index (OI versus HI) from Rock-Eval pyrolysis (Fig. 3.4). Both diagrams can be used with caution to assess petroleum generative potential (Espitalié et al., 1977; Peters et al., 1983; Peters, 1986; Dembicki, 2009). Pyrolysis generally yields results similar to elemental analysis, but requires less sample (~100 mg of cuttings), and is faster and less expensive, thus making it a key method for most source rock studies. However, Rock-Eval HI can underestimate kerogen quality or generative potential because highly oil-prone kerogens yield high atomic H/C, but do not always show correspondingly high pyrolytic yields. For this reason, the atomic H/C of selected kerogens should be used to support Rock-Eval HI measurements of the remaining generative potential of rock samples (Baskin, 2001).

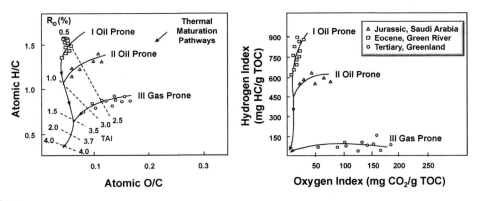

FIGURE 3.4

Van Krevelen (left) and modified van Krevelen (right) diagrams show kerogen composition based on elemental analysis and Rock-Eval pyrolysis, respectively. Type I, II, and III pathways of kerogen thermal maturation (see Appendix) are for descriptive purposes only. Kerogen composition in different source rocks or within the same source rock can be highly variable and many samples plot between, above, or below these pathways (e.g., Type II/III or Type IV). Dashed lines (left) show approximate values of R_o and thermal alteration index (TAI).

Modified from Peters et al. (2005).

3.2.2.2 TOC versus S2 Plots

The best estimate for HI of a rock unit is based on the slope of the TOC versus S2 plot rather than the average HI for individual samples. For example, a plot of TOC versus S2 for cuttings samples from the Shell Worsley 6-34-87-7W6 well (lat 56.584919, long −119.029623) from Alberta (Snowdon, 1997) shows two distinct organofacies corresponding to the Fernie (3216–3422 ft; 980–1043 m) and Nordegg-Montney (3422–3622 ft; 1043–1104 m) formations (Fig. 3.5). Based on the slopes of the best fit lines, the HI for the Fernie section is 255 mg HC/g TOC with a correlation coefficient of 0.998 for six samples, while that for the Nordegg–Montney section is 680 mg HC/g TOC with a correlation coefficient of 0.994 for seven samples. HI calculated for the Fernie and Nordegg–Montney formations based on average HI for samples in each interval (336 and 451 mg HC/g TOC) differs substantially from that based on the TOC versus S2 plot.

The Belloy Formation (3622–4212 ft; 1104–1284 m) in the Shell Worsley well represents a third organofacies in which TOC ranges from 0.04 to 0.69 for 20 samples. These samples are not shown in Fig. 3.5 because they cannot represent source rock. Regardless of the original HI, expulsion efficiencies for rocks originally containing <1 to 2 wt% TOC will be low. This agrees with data suggesting that rocks having <2.5 wt% TOC are incapable of establishing a continuous bitumen network to facilitate expulsion (Lewan, 1987). Therefore, intervals that contain less than ∼2 wt% TOC within a source rock interval should be excluded because they are unlikely to expel petroleum upon thermal maturation. For BPSM, nonsource intervals within a larger source interval may require splitting of the source rock into several vertical subunits to exclude the lean zones and focus on generation from the richer intervals.

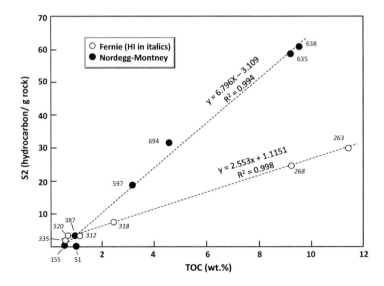

FIGURE 3.5

TOC versus Rock-Eval S2 for two organofacies in the Shell Worsley well from Alberta based on depths for the Fernie and Nordegg–Montney source rocks. Numbers near points are calculated hydrogen indices (HI, mg hydrocarbon/g TOC).

3.2.2.3 Organic Petrography

R_o is commonly used to describe the thermal maturity of source rocks with respect to crude oil or hydrocarbon gas generation (Bostick, 1979; Peters and Cassa, 1994). R_o is also important because it is used to create regional maturity maps of petroleum source rock (Demaison, 1984) and calculate the thickness of lost section due to erosion at unconformities (Armagnac et al., 1989). Measured R_o in a well can be compared to that calculated using EASY%R_o kinetics (Sweeney and Burnham, 1990) to calibrate BPSM (e.g., Schenk et al., 2012).

Two methods are commonly used to prepare rock samples for petrographic measurement of R_o: (1) some very organic-rich rocks, such as coal (>50 wt% TOC), can be prepared using whole rock, and (2) for less organic-rich rocks, including most petroleum source rocks, kerogen is isolated from the rock matrix (see Appendix). The kerogen is then mixed with epoxy, dried, and polished using fine alumina grit (Bostick and Alpern, 1977; Baskin, 1979). A petrographic microscope equipped with a photometer is used to measure the percentage of incident white light (546 nm) reflected from vitrinite phytoclasts in the prepared samples under an oil immersion objective (Taylor et al., 1998). Reported R_o typically represents mean or average values for ~20–100 phytoclasts in each polished slide as subjectively identified by the operator. Some samples, particular oil-prone source rocks, contain fewer than 20 vitrinite phytoclasts, which contributes to uncertainty in the measurement. Analyses of R_o for the same sample by different operators can result in discrepancies because of complex mixtures of macerals and misidentification of vitrinite. Microscopes with rotating stages allow measurement of anisotropy in vitrinite, which becomes significant at $R_o > 1.0\%$, resulting in terms R_{max} and R_{min} for the maximum and minimum reflectance obtained upon rotation of each phytoclast. When rotation of the microscope stage is not possible, R_r or R_m are commonly used to indicate random or mean vitrinite reflectance, respectively.

All measured R_o values for a sample are commonly displayed as a histogram of R_o versus the number of measurements. Operators normally edit polymodal histograms to leave only those R_o values that correspond to their interpretation of indigenous rather than recycled or contaminant phytoclasts (Fig. 3.6). Thus, standard deviations reported with R_o values can be misleading because the operator may have selected an incorrect maceral for measurement rather than the true vitrinite. The mean R_o for each edited histogram is plotted versus depth to generate a reflectance profile that can be used to describe the thermal maturity of the section.

Because of the diversity of phytoclasts in source rocks, identification of vitrinite as opposed to other macerals, such as liptinite or inertinite can be difficult. Fluorescence petrography has proven useful to differentiate macerals (e.g., Crelling, 1983). Liptinite macerals from the waxy parts of plants and some vitrinite macerals fluoresce. Ottenjann (1988) related the fluorescence spectra of vitrinite macerals to various properties of coal. For this reason, most major steel companies employ petrographic laboratories to predict the coking properties of coal. Current methods that combine R_o and fluorescence are based on piecemeal measurements of individual phytoclasts in each sample slide, typically involving no more than a dozen measurements. These measurements are so slow that they are generally not employed by industry. For example, the method of Wilkins et al. (1995) requires 700 s for each fluorescence measurement, i.e., nearly 12 min for each phytoclast.

Interpretation of thermal maturity from R_o measurements as described above is limited by various pitfalls, which include: (1) suppression of R_o (perhydrous vitrinite; Price and Barker, 1985; Wilkins et al., 1992) by samples rich in oil-prone macerals, (2) contamination by particulate drilling

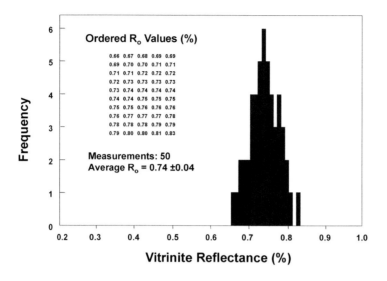

FIGURE 3.6

Example of a vitrinite reflectance histogram for kerogen from a sedimentary rock sample. Fifty reflectance values were determined as the percentage of incident white light reflected from each vitrinite phytoclast. The average R_o is based on all reflectance values in the histogram. The standard deviation (0.04% in this case) can be misleading because it is a measure of the ability of the operator to select what is assumed to be vitrinite rather than other macerals in the slide.

additives or caving of materials from shallower zones in a well, (3) subjective misidentification of vitrinite, and (4) insufficient measurements for statistically valid results. R_o is commonly inaccurate in organic-rich marine source rocks that lack or contain little land plant input. Therefore, R_o for source rock is typically interpolated from measured values in organic-lean units above and below the source rock. T_{max} from Rock-Eval pyrolysis can also be suppressed within source rocks (Fig. 3.3, right column). Vitrinite is absent in Silurian and older rocks, which were deposited prior to the evolution of large land plant communities. Furthermore, vitrinite consists of woody higher-plant remains that contribute little or no oil during burial maturation. For this reason, R_o for the beginning and end of oil generation can vary somewhat depending on kerogen type or structure and kinetics (Fig. 3.1). Approximate R_o values have been assigned to the beginning and end of oil generation (~0.6% and 1.4%, respectively).

3.2.3 INORGANIC GEOCHEMICAL LOGS

Inorganic geochemical logging tools were first introduced in the 1970s for cased-hole evaluation of saturation using carbon to oxygen ratios and qualitative evaluation of lithology based on Si, Ca, and Fe (Culver et al., 1974; Hertzog, 1980). Over the next two decades, focus on elemental spectroscopy logging expanded to include open-hole formation evaluation where technological advances led to new tools to measure elemental concentrations and interpret formation mineralogy and nuclear properties (e.g., Herron and Herron, 1996; Radtke et al., 2012).

The elemental capture spectroscopy (Alexander et al., 2011) tool was the first commercial service designed for the open-hole formation evaluation. It uses a radionuclide source to bombard the formation with neutrons that are captured by nuclei of specific atoms, which emit γ-rays of characteristic energy that are measured and interpreted in terms of formation composition. More modern tools (Pemper et al., 2009; Radtke et al., 2012) use pulsed neutron generator sources that allow simultaneous acquisition of the capture γ-ray spectrum and a spectrum of γ-rays produced by inelastic scattering reactions, which allows measurement of formation carbon and computation of TOC.

Modern neutron-induced γ-ray spectroscopy or elemental spectroscopy logging tools yield concentration logs of important rock-forming elements. For example, LithoScanner reports concentrations of the major elements Si, Ca, Fe, Mg, S, K, Al, Na, and C as well as some minor or trace elements, such as Mn, Ti, and Gd (Radtke et al., 2012; Aboud et al., 2014). This information helps to assess the geomechanical behavior of rock units. For example, Fig. 3.7 shows elevated calcite and dolomite in Zone B (middle member of the Bakken Formation) between organic-rich zones A and C (upper and

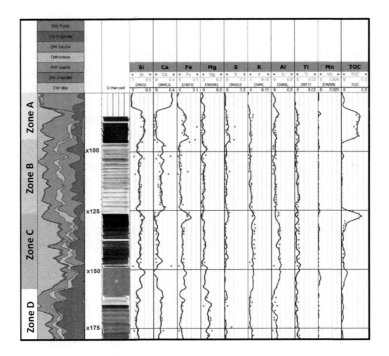

FIGURE 3.7

Inorganic geochemical log from a well in North Dakota shows results for Bakken (zones A–C) and Three Forks (zone D) formations. Black curves compare elemental weight fractions measured by LithoScanner with those derived by laboratory core analysis (red points). Note excellent agreement between LithoScanner TOC and TOC by laboratory filter acidification and combustion (far right; 0.2 on scale = 20 wt% TOC). Colored distributions of mineralogy and core photographs are to the left and right of depth track, respectively. Logging speed was 600 ft/h (183 m/h).

From Radtke et al. (2012).

lower members), which suggests that Zone B is more brittle and amenable to hydraulic fracturing for unconventional petroleum.

Elemental concentration logs can be used to describe TOC, lithology or mineralogy, and matrix properties. The techniques for interpreting mineralogy vary. The most general technique computes mineral groups where total clay is determined from Si, Ca, Fe, and S; carbonate from Ca or calcite, and dolomite from Ca and Mg; anhydrite and/or pyrite from S; and the sum of quartz, feldspar, and mica comprise the remainder (Herron and Herron, 1996). The mineral interpretation technique uses a model-independent mapping function to predict 14 minerals, including individual clay, feldspar, and carbonate minerals (Freedman et al., 2014). Inorganic carbon content in the minerals can be determined based the formation mineralogy. An accurate and continuous TOC log can be then determined by subtracting inorganic carbon from total carbon (Radtke et al., 2012; Al-Salim et al., 2014). The method also indicates open-hole hydrocarbon saturation (Craddock et al., 2013). Several matrix properties, including matrix density and matrix hydrogen content can also be determined directly from elemental concentration logs (Aboud et al., 2014), both of which are used with bulk density and neutron logs to identify gas and to produce accurate porosity logs.

Figures 3.3 and 3.7 can be used independently to rationalize enhanced tight oil production from the middle member of the Bakken Formation (zone B in Fig. 3.7). High TOC and HI from the organic geochemical log (Fig. 3.3) indicates excellent source rock in the upper and lower members, which is supported by high TOC for these units on the inorganic geochemical log (Fig. 3.7). The organic geochemical log indicates oil-prone organic matter (HI > 300 mg hydrocarbon/g TOC) in the upper and lower members (zones A and C). High OSI (Fig. 3.3) and relatively more abundant and brittle calcite and dolomite (Figs 3.3 and 3.7) in the middle member compared with the upper and lower members are favorable indicators of a productive tight oil zone.

Organic and inorganic geochemical logs have distinct advantages and disadvantages. For example, the organic geochemical log in Fig. 3.3 is based on direct measurements of carbonate content and pyrolysis response, but it requires discrete samples at predetermined depth intervals, which are typically analyzed in a laboratory rather than at the drillsite. Inorganic geochemical logs are based on indirect calculation of mineral content and TOC without the need to collect samples. Although they do not determine the quality or thermal maturity of the kerogen (e.g., as determined by Rock-Eval pyrolysis HI or T_{max}, respectively), they provide continuous data with depth and can be quantified in real time during drilling.

3.2.4 FLUID ADSORPTION IN UNCONVENTIONAL RESERVOIRS

Fluid adsorption and absorption influence the estimated volumes of petroleum that might occur in unconventional plays (Table 3.1). Compared with conventional reservoirs, unconventional reservoirs are characterized by small pores and large surface areas (Nelson, 2009). Consequently, adsorption and absorption play more important roles in unconventional than conventional reservoirs. Adsorption is the adhesion of molecules on solid surfaces, unlike absorption in which molecules are taken into the bulk phase of the solid. However, in porous materials such as shale or kerogen, the difference between absorption and adsorption can be unclear. When pore diameters are comparable to molecular diameters, it is difficult to determine whether molecules adhere to the pore wall or simply reside within the pore volume. In this section, we focus on the concept of adsorption, but it should be noted that in some cases adsorption and absorption are not readily distinguished.

Table 3.3 Comparison of Heats of Adsorption (kJ/mol) of Methane, Ethane, and Water on Polar and Nonpolar Surfaces

	Methane	Ethane	Water
Polar aluminosilicate surfaces (zeolite)	20.0 (Smit, 1995)	31.1 (Smit, 1995)	43 (Carmo and Gubulin, 1997)
Nonpolar carbon surfaces (activated carbon)	18.5 (Cruz and Mota, 2009)	31.8 (Cruz and Mota, 2009)	10 (Groszek, 2001)

Adsorption reflects the affinity of small molecules to solid surfaces, which is usually a weak physisorption (<50 kJ/mol) in petroleum reservoirs where chemical bonds do not dissociate. Adsorption affinity largely depends on the polarity and polarizability of surfaces and molecules. Solid surfaces in a nonconventional petroleum reservoir include both kerogen and mineral surfaces, which may be polar or nonpolar. Most mineral surfaces have polar character. Kerogen surfaces become more nonpolar during diagenesis and catagenesis due to the removal of the polar functional groups, such as hydroxyl and carboxylic groups. Polar molecules such as water have strong affinity for polar surfaces and their affinity for nonpolar surfaces is commonly weaker than interactions between polar molecules.

Table 3.3 shows typical adsorption values for polar (water) and nonpolar (methane and ethane) molecules on polar and nonpolar surfaces. The data indicate that mineral surfaces are mainly covered by water in both conventional and unconventional reservoirs, while kerogen surfaces are mainly covered by hydrocarbons. Because of multiple components and heterogeneous surfaces, shale reservoirs are complex adsorption systems. The following discussion addresses the key question related to the contributions of adsorbed hydrocarbons to in-place gas and oil.

3.2.4.1 Quantifying Adsorption under Reservoir Conditions

The amount of adsorbed hydrocarbons is determined by the maximum adsorption capacity and the fractional coverage. The fractional coverage of adsorbed hydrocarbons under reservoir conditions cannot be directly measured, but can be estimated using suitable models calibrated to lab data. There are many models to describe adsorption; each has its own isotherm equation to describe the amount of adsorbed hydrocarbon as a function of pressure at constant temperature. The simplest is the Langmuir model:

$$\frac{n_{ads}}{n_m} = \frac{K(p/p^\circ)}{1 + K(p/p^\circ)} \tag{3.7}$$

where n_{ads} is the amount of adsorbed hydrocarbon for a given rock mass (in mol/g) and n_m is the monolayer adsorption capacity for the rock (mol/g). Consequently, n_{ads}/n_m is the fractional coverage (dimensionless). Strictly, n_{ads} should be the excess amount adsorbed, which is the difference between the amount of adsorbed gas and the gas that would occupy the volume without adsorption. The latter is usually measured by helium absorption, assuming no interaction between helium and solid surfaces. K is the Langmuir coefficient (dimensionless), p is the pressure of adsorptive gas, and p° is the standard pressure (10^5 Pa, 1 bar, or 14.504 psi).

The Langmuir coefficient K is a function of adsorption heat (q, in J/mol) and standard adsorption entropy ($s°$, J/mol K):

$$K = \exp\left(\frac{q}{RT} + \frac{\Delta s°}{R}\right) \quad (3.8)$$

where T is temperature (K) and R is ideal gas constant (8.3144 J/mol K).

The Langmuir isotherm is sometimes written as:

$$\frac{n_{ads}}{n_m} = \frac{p}{p + p_L} \quad (3.9)$$

where p_L is the Langmuir pressure, under which half of the surfaces are covered by adsorbed components. It can be shown that:

$$p_L = p°/K \quad (3.10)$$

where $p° =$ standard pressure (1 bar), and K is the Langmuir coefficient (defined above).

Once values of q, $\Delta s°$, and n_m are known, the amount of adsorbed hydrocarbon can be determined at any temperature and pressure using the Langmuir model. Studies (Zhang et al., 2012; Ji et al., 2012; Gasparik et al., 2014; Hu et al., 2015) have analyzed these values for methane adsorption on organic-rich shale. The adsorption capacity of shale depends on both TOC and thermal maturity. Gasparik et al. (2014) reported adsorption capacity >2 to 4 mmol/g TOC for several gas shale samples ($R_o = 1.3$–2.0%, gas window) having TOC > 2.0%. Such high adsorption capacities indicate that the surface area is ~200–400 m²/g TOC (considering the diameter of methane = 0.4 nm and most pore diameters are <3 to 5 nm), consistent with direct surface area measurements of mature kerogen (Suleimenova et al., 2014). This adsorption capacity is several times higher than that for methane adsorption on coal based on earlier studies (e.g., 1 mmol/g TOC; Clarkson et al., 1997).

Values of q and $\Delta s°$ can be derived from Langmuir isotherms at different temperatures, based on Eq. (3.8). A reasonable value of $\Delta s°$ for physisorption should be close to the loss of translational entropy from three-dimensional free gas to two-dimensional adsorbates, which is about -82 to -87 J/mol K at 273–423 K for methane (Xia and Tang, 2012). Values for q and $\Delta s°$ were determined for naturally mature shale, kerogen, and artificially mature shale samples (Zhang et al., 2012; Gasparik et al., 2014; Hu et al., 2015). The $\Delta s°$ values are between -60 and -120 J/mol K, with a consistent linear variation of adsorption heat (q) with entropy change ($\Delta s°$), indicating their compensating effect on the Langmuir coefficient K (Eq. (3.8)). Because $\Delta s°$ values can be less than the range of two-dimensional adsorption (-82 to -87 J/mol K), some "adsorbed" molecules do not actually adhere to surfaces, but occur as free gas trapped in pores (Xia and Tang, 2012). Measurements on kerogen and artificially mature samples show stronger adsorption ($q > 20$ kJ/mol and $-\Delta s° > 100$ J/mol K), suggesting that some stronger adsorption sites were created during laboratory processing. A value of $\Delta s° = -87$ J/mol K and $q = 18$ kJ/mol is applied for the following calculation.

3.2.5 CONTRIBUTION OF ADSORBED GAS TO GAS-IN-PLACE AND PRODUCTION

Assuming the above parameters ($n_m = 2$–4 mmol/g TOC, $\Delta s° = -87$ J/mol K and $q = 18$ kJ/mol), for a gas shale at 3000–5000 m depth with $T = 80$–180 °C and $p = 30$–50 MPa, the fractional coverage (n/n_m) is in the range 0.7–0.9. The variation of fractional coverage under reservoir temperature and

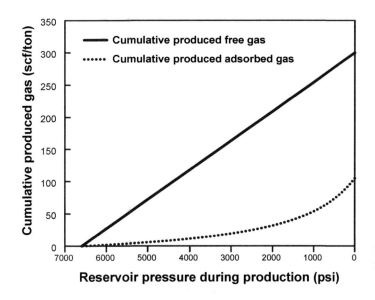

FIGURE 3.8

Contributions of absorbed (free) gas exceed adsorbed gas during the depletion of shale play according to the Langmuir model.

pressure is narrow according to the Langmuir model. Therefore, the main factors controlling the amount of adsorbed gas are TOC and monolayer adsorption capacity. For a shale layer with TOC of 4% and n_m of 2–4 mmol/g TOC, the amount of adsorbed gas is 50–130 standard cubic feet (scf)/ton. Considering a porosity of 5–10% for gas shale plays, "free gas" amounts to ~100–400 scf/ton rock.

One might assume that the amount of adsorbed gas is comparable to free gas in unconventional accumulations. However, the contribution of adsorbed gas to production is not that significant. During reservoir depletion, produced free gas is proportional to depleted pressure (the compressibility z of natural gas is close to 1 at reservoir conditions), but the amount of desorbed gas is insignificant until the reservoir pressure is strongly depleted (Fig. 3.8). Considering a recovery factor of 30% for shale gas, the contribution of desorbed gas is <10% of the total produced gas.

3.2.6 HYDROCARBON GENERATION, EXPULSION, AND RETENTION

Unlike conventional reservoirs, unconventional reservoirs usually contain petroleum that has undergone only short-distance migration within a source rock or adjacent layers. Secondary migration from source to reservoir rock is important for conventional reservoirs, while the generation history and storage capacity of unconventional reservoirs are most important (Fig. 3.9).

Petroleum generation can be described using a collection of first-order reactions having different activation energies. The rate equation for a first-order reaction is:

$$\frac{dy}{dt} = k(1-y) \tag{3.11}$$

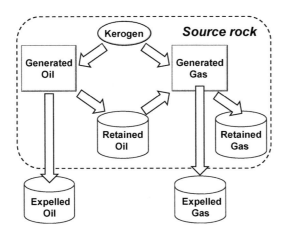

FIGURE 3.9

Scheme depicts the distribution of fluids within a source rock (dashed box) after expulsion. The terms expulsion or primary migration refer to movement of petroleum out of the source rock into adjacent carrier beds. However, petroleum can also be "expelled" from organic-rich intervals in hybrid shale (discussed earlier) into more porous and permeable organic–lean intervals within the source rock. Note that retained gas can originate from primary cracking of kerogen and secondary cracking of retained oil.

where y is the conversion of the precursor (dimensionless), t is time (s), and k is the reaction rate constant (s^{-1}). The Arrhenius equation is another way to express this relationship between reaction rate, activation energy, and temperature:

$$k = A \exp(E_a/RT) \tag{3.12}$$

where A is the preexponential factor (in s^{-1}), E_a is activation energy (in J/mol or kcal/mol; 1 kcal = 4184 J), R is the gas constant (8.314 J/mol K), and T is temperature (in K).

Laboratory pyrolysis data can be used to determine petroleum generation kinetics (e.g., Peters et al., 2015, and references therein). Xia (2014) applied Rock-Eval pyrolysis and field data to calibrate kinetic parameters for gaseous hydrocarbon generation (Table 3.4). These kinetic parameters suggest that the petroleum generative potential of kerogen is depleted at R_o of 1.5–2.0%. Gas retained in source rock with $R_o > 2.0\%$ is mainly the product of earlier maturity or secondary oil cracking.

The amount of petroleum retained in source rock depends mainly on storage capacity. At high maturity, the amount of expelled petroleum likely far exceeds the amount retained. During maturation, fluid previously contained in the pores is replaced by later products. In the oil window, connate water may still occupy some pore volume. Therefore, the storage capacity for oil is determined by porosity and water saturation. With increasing petroleum generation, water is nearly completely expelled from the source rock. Oil is progressively replaced by condensate and natural gas with increasing maturity. Xia et al. (2013) attribute the isotopic "rollover" (discussed below) to cracking of the retained liquid petroleum to gas at high maturity. This process can be modeled using the above kinetic parameters while accounting for porosity and adsorption capacity. Significant

Table 3.4 Best-fit Results for Kinetic Parameters and Precursor Fraction (mg C in Product/mg C in Precursor) for Gaseous Hydrocarbon Generation from Barnett Shale

Products	Kinetic Parameters (Primary Cracking)		Precursor Composition (mol C%)	
	A (10^{13} s^{-1})	E_a (kcal/mol)	Primary Cracking	Secondary Cracking[a]
CH_4	0.03	51.0–55.0	83.29	74.07
C_2H_6	1	52.9–55.7	10.71	22.22
C_3H_8	1	52.5–55.0	3.57	2.96
i-C_4H_{10}	1	53.2–54.8	0.95	0.04
n-C_4H_{10}	1	52.3–54.7	0.95	0.56
i-C_5H_{12}	1	53.0–54.5	0.29	0.04
n-C_5H_{12}	1	52.0–54.4	0.24	0.11

[a] $A = 3.85 \times 10^{16}$ s^{-1} (Behar et al., 2008) and $E_a = 65.5$ kcal/mol for secondary cracking. The ratio of secondary cracking to primary cracking precursors is 1.5×10^{-3} (in carbon weight).

uncertainty remains with respect to the compositional fractionation of oil and gas into different pores during maturation (Xia et al., 2014).

3.2.7 STABLE CARBON ISOTOPE ROLLOVER

Stable carbon isotopic compositions ($\delta^{13}C$) of gaseous hydrocarbons (from methane to pentanes) are initially controlled by the isotopic composition of kerogen, but are later revised by kinetic isotopic fractionation during gas generation, expulsion, and retention (Fig. 3.9). Two distinct patterns of carbon isotopic composition are commonly observed for natural gas from a single precursor (e.g., kerogen or oil):

1. The normal alkanes are enriched in ^{13}C with increasing carbon number (i.e., $\delta^{13}C_{methane}$ < $\delta^{13}C_{ethane}$ < $\delta^{13}C_{proane}$ < $\delta^{13}C_{n\text{-buane}}$). Chung et al. (1988) explained this order based on kinetic isotopic fractionation during natural gas generation from heavier precursors.
2. The $\delta^{13}C$ of n-alkanes become more positive with increasing maturity due to the kinetic isotopic effect (KIE), in which a ^{12}C–^{12}C bond is easier to break than a ^{12}C–^{13}C bond. KIE has been quantified by Lorant et al. (1998) and Tang et al. (2000).

Reversals of the typical carbon isotopic trend with n-alkane carbon number (item 1, above) occur in some rare conventional gas reservoirs and were tentatively attributed to mixing of gas from different sources (Jenden et al., 1993; Dai et al., 2005). A similar reversal of $\delta^{13}C_{methane}$ > $\delta^{13}C_{ethane}$ in Barnett Shale gas was also attributed to mixed gases (Rodriguez and Philp, 2010). Xia et al. (2013) quantitatively demonstrated this mixing mechanism for isotopic rollover, as discussed below.

A reversal or "rollover" of carbon isotopic compositions of ethane and propane with respect to maturity (item 2, above) was reported for Barnett Shale gas (Zumberge et al., 2012). Increasing maturity is also reflected by a decrease in gas wetness (i.e., the volume fraction of ethane through pentane relative to total gaseous hydrocarbons). As shown in Fig. 3.10, $\delta^{13}C_{ethane}$ becomes more

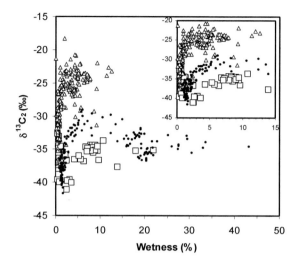

FIGURE 3.10

Wetness-dependent variation of $\delta^{13}C_2$ (ethane) in natural gas from the Ordos Basin (triangles; Dai et al., 2005; Xia, 2000; Hu et al., 2008), the Fort Worth Basin (dots; Zumberge et al., 2012), and the Appalachian Basin (squares; Burruss and Laughrey, 2010). Wetness is the volume fraction of C_{2+} (ethane and heavier) gaseous hydrocarbons in the total hydrocarbon gases. Inset expands part of the full data to emphasize rollover.

From Goddard et al. (2013).

positive with increasing maturity and decreasing wetness at wetness greater than 10%. As wetness decreases further with increasing maturity, $\delta^{13}C_{ethane}$ becomes more negative. This "rollover" is now recognized in many other shale plays having broad maturity distributions, such as the Ordovician Utica and Devonian Marcellus formations in the Appalachian Basin, and the Cretaceous Eagle Ford Formation in South Texas. This phenomenon is not unique to shale plays because similar isotopic reversals with maturity occur in some conventional reservoirs, as in the Ordos Basin (Fig. 3.10).

During exploration and development of shale gas, new data are obtained by systematic analysis of source rock (e.g., TOC and Rock-Eval pyrolysis) and petroleum fluids (e.g., gas and isotopic compositions). The data provide new information related to petroleum generation and expulsion. Unconventional rock units that display a stable carbon isotope rollover are generally overpressured and productive, i.e., the rollover is a useful tool to identify sweet spots. For example, Ferworn et al. (2008) show that wells with high initial production and high stabilized production always occur in zones of ethane isotope rollover. However, Madren (2012) found isotopic rollovers in the Marcellus Shale in western Pennsylvania, but the trends are the same in areas of good and poor production. He concluded that better productivity in the Marcellus Shale is more reliably predicted from porosity and permeability than isotopic rollover.

Xia et al. (2013) quantitatively showed that mixing of primary and secondary gas accounts for the occurrence of isotopic rollover in some shale plays (Fig. 3.11, top). Gas generated by secondary cracking of retained liquid is wetter (enriched in ethane through pentane) and has more negative

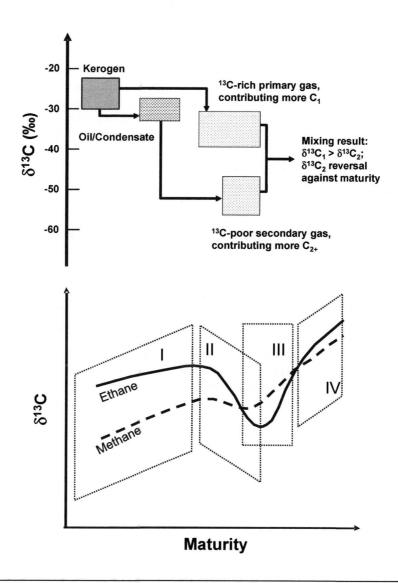

FIGURE 3.11

Top: Scheme shows a $\delta^{13}C$ reversal due to mixing of primary and secondary gas in a closed-system source rock. Bottom: Scheme of complete trend of maturity-dependent $\delta^{13}C_{ethane}$ and $\delta^{13}C_{methane}$. Region I, normal trend; II, $\delta^{13}C_{ethane}$ reversal with respect to maturity trend; III, $\delta^{13}C$ reversal with respect to carbon number ($\delta^{13}C_{methane} > \delta^{13}C_{ethane}$); IV, normal trend. Note that the ratio of early to late products in the retained hydrocarbons depends on expulsion efficiency.

Modified from Goddard et al. (2013).

carbon isotopic composition than the primary gas cracked from kerogen. As thermal maturity increases, oil retained in source rock cracks and causes the $\delta^{13}C_{ethane}$ to decrease.

Figure 3.11 (bottom) illustrates a complete maturity trend of natural gas isotopic compositions with respect to thermal maturity. At low maturity, with little or no contribution of secondary cracked gas, there is a normal $\delta^{13}C$ trend with respect to carbon number and maturity (region I). The contribution of secondary gas increases with maturity and $\delta^{13}C_{ethane}$ becomes more negative (region II). High maturity Barnett Shale gas falls in this region. As maturity continues to increase, a reversal with respect to carbon number ($\delta^{13}C_{ethane} < \delta^{13}C_{methane}$) occurs (region III), as is common in the eastern Sichuan Basin (Dai et al., 2005) and the Western Canada Sedimentary Basin (Tilley et al., 2011). The trend may become normal again (region IV) at extremely high maturity, either due to decreased secondary gas contribution or its enrichment in ^{13}C.

3.2.8 EFFECT OF TRANSIENT FLOW ON GEOCHEMICAL PARAMETERS

Many analyses focus on the gas released from core and cutting samples (e.g., Ferworn et al., 2008; Strąpoć et al., 2006; Zhang et al., 2014). However, the chemical and isotopic compositions of gas collected in these analyses may not represent original reservoir compositions because much of the original gas can be lost during sample collection. Such degassing involves convection, diffusion, and adsorption/desorption. Diffusion and adsorption/desorption are usually accompanied by compositional fractionation.

Xia and Tang (2012) used a continuum flow model to quantify carbon isotopic fractionation during gas release from shale under reservoir or laboratory conditions. Their model links diffusion and adsorption-desorption and shows minimal isotopic fractionation between the free and adsorbed gas phases (0.2–0.5‰ for $\delta^{13}C_{methane}$ at reservoir temperatures). Isotopic fractionation during gas release is mainly caused by diffusion. The instantaneous isotopic composition of any compound in the released gas is a function of both the residual fraction and the diffusivity ratio:

$$\delta^{13}C = \delta^{13}C_{initial} + 1000[1.21 + \ln(1-f)]\ln(D^*/D) \tag{3.13}$$

where f is the recovery factor (ratio of the amount of released to initial gas) and D^*/D is the ratio between the diffusivities of a gas molecule with and without a ^{13}C atom.

Equation (3.13) indicates that the isotopic fractionation is minimal during hydrocarbon migration in geological time and during gas production from an unconventional reservoir because diffusion is not the main mechanism of mass transport and therefore isotopic fractionation is weak ($D^*/D \sim 1$). In addition, the recovery factor f is not particularly close to 1 under geological or reservoir conditions. On the contrary, when gas is released from core or cuttings samples under laboratory conditions, diffusion plays an important role (D^*/D deviates from 1), the recovery factor f can be very close to 1, and significant isotopic fractionation can occur (Xia and Tang, 2012). Figure 3.12 shows modeling of isotopic fractionation during degassing of a coal core sample.

Based on limited data, carbon isotope fractionation appears useful to identify sweet spots vertically with a source rock. For example, carbon isotopic fractionation in methane during shale gas production has been observed by real-time measurements in a Barnett Shale gas well (Fig. 3.13). With increasing production time, the produced gas becomes more enriched of $\delta^{13}C$. This fractionation can be applied to predict production behavior and the production decline curve (Goddard et al., 2013). Furthermore, preliminary work suggests that vertical zones having high adsorbed hydrocarbon gas content show

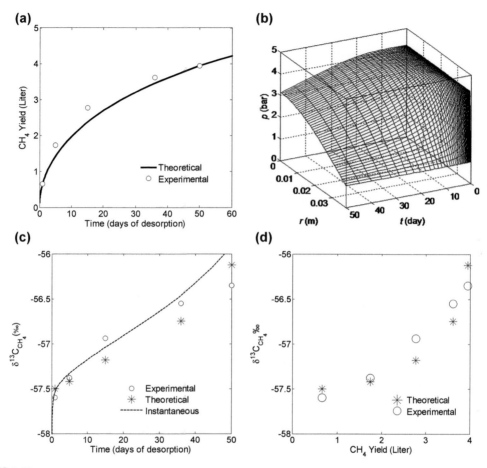

FIGURE 3.12

Calibration of diffusivity using data for coal degassing. (a) Cumulative amount of degassed methane (CH_4) versus time showing theoretical values (see Table 3.4 for parameters) and experimental data; (b) spatial (shown as r, the distance to the axis of the cylinder) and temporal change in methane partial pressure; (c) isotopic composition versus time; (d) isotopic composition versus methane yield. Experimental data are from sample V-3/1 in Strąpoć et al. (2006).

strong isotopic fractionation in methane from headspace to coarse and progressively finer crushed samples of the rock (Y. Tang, personal communication, 2015).

3.2.9 MASS BALANCE AND HYDROCARBON GAS RETENTION EFFICIENCY

Gas retention efficiencies for the various maturation stages of source rock can be calculated as a function of TR using bulk and compositional mass balance models (Horsfield et al., 2010). These calculations are important because they can be used to prioritize gas shale targets for exploitation. The

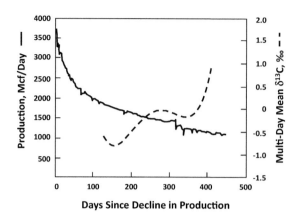

FIGURE 3.13

Production rate and stable carbon isotopic fractionation of methane (difference between real time $\delta^{13}C_{methane}$ and averaged $\delta^{13}C_{methane}$) through time for a Barnett Shale gas well in the Fort Worth Basin. Dashed line is the polynomial regression of the daily mean $\delta^{13}C_{methane}$ averaged from high-resolution field measurements.

Modified from Goddard et al. (2013).

method requires pyrolysis of thermally immature equivalents of the source rock to the TR calculated from mass balance considerations.

$$\text{GIP} = \text{TOC}_o \times \text{TR} \times \text{retention efficiency} \times \text{GOR} \tag{3.14}$$

$$\begin{aligned}\text{PGI} &= (\text{Initial petroleum} + \text{generated petroleum})/\text{Total petroleum potential} \\ &= [(S2_o - S2_m) + S1_o]/(S2_o + S1_o)\end{aligned} \tag{3.15}$$

$$\begin{aligned}\text{PEE} &= \text{Petroleum expelled}/(\text{Initial petroleum} + \text{generated petroleum}) \\ &= [(S2_o + S1_o) - (S2_m - S1_m)]/[(S2_o - S2_m) + S1_o]\end{aligned} \tag{3.16}$$

where

GIP = in-place gas
TOC_o = original TOC
TR = transformation ratio
GOR = gas-oil ratio
PGI = petroleum generation index
$S2_o$ = original Rock-Eval S2
$S2_m$ = measured S2
$S1_o$ = original S1
PEE = petroleum expulsion efficiency.

Gas loss from $S1_m$ can be assessed using microscale sealed vessel pyrolysis.

3.2.10 BASIN AND PETROLEUM SYSTEM MODELING

A new systematic play-based methodology uses computerized BPSM to find and assess sweet spots in unconventional resources (Neber et al., 2012). This methodology identifies sweet spots early in the exploration phase and can be used to predict the quantity and composition of petroleum that remains in the source rock. For example, special functions for gas shale and tight oil modeling in current BPSM software include temperature-pressure dependent Langmuir adsorption modeling, organic porosity modeling, and geomechanics modeling to predict the stress regime through time. BPSM can be used to high-grade parts of shale plays that contain mainly oil, condensate, or hydrocarbon gas prior to drilling. Production trend analysis of the Eagle Ford Shale in southwest Texas shows a progression from black oil to condensate to dry gas toward the southeast (EIA, 2010; Cander, 2012; Cardineaux, 2012). The oil, condensate, and dry gas zones correspond to increasing temperature and depth of the formation and equivalent R_o values of 0.6–1.1%, 1.1–1.4%, and >1.4%, respectively. Operators continue to successfully identify sweet spots in the Eagle Ford Shale by combining these maturity controlled compositional trends with maps of source rock thickness, high-resolution seismic facies maps, mineralogy and geomechanical properties (e.g., carbonate versus clay content), and favorable structural features, such as faults, fractures and sealing stratigraphy. The advantage of the new methodology described above is that it can identify sweet spots prior to drilling, thus significantly reducing delays associated with production trend analysis. Furthermore, by using pressure–temperature dependent Langmuir sorption parameters, BPSM also can be used to predict the relative amounts of gas adsorbed within kerogen pores or on minerals versus free gas in pore space or fractures. Langmuir parameters are best measured on core or outcrop samples of the source rock (Peters et al., in press).

Geomechanical properties of unconventional resources also provide information that can be used to identify areas that are more likely to be naturally fractured due their stress-strain history. The geomechanical model in Neber et al. (2012) includes fluid pressure and rock stress predictions, which are closely coupled. Because the traditional Terzaghi model considers only the vertical stress component, it is limited for predicting rock failure or fluid flow. Developments in BPSM simulators extend this concept to a three-dimensional (3D) poroelasticity rock stress model (Peters et al., in press). The 3D rock stress model considers present-day geomechanical conditions and the evolution of basin-scale geomechanical properties through time. Calculated stress and strain can be used to improve evaluation of seal capacity and fracture orientations or fault properties.

3.2.10.1 SARA Modeling

Predictions of the aromatic and asphaltene content in crude oil and bitumen remaining in source rocks cannot be made using standard published kinetics. A new saturate-aromatic-resin-asphaltene (SARA) kinetic modeling approach includes 11 components (four bitumen, two oil, three hydrocarbon gas, H_2S, and CO_2) and can be used to improve predictions of the quality of migrated oil as well as bitumen remaining in source rock (Fig. 3.14; Peters et al., 2013). Additional features include complex secondary cracking through a multistage reaction network for bitumen oil, oil–gas, and bitumen gas, and an adsorption model for the bitumen components. The 11 components are lumped according to physical and chemical properties in order to minimize processing time. The approach allows prediction of asphaltene flocculation and tar mats as well as H_2S and CO_2 formation.

3.2.11 EXAMPLE OF BPSM MODELING FOR SHALE GAS

Jurassic Posidonia Shale as well as Cretaceous Wealden and Carboniferous shales in northwest-Germany and the Netherlands are potential targets for shale gas exploration. Bruns et al. (2014) used 3D BPSM to assess shale gas prospectivity in this area. Two different tectonic scenarios for basal heat flow and the extent of lost section due to uplift and erosion were incorporated into the model, which was calibrated using geochemical data. The model scenarios provide high-resolution images of regional source rock thermal maturity as well as predicted pressure, temperature, and gas storage capacity in the Posidonia Shale based on source rock thickness maps and experimentally derived

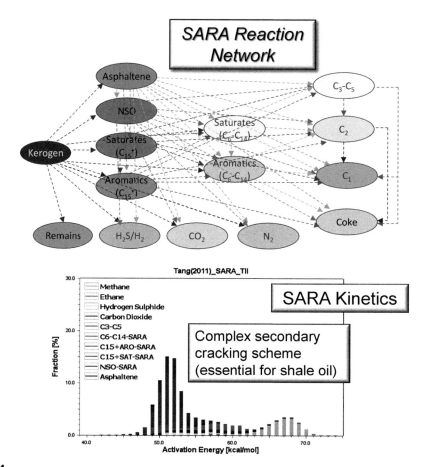

FIGURE 3.14

SARA kinetic modeling includes 11 components (four bitumen, two oil, three hydrocarbon gas, CO_2, and H_2S) and complex secondary cracking through a multistage reaction network (top). Unlike conventional secondary cracking kinetics designed to model expelled oil and gas, SARA kinetics accounts for compositional variation in the source rock with temperature and time, thus improving predictions of the quality of petroleum that might be produced from unconventional targets.

Langmuir sorption parameters. In this discussion, we focus on burial and uplift scenario 1 for the Posidonia Shale, which favors more gas sorption than scenario 2.

Based on scenario 1, Posidonia Shale in the Lower Saxony Basin near Osnabrück was buried up to 10,000 m, resulting in temperatures near 330 °C and very high calculated transformation ratios (Fig. 3.15). Average thermal maturity in the depocenter reached the dry gas stage (>2.3% R_o) and remained in the oil window around the basin margin. Sub-Hercynian uplift and erosion of the Lower Saxony Basin removed up to 8950 m of overburden. Bruns et al. (2014) used Langmuir sorption parameters measured on Posidonia Shale samples of various maturities from the nearby Hils Syncline

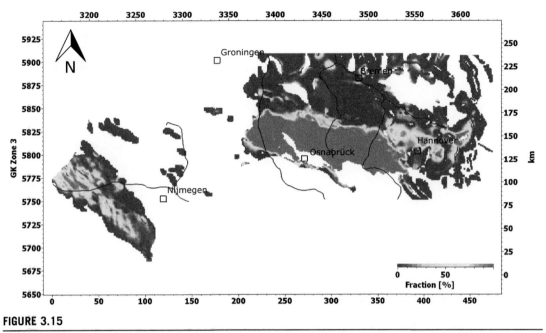

FIGURE 3.15

Calculated present-day transformation ratio (fraction, %) of the Posidonia Shale based on tectonic scenario 1.

Courtesy of Benjamin Bruns (Bruns, submitted for publication).

(Gasparik et al., 2014) to assess Posidonia sorption capacity and sorbed gas contents based on their calibrated burial and thermal history. Bulk adsorption capacities of about 1.3×10^6 tons and gas contents of up to 82 scf/ton rock were predicted for the Posidonia Shale, indicating significant gas potential in specified areas within the Lower Saxony Basin (Fig. 3.16).

3.2.12 KEROGEN ANALYSES FOR STRUCTURAL ELUCIDATION

One of the two objectives of this chapter is to discuss the status of our understanding of kerogen structure. It is hoped that further understanding of kerogen will facilitate identification of unconventional sweet spots. Unfortunately, little of this work has been practically applied toward identifying sweet spots and most of the analytical methods are not amenable to rapid, closely-spaced analyses in actively drilling wells. One notable exception is DRIFTS, as discussed earlier. Modifications of these methods to make them more suitable for characterizing unconventional sweet spots would be a valuable contribution. Nevertheless, these measurements contribute to understanding of the fundamental processes that ultimately play a role in the distribution of sweet spots.

3.2.12.1 Elemental Analysis

In the early 1960s, van Krevelen (1993) determined the atomic H/C and O/C ratios of coals. Coal is defined as sedimentary rock that contains more than 50 wt% TOC. Most European and North American coals are dominated by type III kerogen, but there are also type I and II coals

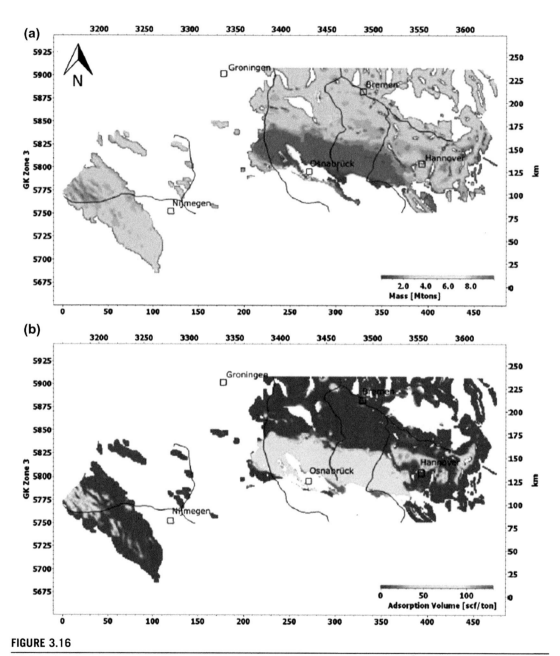

FIGURE 3.16

(a) Calculated present-day total bulk adsorption capacity of the Posidonia Shale (10^6 tons per layer thickness within a grid cell size of 1 km^2), and (b) average volume of methane at standard conditions per mass of rock (scf/ton rock) based on tectonic scenario 1.

Courtesy of Benjamin Bruns; modified from Bruns et al. (2014).

(Peters et al., 2005). The H/C and O/C ratios of kerogen decrease with maturity (van Krevelen, 1993), consistent with a shift from aliphatic to aromatic carbon and loss of oxygen-containing functional groups.

Kerogen is composed of the same elements in approximately the same abundance as asphaltenes. Kerogen typically contains 80–85% carbon by mass depending on type and thermal maturity. Other common elements include hydrogen, nitrogen, oxygen, and sulfur, which decrease in abundance as the carbon content increases. The atomic H/C ratio can be >1.5 for immature type I kerogen and is rarely <0.4 for mature kerogen. The most variable element is sulfur, which can be present at <1% for some kerogens and up to 15% for type IIS kerogens.

3.2.12.2 Nuclear Magnetic Resonance Spectroscopy

NMR spectroscopy is used to measure the carbon backbone in kerogen and bitumen, including carbon types and average molecular parameters. The technique measures deviations in local magnetic field strength for carbon (^{13}C) and hydrogen (^{1}H) nuclei exposed to an external magnetic field. The deviations reflect the local distribution of electrons, which indicates the chemical environment. NMR is a versatile technique used previously to study soluble and insoluble hydrocarbon mixtures, such as coal (Solum et al., 1989), kerogen (Kelemen et al., 2007; Werner-Zwanziger et al., 2005; Mao et al., 2010; Washburn and Birdwell, 2013a; Cao et al., 2013), bitumen (Feng et al., 2013; Solum et al., 2014), petroleum (Korb et al., 2013; Ward and Burnham, 1984), and asphaltenes (Andrews et al., 2011; Lisitza et al., 2009; Dutta Majumdar et al., 2013).

Variations in local magnetic field strength can resolve certain types of carbon, such as aliphatic and aromatic carbon. Greater insight from NMR was made possible by the introduction of techniques such as distortionless enhancement by polarization transfer (DEPT) pulse sequence NMR (Andrews et al., 2011). DEPT involves transfer of magnetization between carbon atoms and covalently bonded protons. Varying the flip angle of the last proton pulse modulates the intensity and phase of the detected carbon signal by an amount that depends on the number of protons covalently bound to the carbon. By performing the measurement at several angles, carbons attached to zero, one, two, and three protons are resolved. The average aromatic cluster size is estimated from the measured molar fraction of bridgehead carbons. Average molecular properties can be determined once the average cluster size is found.

NMR measurements highlight the great diversity in kerogen composition. Some kerogens are dominated by aliphatic carbon (25/75 aromatic/aliphatic), while others are dominated by aromatic carbon (82/18 aromatic/aliphatic; Kelemen et al., 2007). Much of this variability occurs even among kerogens of different maturity from the same formation (Kelemen et al., 2007). The average number of aromatic carbons per cluster can range from 10 to 20, with much of that variability due to maturity differences (Kelemen et al., 2007). These trends result from both cracking of aliphatic chains to form oil and aromatization of naphthenic rings during maturation. NMR spectra have been measured for bitumen samples produced by semi-open pyrolysis of type I Green River oil shale, and their maturity was calculated using EASY%R_o (LeDoan et al., 2013; Sweeney and Burnham, 1990; Feng et al., 2013). Like kerogen, bitumen becomes more aromatic with maturity, with equally dramatic trends—13/87 aromatic/aliphatic for immature bitumen to 74/26 aromatic/aliphatic for mature bitumen. As shown in Fig. 3.17, these trends are consistent with atomic H/C measurements, as hydrogen-rich aliphatic carbon gives way to hydrogen-lean aromatic carbon with maturity. However, the size of the aromatic clusters stays constant in bitumen, while it increases with maturity of kerogen.

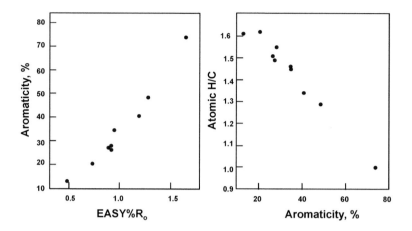

FIGURE 3.17

Aromaticity versus EASY%R$_o$ maturity (left; Sweeney and Burnham, 1990) and atomic H/C ratio (right) for bitumen samples from Green River shale semi-open pyrolysis experiments.

Modified from Feng et al. (2013).

3.2.12.3 Infrared spectroscopy

IR spectroscopy was among the first analytical techniques to be applied to kerogen (Rouxhet et al., 1980). This technique measures vibrational frequencies, which can be assigned to particular chemical groups. Spectra are typically recorded in transmission (Painter et al., 1981), diffuse reflection (Fuller et al., 1982), total attenuated reflection (Washburn and Birdwell, 2013b), or photoacoustic mode (Michaelian and Friesen, 1990) and are analyzed by fitting the peaks to a combination of Lorentzian and Gaussian forms (Painter et al., 1981). Work based on IR imaging enabled measurement of individual macerals (Guo and Bustin, 1998; Chen et al., 2012). In addition to the carbon backbone, this technique provides insight into the chemistry of oxygen in kerogen and bitumen. Unfortunately, standard IR analysis of kerogen or bitumen is too laborious and time-consuming for routine application. However, as discussed earlier, DRIFTS can be applied to core or cuttings and it is fast enough to keep up with the drill bit as long as any oil-based mud contaminants have been removed.

Representative IR spectra of kerogen and bitumen samples from semi-open Green River oil shale pyrolysis are presented in Fig. 3.18. The spectra of kerogen and bitumen are dominated by aliphatic C-H stretch near 2900 cm^{-1} at low maturities, but the intensity of those peaks drops as the kerogen and the bitumen mature. For kerogen, low frequency vibrations associated with C=C and C=O vibrations become dominant, but in bitumen the C–H vibrations remain the most prominent. The aliphatic C–H stretches can be resolved into CH$_2$ and CH$_3$ groups, and the CH$_2$/CH$_3$ ratio typically decreases with maturity in kerogen. This decrease has been attributed to cleavage of aliphatic chains at the position beta to an aromatic ring during maturation, replacing a long chain in immature kerogen with a methyl group in mature kerogen plus an aliphatic chain that partitions into the bitumen or oil phase (Lin and Ritz, 1993; Riboulleau et al., 2000; Lis et al., 2005). On the other hand, the bitumen CH$_2$/CH$_3$ ratio increases during maturation in the range studied here, as bitumen is primarily being created by kerogen

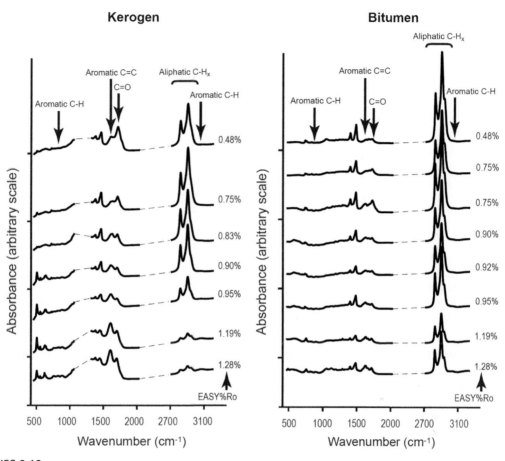

FIGURE 3.18

Infrared spectra of kerogen and bitumen from Green River shale semi-open pyrolysis experiments. Spectra are dashed for wavelengths excluded from the figure.

decomposition, potentially due to the influence of the long chains introduced from kerogen breakdown.

One method that is useful for interpreting these spectra is to calculate the A and C-factors, which are the ratios of aliphatic/aromatic and carbonyl/aromatic bands, respectively (Ganz and Kalkreuth, 1991). These factors are analogous to the hydrogen and oxygen indices from Rock-Eval pyrolysis and Fig. 3.19 presents a pseudo van Krevelen diagram showing these ratios for the same kerogen and bitumen samples. Several interesting trends, many of which are typical of those for other formations, are observed (Painter et al., 1981; Lin and Ritz, 1993; Lis et al., 2005; Ganz and Kalkreuth, 1991; Durand and Espitalié, 1976; Ibarra et al., 1994, 1996). In both kerogen and bitumen, the A-factor decreases with maturity, indicating that the materials become more aromatic and less aliphatic. However, the trend is more dramatic in kerogen, with bitumen appearing less aromatic than kerogen at

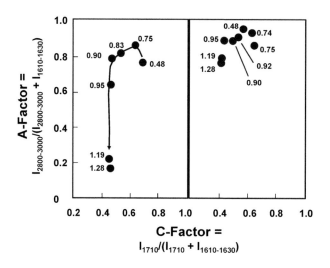

FIGURE 3.19

Pseudo van Krevelen diagrams for type I kerogen (left) and bitumen (right) from Green River Shale semi-open pyrolysis experiments. A and C-factors are aliphatic/aromatic and carbonyl/aromatic ratios from infrared spectra. Numbers near points indicate equivalent vitrinite reflectance (R_o, %).

the highest maturities. In kerogen, the C-factor decreases with maturity, indicating simultaneous decrease in oxygenated components and increase in aromaticity. The C-factor in bitumen initially increases before falling as maturity increases. Thus, the bitumen can appear more oxidized than kerogen in the same rock, which is consistent with other measurements discussed below.

3.2.12.4 X-ray Absorption Near-Edge Structure Spectroscopy

Sulfur content can vary widely and is believed to influence some properties of kerogen, such as kinetics for petroleum generation (Lewan, 1998). Beyond measurement of the total concentration of sulfur, the bonding environment of sulfur can be measured by K-edge X-ray absorption near-edge structures (XANES) spectroscopy. XANES measures the relative abundance of various sulfur-containing functional groups, such as thiophene (sulfur in an aromatic ring), sulfide (sulfur in an aliphatic chain), and sulfoxide (sulfur with a double bond to oxygen). The experiment involves electronic transitions from 1s orbitals on sulfur to vacant molecular orbitals with significant 3p character (Pickering et al., 2001). The energy of that transition depends on the oxidation state of sulfur, which can be mapped to the speciation of the sulfur (Frank et al., 1987; George and Gorbary, 1989). Distinguishing sulfur species in similar oxidation states can be difficult (Behyan et al., 2014; George et al., 2014). The experiments require tunable X-rays of ~2500 eV and are therefore performed at synchrotron facilities. XANES has been used to measure sulfur speciation in coal (George et al., 1991; Hussain et al., 1982; Spiro et al., 1984; Huffman et al., 1991), asphaltenes (Pomerantz et al., 2013; George and Gorbaty, 1989; Waldo et al., 1992), kerogen (Pomerantz et al., 2014; Kelemen et al., 2007; Wiltfong et al., 2005), bitumen (Pomerantz et al., 2014), and petroleum (Waldo et al., 1991; Mitra-Kirtley et al., 1998). While most measurements are for bulk samples, XANES can also be performed in imaging mode with spatial resolution better than 100 nm (Prietzel et al., 2011).

Figure 3.20 shows sulfur XANES spectra for kerogen and bitumen samples from two formations (Pomerantz et al., 2014). NMR and elemental analysis show that kerogen and bitumen typically have similar compositions, with the major difference being molecular weight; kerogen is large and insoluble in organic solvents, while bitumen is smaller and soluble (Salmon et al., 2011). Nonetheless, XANES data indicate a clear difference in composition, in which sulfur-containing moieties in bitumen are enriched in oxidized and polar forms, such as sulfoxides, while sulfur-containing moieties in kerogen are dominated by reduced and nonpolar forms, such as sulfides and thiophenes. Mass balance calculations indicate that about half of bitumen molecules contain a sulfoxide group. Consistently, extraction with less polar solvents results in a sulfoxide-lean bitumen in about half of the yield compared to that from more polar solvents. Bitumen that contains a polar sulfoxide group attached to a relatively nonpolar carbon backbone could act as a naturally-occurring surfactant during migration.

The evolution of sulfur speciation with type and maturity has also been studied in kerogen and bitumen (Pomerantz et al., 2014; Kelemen et al., 2007; Riboulleau et al., 2000; Wiltfong et al., 2005; Sarret et al., 2002). In kerogen, sulfur speciation was found to mimic the carbon chemistry; type III kerogen and highly mature kerogen have more aromatic carbon and aromatic sulfur, while kerogen with more aliphatic carbon also has more aliphatic sulfur (Kelemen et al., 2007; Riboulleau et al., 2000; Wiltfong et al., 2005; Sarret et al., 2002). For Green River oil shale subjected to semi-open pyrolysis, the bitumen is dominated by sulfoxide moieties at low maturity, evolves to more oxidized

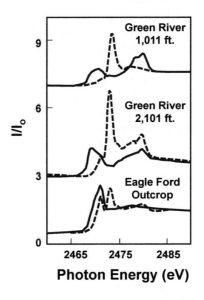

FIGURE 3.20

Sulfur XANES spectra of kerogen (solid lines) and bitumen (dashed lines) from the Green River and Eagle Ford formations (modified from Pomerantz et al., 2014). The x and y-axes represent energy of the excitation photon and normalized fluorescence intensity, respectively. The peak near 2473 eV corresponds to sulfoxide, which dominates the bitumen spectra, but is weak in the kerogen spectra.

sulfone forms at intermediate maturity, and then to reduced thiophene forms at high maturity (LeDoan et al., 2013). A similar maturity evolution of oxidized species has been observed in IR measurements of bitumen.

3.2.12.5 Other Methods
Kerogen and bitumen have been analyzed by many other techniques, including X-ray photoelectron spectroscopy (Kelemen et al., 2007), Raman spectroscopy (Kelemen and Fang, 2001), and ion cyclotron resonance mass spectrometry (Salmon et al., 2011). Measurements of the pore geometry in kerogen using techniques such as small-angle neutron scattering (SANS) have also been performed (Thomas et al., 2014; Clarkson et al., 2013). Future work may involve additional measurements, such as X-ray Raman spectroscopy (Bergmann et al., 2003) and optical spectroscopy (Mullins et al., 1992), which have proven useful for asphaltene analysis. Additionally, imaging experiments studying the chemical heterogeneity of organic matter in shale is of interest in the future. Given the great challenge and importance of kerogen and bitumen chemical analysis, it is clear that we are at only the beginning of a long road to understand their structure and properties.

3.2.13 KEROGEN STRUCTURE THROUGH ASPHALTENE CHEMISTRY
Traditional measurements of kerogen include elemental analysis (e.g., atomic H/C), Rock-Eval pyrolysis, and vitrinite reflectance, as discussed above. Although these techniques can be used to describe the generative potential and thermal maturity of mudrocks, they provide little information on the molecular structure of kerogen. Work has focused on analytical techniques that provide detailed chemical and physical information on kerogen based on asphaltenes. Asphaltenes (Mullins et al., 2007) are defined as soluble in toluene and insoluble in n-heptane, while kerogen is insoluble in both. Differences in the solubility of these materials are largely controlled by molecular weight. Asphaltenes originate from kerogen, but are generally much smaller (<1 kDa). The solubility and comparatively low molecular weight of asphaltenes allows various analytical methods to be employed, while only a subset of these methods can be applied to kerogen. Because asphaltenes and kerogen display similar chemical behavior, asphaltene analyses can be used to infer physicochemical properties of specific kerogens. In addition, the colloidal character of asphaltenes has been intensively studied and provides useful information on the nanostructure of kerogen. The contrast between coal-derived asphaltenes and unaltered petroleum asphaltenes (discussed below) has been useful to identify chemical factors responsible for their nanostructure.

While insolubility of kerogen precludes some measurements that can be performed on soluble asphaltenes, several experimental techniques, particularly solid-state spectroscopy, have been used effectively on kerogen. What emerges is a rich variation in chemical compositions. While asphaltenes are subject to many processes, they have similar structures, as constrained by their solubility (Mullins, 2010; Zuo et al., 2013; Rane et al., 2013). Kerogens from different basins or having different maturities differ in structure, while asphaltenes are similar.

The dominant molecular and stable nanocolloidal structures of asphaltenes have been resolved and codified in the Yen–Mullins model (cf. Fig. 3.21; Mullins, 2010). In solution and crude oil, asphaltenes self assemble to form nanoaggregates with aggregation numbers of about six molecules. For high asphaltene concentrations in heavy oils, asphaltene nanoaggregates form clusters with aggregation

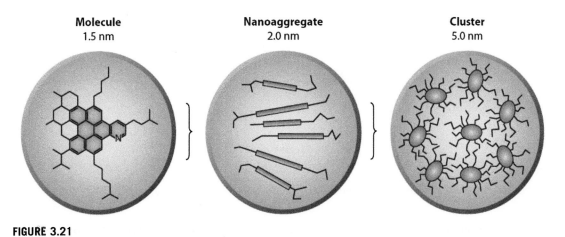

FIGURE 3.21

Schematic shows the Yen–Mullins model for the predominant molecular and colloidal structures of asphaltenes. These structures occur in crude oil and are stable over geologic time.

numbers of about eight. The nanoaggregate has a higher binding energy than the cluster and thus forms at lower concentrations. The polycyclic aromatic hydrocarbons (PAH) in kerogen and asphaltenes are of interest because they have relatively high intermolecular interaction energies that influence nanocolloidal structure, surface interactions, and reaction chemistry. Aromatics interact more strongly than alkanes and much of the heteroatom content in kerogen and asphaltenes occurs in aromatic rings. The disordered PAH stack in Fig. 3.21 shows a favorable energetic configuration for PAH. Steric repulsion of the peripheral alkane chains disrupts stacking and interferes with long-range order.

The Yen–Mullins model has been used with a modified regular solution theory, the Flory–Huggins–Zuo equation of state, to model various crude oil reservoirs, especially to simulate equilibrated gradients of dissolved asphaltenes (Zuo et al., 2013). In addition, this nanostructure model has been linked with the Langmuir equation of state to understand the impact of asphaltenes on interfacial properties at the oil-water interface (Rane et al., 2013). All of these developments aid in understanding asphaltene intermolecular interactions, especially the central role of asphaltene PAH. Below, we review the chemical properties of asphaltenes that may enhance understanding of kerogen.

Asphaltene molecular weight is primarily in the range 600–750 Da based on mass spectroscopy and molecular diffusion measurements (Mullins et al., 2007; Mullins, 2010; Groenzin and Mullins, 1999; Pomerantz et al., 2009; Pomerantz et al., 2015). Molecular weights vary from 400 to 1000 Da with low and high mass tails illustrating the polydispersity of asphaltenes (Fig. 3.22, Pomerantz et al., 2009).

Molecular diffusion (Groenzin and Mullins, 1999) and two-step laser mass spectrometry (L^2MS) measurements (Sabah et al., 2011, 2012; Pomerantz et al., 2009; Pomerantz et al., 2015) typically show only one PAH per asphaltene molecule. The diffusion measurements show that smaller PAH in asphaltenes rotationally diffuse 10 times faster than the larger PAH; thus, they are not attached to each other (Groenzin and Mullins, 1999). L^2MS has been used to evaluate laser induced decomposition of asphaltenes and 23 model compounds with one or more PAH per molecule (Sabah et al., 2011, 2012).

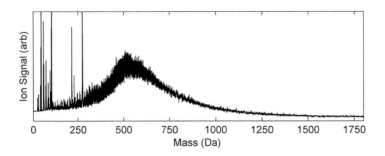

FIGURE 3.22

Laser desorption, laser ionization mass spectrometry of asphaltenes (Pomerantz et al., 2009).

These experiments show that asphaltenes and model compounds with one PAH resist decomposition, while model compounds with two or more PAH decompose at higher laser power. This supports the interpretation that asphaltenes mainly contain a single PAH. Kerogens differ from asphaltenes because they contain cross-linked PAH. Nevertheless, the PAH in asphaltenes and kerogens are likely similar.

Various measurements show that asphaltene PAH have on average seven fused rings, but the number may vary from four to 15 fused rings. Direct molecular imaging of asphaltenes shows a mean of about seven rings, but the images are difficult to interpret (Zajac et al., 1994). NMR shows that asphaltenes have on average seven-ring PAH (Andrews et al., 2011; Dutta Majumdar et al., 2013).

Andrews et al. (2011) and Mullins et al. (2012) estimated the number of fused rings in PAH using single pulse excitation (SPE) to obtain ^{13}C NMR spectra. A spectral cut-off between protonated and nonprotonated aromatic carbon (130 ppm) was used to estimate bridgehead versus peripheral aromatic carbon. Information on aromatic carbon bonded to hydrogen was obtained using DEPT NMR. Aromatic carbon with hydrogen can be determined by comparing laboratory DEPT ^{13}C NMR with SPE spectra. Fig. 3.23 shows the DEPT and SPE spectra for coal-derived asphaltenes. The study concluded that petroleum asphaltene PAH contain roughly seven-ring PAH, while coal asphaltenes have six-ring PAH. Similar results for petroleum asphaltenes were obtained in another NMR study (Dutta Majumdar et al., 2013).

The extent of the PAH distribution was obtained by optical spectroscopy. Each π-electron has oscillator strength of one. Thus, all π-electron transitions must appear. Larger PAH have longer wavelength transitions in accord with a quantum "particle-in-a-box". Carbon X-ray Raman spectroscopy revealed that asphaltenes contain mainly the more stable aromatic sextet carbon as opposed to isolated double-bond carbon (Bergmann et al., 2003). With this understanding, the asphaltene optical absorption spectrum can be modeled using a sum of spectra from many contributing PAH. The modeled and measured asphaltene spectra can be compared and the presumed asphaltene PAH distribution can be adjusted to obtain a match (Ruiz-Morales and Mullins, 2009). An exhaustive study employing 523 PAH was used to obtain synthetic asphaltene optical spectra that closely matched observations (Fig. 3.24; Ruiz-Morales and Mullins, 2009).

The relationship between PAH in asphaltenes versus kerogen can be investigated by analysis of optical spectra or color (Fig. 3.25). Darker kerogen color generally indicates higher maturity. The low-energy electronic absorption edge depicted in Fig. 3.25 is linear when optical density and photon

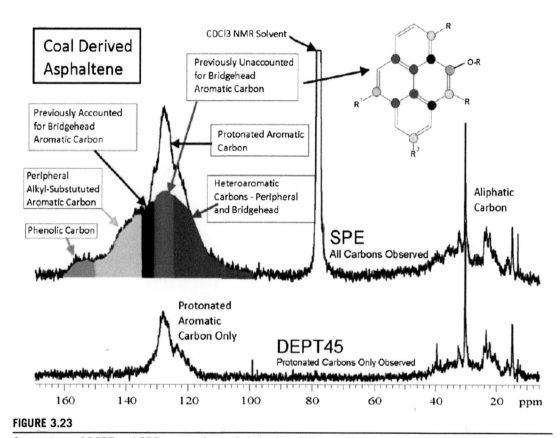

FIGURE 3.23

Comparison of DEPT and SPE spectra for coal-derived asphaltenes (Andrews et al., 2011; Mullins et al., 2012). The DEPT spectrum identifies protonated aromatic carbon, while the SPE spectrum shows all aromatic carbon. Bridgehead carbon dominates the difference.

energy are plotted on log and linear scales, respectively. This linear relationship is called the Urbach tail, as in semiconductor physics, and it corresponds to an exponential decrease of absorption at increasing wavelengths. Urbach tails are associated with thermal processes. This corresponds to a rapid decrease in the larger PAH during catagenesis of asphaltenes. Both crude oil and asphaltenes exhibit the Urbach tail, but lighter, more mature oil exhibits this at shorter wavelengths and thus small PAH size, as seen in Fig. 3.25 (Mullins et al., 1992), thereby giving the centroid and width of the PAH distribution.

Asphaltene PAH are sites of attraction due mainly to high polarizability, e.g., the Hildebrand solubility parameter for asphaltenes is 20.4 $MPa^{1/2}$. The Hildebrand solubility parameter provides a measure of the degree of interaction between materials and can be a good indicator of solubility (Burke, 1984). The projection of this solubility parameter into the Hansen solubility parameters is dispersion or polarizability, 19.5 ($MPa^{1/2}$); dipole 4.7 ($MPa^{1/2}$); and H-bonding, 4.2 ($MPa^{1/2}$; Acevedo et al., 2010). This finding is consistent with most intermolecular interaction among asphaltenes

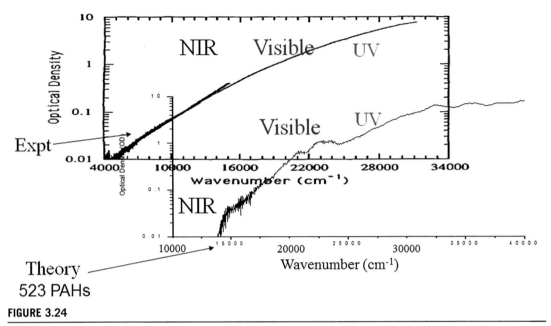

FIGURE 3.24

Measured asphaltene optical spectrum (top). Modeled asphaltene optical spectrum using 523 PAH spectra obtained from molecular orbital calculations (bottom) (Ruiz-Morales and Mullins, 2009).

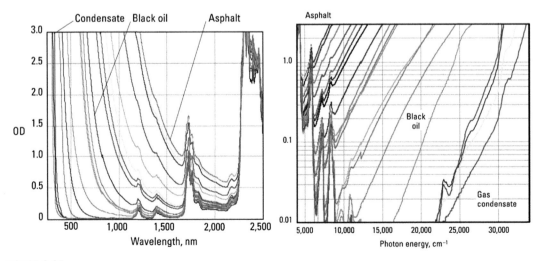

FIGURE 3.25

The ultraviolet (UV)–visible–near IR (NIR) spectra of crude oils vary from condensate to heavy crude oil (indicated by asphalt; Mullins et al., 1992). The optical absorption edge is linear when plotted on a log-linear plot versus photon energy, which implies decreased concentrations of larger PAH in crude oil due to thermally produced PAH distributions. Light crude oil lacks large PAH, unlike heavy oil.

through PAH rather than heteroatom chemistry, which would have stronger polarity and hydrogen bonding forces. For comparison, Hildebrand solubility for *n*-heptane is 15.3 (MPa^{½}), which projects into the Hansen polarizability parameter as dispersion or polarizability, 15.3 (MPa^{½}); dipole, 0 (MPa^{½}); and hydrogen bonding, 0 (MPa^{½}).

A consequence of the above is that asphaltene nanoaggregates contain a stack of PAH in the nanoaggregate (Fig. 3.21, center), as illustrated by a combined study of small-angle X-ray scattering (SAXS) and SANS. X-rays scatter preferentially from carbon in asphaltenes, which is enriched in the aromatic cores, while neutrons scatter preferentially from hydrogen, which is mainly in peripheral alkane groups. The divergence of the SAXS versus SANS cross-sections in Fig. 3.26 demonstrates that aromatic and saturate carbon atoms are distributed in asphaltenes at the 1.4 nm length scale; the nanoaggregate (Eyssautier et al., 2011). These data suggest that there is an aromatic carbon core surrounded by alkane chains in asphaltene nanoaggregates. This length scale yields an aggregation number of ~10.

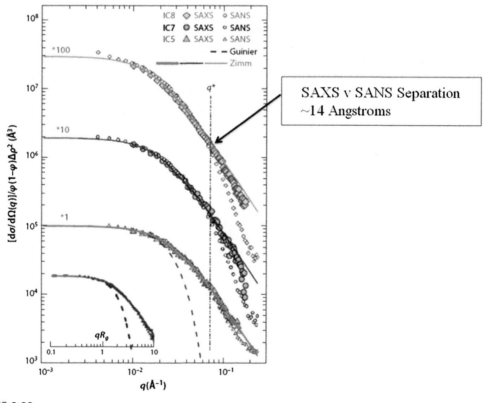

FIGURE 3.26

Comparison of cross-sections for asphaltenes from small-angle X-ray scattering versus small-angle neutron scattering shows that asphaltene nanoaggregates have a core of PAH with peripheral alkanes having an aggregation number of ~10 (with permission from Barré et al., 2009).

Various other measurements reinforce small aggregation numbers for asphaltenes. Surface enhanced laser desorption ionization mass spectrometry coupled with L^{2}MS suggest asphaltene aggregation numbers of ~8 (Wu et al., 2014; Pomerantz et al., 2015). Direct current (DC)-conductivity and centrifugation and NMR studies yield similar results for aggregation number (Mullins, 2010). Many studies, including high-Q ultrasonics, DC-conductivity, centrifugation and NMR, yield similar values for the critical nanoaggregate concentration in toluene of about 10^{-4} mass fraction (Mullins, 2010) At higher concentrations of 10^{-3} mass fraction in toluene, asphaltenes form clusters of nanoaggregates with an aggregation number of ~8, as supported by NMR (Dutta Majumdar et al., 2013), SAXS, and SANS studies (Eyssautier et al., 2011). Consequently, cluster binding energies are weaker than the nanoaggregate. To understand kerogen, strong nanoaggregate binding is most relevant.

The heteroatom chemistry of asphaltenes must be considered as potentially important even though asphaltene PAH are the main source of intermolecular interaction. Sulfur is typically the most important heteroatom, commonly representing several percent by mass, while nitrogen and oxygen occur in lower concentrations. Studies of the heteroatom chemistry of kerogen overlap with those of asphaltene heteroatom chemistry.

Asphaltene sulfur speciation can be determined using sulfur–XANES. Sulfur occurs mainly in sulfide and thiophene groups, which are rather non-polar. Sulfoxide groups can occur and are highly polar (up to ~4 Debye; double the dipole moment of water). Nitrogen–XANES indicates that the dominant asphaltene nitrogen moieties are pyrrolic and pyridinic nitrogen in PAH. Pyridinic nitrogen is basic and pyrrolic nitrogen is acidic, causing some charge separation within asphaltene PAH. Asphaltene oxygen moieties can be studied by well-known IR and NMR methods. There is a small carboxylic acid component in asphaltenes. Other oxygen-containing groups include ethers, phenols, and sulfoxides. Vanadium and nickel occur in very low concentrations (less than ppt), mainly in metalloporphyins.

3.3 CONCLUSIONS

Unconventional rock or sediment units require stimulation to produce retained petroleum due to the combined effects of petroleum viscosity and matrix permeability. Current methods to identify unconventional sweet spots in map view and vertical sections are mainly empirical and rely on independent organic and inorganic geochemical measurements that are compared with those of productive rock units. Many of these measurements are time-consuming, cannot be completed while drilling, and are not at the proper sampling scale to most effectively identify sweet spots. Improved workflows are needed that include rapid, properly scaled geochemical and geomechanical measurements that can be completed while drilling, coupled with fully integrated 3D basin and petroleum system models that predict unconventional targets through time. Nevertheless, some significant recent advances include 3D poroelasticity, SARA, and adsorption modeling to predict rock stress, susceptibility to fracture, and the amount and composition of petroleum that remains in unconventional rock units.

Organic and inorganic geochemical logs offer independent means to identify sweet spots. Most organic geochemical logs are based on direct measurements of carbonate content and organic pyrolysis response, but they require discrete samples at predetermined depth intervals, which are typically analyzed in a laboratory rather than at the drillsite. Inorganic geochemical logs are based on indirect calculation of mineral content and TOC without the need to collect samples. Although inorganic geochemical logs do not determine the quality or thermal maturity of the kerogen, they provide continuous data with depth and can be quantified in real time during drilling. For example, DRIFTS has emerged as a robust method to determine TOC, maturity, and mineralogy from cuttings during drilling of horizontal wells.

The main factors controlling the amount of adsorbed hydrocarbon gas in unconventional gas shales are TOC and monolayer adsorption capacity. The amount of adsorbed hydrocarbon can be determined at any temperature and pressure using the Langmuir equation. During production, the amount of adsorbed gas is low compared to free gas until the reservoir pressure is strongly depleted. Calibrated kinetic parameters for light hydrocarbons suggest that petroleum generative potential of kerogen is depleted at 1.5–2.0% R_o and that retained in source rock at $R_o > 2.0\%$ is mainly the product of early maturity or secondary oil cracking.

As discussed above, recent work has focused on analytical techniques that provide detailed chemical and physical information on kerogen and asphaltenes. These measurements contribute to understanding of processes that play a role in the distribution of sweet spots, but most cannot be practically applied to identify sweet spots in actively drilling wells. Modifications of these methods to make them more suitable for characterizing unconventional sweet spots would be a valuable contribution.

3.4 APPENDIX: KEROGEN TYPES AND PREPARATION
3.4.1 KEROGEN TYPES

Four general types of kerogens in thermally immature coal and sedimentary rock are defined by the van Krevelen diagram based on original atomic H/C or hydrogen index (HI; e.g., Peters and Cassa, 1994):

1. Oil-prone kerogen dominated by liptinite macerals; high atomic H/C (≥ 1.5) and HI (>600 mg hydrocarbon/g TOC) and low O/C (≤ 0.1).
2. Oil-prone kerogen shows moderate atomic H/C (1.2–1.5) and HI (300–600 mg hydrocarbon/g TOC), and low O/C compared with types III and IV. Some kerogens of this type also contain abundant organic sulfur (8–14 wt% sulfur; atomic S/C ≥ 0.04) and are designated as type IIS. (Note that sulfur-rich type IS and IIIS kerogens also occur, but are less common.) Type II/III kerogen shows atomic H/C (1.0–1.2) and HI of 200–300 mg hydrocarbon/g TOC.
3. Gas-prone kerogen shows low atomic H/C (<1.0) and HI (50–200 mg hydrocarbon/g TOC), and high O/C (≤ 0.3). The term "gas-prone" is misleading because type III kerogen typically yields less gas than type I or II kerogen.
4. "Dead carbon" dominated by inertinite macerals that generate little or no petroleum. Kerogens of this type show very low atomic H/C (~ 0.5–0.6) and HI (<50 mg hydrocarbon/g TOC) and variable O/C (≤ 0.3).

Kerogens are mixtures of macerals. Intermediate compositions between types I, II, III, and IV are common on van Krevelen diagrams. For example, a type III kerogen might be composed dominantly of gas-prone vitrinite macerals or it could be a mixture of type II and IV and thus have significant oil-prone character. Furthermore, kerogen types can vary within any depositional setting. For example, a lacustrine source rock might contain type I, II, III, and IV kerogen depending on location within the basin and the depositional setting for the organic matter. Other common misconceptions are that all type I kerogens originated from lacustrine source rocks, such as the Mahogany Ledge unit in the Green River Formation, and that all type II kerogens are from marine source rocks, such as the Toarcian Shale in the Paris Basin.

3.4.2 KEROGEN PREPARATION

Kerogen consists of the insoluble organic matter in sedimentary rocks (Durand and Nicaise, 1980; Groenzin and Mullins, 1999). Pyrobitumen and coke are also insoluble and in some cases can comprise a significant proportion of the kerogen. Bitumen is the soluble organic matter isolated using solvents, such as dichloromethane, pyridine, or toluene. Free oil and gas are defined as having escaped the rock by volatilization or depressurization, so all residual soluble components are categorized as bitumen. Bitumen can be extracted from shale using organic solvents, as in a Soxhlet extractor. A moderately polar solvent is typically used because low polarity solvents can result in lower yields and different compositions of bitumen (Salmon et al., 2011; Pomerantz et al., 2014).

Kerogen can be isolated from rock by solvent extraction to remove soluble components and acid treatment to remove most minerals. Demineralization typically involves hydrochloric, hydrofluoric, and boric acids to remove carbonates, silicates, and neoformed fluorides, respectively (Robinson and Taulbee, 1995). This method removes nearly all minerals except pyrite. Additional treatment with chromous chloride can remove residual pyrite (Acholla and Orr, 1993; Ibrahimov and Bissada, 2010). Acid demineralization has been shown by various spectroscopy tools to preserve the chemical composition of kerogen, allowing the method to be used to prepare samples for chemical analysis (Pomerantz et al., 2014; Durand and Nicaise, 1980). Because typical acid demineralization procedures alter the pore geometry in kerogen, a new method that includes critical point drying can be used to preserve kerogen physical structure (Suleimenova et al., 2014).

ACKNOWLEDGMENTS

We thank Ian Bryant, Thomas Hantschel, Susan Herron, Oliver Schenk, Rodney Warfford, and Bjorn Wygrala (Schlumberger) for encouragement and useful discussions and the following reviewers, whose comments improved the manuscript: Christina Calvin, Zee Ma, and Daniel Palmowski.

REFERENCES

Aboud, M.R., Badry, R., Grau, J., Herron, S., Hamichi, F., Horkowitz, J., Hemingway, J., MacDonald, R., Saldungaray, P., Stachiw, D., Stellor, C., Williams, R.E., 2014. High-definition spectroscopy—determining mineralogic complexity. Oilfield Review 20, 34–50.
Acevedo, S., Castro, A., Vasquez, E., Marcano, F., Ranaudo, M.A., 2010. Investigation of physical chemistry properties of asphaltenes using solubility parameters of asphaltenes and their fractions A1 and A2. Energy & Fuels 24 (11), 5921–5933.
Acholla, F.V., Orr, W.L., 1993. Pyrite removal from kerogen without altering organic matter: the chromous chloride method. Energy & Fuels 7, 406–410.
Alexander, T., Baihly, J., Boyer, C., Clark, B., Waters, G., Jochen, V., Le Calvez, J., Lewis, R., Miller, C.K., Thaeler, J., Toelle, B.E., 2011. Shale gas revolution. Oilfield Review 23, 40–55.
Allix, P., Burnham, A., Fowler, T., Herron, M., Kleinberg, R., Symington, W., 2011. Coaxing oil from shale. Oilfield Review 22, 4–15.

REFERENCES

Al-Salim, A., Meridji, Y., Musharfi, N., Al-Waheed, H., Saldungaray, P., Herron, S., Polyakov, M., 2014. Using a new spectroscopy tool to quantify elemental concentrations and TOC in an unconventional shale gas reservoir: case studies from Saudi Arabia. Society of Petroleum Engineers. http://dx.doi.org/10.2118/172176-MS. Supplemental info: SPE Saudi Arabia Section Technical Symposium and Exhibition, April 21–24, Al-Khobar, Saudi Arabia SPE-172176-MS.

Altindag, R., Guney, A., 2010. Predicting the relationships between brittleness and mechanical properties (UCS, TS and SH) of rocks. Scientific Research Essays 5, 2107–2118.

Andrews, A.B., Edwards, J.C., Pomerantz, A.E., Mullins, O.C., Nordlund, D., Norinaga, K., 2011. Comparison of coal-derived and petroleum asphaltenes by ^{13}C nuclear mgnetic resonance, DEPT, and XRS. Energy & Fuels 25, 3068–3076.

Armagnac, C., Bucci, J., Kendall, C.G.St.C., Lerche, I., 1989. Estimating the thickness of sediment removed at an unconformity using vitrinite reflectance data. In: Naeser, N.D., et al. (Eds.), Thermal History of Sedimentary Basins. Springer-Verlag, New York, pp. 217–238.

Aubertin, M., Gill, D.E., Simon, R., 1994. On the use of the brittleness index modified (BIM) to estimate the post-peak behaviour of rocks. In: Proceedings of the First North American Rock Mechanics Symposium, Balkema, pp. 945–952.

Barré, L., Jestin, J., Morisset, A., Palermo, T., Simon, S., 2009. Relation between nanoscale structure of asphaltene aggregates and their macroscopic solution properties: Oil & Gas Science and Technology – Review Institut Français du Pétrole 64, 617–628.

Baskin, D.K., 1979. A method of preparing phytoclasts for vitrinite reflectance analysis. Journal of Sedimentary Petrology 49, 633–635.

Baskin, D.K., 2001. Comparison between atomic H/C and Rock-Eval hydrogen index as an indicator of organic matter quality. In: Isaacs, C.M., Rullkötter, J. (Eds.), The Monterey Formation – From Rocks to Molecules. Columbia University Press, New York, pp. 230–240.

Behar, F., Lorant, F., Lewan, M.D., 2008. Elaboration of a new compositional kinetic schema for oil cracking. Organic Geochemistry 39, 764–782.

Behyan, S., Hu, Y., Urquhart, S.G., 2014. Chemical sensitivity of sulfur 1s NEXAFS spectroscopy I: speciation of sulfoxides and sulfones. Chemical Physics Letters 592, 69–74.

Bergmann, U., Groenzin, H., Mullins, O.C., Glatzel, P., Fetzer, J., Cramer, S.P., 2003. Carbon K-edge X-ray Raman spectroscopy supports simple yet powerful description of aromatic hydrocarbons and asphaltenes. Chemical Physics Letters 369, 184–191.

Bjørlykke, K., 1998. Clay mineral diagenesis in sedimentary basins—a key to the prediction of rock properties. Examples from the North Sea Basin. Clay Minerals 33, 15–34.

Bostick, N.H., 1979. Microscopic measurement of the level of catagenesis of solid organic matter in sedimentary rocks to aid exploration for petroleum and to determine former burial temperatures – a review. In: Society of Economic Paleontologists and Mineralogists Special Publication, vol. 26, pp. 17–43.

Bostick, N.H., Alpern, B., 1977. Principles of sampling, preparation and constituent selection for microphotometry in measurement of maturation of sedimentary organic matter. Journal of Microscopy 109, 41–47.

Bruns, B. Unconventional Petroleum Systems in NW-Germany and the Netherlands: A 3D Numerical Basin Modeling and Organic Petrography Study (Dissertation). RWTH Aachen University, Aachen, Germany, submitted for publication.

Bruns, B., Littke, R., Gasparik, M., van Wees, J.-D., Nelskamp, S., 2014. Thermal evolution and shale gas potential estimation of the Wealden and Posidonia Shale in NW-Germany and the Netherlands: a 3D basin modeling study. Basin Research. http://onlinelibrary.wiley.com/doi/10.1111/bre.12096/abstract (accessed 20.12.14.). Supplemental info: Published online before inclusion in an issue.

Burke, J., 1984. Solubility parameters: theory and application. In: Jensen, C. (Ed.), AIC (American Institute for Conservation) Book and Paper Group Annual, 3, pp. 13–58.

Burruss, R.C., Laughrey, C.D., 2010. Carbon and hydrogen isotopic reversals in deep basin gas: evidence for limits to the stability of hydrocarbons. Organic Geochemistry 41, 1285–1296.

Cao, X., Birdwell, J.E., Chappell, M.A., Li, Y., Pignatello, J.J., Mao, J., 2013. Characterization of oil shale, isolated kerogen, and postpyrolysis residues using advanced ^{13}C solid-state nuclear magnetic resonance spectroscopy. American Association of Petroleum Geologists Bulletin 97, 421–436.

Cardineaux, A.P., 2012. Mapping of the Oil Window in the Eagle Ford Shale Play of Southwest Texas Using Thermal Modeling and Log Overlay Analysis (Masters thesis). Department of Geology and Geophysics, Louisiana State University, 74 p.

Cander, H., 2012. Sweet spots in shale gas and liquids plays: prediction of fluid composition and reservoir pressure. AAPG Search and Discovery. Article #40936. Supplemental info: Adapted from oral presentation at AAPG Annual Convention and Exhibition, Long Beach, California, April 22–25, 2012.

Carmo, M.J., Gubulin, J.C., 1997. Ethanol-water adsorption on commercial 3A zeolites: kinetic and thermodynamic data. Brazilian Journal of Chemical Engineering 14 (3).

Charsky, A., Herron, S., 2013. Accurate, direct total organic carbon (TOC) log from a new advanced geochemical spectroscopy tool: comparison with conventional approaches for TOC estimation. AAPG Search and Discovery. Article #41162. Supplemental info: AAPG Search and Discovery Article #90163©2013AAPG 2013 Annual Convention and Exhibition, Pittsburgh, Pennsylvania, May 19–22, 2013.

Chen, Y., Mastalerz, M., Schimmelmann, A., 2012. Characterization of chemical functional groups in macerals across different coal ranks via micro-FTIR spectroscopy. International Journal of Coal Geology 104, 22–33.

Chung, H.M., Gormly, J.R., Squires, R.M., 1988. Origin of gaseous hydrocarbons in subsurface environments: theoretical considerations of carbon isotope distribution. Chemical Geology 71, 97–103.

Clarkson, C.R., Bustin, R.M., Levy, J.H., 1997. Adsorption of the mono/multilayer and adsorption potential theories to coal methane adsorption isotherms at elevated temperature and pressure. Carbon 35, 1689–1705.

Clarkson, C.R., Solano, N., Bustin, R.M., Bustin, A.M.M., Chalmers, G.R.L., He, L., Melnichenko, Y.B., Radliński, A.P., Black, T.P., 2013. Pore structure characterization of North American shale gas reservoirs using USANS/SANS, gas adsorption, and mercury intrusion. Fuel 103, 606–616.

Craddock, P., Herron, S.L., Badry, R., Swager, L.I., Grau, J.A., Horkowitz, J.P., Rose, D., 2013. Hydrocarbon saturation from total organic carbon logs derived from inelastic and capture nuclear spectroscopy. In: Society of Petroleum Engineers, SPE 166297 Presented at the SPE Annual Technical Conference and Exhibition, New Orleans, Louisiana, USA, 30 September–2 October.

Crelling, J.C., 1983. Current uses of fluorescence microscopy in coal petrology. Journal of Microscopy 132, 251–266.

Cruz, J.A.L., Mota, J.P.B., 2009. Thermodynamics of adsorption of light alkanes and alkenes in single-walled carbon nanotube bundles. Physics Review B 79, 1654261–16542614.

Culver, R.B., Hopkinson, E.C., Youmans, A.H., October 1974. Carbon/oxygen (C/O) logging instrumentation. Society of Petroleum Engineers Journal 14, 463–470. SPE-4640-PA.

Dahl, J., Moldowan, J.M., Walls, J., Nur, A., DeVito, J., 2012. Creation of porosity in tight shales during organic matter maturation. AAPG Search and Discovery. Article #40979. Supplemental info: Adapted from oral presentation at AAPG Annual Convention and Exhibition, Long Beach, California, USA, April 22–25, 2012.

Dai, J., Li, J., Luo, X., Zhang, W., Hu, G., Ma, C., Guo, J., Ge, G., 2005. Stable carbon isotope compositions and source rock geochemistry of the giant gas accumulations in the Ordos Basin, China. Organic Geochemistry 36, 1617–1635.

Demaison, G.J., 1984. The generative basin concept. In: Demaison, G.J., Murris, R.J. (Eds.), Petroleum Geochemistry and Basin Evaluation, American Association of Petroleum Geologists Memoir, 35, 1–14.

Dembicki, H., 2009. Three common source rock evaluation errors made by geologists during prospect or play appraisals. American Association of Petroleum Geologists Bulletin 93, 341–356.

Dewhurst, D.N., Yang, Y., Aplin, A.C., 1999. Permeability and fluid flow in natural mudstones. Geological Society of London Special Publications 158, 32–43.

Durand, B., Espitalié, J., 1976. Geochemical studies on the organic matter from the Douala Basin (Cameroon)–II. Evolution of kerogen. Geochimica et Cosmochimica Acta 40, 801–808.

REFERENCES

Durand, B., Nicaise, G., 1980. Procedures for kerogen isolation. In: Durand, B. (Ed.), Kerogen: Insoluble Organic Matter from Sedimentary Rocks. Editions Technip, Paris.

Dutta Majumdar, R., Gerken, M., Mikula, R., Hazendonk, P., 2013. Validation of the Yen–Mullins model of Athabasca oil-sands asphaltenes using solution-state 1H NMR relaxation and 2D HSQC spectroscopy. Energy & Fuels 27, 6528–6538.

EIA, 2010. Eagle Ford Shale Play Map. At: www.eia.gov (accessed 2.01.15.).

Elias, R., Duclerc, D., Le-Van-Loi, R., Gelin, F., Dessort, D., 2013. A new geochemical tool for the assessment of organic-rich shales. In: Unconventional Resources Technology Conference, URTeC 1581024.

Engleman, E.E., Jackson, L.L., Norton, D.R., Fischer, A.G., 1985. Determinations of carbonate carbon in geological materials by coulometric titration. Chemical Geology 53, 125–128.

Espitalié, J., Madec, M., Tissot, B., Menning, J.J., Leplat, P., 1977. Source rock characterization methods for petroleum exploration. In: Proceedings of the 1977 Offshore Technology. Conference, vol. 3, pp. 439–443.

Eyssautier, J., Levitz, P., Espinat, D., Jestin, J., Gummel, J., Brillo, I., Barré, L., 2011. Insight into asphaltene nanoaggregate structure inferred by small angle neutron and X-ray scattering: Journal of Physical Chemistry B 115, 6827–6837.

Feng, Y., Le Doan, T.V., Pomerantz, A.E., 2013. The chemical composition of bitumen in pyrolyzed Green River oil shale: characterization by ^{13}C NMR spectroscopy. Energy & Fuels 27, 7314–7323.

Fertl, W.H., Chilingar, G.V., 1988. Total organic carbon content determined from well logs. In: Society of Petroleum Engineers, vol. 3. SPE-15612-PA.

Ferworn, K., Zumberge, J., Brown, S., 2008. Gas character Anomalies Found in Highly Productive Gas Shale Wells. At: https://www.google.com/?gws_rd=ssl#q=ferworn+isotopic+rollover (accessed 19.12.14.).

Frank, P., Hedman, B., Carlson, R.M.K., Tyson, T.A., Roe, A.L., Hodgson, K.O., 1987. A large reservoir of sulfate and sulfonate resides within plasma cells from Ascidia ceratodes revealed by X-ray absorption near-edge structure spectroscopy. Biochemistry 26, 4975–4979.

Freedman, R., Herron, S., Anand, V., Herron, M.M., May, D.H., Rose, D.A., 2014. New method for determining mineralogy and matrix properties from elemental chemistry measured by gamma ray spectroscopy logging tools. Society of Petroleum Engineers. http://dx.doi.org/10.2118/170722-MS. Supplemental info: SPE Annual Technical Conference and Exhibition, October 27–29, Amsterdam, The Netherlands SPE-170722-MS.

Fuller, M.P., Hamadeh, I.M., Griffiths, P.R., Lowenhaupt, D.E., 1982. Diffuse reflectance infrared spectrometry of powdered coals. Fuel 61, 529–536.

Ganz, H.H., Kalkreuth, W., 1991. IR classification of kerogen type, thermal maturation, hydrocarbon potential and lithological characteristics. Journal of Southeast Asian Earth Science 5, 19–28.

Goddard III, W.A., Tang, Y., Wu, S., Deev, A., Ma, Q., Li, G., 2013. Novel Gas Isotope Interpretation Tools to Optimize Gas Shale Production. Research Partnership to Secure Energy for America (RPSEA). Report No. 08122.15.final, 90 p.

Gasparik, M., Ghanizadeh, A., Bertier, P., Gensterblum, Y., Bouw, S., Krooss, B.M., 2012. High-pressure methane sorption isotherms of black shales from the Netherlands. Energy & Fuels 26, 4995–5004.

Gasparik, M., Bertier, P., Gensterblum, Y., Ghanizadeh, A., Krooss, B.M., Littke, R., 2014. Geological controls on the methane storage capacity in organic-rich shales. International Journal of Coal Geology 123, 34–51.

George, G.N., Gorbaty, M.L., 1989. Sulfur K-edge X-ray absorption spectroscopy of petroleum asphaltenes and model compounds. Journal of the American Chemical Society 111, 3182–3186.

George, G.N., Gorbaty, M.L., Kelemen, S.R., Sansone, M., 1991. Direct determination and quantification of sulfur forms in coals from the Argonne Premium Sample Program. Energy & Fuels 5, 93–97.

George, G.N., Hackett, J.J., Sansone, M., Gorbaty, M.L., Kelemen, S.R., Prince, R.C., Harris, H.H., Pickering, I.J., 2014. Long-range chemical sensitivity in the sulfur K-edge X-ray absorption spectra of substituted thiophenes. The Journal of Physical Chemistry A 118, 7796–7802.

Gonzalez, J., Lewis, R., Hemingway, J., Grau, J., Rylander, R., Schmitt, R., 2013. Determination of formation organic carbon content using a new neutron-induced gamma ray spectroscopy service that directly measures carbon. In: 54 Annual Logging Symposium, June 22–26, 2013. Society of Petrophysicists and Well Log Analysts (SPWLA), pp. 1–15.

Groenzin, H., Mullins, O.C., 1999. Asphaltene molecular size and structure. Journal of Physical Chemistry A 103, 11237–11245.

Groszek, A.J., 2001. Heats of water adsorption on microporous carbons from nitrogen and methane carriers. Carbon 39, 1857–1862.

Guo, Y., Bustin, R.M., 1998. Micro-FTIR spectroscopy of liptinite macerals in coal. International Journal of Coal Geology 36, 259–275.

Herron, S.L., Herron, M.M., 1996. Quantitative lithology: an application for open and cased-hole spectroscopy. In: Proceedings 37th SPWLA Annual Logging Symposium, New Orleans, Louisiana, USA, 16–19 June, Paper E.

Herron, M.M., Loan, M., Charsky, A., Herron, S.L., Pomerantz, A.E., Polyakov, M., 2014. Kerogen content and maturity, mineralogy and clay typing from DRIFTS. Analysis of Cuttings or Core. Petrophysics 55, 435–446.

Hertzog, R.C., 1980. Laboratory and field evaluation of an inelastic neutron scattering and capture gamma ray spectrometry tool. Society of Petroleum Engineers Journal 20, 327–340.

Hood, A., Gutjahr, C.C.M., Heacock, R.L., 1975. Organic metamorphism and the generation of petroleum. American Association of Petroleum Geologists Bulletin 59, 986–996.

Horsfield, B., Littke, R., Mann, U., Bernard, S., Vu, T.A.T., di Primio, R., Schulz, H.-M., 2010. Shale gas in the Posidonia Shale, Hils area, Germany. AAPG Search and Discovery. Article #110126. Supplemental info: Adapted from oral presentation at session, Genesis of Shale Gas–Physicochemical and Geochemical Constraints Affecting Methane Adsorption and Desorption, at AAPG Annual Convention, New Orleans, LA, April 11–14, 2010.

Hu, A.P., Li, J., Zhang, W.J., Li, Z.F., Hou, L., Liu, Q.Y., 2008. Geochemical characteristics and origin of gases from the Upper, Lower Paleozoic and the Mesozoic reservoirs in the Ordos Basin, China. Science in China Series D-Earth Science 51, 183–194.

Hu, H., Zhang, T., Wiggins-Camacho, J.D., Ellis, G.S., Lewan, M.D., Zhang, X., 2015. Experimental investigation of changes in methane adsorption of bitumen-free Woodford Shale with thermal maturation induced by hydrous pyrolysis. Marine and Petroleum Geology 59, 114–128.

Huffman, G.P., Mitra-Kirtley, S., Huggins, F.E., Shah, N., Vaidya, S., Lu, F., 1991. Quantitative analysis of all major forms of sulfur in coal by X-ray absorption fine structure spectroscopy. Energy & Fuels 5, 574–581.

Hussain, Z., Umbach, E., Shirley, C.A., Stöhr, J., Feldhaus, J., 1982. Performance and application of a double crystal monochrometer in the energy region $800 < h\nu < 4500$ eV. Nuclear Instruments and Methods 195, 115–131.

Ibarra, J.V., Moliner, R., Bonet, A.J., 1994. FT-i.r. investigation on char formation during the early stages of coal pyrolysis. Fuel 73, 918–924.

Ibarra, J.V., Munoz, E., Moliner, R., 1996. FTIR study of the evolution of coal structure during the coalification process. Organic Geochemistry 24, 725–735.

Ibrahimov, R.A., Bissada, K.K., 2010. Comparative analysis and geological significance of kerogen isolated using open-system (palynological) versus chemically and volumetrically conservative closed-system methods. Organic Geochemistry 41, 800–811.

Issler, D.R., Hu, K., Block, J.D., Katsube, T.J., 2002. Organic Carbon Content Determined from Well Logs: Examples from Cretaceous Sediments of Western Canada. Geological Survey of Canada. Open File Report 4362, 19 p.

Janik, L.J., Skjemstad, J.O., 1995. Characterization and analysis of soils using mid-infrared partial least-squares. II. Correlations with some laboratory data. Australian Journal of Soil Research 33, 637–650.

Jarvie, D.M., Claxton, B.L., Henk, F., Breyer, J.T., 2001. Oil and shale gas from the barnett shale, fort Worth Basin, Texas. American Association of Petroleum Geologists Bulletin 85, A100. Abstract.

Jarvie, D.M., 2012a. Shale resource systems for oil and gas: Part 1-Shale-gas resource systems. In: Breyer, J.A. (Ed.), Shale Reservoirs—Giant Resources for the 21st Century, American Association of Petroleum Geologists Memoir, 97, 69–87.

Jarvie, D.M., 2012b. Shale resource systems for oil and gas: Part 2—Shale-oil resource systems. In: Breyer, J.A. (Ed.), Shale Reservoirs—Giant Resources for the 21st Century, American Association of Petroleum Geologists Memoir, 97, 89–119.

Jarvie, D.M., Hill, R.J., Ruble, T.E., Pollastro, R.M., 2007. Unconventional shale-gas systems: the Mississippian Barnett Shale of North-Central Texas as one model for thermogenic shale-gas assessment. American Association of Petroleum Geologists Bulletin 91, 475–499.

Jenden, P.D., Drazan, D.J., Kaplan, I.R., 1993. Mixing of thermogenic natural gases in northern Appalachian basin. American Association of Petroleum Geologists Bulletin 77, 980–998.

Ji, L., Zhang, T., Milliken, K.L., Qu, L., Zhang, X., 2012. Experimental investigation of main controls to methane adsorption in clay-rich rocks. Applied Geochemistry 27, 2533–2545.

Jones, R.W., 1987. Organic facies. In: Brooks, J., Welte, D. (Eds.), Advances in Petroleum Geochemistry. Academic Press, New York, pp. 1–90.

Kelemen, S.R., Fang, H.L., 2001. Maturity trends in Raman spectra from kerogen and coal. Energy & Fuel 15, 653–658.

Kelemen, S.R., Walters, C.D., Ertas, D., Freund, H., Curry, D.J., 2006. Petroleum expulsion. Part 3. A model of chemically driven fractionation during expulsion of petroleum from kerogen. Energy & Fuels 20, 309–319.

Kelemen, S.R., Afeworki, M., Gorbaty, M.L., Sansone, M., Kwiatek, P.J., Walters, C.C., Freund, H., Siskin, M., Bence, A.E., Curry, D.J., Solum, M., Pugmire, R.J., Vandenbroucke, M., Leblond, M., Behar, F., 2007. Direct characterization of kerogen by X-ray and solid-state ^{13}C nuclear magnetic resonance methods. Energy & Fuels 21, 1548–1561.

Korb, J.P., Louis-Joseph, A., Benamsili, L., 2013. Probing structure and dynamics of bulk and confined crude oils by multiscale NMR spectroscopy, diffusometry, and relaxometry. Journal of Physical Chemistry B 117, 7002–7014.

van Krevelen, D.W., 1993. Coal: Typology – Chemistry – Physics – Constitution. Elsevier Science, 1002 p.

Lafargue, E., Espitalié, J., Marquis, F., Pillot, D., 1998. Rock-eval 6 applications in hydrocarbon exploration, production and in soil contamination studies. Revue de l'Institut Français du Pétrole 53 (4), 421–437.

LeDoan, T.V., Bostrom, N.W., Burnham, A.K., Kleinberg, R.L., Pomerantz, A.E., Allix, P., 2013. Green River oil shale pyrolysis: semi-open conditions. Energy & Fuels 27, 6447–6459.

Lewan, M.D., 1987. Petrographic study of primary petroleum migration in the Woodford Shale and related rock units. In: Doligez, B. (Ed.), Migration of Hydrocarbons in Sedimentary Basins. Editions Technip, Paris, pp. 113–130.

Lewan, M.D., 1998. Sulphur-radical control on petroleum formation rates. Nature 391, 164–166.

Lin, R., Ritz, G.P., 1993. Studying individual macerals using i.r. microspectroscopy, and implications on oil versus gas/condensate proneness and "low-rank" generation. Organic Geochemistry 20, 695–706.

Lis, G.P., Mastalerz, M., Schimmelmann, A., Lewan, M.D., Stankiewicz, B.A., 2005. FTIR absorption indices for thermal maturity in comparison with vitrinite reflectance R_o in type-II kerogens from Devonian black shales. Organic Geochemistry 36, 1533–1552.

Lisitza, N.V., Freed, D.E., Senand, P.N., Song, Y.-Q., 2009. Study of asphaltene nanoaggregation by nuclear magnetic resonance (NMR). Energy & Fuels 23, 1189–1193.

Lorant, F., Prinzhofer, A., Behar, F., Huc, A.-Y., 1998. Carbon isotopic and molecular constraints on the formation and the expulsion of thermogenic hydrocarbon gases. Chemical Geology 147, 240–264.

Lüning, S., Kolonic, S., 2003. Uranium spectral gamma-ray response as a proxy for organic richness in black shales: applicability and limitations. Journal of Petroleum Geology 26, 153–174.

Mackenzie, A.S., Quigley, T.M., 1988. Principles of geochemical prospect appraisal. American Association of Petroleum Geologists Bulletin 72, 399–415.

Madren, J., 2012. Stable carbon isotope reversal does not correlate to production in the Marcellus Shale. AAPG Search and Discovery. Article #80233. Supplemental info: Adapted from oral presentation at AAPG Annual Convention and Exhibition, Long Beach, California, April 22–25, 2012.

Mao, J., Fang, X., Lan, Y., Schimmelmann, A., Mastalerz, M., Xu, L., Schmidt-Rohr, K., 2010. Chemical and nanometer-scale structure of kerogen and its change during thermal maturation investigated by advanced solid-state ^{13}C NMR spectroscopy. Geochimica et Cosmochimica Acta 74, 2110–2127.

McKenzie, D., 1978. Some remarks on the development of sedimentary basins. Earth and Planetary Science Letters 40, 25–32.

Michaelian, K.H., Friesen, W.I., 1990. Photoacoustic FT-IR spectra of separated western Canadian coal macerals: analysis of the CH stretching region by curve-fitting and deconvolution. Fuel 69, 1271–1275.

Mitra-Kirtley, S., Mullins, O.C., Ralston, C.Y., Sellis, D., Pareis, C., 1998. Determination of sulfur species in asphaltene, resin, and oil fractions of crude oils. Applied Spectroscopy 52, 1522–1525.

Mullins, O.C., 2010. The modified Yen model. Energy & Fuels 24, 2179–2207.

Mullins, O.C., Mitra-Kirtley, S., Zhu, Y., 1992. The electronic absorption edge of petroleum. Applied Spectroscopy 46, 1405–1411.

Mullins, O.C., Sheu, E.Y., Hammami, A., Marshall, A.G., 2007. Asphaltenes, Heavy Oils, and Petroleomics. Springer, New York, 669 p.

Mullins, O.C., Sabbah, H., Eyssautier, J., Pomerantz, A.E., Barré, L., Andrews, A.B., Ruiz-Morales, Y., Mostowfi, F., McFarlane, R., Goual, L., Lepkowicz, R., Cooper, T., Orbulescu, J., Leblanc, J.M., Edwards, J., Zare, R.N., 2012. Advances in asphaltene science and the Yen-Mullins model. Energy & Fuels 26, 3986–4003.

Neber, A., Cox, S., Levy, T., Schenk, O., Tessen, N., Wygrala, B., et al., 2012. Systematic evaluation of unconventional resource plays using a new play-based methodology. Society of Petroleum Engineers (SPE) 158571, 15.

Nelson, P.H., 2009. Pore-throat sizes in sandstones, tight sandstones, and shales. American Association of Petroleum Geologists Bulletin 93, 329–340.

Ottenjann, K., 1988. Fluorescence alteration and its value for studies of maturation and bituminization. Organic Geochemistry 12, 309–321.

Painter, P.C., Snyder, R.W., Starsinic, M., Coleman, M.M., Kuehn, D., Davis, A., 1981. Concerning the application of FT-IR to the study of coal: a critical assessment of band assignments and the application of spectral analysis programs. Applied Spectroscopy 35, 475–485.

Passey, Q.R., Creaney, S., Kulla, J.B., Moretti, F.J., Stroud, J.D., 1990. A practical model for organic richness from porosity and resistivity logs. American Association of Petroleum Geologists 74, 1777–1794.

Passey, Q.R., Bohacs, K.M., Esch, W.L., Klimentidis, R., Sinha, S., 2010. From Oil-prone Source Rock to Gas-producing Shale Reservoir – Geologic and Petrophysical Characterization of Unconventional Shale-gas Reservoirs. Society of Petroleum Engineers. SPE 131350, 29 p.

Pemper, R., Han, X., Mendez, F., Jacobi, D., LeCompte, B., Bratovich, M., Feuerbacher, G., Bruner, M., Bliven, S., 2009. The direct measurement of carbon in wells containing oil and natural gas using a pulsed neutron mineralogy tool. In: SPE 124234, SPE Annual Technical Conference and Exhibition, New Orleans, Louisiana, October 30–November 2.

Peters, K.E., 1986. Guidelines for evaluating petroleum source rock using programmed pyrolysis. American Association of Petroleum Geologists Bulletin 70, 318–329.

Peters, K.E., Simoneit, B.R.T., 1982. Rock-eval pyrolysis of Quaternary sediments from Leg 64, sites 479 and 480, gulf of California. Initial Reports of the Deep Sea Drilling Project 64, 925–931.

Peters, K.E., Cassa, 1994. Applied source-rock geochemistry. In: Magoon, L.B., Dow, W.G. (Eds.), The Petroleum System—From Source to Trap, American Association of Petroleum Geologists Memoir, 60, 93–120.

Peters, K.E., Whelan, J.K., Hunt, J.M., Tarafa, M.E., 1983. Programmed pyrolysis of organic matter from thermally altered Cretaceous black shales. American Association of Petroleum Geologists Bulletin 67, 2137–2146.

Peters, K.E., Walters, C.C., Moldowan, J.M., 2005. The Biomarker Guide, second ed. Cambridge University Press, Cambridge, U.K. 1155 p.

Peters, K.E., Magoon, L.B., Bird, K.J., Valin, Z.C., Keller, M.A., 2006. North Slope, Alaska: source rock distribution, richness, thermal maturity, and petroleum charge. American Association of Petroleum Geologists Bulletin 90, 261–292.

Peters, K.E., Hantschel, T., Kauerauf, A.I., Tang, Y., Wygrala, B., 2013. Recent advances in petroleum system modeling of geochemical processes: TSR, SARA, and biodegradation. AAPG Search and Discovery. Article #90163. At: http://www.searchanddiscovery.com/abstracts/html/2013/90163ace/abstracts/pete.htm (accessed 23.12.14.). Supplemental info: AAPG Annual Convention and Exhibition, Pittsburgh, Pennsylvania, May 19–22, 2013.

Peters, K.E., Burnham, A.K., Walters, C.C., 2015. Petroleum generation kinetics: single versus multiple heating-ramp open-system pyrolysis. American Association of Petroleum Geologists Bulletin 99, 591–616.

Peters, K.E., Schenk, O., Hosford Scheirer, A., Wygrala, B., Hantschel, T. Basin and petroleum system modeling of conventional and unconventional petroleum resources: In: Hsu, C.S., Robinson, P. (Eds.), Practical Advances in Petroleum Production and Processing. Springer, New York, in press.

Pickering, I.J., George, G.N., Yu, E.Y., Brune, D.C., Tuschak, C., Overmann, J., Beatty, J.T., Prince, R.C., 2001. Analysis of sulfur biochemistry of sulfur bacteria using X-ray absorption spectroscopy. Biochemistry 40, 8138–8145.

Pitman, J.K., Price, L.C., LeFever, J.A., 2001. Diagenesis and Fracture Development in the Bakken Formation, Williston Basin: Implications for Reservoir Quality in the Middle Member. U.S. Geological Survey. Professional Paper 1653, 19 p.

Pomerantz, A.E., Hammond, M.R., Morrow, A.L., Mullins, O.C., Zare, R.N., 2009. Asphaltene molecular weight distribution determined by two-step laser mass spectrometry. Energy & Fuels 23, 1162–1168.

Pomerantz, A.E., Seifert, D.J., Bake, K.D., Craddock, P.R., Mullins, O.C., Kodalen, B.G., Mitra-Kirtley, S., Bolin, T.B., 2013. Sulfur chemistry of asphaltenes from a highly compositionally graded oil column. Energy & Fuels 27, 4604–4608.

Pomerantz, A.E., Bake, K.D., Craddock, P.R., Kurzenhauser, K.W., Kodalen, B.G., Mitra-Kirtley, S., Bolin, T., 2014. Sulfur speciation in kerogen and bitumen from gas and oil shales. Organic Geochemistry 68, 5–12.

Pomerantz, A.E., Wu, Q., Mullins, O.C., Zare, R.N., 2015. Laser-based mass spectroscopic assessment of asphaltene molecular weight, molecular architecture and nanoaggregate weight. Energy & Fuels 29, 2833–2842.

Price, L.C., Barker, C.E., 1985. Suppression of vitrinite reflectance in amorphous rich kerogen—a major unrecognized problem. Journal of Petroleum Geology 8, 59–84.

Prietzel, J., Kögel-Knabner, I., Thieme, J., Paterson, D., McNulty, I., 2011. Microheterogeneity of element distribution and sulfur speciation in an organic surface horizon of a forested Histosol as revealed by synchrotron-based X-ray spectromicroscopy. Organic Geochemistry 42, 1308–1314.

Radtke, R.J., Lorente, M., Adolph, R., Berheide, M., Fricke, S., Grau, J., Herron, S., Horokowitz, J., Jorion, B., Madio, D., May, D., Miles, J., Perkins, L., Philip, O., Roscoe, B., Rose, D., Stoller, C., 2012. A new capture and inelastic spectroscopy tool takes geochemical logging to the next level. In: SPWLA 53 Annual Logging Symposium, June 16–20, 2012, pp. 1–16.

Rane, J.P., Pauchard, V., Couzis, A., Banerjee, S., 2013. Interfacial rheology of asphaltenes at oil-water interfaces and interpretation of the equation of state. Langmuir 29, 4750–4759.

Ribacchi, R., 2000. Mechanical tests on pervasively jointed rock material: insight into rock mass behaviour. Rock Mechanics and Rock Engineering 33, 243–266.

Riboulleau, A., Derenne, S., Sarret, G., Largeau, C., Baudin, F., Connan, J., 2000. Pyrolytic and spectroscopic study of a sulphur-rich kerogen from the "Kashpir oil shales" (Upper Jurassic, Russian platform). Organic Geochemistry 31, 1641–1661.

Rickman, R., Mullen, M., Petre, E., Grieser, B., Kundert, D., 2008. A Practical Use of Shale Petrophysics for Stimulation Design Optimization: All Shale Plays Are Not Clones of the Barnett Shale. http://dx.doi.org/10.2118/115258-MS. SPE 115258.

Robinson, L.R., Taulbee, D.N., 1995. Demineralization and kerogen maceral separation and chemistry. In: Snape, C. (Ed.), Composition, Geochemistry and Conversion of Oil Shales, vol. 455. Kluwer Academic Publishers, NATO ASI Series, pp. 35–50.

Rodriguez, N.D., Philp, R.P., 2010. Geochemical characterization of gases from the Mississippian Barnett Shale, Fort Worth Basin, Texas. American Association of Petroleum Geologists Bulletin 94, 1641–1656.

Rosen, P., Vogel, H., Cunningham, L., Hahn, A., Hausmann, S., Pienitz, R., Zolitschka, B., Wagner, B., Persson, P., 2011. Universally applicable model for the quantitative determination of lake sediment composition using Fourier transform infrared spectroscopy. Environmental Science & Technology 45, 8858–8865.

Rouxhet, P.G., Robin, P.L., Nicaise, G., 1980. Characterization of kerogens and their evolution by infrared spectroscopy. In: Durand, B. (Ed.), Kerogen: Insoluble Organic Matter from Sedimentary Rocks. Editions Technip, Paris.

Ruiz-Morales, Y., Mullins, O.C., 2009. Simulated and measured optical absorption spectra of asphaltenes. Energy & Fuels 23, 1169–1177.

Sabah, H., Morrow, A.L., Pomerantz, A.E., Zare, R.N., 2011. Evidence for island structures as the dominant architecture of asphaltenes. Energy & Fuels 25, 1597–1604.

Sabah, H., Pomerantz, A.E., Wagner, M., Müllen, K., Zare, R.N., 2012. Laser desorption single-photon ionization of asphaltenes: mass range, compound sensitivity, and matrix effects. Energy & Fuels 26, 3521–3526.

Salmon, E., Behar, F., Hatcher, P.G., 2011. Molecular characterization of Type I kerogen from the Green River Formation using advanced NMR techniques in combination with electrospray ionization/ultrahigh resolution mass spectrometry. Organic Geochemistry 42, 301–315.

Sarg, J.F., 2012. The Bakken—An unconventional petroleum reservoir system. In: Final Scientific/Technical Report, September 18–December 31, 2011. Office of Fossil Energy, National Energy Technology Laboratory, p. 65.

Sarret, G., Mongenot, T., Conna, J., Derenne, S., Kasrai, M., Bancroft, G.M., Largeau, C., 2002. Sulfur speciation in kerogens of the Orbagnous deposit (Upper Kimmeridgian, Jura) by XANES spectroscopy and pyrolysis. Organic Geochemistry 33, 877–895.

Schenk, O., Bird, K.J., Magoon, L.B., Peters, K.E., 2012. Petroleum system modeling of northern Alaska. In: Peters, K.E., Curry, D., Kacewicz, M. (Eds.), Basin Modeling: New Horizons in Research and Applications, vol. 4. American Association of Petroleum Geologists Hedberg Series, pp. 317–338.

Schmoker, J.W., 1981. Determination of organic-matter content of Appalachian Devonian shales from gamma-ray logs. American Association of Petroleum Geologists Bulletin 65, 1285–1298.

Schmoker, J.W., Hester, T.C., 1983. Organic carbon in Bakken Formation, United States portion of Williston Basin. American Association of Petroleum Geologists Bulletin 67, 2165–2174.

Shanley, K.W., Cluff, R.M., Robinson, J.W., 2004. Factors controlling prolific gas production from low-permeability sandstone reservoirs: implications for resource assessment, prospect development, and risk analysis. American Association of Petroleum Geologists Bulletin 88, 1083–1121.

Smit, B., 1995. Simulating the adsorption isotherms of methane, ethane, and propane in the zeolite silicalite. Journal of Physical Chemistry 99, 5597–5603.

Snowdon, L.R., 1997. Rock-Eval/TOC Data for Six Wells in the Worsley Area of Alberta (Townships 80 to 87 and Ranges 3W6 to 10W6). Geological Survey of Canada. Open File 2492. http://geogratis.gc.ca/api/en/nrcan-rncan/ess-sst/6aa6b720-3ccc-5d29-8f9b-8023d1fa5b29.html (accessed 12.12.14.).

Solum, M.S., Pugmire, R.J., Grant, D.M., 1989. ^{13}C solid-state NMR of Argonne premium coals. Energy & Fuels 3, 187–193.

Solum, M.S., Mayne, C.L., Orendt, A.M., Pugmire, R.J., Adams, J., Fletcher, T.H., 2014. Characterization of macromolecular structure elements from a Green River oil shale, I. Extracts. Energy & Fuels 28, 453–465.

Spiro, C.L., Wong, J., Lytle, F.W., Greegor, R.B., Maylotte, D.H., Lamson, S.H., 1984. X-ray absorption spectroscopic investigation of sulfur sites in coal: organic sulfur indentification. Science 226, 48–50.

Strąpoć, D., Schimmelmann, A., Mastalerz, M., 2006. Carbon isotopic fractionation of CH_4 and CO_2 during canister desorption of coal. Organic Geochemistry 37, 152–164.

Suleimenova, A., Bake, K.D., Ozkan, A., Valenza, J.J., Kleinberg, R.L., Burnham, A.K., Ferralis, N., Pomerantz, A.E., 2014. Acid demineralization with critical point drying: a method for kerogen isolation that preserves microstructure. Fuel 135, 492–497.

Sweeney, J.J., Burnham, A.K., 1990. Evaluation of a simple model of vitrinite reflectance based on chemical kinetics. American Association of Petroleum Geologists Bulletin 74, 1559–1570.

Tang, Y., Perry, J.K., Jenden, P.D., Schoell, M., 2000. Mathematical modeling of stable carbon isotope ratios in natural gases. Geochimica et Cosmochimica Acta 64, 2673–2687.

Tilley, B., McLellan, S., Hiebert, S., Quartero, B., Veilleux, B., Muehlenbachs, M., 2011. Gas isotope reversals in fractured gas reservoirs of the western Canadian Foothills: mature shale gases in disguise. American Association of Petroleum Geologists Bulletin 95, 1399–1422.

Taylor, G.H., Teichmüller, M., Davis, A., Diessel, C.F.K., Littke, R., Robert, P., 1998. Organic Petrology. Gebrüder Borntraeger, Berlin, 704 p.

Tegelaar, E.W., de Leeuw, J.W., Derenne, S., Largeau, C., 1989. A reappraisal of kerogen formation. Geochimica et Cosmochimica Acta 53, 3103–3106.

Tegelaar, E.W., Noble, R.A., 1994. Kinetics of hydrocarbon generation as a function of the molecular structure of kerogen as revealed by pyrolysis-gas chromatography. Organic Geochemistry 22, 543–574.

Thomas, J.J., Valenza, J.J., Craddock, P.R., Bake, K.D., Pomerantz, A.E., 2014. The neutron scattering length density of kerogen and coal as determined by CH_3OH/CD_3OH exchange. Fuel 117, 801–811.

Tian, H., Wang, Z., Xiao, Z., Li, X., Xiao, X., 2006. Oil cracking to gases: kinetic modeling and geological significance. Chinese Science Bulletin 51, 2763–2770.

Tissot, B.P., Welte, D.H., 1978. Petroleum Formation and Occurrence. Springer-Verlag, Berlin, Germany, 538 p.

Tissot, B.P., Welte, D.H., 1984. Petroleum Formation and Occurrence. Springer-Verlag, Berlin, 699 p.

Vernik, L., Chi, S., Khadeeva, Y., 2012. Rock physics of organic shale and its applications. In: 2012 Society of Exploration Geophysicists (SEG) Annual Meeting 4–9 November, Las Vegas, SEG-2012-0184. Society of Exploration Geophysicists.

Vogel, H., Rosen, P., Wagner, B., Melles, M., Persson, P., 2008. Fourier transform infrared spectroscopy, a new cost-effective tool for quantitative analysis of biogeochemical properties in long sediment records. Journal of Paleolimnolpgy 40, 689–702.

Wang, F.P., Gale, J.F.W., 2009. Screening criteria for shale-gas systems. Gulf Coast Association of Geological Societies (GCAGS) Transactions 59, 779–793.

Waples, D.W., Marzi, R.W., 1998. The universality of the relationship between vitrinite reflectance and transformation ratio. Organic Geochemistry 28, 383–388.

Waldo, G.S., Carlson, R.M.K., Moldowan, J.M., Peters, K.E., Penner-Hahn, J.E., 1991. Sulfur speciation in heavy petroleums: information from X-ray absorption near-edge structure. Geochimica et Cosmochimica Acta 55, 801–814.

Waldo, G.S., Mullins, O.C., Penner-Hahn, J.E., Cramer, S.P., 1992. Determination of the chemical environment of sulphur in petroleum asphaltenes by X-ray absorption spectroscopy. Fuel 71, 53–57.

Ward, R.L., Burnham, A.K., 1984. Identification by ^{13}C n.m.r of carbon types in shale oil and their relation to pyrolysis conditions. Fuel 63, 909–914.

Washburn, K.E., Birdwell, J.E., 2013a. Updated methodology for nuclear magnetic resonance characterization of shales. Journal of Magnetic Resonance 233, 17–28.

Washburn, K.E., Birdwell, J.E., 2013b. Multivariate analysis of ATR-FTIR spectra for assessment of oil shale organic geochemical properties. Organic Geochemistry 63, 1–7.

Werner-Zwanziger, U., Lis, G., Mastalerz, M., Schimmelmann, A., 2005. Thermal maturity of type II kerogen from the New Albany Shale assessed by ^{13}C CP/MAS NMR. Solid State NMR 27, 140–148.

Wilkins, R.W.T., Wilmshurst, J.R., Russell, N.J., Hladky, G., Ellacott, M.V., Buckingham, C., 1992. Fluorescence alteration and the suppression of vitrinite reflectance. Organic Geochemistry 18, 629–640.

Wilkins, R.W.T., Wilmshurst, J.R., Hladky, G., Ellacott, M.V., Buckingham, C.P., 1995. Should fluorescence alteration replace vitrinite reflectance as a major tool for thermal maturity deterimination in oil exploration? Organic Geochemistry 22, 191–209.

Wiltfong, R., Mitra-Kirtley, S., Mullins, O.C., Andrews, A.B., Fujisawa, G., Larsen, J.W., 2005. Sulfur speciation in different kerogens by XANES spectroscopy. Energy & Fuels 19, 1971–1976.

Wimberley, J.W., 1969. A rapid method for the analysis of total organic carbon in shale with a high-frequency combustion furnace. Analytica Chimica Acta 48, 419–423.

Wu, Z., Pomerantz, A.E., Mullins, O.C., Zare, R.N., 2014. Laser-based mass spectrometric determination of aggregation numbers for petroleum- and coal-derived asphaltenes. Energy & Fuels 28, 475–482.

Xia, X., 2000. Hydrocarbon Potential of Carbonates and Source Rock Correlation of the Changqing Gas Field. Petroleum Industry Press (in Chinese), Beijing, 1–164 p.

Xia, D., 2014. Kinetics of gaseous hydrocarbon generation with constraints of natural gas composition from the Barnett Shale. Organic Geochemistry 74, 143–149. http://dx.doi.org/10.1016/j.orggeochem.2014.02.009.

Xia, X., Tang, Y., 2012. Isotope fractionation of methane during natural gas flow with coupled diffusion and adsorption/desorption. Geochimica et Cosmochimica Acta 77, 489–503.

Xia, X., Chen, J., Braun, R., Tang, Y., 2013. Isotopic reversals with respect to maturity trends due to mixing of primary and secondary products in source rocks. Chemical Geology 339, 205–212.

Xia, X., Guthrie, J.M., Burke, C., Crews, S., Tang, Y., 2014. Predicting hydrocarbon composition in unconventional reservoirs with a compositional generation/expulsion model. Search and Discovery article #80397. http://www.searchanddiscovery.com/documents/2014/80397xia/ndx_xia.pdf (accessed 19.12.14.). Supplemental info: Adapted from oral presentation given at 2014 AAPG Annual Convention and Exhibition, Houston, Texas, April 6–9, 2014.

Zajac, G.W., Sethi, N.K., Joseph, J.T., 1994. Molecular imaging of asphaltenes by scanning tunneling microscopy: verification of structure from ^{13}C and proton NMR data. Scanning Microscopy 8, 463–470.

Zhang, T., Ellis, G.S., Ruppel, S.C., Milliken, K., Yang, R., 2012. Effect of organic-matter type and thermal maturity on methane adsorption in shale-gas systems. Organic Geochemistry 47, 120–131.

Zhang, T., Yang, R., Milliken, K.L., Ruppel, S.C., Pottorf, R.J., Sun, X., 2014. Chemical and isotopic composition of gases released by crush methods from organic rich mudrocks. Organic Geochemistry 73, 16–28.

Zumberge, J., Ferworn, K., Brown, S., 2012. Isotopic reversal ('rollover') in shale gases produced from the Mississippian Barnett and Fayetteville formations. Marine and Petroleum Geology 31, 43–52.

Zuo, J.Y., Mullins, O.C., Freed, D.E., Dong, C., Elshahawi, H., Seifert, D.J., 2013. Advances of the Flory-Huggins-Zuo equation of state for asphaltene gradients and formation evaluation. Energy & Fuels 27, 1722–1735.

CHAPTER 4

PORE-SCALE CHARACTERIZATION OF GAS FLOW PROPERTIES IN SHALE BY DIGITAL CORE ANALYSIS

Jingsheng Ma
Institute of Petroleum Engineering, Heriot-Watt University, Edinburgh, UK

4.1 INTRODUCTION

The characterization of flow properties for shale gas reservoirs is of importance for assessing shale gas potentials, gas recoverability, and the economics of recovering gas in place. Unlike in a conventional reservoir, an appropriate characterization of shale gas reservoirs must take into account nonideal gas flow behavior in pore space that is predominated in submicron-sized pores. In such pores, gas flow is no longer in Darcy nor slip flow regimes, but falls primarily within transition flow regimes, while in nanometer pores gas molecules are transported through Knudsen and surface diffusions. These gas behaviors are interwoven with gas adsorption that takes place at pore surfaces. The surface concentration of the gas molecules depends on pore geometry, surface morphology, and chemical compositions of solid phases, as well as in-situ pore pressures and temperatures. Each of these factors can impact free gas flow, and therefore the prediction of flow properties, like gas permeability.

It is not trivial to characterize gas flow properties for shale samples. The experimental approach to estimating gas flow properties is fraught with difficulties; the small-sized pores make it too time-consuming to measure those properties by conventional laboratory experiments; and, even with newer and more efficient laboratory experimental techniques, how they really work is not well understood. The digital core analysis (DCA) approach, on the other hand, offers a set of complementary, if not replacing, techniques for lab-based petrophysical analysis. DCA involves: digital characterization of porous samples, reconstruction of pore-scale models, numerical simulations of pore-scale physicochemical processes on reconstructed models, and property determination. This chapter highlights the importance of considering the full range of gas behaviors in small pores for characterizing gas flow properties.

The first section of this chapter reviews the nature of shale pore space and the DCA approach to characterization. The second section highlights behaviors of gas flow in shale pores and models that have been developed for nonDarcy flows and Knudsen diffusions of nonideal gas. The third section considers the potential impact of gas adsorption on free gas flow using a simple effective multilayer adsorption model. The last section examines the aggregated effects of nonDarcy flow and gas adsorption on the prediction of the gas permeability for a realistic shale model that was previously characterized following our DCA workflow. The chapter ends with conclusions.

4.2 GAS SHALE CHARACTERIZATION BY DCA

Shales are composed of very fine-grained sediments that may be internally heterogeneous in terms of grain textures and chemical compositions within individual depositional units (e.g., the form of their interbedding/lamination), which may be modified by mechanical deformations (e.g., folding, fracturing, and faulting processes). Fine-grained sediments that comprise shales typically contain grains much smaller than 2 μm in diameter, while the shale pores, which exist between and within grains (Loucks et al., 2009), may vary by several orders of magnitude and are usually much smaller than those of very tight-sandstones (Chalmers et al., 2012). The permeability of intact shales is typically at a range of hundreds of nano-Darcy or less (Heller et al., 2014). Gas shales are rich in organic matter and kerogen which exist in the form of individual grains or more typically in clumps. Kerogens may contain additional pores of submicron/nano scales, due probably to hydrocarbon generation (Curtis et al., 2012). These organic pores, along with inorganic pores and fissures to which the organic pores connect, provide storage space for gas in the "matrix" rock. When the pores are connected to natural and induced fractures, the free gas can flow to the well for production.

However, it is very difficult to characterize the petrophysical properties of gas shales using traditional lab-based analytical techniques because of the small sizes and heterogeneities of grains and pores. Chalmers et al. (Chalmers et al., 2012) analyzed gas shale samples from all major United States shale gas fields and showed the diversities of types, sizes, and shapes of shale grains and pores and their spatial distributions. Figure 4.1 shows the relationships between differential pore volume and pore size

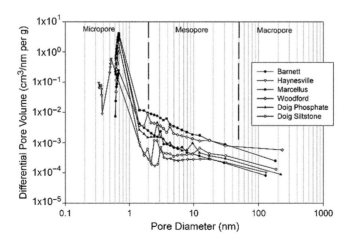

FIGURE 4.1

The relationship between differential pore volume and pore size distribution from gas adsorption measurements.

Adapted from Chalmers et al. (2012) under the terms of the fair use of AAPG published images.

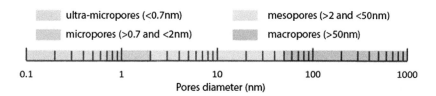

FIGURE 4.2

IUPAC pore size classification.

distribution from gas adsorption measurements on shale samples of selected United States gas shale formations. It is clear that for all samples, their pore diameters vary by over four orders of magnitude and that there is a larger proportion of pores whose pore diameters are less than 2 nm. Pores below that size are referred to as micropores or ultramicropores according to the pore size classification of International Union of Pure and Applied Chemistry (IUPAC) (Sing et al., 1985), as illustrated in Fig. 4.2.

It is well known that many standard analytical techniques for core analysis of conventional reservoirs rocks cannot be applied to shale cores in an efficient manner. For example, steady-state flow measurement on a shale core is likely to take at least 100 times longer than on a sandstone core and is, therefore, not acceptable for practical use. For this reason, new unsteady-state flow measurement procedures have been devised, such as the Gas Research Institute technique and pressure pulse-decay on crushed and noncrushed cores (Luffel et al., 1993; Cui et al., 2009). Although these techniques can make measurements considerably faster, the reliability and robustness of the predicted properties are not well known. Moreover, for these new techniques, the measurements need to be interpreted using numerical models that are often inadequate to capture the full spectrum of physical processes that occur at different spatial and time scales in a full range of pores (see the next section for discussion).

DCA comprises a number of techniques including imaging porous samples, pore-scale characterization and reconstruction of samples, numerical simulations of pore-scale physicochemical processes on reconstructed models, and determination of macroscopic properties. It has been touted as a promising technology to complement, if not to replace, lab-based petrophysical techniques. Figure 4.3 illustrates the DCA workflow (Ma et al., 2014c) and its key elements are explained briefly below.

4.2.1 IMAGING POROUS SAMPLES

Advances in imaging technology have made it possible to obtain images of samples at a high resolution that resolve submicron grains and pores. For imaging shale samples, the combined Broad/Focused Ion Beam milling and Scanning Electron Microscopy (B/FIB/SEM) approach, which creates one or a series of image slices, is the most widely used technology. It may be capable of resolving grains and pores around 5 nm or less in diameter on each milled surface that may be 10 nm from the next milled surface (Bushby et al., 2011). The size of the field of view can reach thousands of micrometers or hundreds of micrometers in each dimension on a two-dimensional (2D) slice or a three-dimensional (3D) volume of stacked 2D slices with BIB/SEM and FIB/SEM (Bushby et al., 2011; Desbois et al., 2013). Those instruments enable the identification of the chemical composition of grains through Energy Dispersive Spectroscopy at the same time. B/FIB/SEM has been the main workhorse for shale imaging for pore-grain characterization (Loucks et al., 2009; Curtis et al., 2010; Klaver et al., 2012; Bai et al., 2013).

130 CHAPTER 4 PORE-SCALE CHARACTERIZATION

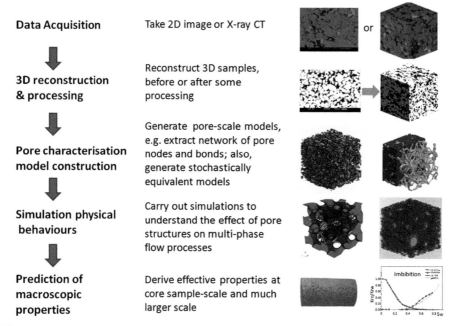

FIGURE 4.3

Illustration of the workflow of digital core analysis (DCA).

Adapted from Fig. 4.1 of Ma et al. (2014c).

4.2.2 PORE-SCALE CHARACTERIZATION AND RECONSTRUCTION

One purpose of pore-scale characterization is to develop representative models for a sample. Such a model does not need to represent all of the geometric, topologic, and chemical compositional aspects of a sample, but needs to capture selected aspects that have the most influence on pore-scale physicochemical processes that will be simulated on that reconstructed model. For simulating gas flow in gas shale pores, it is necessary to identify organic and inorganic grains and pore space associated with them, while knowing the distribution of chemical composition is useful and sometimes critical to account for surface-related processes, e.g., gas adsorption and surface diffusion.

Given a 3D image of a sample, pore-scale characterization is concerned with different types of grains and pores, and analyzing their geometry and connectivity. With respect to different simulation purposes, appropriate simplification of the components is needed, for example, by aggregating small grains and pores, for constructing models suitable for the intended simulations.

An alternative to the slice-reconstruction method is to generate stochastic 3D models from 2D images. Given a set of 2D images, for example, three images taken from regions on orthogonal surfaces, the pore-scale characterization involves the reconstruction of one or more 3D volumes from the 2D images. Wu et al. (2006) proposed a Markov Chain Monte Carlo technique to reconstruct a 3D volume using three orthogonal 2D binary images, which depict the solid and pore locations. The input images are used to train or construct transitional probability matrixes of a Markov Chain. Those authors devised a single-scan algorithm that only needs to determine 348 conditional probabilities for a binary system using a 3D template containing 15 neighbors, leading to a significant reduction in training and simulation time. That technique has been demonstrated to be able to reconstruct realistic 3D volumes that well capture the heterogeneity of grain-pore distributions in 2D images for samples of sandstones, siltstones, shales, and carbonate rocks, and can be extended for grains of different compositions.

Characterizing pore space is a critical part of the pore-scale analysis, especially when fluid flow and transport are of the primary concern. Because the pore structure of natural geomaterials is often very complicated in terms of geometry and topology, this task is not trivial and calls for advanced techniques. Broadly speaking, there are two approaches to pore-space characterization: using local statistics computed on 3D binary images (Biswal and Hilfer, 1999), or treating the pore space as a network of pore elements, with an accompanying analysis of the network properties (Jiang et al., 2007; Dong and Blunt, 2009). The latter approach is advantageous in that a resultant network is readily available to generate simplified pore network models on which flow and transport simulations can be carried out efficiently, although this incurs implicitly or explicitly a degree of arbitrariness in defining the criteria of partitioning the pore space.

Jiang et al. (2007) developed a technique to partition the pore space of a binary 3D image into a topology-preserved network of nodes and bonds. Firstly, it calculates the Euclidean distance map and extracts the medial axial skeleton from the pore space. Then it identifies all junction voxels on the skeleton and, on each of them, it centers the maximal inscribed sphere and labels all skeleton voxels inside that volume as belonging to a node. All remaining skeleton voxels are divided into segments of skeleton voxels, and for each segment it labels them as a bond. Finally it allocates each remaining pore voxel, in turn, to the closet node or bond according to its geodesic distance. On such a network, the geometry of every node and bond, as well as their topology, can be computed and measured. This technique has been used to characterize the pore space of sandstones, siltstones, shales and carbonate rocks, and even fractured rocks with an extension. It can also produce simplified pore networks for simulating single and multiphase fluid flow, and has been extended to generate multiscale stochastic networks from more than one network (Wu et al., 2007; Jiang et al., 2013).

4.2.3 MODELING PORE-SCALE PHYSICOCHEMICAL PROCESSES

Physicochemical processes may be simulated numerically on 3D images directly, or using simplified forms of the pore space, such as a pore network of nodes and bonds. As an image-based 3D model offers the full information about geometric, topological, and compositional complexity of the model, image-based simulation models can be implemented to simulate processes, in addition to fluid flow within pores, to account for mechanical deformation, surface reactive and diffusive processes, or other transport phenomena (electrical current or nuclear magnetic resonance responses). For modeling single and multiphase flow processes, image-based simulators have been developed by solving Navier–Stokes/Stokes equations coupled with equations describing phase relationships (Cahn and Hilliard, 2013) using numerical techniques and Boltzmann equations with lattice Boltzmann (LB) models. Ma et al. (Ma et al., 2010) developed a simulation scheme for an LB model which reduces the requirement for computer storage space and exploits the connectivity of pore voxels to speed up the computation, in particular for low-porosity rocks. Zaretskiy et al. (2012) developed a Stokes solver taking a 3D binary image as an input although it does not work exactly on the voxels of an image but on a mesh constructed from that image, as most of other image-based NS solvers do.

The image-based simulation can be computationally intensive because of the sheer number of voxels in 3D images and the multiphysics to be considered in a simulation; thus, this approach may be so inefficient that it does not produce results in a reasonable time frame. Therefore, reducing the complexity of a 3D image is not only desirable but also necessary in many cases, especially when more physics are to be considered. Many numerical models were developed assuming the simulation domain is much simpler. For simulating fluid flow, the pore-network model (PNM) pioneered by Fatt (1956) is one of the most well-known examples of this kind. The model was designed originally to simulate capillary-dominated two-phase flow over a pore network of connected simple cylindrical elements on a regular lattice but with a set of prescribed physics (see (Blunt, 2001) for review) and has drawn considerable attention from those who are interested in modeling pore-scale fluid flow. PNM has evolved to being able to take geometrically more-complex elements and to model multiphase reactive or nonreactive flow driven by capillary, viscous and gravity force. There is a large collection of literature on PNM, including those contributed from the author's institution (McDougall and Sorbie, 1997; Bondino et al., 2007, 2011; Ezeuko et al., 2010; van Dijke et al., 2004, 2007; van Dijke and Sorbie, 2006; Ma et al., 2014a).

These advances in PNM simulation capability motivate the development of methods that lead to construction of simplified pore-space models from 3D images. The technique developed by Jiang et al. (2007) can generate topology-preserved pore network models containing nodes and bonds with circular, triangular and rectangular cross-sections. Pore network flow codes have been developed to work on such "true" pore network configurations (Ryazanov et al., 2009; Valvatne and Blunt, 2004).

4.2.4 DETERMINATION OF MACROSCOPIC PROPERTIES FOR A SAMPLE

By simulating physicochemical processes, certain quantities may be determined to estimate macroscopic properties of interest for a sample, if the digital rock model is a representative element volume. For a typical core sample at a size of cm in each dimension, a reconstructed model may at best represent a fractional volume of the sample due to the mutual constraint of the size of the field of view (FOV) of an imaging instrument, and its resolution. In order to obtain a useful image of the pore space, the FOV needs to be selected so that pore-space components are resolved into more than one pixel/voxel. This constraint is particularly challenging for carbonate rocks and shales as they contain pores whose sizes may vary by four orders of magnitude or more.

To overcome this issue, there is active research that seeks to develop multiscale approaches to pore-scale characterization on a larger volume. This topic clearly demands thoughtful technical integration of imaging, characterization and simulation under a practical framework. Knackstedt et al. (2006) demonstrated an integrated technique that uses X-ray tomography (XRT) and SEM to capture vuggy pores and large grains in 3D and micrometer or even smaller pores in 2D on a carbonate rock sample, respectively, and then to "impregnate" the latter over the former to achieve two-scale imaging by image registration. Although it should be possible to make use of more than one SEM image in this technique instead, it is not clear how this may be done systematically. Another weakness of this technique is that it yields a monolithic model in the 3D image that can be colossal to carry out any characterization and simulation. Both limitations obstruct potential application of that technique to shale.

Ma et al. (2014c) proposed a multiscale DCA framework for shale gas flow characterization. Under this framework, a shale sample is characterized through a process of multiple steps: (1) take XRT imaging a shale sample at a volume of $2 \times 2 \times 2$ mm^3; (2) virtually subdivide the image into $20 \times 20 \times 20$ μm^3 and classify these "coarse" textures into groups (Ma et al., 2014c); (3) select one or more subvolumes in each group and carry out FIB/SEM on them; (4) generate stochastic pore networks from each FIB/SEM image (Jiang et al., 2012) to populate all subvolumes; (5) do two-scale simulation (Ma and Couples, 2004) on the XRT image using gray lattice Boltzmann methods (Zhu and Ma, 2013) and on individual networks in parallel using a pore network model developed for modeling shale gas (Ma et al., 2014c). Figure 4.4 highlights key elements of that framework where the top row indicates types of models while the bottom row illustrates methods to be employed. Note that the XRT gray-scale image is the coarse "gray" model that contains **s**olids, unresolved **a**ggregates of pores and solids, and resolved **p**ores, and therefore is referred to as SAP. A resolved solid-pore model at the fine scale by FIB/SEB is referred to as SP. Also note that if pores less than 10 nm on an FIB-SEM volume could be resolved by TEM, the technique of Wu et al. (2007) can then be used to generate stochastic 3D models. Then two-scale pore networks can be integrated into a large and more representative network using a technique (Jiang et al., 2013) if a good estimate of the interconnectivity between resolved pores and aggregates can be established.

Even if this framework can be implemented in a practical workflow, the estimated properties calculated for samples at mm scales may not be appropriate and of great use if the aim is to perform reservoir scale simulation, where the properties at this scale need to be upscaled to a suitable size. It is well known that upscaling must account for the geological heterogeneities observed in well-logs, outcrops, and seismic data. The concept of geological-based upscaling and associated techniques has been developed primarily since the 1990s for reservoirs (see (Pickup et al., 2005) for a review) and has been expanded since then for organic-lean shales within different genetic units (Aplin et al., 2012; Ma et al., 2013) in the Caprocks project (Caprock Project, 2004–2013) from millimeters to basin scales. A further extension to include the pore scale via DCA has been proposed for organic-rich gas shales by incorporating DCA (Ma et al., 2013). Figure 4.5 illustrates a genetic unit-based upscaling framework for shale.

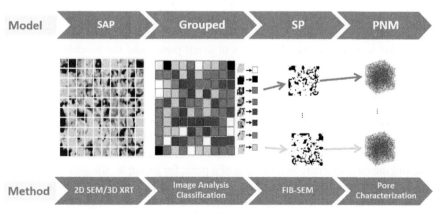

FIGURE 4.4

A multiscale DCA framework for shale gas flow and transport characterization.

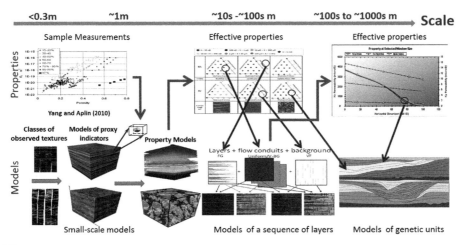

FIGURE 4.5

A multiscale upscaling framework for shale.

From Ma et al. (2013).

In summary, this section highlights the challenges in characterizing gas flow and transport properties of shale, the key components of DCA and developments of a multiscale DCA framework for shale as well as DCA complementing the multiscale upscaling framework for shale.

4.3 GAS FLOW BEHAVIORS IN SHALE PORES AND PORE-NETWORK MODELS

Gas flow has long been known to behave differently in confined spaces due to increased interactions between gas molecules and the pore-wall surface (Knudsen, 1909; Kennard, 1938), and it has been studied experimentally and numerically at microscopic scales (see (Nicholson and Bhatia, 2009) for a review). This section summarizes key behaviors in small pores, and the models, as well as the links between flow behaviors and estimated flow properties, such as apparent gas permeability.

4.3.1 GAS FLOW REGIMES

Gas flow may be broadly classified into the following regimes: continuum, slip, transition flow, and Knudsen diffusion (Roy et al., 2003; Xiao and Wei, 1992), according to Knudsen number, Kn. Kn is the ratio between the mean free path that a gas molecule travels before colliding with another molecule or the pore walls, and the characteristic length scale of a pore, for example, the pore diameter. For a given pore, Kn is inversely proportional to the product of gas pressure and the pore diameter and proportional to the temperature. With respect to the order of magnitude of Kn, those regimes are corresponding to Kn (continuum) < 0.01, $0.01 <$ Kn (slip) < 0.1, $0.1 <$ Kn (transition) < 10, and Kn (Knudsen) > 10. Shale gas is thought to mostly fall in the transition flow regime (Freeman et al., 2011).

Mathematical models of gas flow have been developed from both molecular and continuum approaches. In the continuum regime, models are based on Euler and Navier–Stokes (NS) equations, while in the slip regime the NS models were extended to allow nonzero gas velocity on the pore surfaces, which gives rise to the well-known Klinkenberg effect (Klinkenberg, 1941). Also well-known are first-order Maxwellian slip-wall models (Kennard, 1938; Arya et al., 2003) and second-order Maxwellian slip-wall models (Beskok and Karniadakis, 1999). Those extended continuum models are phenomenological in nature (Zhang et al., 2012). In the transition flow regime, the NS equation ceases to be valid and higher-order hydrodynamics are required due to increasing collisions among individual molecules and between molecules and pore surfaces. Burnet (Burnett, 1935) derived the first higher-order hydrodynamic equation, referred to as the Burnett equation in literature, using Chapman's solution of the Boltzmann equation (Chapman and Cowling, 1991), and a few other Burnet-type equations have been derived since then. However, Burnett-type models suffer from intrinsic numerical instability and inconsistency in predicting gas rarefaction (Lockerby et al., 2005). There is a growing effort to develop suitable kinetic models from the molecular approach, such as using Boltzmann equations (Meng et al., 2011). In the Knudsen flow regime, the collisions between molecules occur less frequently than the collisions between molecules and pore surfaces, and this enhances gas diffusive flow. The collisionless Boltzmann equation properly captures the Knudsen diffusion. Under the continuum approach, the Knudsen diffusion can be captured using correct diffusivity that can be obtained from the kinetic theory of gases (Knudsen, 1909) and is proportional to the mean velocity with a coefficient. That coefficient depends on the geometry of the pore space (Gruener and Huber, 2008; Evans et al., 1961) and the particular gas models assumed (Mason and Malinauskas, 1983). The continuum approach yields models that are more efficient computationally.

Continuum models have drawn considerable attention because they are needed to invert laboratory experimental results (Civan, 2010; Cui et al., 2009; Darabi et al., 2012). The models also play more crucial roles for estimating the macroscopic properties of gas shales because laboratory experiments are difficult to setup and to undertake under in situ conditions to make robust and reliable property estimation. It is extremely difficult to preserve gas shale cores and any change to the pressure and temperature could lead to the deformation of pore space and other artifacts. For example, trapped gas may crack the shale matrix by gas expansion when the cores are being lifted to the surface (Chenevert and Amanullah, 2001). Recent advanced high-resolution imaging enables characterization of pores and grains, as well as the details of the pore structures of a porous material, and therefore there is a greater potential for DCA to complement and, perhaps, even to replace unreliable laboratory procedures. In principle, the mentioned artifacts introduced into samples could be identified in images and eliminated from the DCA, allowing a "pristine" model to be used to calculate the properties.

Due to the computational efficiency, pore network models for shale gas flow have been developed under the continuum approach. Given the difficulty in modeling the transition flow, most, if not all, proposed continuum models for gas flow are basically developed using some form of superposition onto the continuum model, for example, a first-order Maxwellian slip-wall model (Kennard, 1938; Arya et al., 2003), and a model for Knudsen diffusion (Gruener and Huber, 2008; Kast and Hohenthanner, 2000; Mason, 1983) such as that proposed by Javadpour (2009). Sakhaee-Pour and Bryant developed a pore network model that considers the slip flow and gas adsorption/desorption (see discussion in next section), but not the Knudsen diffusion, for ideal gas only (Sakhaee-Pour and Bryant, 2012). Mehmani et al. (2013) developed an extended pore network model based on Javadpour's model (Javadpour, 2009) that takes into account the slip and Knudsen phenomena as well as nonideal PVT properties for simulating single-phase gas flow, and they applied that model to study the impacts of pore geometry (e.g., cylinder pore diameter and length) constraints on the gas flow in a shale matrix. Independent from that work, Ma et al. (2014a) developed a similar extended pore network model. That model can simulate gas and gas–liquid flow on a pore network that contains pore bonds with circular, triangular and square cross-sections. The authors showed nonDarcy effects could cause gas flow to depart markedly from the continuum flow prediction, and therefore apparent gas permeability to deviate from the normal Darcy permeability. That model was applied to simulate gas–water flow in a microporous layer, a nanopore component in proton exchange membrane fuel cell (Ma et al., 2014b).

Equation (4.1) defines the apparent gas permeability for each network element where the effects of the slip flow and the Knudsen diffusion are accounted in the second and third terms inside the brace bracket, respectively. The equation was developed for an element with a circular cross-section first, and then adapted for an element with a noncircular cross-section through a correction factor C_g. The nonideal gas (NIG) effect was expressed using van der Waals's two-parameter principle of corresponding states in coefficient terms NIG. For a brief explanation of nomenclature in the equation see Table 4.1. The reader is also referred to the original publication (Ma et al., 2014a) for details of the definition for every coefficient in Equation (4.1).

$$K_{\text{NIG}} = \frac{C_g R_h^2 \rho_2}{\mu_2} \left\{ \frac{\mu_2}{\mu_0} \text{NIG}_c + \frac{\mu_2}{\mu_2^{ig}} \frac{8}{3} \left(\frac{2 - \text{TMAC}}{\text{TMAC}} \right) k_{n,2h} + \frac{\mu_2}{\mu_2^{ig}} \frac{64}{9 C_g \pi} \text{NIG}_k k_{n,2h} \right\} \quad (4.1)$$

Table 4.1 Nomenclature for Eq. (4.1)

	Description
C_g	A correction factor for a cross-section of circle, square, or equilateral triangle
R_h [1/m]	Hydraulic radius—the perimeter over the area of a cross-section
ρ_2 [kg/m³]	Density of gas at the outlet of an element
μ_2 [Pa.s]	Viscosity of nonideal gas at the outlet of an element
μ_0 [Pa.s]	Reference viscosity of gas typically taken at 1 atm
μ_2^{ig} [Pa.s]	Viscosity of ideal gas at the outlet of an element
NIG_c	The nonideal gas coefficient for the continuum term
NIG_k	The nonideal gas coefficient for the Knudsen term
TMAC	Tangential momentum accommodation coefficient (TMAC) at pore wall
$k_{n,2h}$	Knudsen number

4.3.2 BEHAVIORS OF APPARENT GAS PERMEABILITY

From Eq. (4.1), it is clear that slip flow and Knudsen diffusion influence apparent gas permeability. This can be highlighted by analyzing the ratio of the apparent gas permeability, Kapp, and the fluid permeability, Kd, without considering slip flow and Knudsen diffusion, for a cylindrical pore. Tangential momentum accommodation coefficient (TMAC) falls within [0, 1] and determines how much gas slippage occurs at the pore wall. Figure 4.6 (a) shows the relationship between Kapp/Kd and

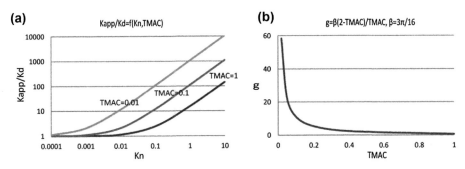

FIGURE 4.6

Impact of slippage and gas rarefaction on the apparent permeability (a) and the relative importance of slip flow and Knudsen diffusion (b).

Adapted from Ma et al. (2014a).

Kn for TMAC equal to 0.01, 0.1, and 1. For a given TMAC or a degree of slippage, Kapp/Kd increases with the increase of Kn, nonlinearly: that is, the greater the gas rarefaction is, the greater Kapp/Kd. At a given Kn, the smaller TMAC is, and the stronger is the slippage and the greater is Kapp/Kd. Figure 4.6 (b) shows g, the ratio of the coefficient of the slip term and the coefficient of the Knudsen term at a given Kn with respect to TMAC. It is clear that as TMAC approaches zero, the ratio increases exponentially, that is, the slip flow becomes more dominant than Knudsen diffusion, although they both are of the same order in terms of the Knudsen number, Kn.

Another key character of the model (Eq. (4.1)) is that it deals with nonideal gas. Figure 4.7 shows the relative differences between coefficients NIG and IG for the Darcy and Knudsen terms in Eq. (4.1) when gas is considered to be nonideal or ideal gas, respectively. For both terms, the relative differences may reach as high as 80%, and 150% and -80% around the critical states (see Fig. 4.7(a) and (c)), respectively, and 30% and +20% and −60% even when the states are in the middle range of shale gas operational conditions for methane (see Fig. 4.7(b) and (d)), respectively. These differences suggest that the assumption of an ideal gas would lead to over or underestimation of the apparent permeability and the flow conductance at different states. Because continuum models of this general type are used to invert laboratory measurements, the discrepancies identified can lead to significant errors in the reported properties if the issues noted are not recognized. At a higher temperature and higher pressure, the variation of viscosity ratios in the coefficients may become big even in a short pore, resulting in additional discrepancy. This analysis shows the importance of a pore network model being able to treat nonideal gas appropriately. For details, the reader is referred to (Ma et al., 2014a).

4.4 SURFACE ADSORPTION/DESORPTION AND AN EFFECTIVE MULTILAYER ADSORPTION MODEL

Under reservoir conditions, a large amount of gas molecules may be adsorbed on organic pore surfaces, and when gas pressure reduces, for example, due to gas production, some of them desorb off the pore surfaces into pores as free gas, which can flow to the producing wells. This represents a gas state-dependent process in which the pore space available to free gas and complex gas flow, as discussed in

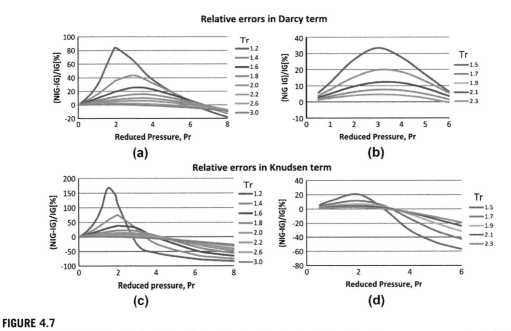

FIGURE 4.7

Relative differences between NIG and IG for Darcy and Knudsen terms for a full range of pressure and temperature conditions (a) and (c) and methane production conditions (b) and (d).

From Ma et al. (2014a).

the previous section, is coupled with gas sorption. Therefore, understanding gas adsorption and desorption behaviors in gas shale is critical for not only estimating gas reserves, and ultimate gas recovery and developing a maximizing recovery strategy, but also predicting flow properties like permeability (Sakhaee-Pour and Bryant, 2012; Sondergeld et al., 2010; Yu and Sepehrnoori, 2014).

4.4.1 BASICS OF ADSORPTION AND DESORPTION

Adsorption is a physicochemical process that adheres one substance, adsorbate, on the surface of another substance, adsorbent, by surface force, or bonding of ions, referred to as physical sorption or chemical sorption, respectively. In the physical sorption, van der Waals force is the fundamental interacting force, originated from the interactions between induced, permanent or transient electric dipoles, to give rise a weak interaction energy. In the chemical sorption, there are changes taking place in electronic structure of bonding molecules in the form of covalent or ionic bonds, and the binding energy of the adsorbate and the adsorbent is much higher than that in the physical sorption. In gas shale, the physical sorption is the predominant cause of adsorption.

Estimating shale gas adsorption capacity is of the most importance in shale gas evaluation, but it is also challenging. The amount of adsorbate being adsorbed on adsorbent, or the concentration as a function of pressure at a given temperature, is referred to as an adsorption isotherm. Many different experimental procedures have been devised, including gas volumetric and gravimetric methods, which measure the amount of gas being removed from the gas phase or mass increase, respectively, and static or dynamic methods, which measures the volumetric difference of admitted gas and gas required to fill the space around the adsorbent (Sing et al., 1985).

The static or dynamic gas volumetric methods are widely used in routine experiments because their procedures are simpler than their counterpart gravimetric methods, and they do not require complex calibrations (Sing et al., 1985). A simple setup for volumetric determination may contain two thermostatic cells, one for hosting adsorptive gas and another the adsorbent of a confined volume, connected through a dosing valve for admitting gas of a fixed quantity (static) or at a slow and constant rate (dynamic). The gas hosting cell has a pair of gas inlet and outlet valves, and is attached to pressure transducers to record pressure decrease toward equilibrium or quasiequilibrium. An adsorption isotherm can be constructed point-by-point with incremental gas admission. The amount of gas adsorbed at the equilibrium is the difference between the amount of admitted gas and the amount of gas required to fill the space of the adsorbent, which can be determined using the same setup with nonadsorptive gas (e.g., helium) or nitrogen at the temperature of the boiling point of nitrogen at ambient atmospheric pressure (Sing et al., 1985).

According to IUPAC classifications for the physical sorption (Sing et al., 1985), adsorption isotherms of porous media may fall into one of six types as illustrated in Fig. 4.8. Type I isotherms characterize microporous adsorbents. Types II and III describe adsorption on macroporous adsorbents with strong and weak adsorbate–adsorbent interactions, respectively. Types IV and V represent adsorption isotherms with hysteresis. Type VI has steps and represents stepwise multilayer adsorption on a uniform nonporous surface.

Note that even the IUPAC classification does not cover a full range of isotherms, but a subrange at subcritical temperatures. Also adsorption isotherms may not always be monotonic functions of pressure as shown by Donohue and Aranovich (1999), who also presented a new classification.

4.4.2 GAS–SOLID ADSORPTION MODELS

Langmuir (1918) developed one of earliest adsorption isotherm models by treating adsorption and desorption as kinetic processes. He assumed the following: (1) there is a fixed number of vacant or adsorption sites of the same size and shape available on an open and flat surface of an adsorbent; (2) each vacant site attracts one gaseous molecule only by releasing a constant amount of heat energy during this process; and (3) a dynamic equilibrium is reached between adsorbed gaseous molecules

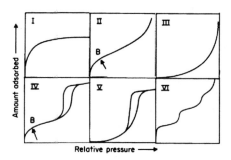

FIGURE 4.8

An illustration of IUPAC classification of isotherms.

and the free gaseous molecules. The Langmuir isotherm model has been discussed in the context of shale gas application (Boyer et al., 2006).

However, the Langmuir model has several limitations. It is of less practical use to estimate the surface area available because the monolayer formation assumption may break down at supercritical conditions (Rexer et al., 2013). When experimentally obtained adsorption data is fitted to the Langmuir equation, it tends to overestimate the surface area, especially for those experimental isotherms that exhibit a nonzero slope toward high pressure. The inflection points indicate that adsorption occurs in more than one distinct phase: homogeneous adsorption at the sorbent surface sites, and adsorption in layers beyond the surface of the adsorbent. This suggests that multilayer adsorption is more appropriate. In addition, adsorbed gas no longer behaves as an ideal vapor phase and the gas molecules attract each other. The Langmuir model has been adapted to address the limitations. Brunauer et al. (1938) developed a multilayer model, referred to as the Brunauer–Emmett–Teller (BET) model, by introducing a number of simplifying assumptions, namely (1) all layers, except the first, have the same energy of adsorption of liquefaction, and (2) the number of layers increases to infinity as the pressure approaches the saturation pressure and the adsorbed phase becomes a liquid.

Although the Langmuir and BET models have been widely used for estimating pore surface areas for shales, they are developed by assuming that pores have ideal flat surfaces. This is clearly not the case in most natural porous media, in particular in shales. Therefore, a more appropriate model is needed to account for the structure of pores. The Dubinin–Radushkevich isotherm was developed from a pore-filling theory of adsorption that overcomes the limitation of layer theories in explaining adsorption in highly microporous media. Several factors may alter the preferences of pore filling (adsorption) mechanisms as pressure increases. There are several key factors that can affect the shapes of adsorption isotherms, including sorbent chemical composition, porosity, pore size distribution, temperature, and pressure. Micropores will primarily be subject to physical sorption only at a pore-filling stage, limited mostly to a single layer or very few condensed molecular layers, whereas meso and macropores will experience two stages of adsorption, single and multi-layer adsorption, and capillary condensation that may lead to denser adsorbates than liquid (Lowell, 2004; Nguyen and Do, 2001).

4.4.3 HETEROGENEOUS MULTILAYER GAS ADSORPTION AND FREE GAS FLOW

This section assesses the impact of gas adsorption on free gas flow as a result of the reduction of the effective size of pore space and the permeability by adsorbed gas molecules. Here we consider not only homogenous multilayer adsorption, where each layer of molecules covers the full sites of the previous layer that is closer to the pore surface, but also heterogeneous multilayer adsorption, where each layer covers just part of sites of the previous layer. Results from molecular dynamic and Monte Carlo simulations on surface adsorption show that heterogeneous multilayer adsorption does occur under a wide range of pressure and temperature conditions (Bojan and Steele, 1993; Xiang et al., 2014; Gensterblum et al., 2014; Lee et al., 2012).

Our purpose is not to study what kind of heterogeneous multilayers may form for a given pore functional group and a specific gas (this will be considered in a separate publication), but to simply answer what impact that either homogenous or heterogeneous multilayers may have on pore structures for the free gas. For this reason, we define a "pseudo" heterogeneous multilayer adsorption model. Rather than working directly on the distribution of occupied sites and the number of molecule layers, we consider that there is an effective layer with a thickness across the surface of pores, equal to $\beta\theta d$

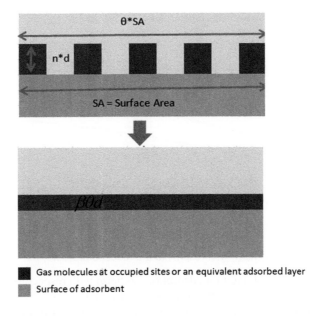

FIGURE 4.9

Determination of an effective layer.

From Houben et al. (2013).

where β is a factor of the effective thickness of the gas layer, θ is the fraction of gas occupied sites, and d is the diameter of a single gas molecule (Fig. 4.9). By so doing, we actually assume that there is a higher probability at each site to attract a gas molecule than an adsorbed gas molecule to attract another one, and for each adsorbed gas molecule closer to the corresponding site, to attract another one than one further away from that site. Experimental simulation results (Nguyen et al., 2013) seem to support the postulation that the attraction decays away from each site.

θ is the fraction of occupied sites on the pore surface in the Langmuir equation (Eq. (4.2)).

$$\theta = \frac{kp}{(1+kp)} \qquad (4.2)$$

where k is the ratio of adsorbing and desorbing rates at equilibrium, and p is the gas pressure. Note that in this work we do not use the k value of the reaction ratio directly, instead take α as a parameter encapsulating k and the critical pressure for any given gas. That is $\alpha = kp_c$ where p_c is the critical pressure of a gas. Now we have a modified Langmuir equation in Eq (4.3), where $p_r = p/p_c$ is the reduced pressure.

Note that we use the reduced pressure to align with the van der Waals's EOS (Klein, 1974) that is defined in terms of reduced pressure and temperature, and implemented in the pore network code used in this work.

$$\theta = \frac{\alpha p_r}{(1+\alpha p_r)} \qquad (4.3)$$

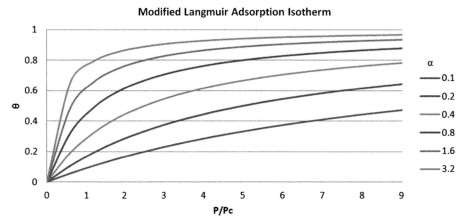

FIGURE 4.10

A modified Langmuir adsorption isotherm with respect to the reduced pressure, p_r.

Figure 4.10 shows a modified Langmuir adsorption isotherm with respect to the p_r for a range of α. It is clear that the larger α is, the higher the fraction of adsorbed gas is.

To explore the behaviors of the new "pseudo" multilayer model in terms of the adsorption effect on the apparent gas permeability, numerical experiments are carried out for a single cylindrical pore. Two gases, methane and nitrogen, are used and their properties are shown in Table 4.2. The effects on the apparent gas permeability, by three parameters α, β, and TMAC, are examined.

For the cylindrical pore model with R being its radius, the actual radius r of the pore space is defined as follows:

$$r = R - \beta\theta d \tag{4.4}$$

The relationship between the ratio of K'app and Kapp, (K'app and Kapp denoted permeability values calculated with and without gas adsorption, respectively) and the reduced pressure are analyzed below. Both K'app and Kapp are calculated using Eq. (4.1) for any given set of α, β and TMAC, and p_r. Using Eqs (4.1), (4.3), and (4.4), one can find that the ratio is a nonlinear function of p_r shown in Eq. (4.5), and note that constant coefficients a_1 to a_3 can be determined precisely:

$$\frac{\text{K'app}}{\text{Kapp}} \sim \frac{a_1 r p_r + a_2(\text{TMAC}) + a_3}{a_1 R p_r + a_2(\text{TMAC}) + a_3} = 1 - \left(\frac{a_1 \beta d \alpha p_r}{(a_1 R p_r + a_2(\text{TMAC}) + a_3)(1 + \alpha p_r)}\right) \tag{4.5}$$

Table 4.2 Gas properties of Methane and Nitrogen

Description	Unit	Methane	Nitrogen
Molar mass	Kg/mol	0.01604	0.02801
Collision diameter	m	3.8E-10	3.75E-10
Critical pressure	MPa	4.696	3.3978
Critical temperature	Kelvin	190.5	126.19

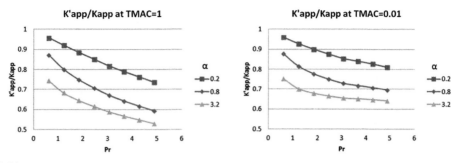

FIGURE 4.11

Plots of the ratio of apparent gas permeability with a monolayer of gas molecules (K'app) and without gas adsorption (Kapp) for nitrogen at TMAC equal to 1 (left) and 0.01 (right).

Equation (4.5) shows that: (1) K'app/Kapp decreases with the increase of the reduced pressure; (2) a thicker layer (larger β) and a high adsorption rate (larger α) lead to the ratio further deviated from (1); and (3) the stronger is the gas slippage, i.e., the smaller TMAC and therefore larger a_2 (TMAC), the closer is the ratio to 1. To show the numerical values of the ratio, we set R equal to 1 nm and calculated K'app with a monolayer (i.e., $\beta = 1$) of gas molecules for nitrogen and Kapp without gas adsorption (i.e., $\beta = 0$). We set α to take 0.2, 0.8 and 3.2, and TMAC to take 1, 0.1 and 0.01. Figure 4.11 shows the selected results for K'app/Kapp versus the reduced pressure and the relationships are consistent with the analysis given above. Note that K'app can be as small as two-thirds of Kapp, and that means that even a single-layer adsorption can have a noticeable impact on the permeability for a small pore.

4.5 AGGREGATED EFFECT ON THE PREDICTED GAS PERMEABILITY

In previous sections, we highlighted the importance of considering nonDarcy flow regimes and gas adsorption in estimating gas permeability. In this section, we recap some results we derived previously to emphasize this importance for a real shale sample following the DCA procedure. We start from reconstructing a realistic 3D shale model from 2D images of that sample. In our previous work (Ma et al., 2014a), SEM images, which were obtained from an Opalinus Clay of the shaly facies from the Mont Terri Rock Laboratory at Switzerland (Houben et al., 2013) by BIB milling at RWTH Aachen University (Klaver et al., 2012), were used. The gray-scale SEM images of the sample were binarized after segmentation, where the original and binary images are shown in Fig. 4.12.

The 2D binary images were used to reconstruct a 3D binary image mode using the Markov Chain Monte Carlo method (Wu et al., 2006). The pore space of that 3D binary image was analyzed and subsequently the pore network of the pore space was extracted (Jiang et al., 2007) to produce a simplified version of the pore network for simulating gas flow. Figure 4.13 shows the 3D shale model and extracted pore network model. The porosity of the extracted network is about 2.9% with the average coordination number of the pore network being slightly less than 3. In terms of pore shapes, 53% of all bonds are found to have a shape close to a square, 45% of them are close to a triangle, and the remaining 2% are close to a circle. Then, the modified pore network flow model, as described

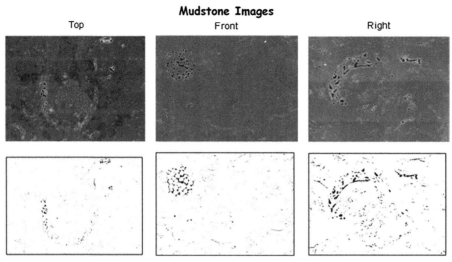

FIGURE 4.12

Portions of original gray-scale SEM images of mudstone (the top row) and their binary images (the bottom row) after segmentations.

From Ma et al. (2014a).

FIGURE 4.13

A shale model and a pore network extracted from it.

From Ma et al. (2014a).

above, was applied to the network to calculate permeability Kapp and Kd. Without considering nonDarcy effects, the calculated water permeability values, along all three directions on the pore network, are in good agreement with those calculated on the 3D binary model using an LB code (Ma et al., 2010). This shows that the extracted network is hydraulically representative.

One of the main results from the previous work (Ma et al., 2014a) is the relationship between Kapp/Kd and Pr for different TMAC values for the methane being treated as ideal gas, and the relative difference between Kapp for the methane being treated as nonideal gas and Kapp for the methane being treated as ideal gas (Fig. 4.14). It is clear that Kapp/Kd increases nonlinearly with the decrease of Pr toward 0.5, and the maximum ratio increases at a rate of the same order of magnitude with respect to the decreasing TMAC. Note that under the simulation conditions the Knudsen number is around 0.1 only, and the Knudsen diffusion has much less impact on Kapp than that of the slip flow. For a given TMAC, the relative difference is dominated by the difference in the Darcy term, and its magnitude decreases when the contribution of the slip flow increases, and reaches just below 50% and just over 20% at temperatures of 340 and 400 K, respectively, at TMAC equal to 1. Therefore, underestimating gas permeability is likely to occur at real pressures between 15 and 20 MPa if the methane is assumed to behave like an ideal gas.

The impact of gas adsorption on the gas permeability for this model was assessed (Couples et al., 2014). Figure 4.15 shows K'app/Kapp, with a monolayer ($\beta = 1$), and with three layers of methane

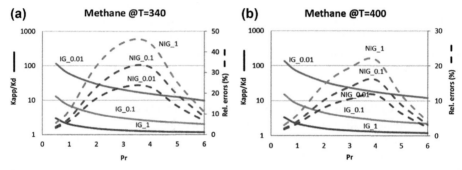

FIGURE 4.14

Kapp/Kd versus Pr to the left axis, and the relative difference to the right axis, at temperature of 340 K (a) and 400 K (b) for methane.

From Ma et al. (2014a).

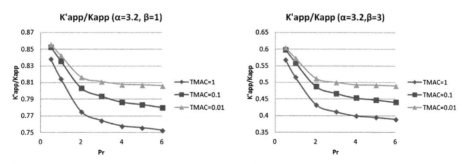

FIGURE 4.15

Plots of K'app/Kapp with a monolayer (left) and three layers (right) of methane molecules at TMAC equal to 1, 0.1, and 0.01 versus the reduced pressure. Methane is treated as nonideal gas.

molecules ($\beta = 3$), at TMAC equal to 1, 0.1, and 0.01, and $\alpha = 3.2$, corresponding to a high adsorption isotherm. The methane is treated as a nonideal gas. Based on the same reasoning for the cylindrical pore, K'app/Kapp decreases with the increase of the gas pressure for every TMAC in both cases. But the rate of the decrease is higher between zero and two of the reduced pressure than between two and six for every TMAC, in both cases. The initial faster decrease in K'app than in Kapp is likely due to the fact that both gas slippage and Knudsen diffusion effects are stronger in the smaller pores. The slower decrease is likely due to the Knudsen diffusion being switched off for some pores when the Knudsen number becomes smaller than 0.1. The decreasing rates are smaller in the monolayer case than in the three-layer case because all the effects mentioned above are stronger for high gas adsorption ($\beta = 3$) than for low ones ($\beta = 1$). If the gas adsorption is not considered, there will an overestimation of gas permeability by 60% of Kapp.

4.6 CONCLUSIONS

This chapter firstly discussed the DCA approach to predicting petrophysical properties for a rock sample, and associated techniques. Those techniques include imaging, the reconstruction of pore-grain space, the characterization of pore space, pore-network simulation, and the prediction of macroscopic petrophysical properties. It then reviewed key natures of nonDarcy flow regimes and a pore-network flow model developed for simulating gas flow and gas adsorptions. It highlighted their impact on free gas flow: analytically on a single-cylinder pore, and then on a realistic shale model. It has shown that:

1. Impact of nonDarcy flow regimes on estimating gas permeability can be significant, in particular, when the gas slippage is the strongest and/or the Knudsen number becomes much larger than 0.1;
2. Treating gas as nonideal gas is important when coming to assess the validity of laboratory measurements and to interpret them to infer the properties of a sample at in situ conditions;
3. Gas adsorption can have strong impact on free gas flow and therefore gas permeability, and needs to be considered in organic pores when the multilayer adsorption of gas molecules occurs.

ACKNOWLEDGMENTS

The author thanks Gary D. Couples, at Heriot–Watt University, for his constructive suggestions on an earlier version of the manuscript, and a reviewer and the editor for commenting on the manuscript, which all helped improve its quality.

REFERENCES

Aplin, A., et al., 2012. Multi-scale effective flow properties of heterogeneous mudstones. In: Petroleum Systems: Modeling the Past, Planning the Future; AAPG Hedberg Conference.
Arya, G., Chang, H.-C., Maginn, E.J., 2003. Molecular simulations of Knudsen Wall-slip: effect of Wall morphology. Molecular Simulation 29 (10–11), 697–709.
Bai, B., et al., 2013. Rock characterization of Fayetteville shale gas plays. Fuel 105, 645–652.
Beskok, A., Karniadakis, G.E., 1999. A model for flows in channels, pipes, and ducts at micro and nano scales. Microscale Thermophysical Engineering 3 (1), 43–78.

Biswal, B., Hilfer, R., 1999. Microstructure analysis of reconstructed porous media. Physica A: Statistical Mechanics and Its Applications 266 (1–4), 307–311.
Blunt, M.J., 2001. Flow in porous media—pore-network models and multiphase flow. Current Opinion in Colloid and Interface Science 6 (3), 197–207.
Bojan, M.J., Steele, W., 1993. Computer simulation studies of the adsorption of Kr in a pore of triangular cross-section. In: Motoyuki, S. (Ed.), Studies in Surface Science and Catalysis. Elsevier, pp. 51–58.
Bondino, I., et al., 2007. Investigation of Gravitational Effects in Solution Gas Drive via Pore Network Modelling: Results from Novel Core-Scale Simulations.
Bondino, I., McDougall, S.R., Hamon, G., 2011. Pore-scale modelling of the effect of viscous pressure Gradients during Heavy oil Depletion experiments. Journal of Canadian Petroleum Technology 50 (2), 45–55.
Boyer, C., et al., 2006. Producing gas from its source. Oilfield Review 18 (3), 36–49.
Brunauer, S., Emmett, P.H., Teller, E., 1938. Adsorption of gases in multimolecular layers. Journal of the American Chemical Society 60 (2), 309–319.
Burnett, D., 1935. The distribution of velocities in a slightly non-uniform Gas. Proceedings of the London Mathematical Society s2-39 (1), 385–430.
Bushby, A.J., et al., 2011. Imaging three-dimensional tissue architectures by focused ion beam scanning electron microscopy. Nature Protocols 6 (6), 845–858.
Cahn, J.W., Hilliard, J.E., 2013. Free energy of a nonuniform system. I. Interfacial free energy. In: The Selected Works of John W. Cahn. John Wiley & Sons, Inc., pp. 29–38
Caprock Project, 2004–2013. http://research.ncl.ac.uk/caprocks/project.htm.
Chalmers, G.R., Bustin, R.M., Power, I.M., 2012. Characterization of gas shale pore systems by porosimetry, pycnometry, surface area, and field emission scanning electron microscopy/transmission electron microscopy image analyses: examples from the Barnett, Woodford, Haynesville, Marcellus, and Doig units. AAPG Bulletin 96 (6), 1099–1119.
Chapman, S., Cowling, T., January 1991. Foreword by Cercignani, C., p. 447. The Mathematical Theory of Non-uniform Gases, vol. 1. Cambridge University Press, Cambridge, UK, ISBN 052140844X.
Chenevert, M., Amanullah, M., 2001. Shale preservation and testing techniques for borehole-stability studies. SPE Drilling and Completion 16 (3), 146–149.
Civan, F., 2010. Effective correlation of apparent gas permeability in tight porous media. Transport in Porous Media 82 (2), 375–384.
Couples, G., Zhao, X., Ma, J., 2014. Pore-scale modelling of shale gas permeability considering shale gas adsorption. In: ECMOR XIV-14th European Conference on the Mathematics of Oil Recovery.
Cui, X., Bustin, A.M.M., Bustin, R.M., 2009. Measurements of gas permeability and diffusivity of tight reservoir rocks: different approaches and their applications. Geofluids 9 (3), 208–223.
Curtis, M., et al., 2010. Structural characterization of gas shales on the micro-and nano-scales. In: Canadian Unconventional Resources and International Petroleum Conference.
Curtis, M.E., et al., 2012. Development of organic porosity in the Woodford Shale with increasing thermal maturity. International Journal of Coal Geology 103, 26–31.
Darabi, H., et al., 2012. Gas flow in ultra-tight shale strata. Journal of Fluid Mechanics 710, 641.
Desbois, G., et al., 2013. Argon broad ion beam tomography in a cryogenic scanning electron microscope: a novel tool for the investigation of representative microstructures in sedimentary rocks containing pore fluid. Journal of Microscopy 249 (3), 215–235.
Dong, H., Blunt, M.J., 2009. Pore-network extraction from micro-computerized-tomography images. Physical Review E 80 (3), 036307.
Donohue, M.D., Aranovich, G.L., 1999. A new classification of isotherms for Gibbs adsorption of gases on solids. Fluid Phase Equilibria 158–160, 557–563.
van Dijke, M.I., et al., 2004. Free energy balance for three fluid phases in a capillary of arbitrarily shaped cross-section: capillary entry pressures and layers of the intermediate-wetting phase. Journal of Colloid and Interface Science 277 (1), 184–201.

van Dijke, M.I.J., et al., 2007. Criteria for three-fluid configurations including layers in a pore with nonuniform wettability. Water Resources Research 43 (12), W12S05.
van Dijke, M.I., Sorbie, K.S., 2006. Existence of fluid layers in the corners of a capillary with non-uniform wettability. Journal of Colloid and Interface Science 293 (2), 455–463.
Evans, R.B., Watson, G.M., Mason, E.A., 1961. Gaseous diffusion in porous Media at uniform pressure. The Journal of Chemical Physics 35 (6), 2076–2083.
Ezeuko, C.C., et al., 2010. Dynamic pore-network simulator for modeling buoyancy-driven migration during depressurization of oil-saturated systems. SPE Journal 15 (04), 906–916.
Fatt, I., 1956. The Network Model of Porous Media. Society of Petroleum Engineers.
Freeman, C.M., Moridis, G.J., Blasingame, T.A., 2011. A numerical study of Microscale flow behavior in tight Gas and shale Gas reservoir systems. Transport in Porous Media 90 (1), 253–268.
Gensterblum, Y., Busch, A., Krooss, B.M., 2014. Molecular concept and experimental evidence of competitive adsorption of H_2O, CO_2 and CH_4 on organic material. Fuel 115, 581–588.
Gruener, S., Huber, P., 2008. Knudsen diffusion in Silicon Nanochannels. Physical Review Letters 100 (6), 064502.
Heller, R., Vermylen, J., Zoback, M., 2014. Experimental investigation of matrix permeability of gas shales. AAPG Bulletin 98 (5), 975–995.
Houben, M.E., Desbois, G., Urai, J.L., 2013. Pore morphology and distribution in the Shaly facies of Opalinus Clay (Mont Terri, Switzerland): insights from representative 2D BIB–SEM investigations on mm to nm scale. Applied Clay Science 71, 82–97.
Javadpour, F., 2009. Nanopores and apparent permeability of gas flow in mudrocks (shales and siltstone). Journal of Canadian Petroleum Technology 48 (8), 16–21.
Jiang, Z., et al., 2007. Efficient extraction of networks from three-dimensional porous media. Water Resources Research 43 (12).
Jiang, Z., et al., 2012. Stochastic pore network generation from 3D rock images. Transport in Porous Media 94 (2), 571–593.
Jiang, Z., et al., 2013. Representation of multiscale heterogeneity via multiscale pore networks. Water Resources Research 49 (9), 5437–5449.
Kast, W., Hohenthanner, C.R., 2000. Mass transfer within the gas-phase of porous media. International Journal of Heat and Mass Transfer 43, 807–823.
Kennard, E.H., 1938. Kinetic Theory of Gases. McGraw-Hill, New York.
Klaver, J., et al., 2012. BIB-SEM study of the pore space morphology in early mature Posidonia Shale from the Hils area, Germany. International Journal of Coal Geology 103, 12–25.
Klein, M.J., 1974. The historical origins of the Van der Waals equation. Physica 73 (1), 28–47.
Klinkenberg, L., 1941. The permeability of porous media to liquids and gases. Drilling and Production Practice 200–213.
Knackstedt, M., et al., 2006. 3D imaging and flow characterization of the pore space of carbonate core samples. In: SPWLA 47th Annual Logging Symposium.
Knudsen, M., 1909. The law of the molecular flow and viscosity of gases moving through tubes. Annalen der Physik 28, 75–130.
Langmuir, I., 1918. The Adsorption of Gases on Plane Surfaces of Glass, Mica and Platinum. Journal of the American Chemical Society 40 (9), 1361–1403.
Lee, E., et al., 2012. Effect of pore geometry on gas adsorption: Grand Canonical Monte Carlo Simulation Studies. Bulletin of the Korean Chemical Society 33 (3), 901–905.
Lockerby, D.A., Reese, J.M., Gallis, M.A., 2005. The usefulness of higher-order constitutive relations for describing the Knudsen layer. Physics of Fluids (1994-present) 17 (10), 100609.

REFERENCES

Loucks, R.G., et al., 2009. Morphology, genesis, and distribution of nanometer-scale pores in Siliceous mudstones of the Mississippian Barnett shale. Journal of Sedimentary Research 79 (12), 848–861.

Lowell, S., 2004. Characterization of Porous Solids and Powders: Surface Area, Pore Size and Density, vol. 16. Springer Science & Business Media.

Luffel, D.L., Hopkins, C.W., Schettler Jr., P.D., 1993. Matrix Permeability Measurement of Gas Productive Shales. Society of Petroleum Engineers.

Ma, J., Couples, G., 2004. A finite element upscaling technique based on the heterogeneous multiscale method. In: 9th European Conference on the Mathematics of Oil Recovery.

Ma, J., et al., 2010. SHIFT: an implementation for lattice Boltzmann simulation in low-porosity porous media. Physical Review E 81 (5), 056702.

Ma, J., et al., 2013. Determination of effective flow properties of caprocks in presence of multi-scale heterogeneous flow elements. In: 6th International Petroleum Technology Conference.

Ma, J., et al., 2014a. A pore network model for simulating non-ideal gas flow in micro- and nano-porous materials. Fuel 116, 498–508.

Ma, J., et al., 2014b. Flow properties of an intact MPL from nano-tomography and pore network modelling. Fuel 136, 307–315.

Ma, J., et al., 2014c. A multi-scale framework for digital core analysis of Gas shale at Millimeter scales. In: Unconventional Resources Technology Conference. Society of Petroleum Engineers, Denver, CO, USA.

Mason, E.A., Malinauskas, A., 1983. Gas Transport in Porous Media: The Dusty-gas Model. Elsevier, Amsterdam.

Mason, E.A., 1983. Gas Transport in Porous Media. In: Chemical Engineering Monographs, vol. 17.

McDougall, S.R., Sorbie, K.S., 1997. The application of network modelling techniques to multiphase flow in porous media. Petroleum Geoscience 3 (2), 161–169.

Mehmani, A., Prodanović, M., Javadpour, F., 2013. Multiscale, multiphysics network modeling of shale matrix gas flows. Transport in Porous Media 99 (2), 377–390.

Meng, J., Zhang, Y., Shan, X., 2011. Multiscale lattice Boltzmann approach to modeling gas flows. Physical Review E 83 (4), 046701.

Nguyen, C., Do, D.D., 2001. The Dubinin–Radushkevich equation and the underlying microscopic adsorption description. Carbon 39 (9), 1327–1336.

Nguyen, P.T.M., Do, D.D., Nicholson, D., 2013. Pore connectivity and hysteresis in gas adsorption: a simple three-pore model. Colloids and Surfaces A: Physicochemical and Engineering Aspects 437, 56–68.

Nicholson, D., Bhatia, S.K., 2009. Fluid transport in nanospaces. Molecular Simulation 35 (1–2), 109–121.

Pickup, G., et al., 2005. Multi-stage upscaling: selection of suitable methods. Transport in porous media 58 (1–2), 191–216.

Rexer, T.F.T., et al., 2013. Methane adsorption on shale under simulated geological temperature and pressure conditions. Energy and Fuels 27 (6), 3099–3109.

Roy, S., et al., 2003. Modeling gas flow through microchannels and nanopores. Journal of Applied Physics 93 (8), 4870–4879.

Ryazanov, A., van Dijke, M.I.J., Sorbie, K.S., 2009. Two-phase pore-network modelling: existence of oil layers during water invasion. Transport in Porous Media 80 (1), 79–99.

Sakhaee-Pour, A., Bryant, S., 2012. Gas permeability of shale. SPE Reservoir Evaluation and Engineering 15 (4), 401–409.

Sing, K.S.W., et al., 1985. Reporting physisorption data for gas/solid systems with special reference to the determination of surface area and porosity (Recommendations 1984). Pure and Applied Chemistry 57 (4), 603–619.

Sondergeld, C.H., et al., 2010. Petrophysical Considerations in Evaluating and Producing Shale Gas Resources. Society of Petroleum Engineers.

Valvatne, P.H., Blunt, M.J., 2004. Predictive pore-scale modeling of two-phase flow in mixed wet media. Water Resources Research 40 (7).

Wu, K., et al., 2006. 3D stochastic modelling of heterogeneous porous media–applications to reservoir rocks. Transport in Porous Media 65 (3), 443–467.

Wu, K., et al., 2007. Reconstruction of multi-scale heterogeneous porous media and their flow prediction. In: International Symposium of the Society of Core Analysts Held in Calgary, Canada (sn).

Xiang, J., et al., 2014. Molecular simulation of the $CH_4/CO_2/H_2O$ adsorption onto the molecular structure of coal. Science China Earth Sciences 57 (8), 1749–1759.

Xiao, J., Wei, J., 1992. Diffusion mechanism of hydrocarbons in zeolites—I. Theory. Chemical Engineering Science 47 (5), 1123–1141.

Yu, W., Sepehrnoori, K., 2014. Simulation of gas desorption and geomechanics effects for unconventional gas reservoirs. Fuel 116, 455–464.

Zaretskiy, Y., Geiger, S., Sorbie, K., 2012. Direct numerical simulation of pore-scale reactive transport: applications to wettability alteration during two-phase flow. International Journal of Oil, Gas and Coal Technology 5 (2), 142–156.

Zhang, W.-M., Meng, G., Wei, X., 2012. A review on slip models for gas microflows. Microfluidics and Nanofluidics 13 (6), 845–882.

Zhu, J., Ma, J., 2013. An improved gray lattice Boltzmann model for simulating fluid flow in multi-scale porous media. Advances in Water Resources 56, 61–76.

CHAPTER 5

WIRELINE LOG SIGNATURES OF ORGANIC MATTER AND LITHOFACIES CLASSIFICATIONS FOR SHALE AND TIGHT CARBONATE RESERVOIRS

Y. Zee Ma[1], W.R. Moore[1], E. Gomez[1], B. Luneau[1], P. Kaufman[1], O. Gurpinar[1], David Handwerger[2]

Schlumberger, Denver, CO, USA[1]; Schlumberger, Salt Lake City, UT, USA[2]

5.1 INTRODUCTION AND OVERVIEW
5.1.1 LITHOFACIES IN SHALE RESERVOIRS

In the early exploration and production of shale reservoirs, researchers initially thought that analysis of lithofacies and rock type was not important as they thought "it is all about shale!" Historically the term "shale" has two connotations: lithological (or compositional) and geological formation-wise (often depositional). In unconventional reservoirs, shale is generally used as a geological formation or perhaps even more accurately a facies; thus, it is important to distinguish shale from clay. Shale describes fine-grained rocks that may contain a number of lithological components: clay, quartz, feldspar, heavy minerals, etc. (Passey et al., 2010), and clay is only one component of the typical shale (Fig. 5.1(a)). Given the grain-size connotation, shale should more properly be referred to as a mudstone in the context of unconventional reservoirs (Jarvie, 2012) that has undergone a certain amount of compaction. As the composition of mudstone is typically more complex than sandstone, shale often has complex lithofacies. The Barnett shale, for example, has a number of lithofacies identifiable using thin-section and core descriptions (Hickey and Henk, 2007; Loucks and Ruppel, 2007). Many well-known shale resource plays are quite different from "traditional" shale from a viewpoint of mineral compositional analysis and organic content (Allix et al., 2011; Gamero-Diaz et al., 2013). Moreover, as with the heterogeneity of rock formations, the variation in mineral composition is especially strong in the vertical direction (Fig. 5.1(b)), which is important to describe for reservoir quality evaluation, and for completion target selection (Ajayi et al., 2013).

Lithofacies are composite variables that incorporate lithology and facies characteristics, and represent an intermediate scale of reservoir properties in the hierarchy of subsurface heterogeneities (Ma et al., 2008). On one hand, lithofacies are characteristics of stratigraphic formations as they are sedimentary depositional features (Passey et al., 2010). On the other hand, they govern the characteristics of petrophysical properties because different lithofacies often have different porosity and

152 CHAPTER 5 WIRELINE LOG SIGNATURES OF ORGANIC MATTER

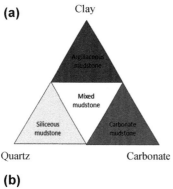
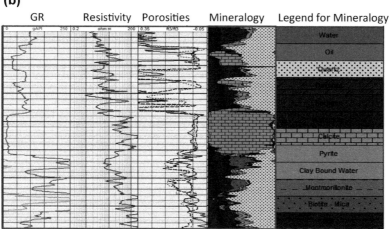

FIGURE 5.1

(a) Ternary plot showing four main lithofacies based on three-mineral compositions for shale. (b) Example of compositional heterogeneity along a vertical wellbore.

permeability ranges. Thus, lithofacies have a duality of being a reservoir attribute property and a modeling reference class in describing multiscaled reservoir heterogeneities (Ma et al., 2009). Yet, lithofacies data are generally limited to interpretations from cores because they are not directly measured. Some newer logging methods, such as nuclear spectroscopy, give mineral compositions that can describe the lithology of the rock formation (Alexander et al., 2011), which can be used for lithofacies or rock type classification (Gamero-Diaz et al., 2013).

Reservoir rocks can be organic-rich source rocks, often mudstones, or other lithofacies juxtaposed to the source rocks (Jarvie, 2012). For example, the most prolific and productive intervals in the Bakken and Three Forks formations are dolomitic siltstone, silty dolostone, fine-grain sandstone, or limestone juxtaposed to the organic-rich source rocks (Pitman et al., 2001; Theloy and Sonnenberg, 2013). Even when the source rocks are the reservoir, other considerations for integrated reservoir studies make lithofacies analysis important, including mineral compositional analysis (Gamero-Diaz et al., 2013), subclassification of lithofacies into rock types, and the relationship between reservoir and nonreservoir lithofacies in the formations.

5.1.2 OVERVIEW OF WIRELINE LOG RESPONSES TO ORGANIC MATTER

As shale reservoirs are composed of fine-grained rocks and are associated with low porosity and extremely low permeability, log responses to these rocks are not the same as to conventional coarser-grained rocks. Gamma rays (GR) are often high because of the various radioactive minerals in the fine-grain rocks (Ma et al., 2014a). GR can be increased dramatically by the presence of uranium while clay content does not necessarily increase or can even decrease. Resistivity tends to be higher because of the resistive nature of the oil-wet organic materials (i.e., lower water content) and/or hydrocarbon generation during the maturation of organic matters (Passey et al., 2010; Lu et al., 2015). Passey et al. (1990) proposed a method of estimating total organic carbon (TOC) using resistivity and sonic logs. Bulk density generally tends to be lower because of the presence of lighter organic materials, such as kerogen, but it could be counterbalanced, to a certain extent, by the presence of heavy minerals, such as pyrite. Apparent neutron porosity tends to be higher because it is impacted by the hydrogen in the system that exists in clays, kerogen, and fluids in the pores. However, neutron typically responds less strongly than the bulk density and sonic (DT) as organic shales tend to have significantly lower density and slower sonic velocity. It is noteworthy that sonic data are impacted by many factors and are more difficult to calibrate to reservoir properties than neutron and density logs. In summary, each of these logging measurements responds to the organic-rich rock differently than to nonorganic rocks, and, thus their responses can be used to interpret material properties of the rock. Table 5.1 summarizes the main response characteristics of several common logs to organic-rich rocks.

Note that the measurements may not always have the typical response listed in Table 5.1 because of the effect of other variables. For example, resistivity may not be high for high TOC formations as a result of wettability change and other petroleum-generating mechanism (Al Duhailan and Cumella, 2014).

One of the major differences in log responses between conventional reservoirs and shale reservoirs is the relationship between GR and fractional volume of clay (Vclay). In conventional reservoirs, Vclay is generally correlated strongly and positively with GR; the correlation coefficient commonly is greater than 0.7, and high GR values commonly indicate low hydrocarbon potential. Geoscientists, while knowing that shale is not clay, have often equated the two in practice. It is quite common to

Table 5.1 Common Logging Measurement Responses in Shale Reservoirs	
Measurement	**Response**
Natural GR	Type II kerogen has anomalously high uranium content
Total GR	High concentration of radioactive materials in organic matters lead to high GR reading
Bulk density	Organic matter is less dense than matrix minerals, leading to low bulk density
Neutron	Organic matter increases apparent neutron porosity, and the degree of maturation can impact the hydrogen index which neutron porosity (NPHI) should be sensitive to
Sonic	Organic matter is less dense, has different textures and increases transit time of acoustic log
Resistivity	Organic materials are usually nonconductive; resistivity reads higher, and can increase dramatically with kerogen conversion into hydrocarbons, which displaces pore water. However, if overmature, organic matter may become graphitic and the resistivity reading is reduced.

Modified after Passey et al. (1990) and Sondergeld et al. (2010).

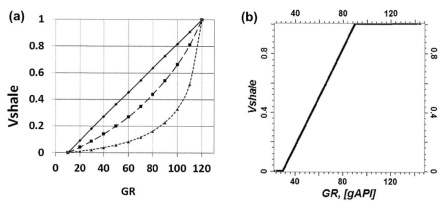

FIGURE 5.2

Vclay (or Vshale, VCL)–GR cross-plots. (a) Common models in conventional formations: linear, Clavier (slightly curved), and Steiber (highly bent curve). (b) Example of deriving Vshale from GR using a linear transform with cutoffs for a conventional clastic reservoir. Correlation coefficient is 0.795.

calculate Vclay from GR data (Bhuyan and Passey, 1994); sometimes, researchers simply compute the fractional volume of shale (Vshale) instead of Vclay (Szabo and Dobroka, 2013). Figure 5.2(a) shows three transforms to generate Vclay from GR (Steiber, 1970; Clavier et al., 1971). In practice, a combination of applying GR cutoffs and a linear model can also be used to derive Vshale while mitigating the overestimation bias, as shown in Fig. 5.2(b). A high Vclay indicates low net-to-gross ratio, and low reservoir quality in conventional reservoirs.

In unconventional reservoirs, however, Vclay may be correlated weakly with GR, and sometime even negatively (an example will be shown in Section 5.4). In source-rock reservoirs, high GR often indicates high TOC and kerogen. In some tight-gas sandstone reservoirs, hydrocarbon-bearing sandstones have abnormally high GR due to high radioactive content (Ma et al., 2014a). This change of correlation between commonly used wireline logs in conventional formations to a different correlation between those same logs in unconventional formations is a manifestation of the Simpson's paradox—a counterintuitive statistical phenomenon in data analysis (Ma et al., 2014a).

5.1.3 SCOPE

Based on the responses of wireline logs to organic matters and various lithofacies, this chapter presents an integrated methodology for classifying lithofacies for shale and tight-carbonate reservoirs. Some statistical and artificial neural network methods were proposed for lithofacies classification in shale reservoirs (Eslinger and Everett, 2012; Wang and Carr, 2012). However, those proposed methods are focused on automatic classification with a minimum of geological and petrophysical interpretations. In fact, other lithofacies classification researches have already proved that automatic classification methods with little use of subject matter knowledge often fail to accurately classify the lithofacies (Ma, 2011; Ma and Gomez, 2015). We first review lithofacies classification methods for conventional formations that are supported by laboratory experiments and accepted by the industry. We then discuss how to extend these classification methods of evaluating conventional formations to the evaluation of shale and tight carbonate reservoirs. Other types of unconventional formations, such as tight-gas sandstones, while sharing some similarities in terms of the classification of lithofacies, are discussed in a separate chapter (Ma et al., 2015).

One common problem in clustering lithofacies is the number of input wireline logs. Using a single log generally requires applying cutoffs, and using a very large number of logs causes inconsistencies and confusions when attempting to understand the contributions of each log in relation to the clusters. The latter is sometimes termed the curse of (high) dimensionality (COD). Principal component analysis (PCA) can be used to overcome the COD, and it has other advantages, such as enabling the geological and petrophysical interpretation of the principal components (PCs), overcoming the correlation reversals (i.e., Simpson's paradox) that occur in relationships among various wireline logs, and rotation of the components for the optimal discrimination of lithofacies.

We present three broad categories: (1) tight carbonate reservoirs with the presence of claystones, but without the presence of organic shale, (2) shale reservoirs with the presence of carbonate lithofacies, and (3) shale reservoirs with a mixture of clayey, siliceous, and carbonate lithofacies. We then propose a multilevel clustering scheme for lithofacies and rock types.

5.2 REVIEW OF LITHOFACIES CLASSIFICATION IN CONVENTIONAL FORMATION EVALUATION

Several methods have been proposed for clustering lithofacies from wireline logs for conventional formations (Ma, 2011). Early methods used one or two logs and applied cutoffs to derive lithofacies; they are based on heuristics and are not optimal. Newer approaches include statistical methods (Tang and White, 2008; Ma et al., 2014b) and artificial neural networks, or ANN (Wang and Carr, 2012). Although neural networks and statistical methods are highly useful, they have a number of pitfalls that can lead to suboptimal classifications (Ma and Gomez, 2015).

For conventional clastic reservoirs, the GR log is commonly used to classify sand, shaly sand, sandy shale, and shale. Alternatively the volume proportion of shale (Vshale) or sand (Vsand) is used to separate those lithofacies. Cutoff values are typically used in the clustering. For carbonate or a mixture of clastic and carbonate reservoirs, benchmark charts using neutron, sonic, density and/or photoelectric logs based on laboratory experiments and field data are generally preferred for interpreting lithofacies (Dewan, 1983; Schlumberger, 1989). Statistical and/or ANN methods have also been used (Wolff and Pelissier-Combescure, 1982; Rogers et al., 1992), but using these methods without the guidance of benchmark references frequently gives poor results (Ma, 2011).

Two commonly used benchmark cross-plotting charts are shown in Fig. 5.3. ANN has been used for clustering lithofacies in both conventional and unconventional formations (Ma, 2011; Wang and Carr, 2012). A back-propagation ANN with a single hidden layer and 10 nodes is illustrated using an example of classifying three lithofacies by two wireline logs in a mixture of limestone, dolomite, and sand for a conventional reservoir (Fig. 5.3(c)). Both ANN alone and ANN in combination with PCA (a tutorial on PCA is given in Appendix A) without the guidance of petrophysical analysis gave poor classification results (Fig. 5.3(d) and (e)).

In applying PCA, it is a common practice to select the major PC(s) for clustering (Wolff and Pelissier-Combescure, 1982; Everitt and Dunn, 2002). An example for lithofacies clustering for a conventional reservoir with a mixture of sand, limestone and dolomite is presented here. The lithofacies clustered by ANN using the first PC (PC1), however, is completely inconsistent with the well-

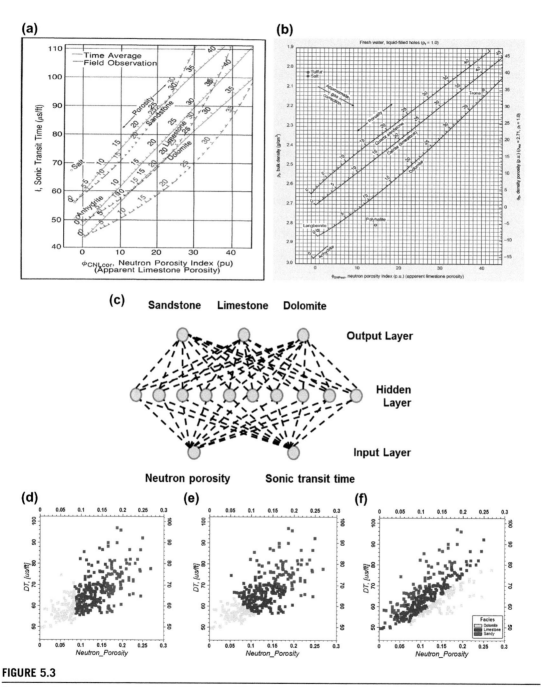

FIGURE 5.3

Benchmark cross-plotting charts for lithology or lithofacies determination. (a) Neutron–density. (b) Neutron–sonic cross-plot; adapted from Schlumberger (1989). (c) ANN design. (d) Neutron–sonic cross-plot overlain by ANN clusters. (e) Neutron–sonic cross-plot overlain by PCA-ANN clusters. (f) Neutron–sonic cross-plot overlain by PCA–ANN clusters using benchmark reference guidance.

known neutron–sonic lithology chart (Fig. 5.3(e)). The overwhelming practice of using the major PC(s) for clustering is based on the mathematical principle, in which the major PC(s) contain more information than a minor PC. In this case, however, PC1 mainly represents porosity, and it does not reveal much about the lithofacies; the second component (PC2 or the minor component in this case) carries the essential information for discriminating the lithofacies, which explains why the lithofacies classification by ANN with the two original logs directly or with their PC1 is very poor. In a later example of shale formation (Section 5.6), PC1 is useful for lithofacies classification because of the effect of organic matter.

In fact, two important properties are interpretable from the neutron–sonic benchmark chart: porosity and lithology. They are approximately orthogonal; porosity is aligned with the maximum axis of information in the cross-plot while lithology is approximately perpendicular to it. A good classification of lithofacies is achieved when PCA and ANN are integrated with this petrophysical analysis. Specifically, when the PC2 is used, without using PC1, the lithofacies clustered by ANN are similar to the benchmark chart, as shown in Fig. 5.3(f).

5.3 TIGHT CARBONATE RESERVOIRS WITHOUT PRESENCE OF ORGANIC SHALE

Tight carbonates have received increasing attention as an unconventional resource. In these subsurface formations, carbonate rocks tend to be the dominant lithofacies, generally with the presence of clay, but with or without the presence of organic mudstones. Benchmark charts (Fig. 5.3(a) and (b)) are commonly useful for the lithofacies prediction using wireline logs in such a setting. In this section, we discuss the lithofacies classification in a mixture of claystones and tight carbonates without the presence of organic mudstones.

As a result of the presence of argillaceous rocks, determining the lithofacies is generally more complex than for the three lithologies shown in the benchmark charts (Fig. 5.3(a) and (b)). GR and/or Vclay are often important for separating clayey lithofacies from other lithofacies. Neutron, sonic, and density logs can be integrated with Vclay, GR, and possibly other wireline logs for lithofacies classification. Here we discuss how to use these wireline logs for lithofacies classification in formations with a mixture of claystones and carbonates. Two different workflows are compared to classify lithofacies in a mixture of claystone, limestone, dolomite, and anhydrite. The GR log is the main discriminator for separating claystone in both workflows.

In the first workflow, neutron porosity (NPHI) and sonic transit time (DT) are used for separating dolomite from limestone (Fig. 5.4(a)) based on the benchmark chart (Fig. 5.3(a)), and density (RHOB) is used to separate anhydrite (Fig. 5.4(b) or (c)). As discussed in Section 5.2, the method for using the benchmark charts is to perform PCA from the wireline logs and cluster lithofacies using PC2 while ignoring PC1. However, because density is not used in classifying dolomite and limestone, these two lithofacies are not well separated on the NPHI–RHOB and DT–RHOB cross-plots (Fig. 5.4(b) and (c)).

In the second workflow, PCA was performed using all the three porosity-measuring wireline logs, NPHI, DT, and RHOB. Correlations between any two variables among these three logs and their PCs are listed in Table 5.2. PC1 is highly correlated to all of the three logs, especially to DT and NPHI. This is not surprising because PC1 mainly represents porosity when those three logs are used and generally has little information for discriminating the major lithofacies in such a setting (Ma, 2011). On the other hand, both PC2 and PC3 contain a significant amount of information for lithofacies classification.

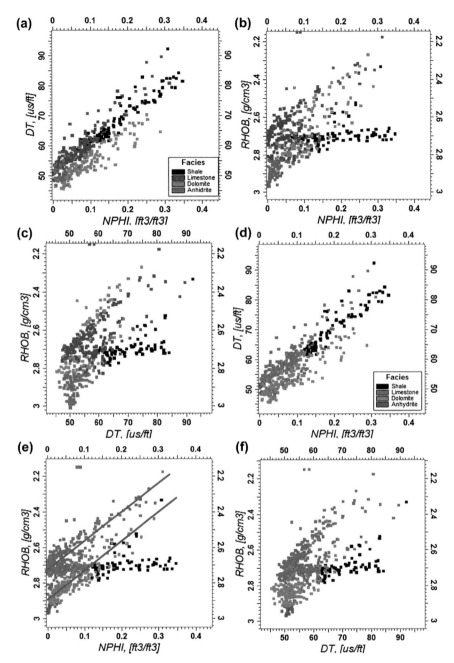

FIGURE 5.4

Cross-plots between two wireline logs in a tight-carbonate reservoir (single well). (a) to (c) are NPHI–DT, NPHI–RHOB, and DT–RHOB cross-plots, respectively, overlain by lithofacies clustered using GR, NPHI, and DT with PCA; density was used only for clustering anhydrite. Shale is nonorganic or claystone. (d) to (f) are NPHI–DT, DT–RHOB, and NPHI-RHOB cross-plots, respectively, overlain by lithofacies clustered using GR, NPHI, DT, and RHOB with a rotated PCA. (g) Cross-plot between PC2 and PC3 overlain by limestone and dolomite. (h) Histogram showing the proportion of each clustered lithofacies. 0, claystone (nonorganic shale); 1, limestone; 2, dolomite; and 3, anhydrite. (i) NPHI–RHOB cross-plot overlain by the ANN-clustered lithofacies. (j) Histogram and its cumulative of Vclay, showing the method of finding a cutoff value that corresponds to 10% claystones, and 90% other lithofacies. The 10% claystone corresponds to 28% of Vclay as the cutoff value.

5.3 TIGHT CARBONATE RESERVOIRS

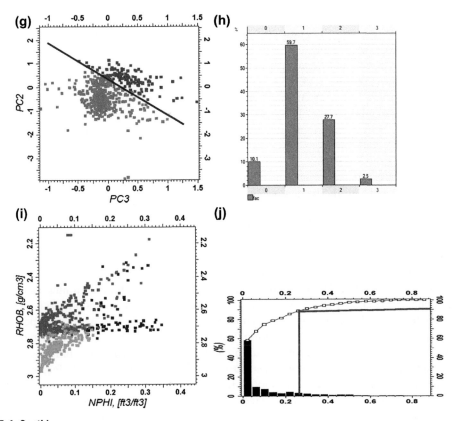

FIGURE 5.4 Cont'd

Table 5.2 Correlation Matrix between Pairs of Eight Variables: Neutron Porosity, Sonic Transit Time, Density, Their PCs (PC1, PC2, and PC3), a Rotated Component, PC2_3, and Vclay (Vclay Was Not Part of the PCA)

	NPHI	DT	RHOB	PC1	PC2	PC3	PC2_3	Vclay
NPHI	1							
DT	0.898	1						
RHOB	−0.454	−0.501	1					
PC1	0.929	0.945	−0.712	1				
PC2	0.230	0.234	0.702	0	1			
PC3	0.218	−0.227	−0.018	0	0	1		
PC2_3	0.369	0.086	0.595	0	0.860	0.510	1	
Vclay	0.767	0.780	−0.098	0.646	0.628	−0.097	0.491	1

By linearly combining PC2 and PC3 (Ma, 2011), a new component was generated, and it has a correlation of 0.86 to PC2 and a correlation of 0.51 to PC3. This enables separating dolomite from limestone more accurately. The clustered lithofacies are generally consistent with all the benchmark charts (compare Fig. 5.4(d) and (e) with Fig. 5.3(a) and (b)). The separation between the two lithofacies is clearly shown in the PC2–PC3 cross-plot (Fig. 5.4(g)), in which the separating line is the new component created from PC2 and PC3. Statistically, this new component can be interpreted as a rotated component in the PC2 and PC3 plan. The correlations between this rotated component and each of the three logs show that density contributes the most in separating limestone from dolomite. This also explains why the first workflow does not work well since the former does not use the density log in separating dolomite from limestone.

Claystone in tight carbonate formations often has higher sonic (DT) readings than limestone and dolomite (Fig. 5.4(d) and 5.4(f)). The neutron response of claystone is generally high in absolute value, but is lower relative to the trend line on NPHI–DT and NPHI density charts (Fig. 5.4(d) and 5.4(e)). When Vclay or GR is available, it is straightforward to classify nonorganic claystone from the other lithofacies. In this example, the mean value of Vclay is about 10%; a cutoff value of 28% Vclay can be easily defined to classify 10% nonorganic claystone (Fig. 5.4(j)). The proportions of all the lithofacies clustered are shown in Fig. 5.4(h). Note also that Vclay was not part of the PCA in this example. If Vclay and GR are not available, DT or NPHI would be the log of choice for discriminating claystone or nonorganic shale, as DT and NPHI have the highest or nearly highest correlation to Vclay (Table 5.2).

This example shows the importance of combining statistical methods with geologic and petrophysical interpretations for lithofacies clustering. Using PCA as a preprocessor for clustering enables not only synthesizing the information from multiple logs, but more importantly, introducing geologic and petrophysical interpretations in lithofacies clustering. In this dataset, the lithofacies clustered by ANN without using PCA is very poor even though the separations of the different clusters are apparent (Fig. 5.4(i)). The poor classification lies in the inconsistency of the clusters with the physical models or experimental results (compare Fig. 5.4(i) with Fig. 5.3(b)). This makes it difficult to assign a cluster to a lithofacies, especially between dolomite and limestone. Moreover, the proportion of each lithofacies is very unrealistic; for example, anhydrite has over 20% proportion instead of 2.5% in the previous classification. On the other hand, the use of PCA in the previous classification enables relating the PCs to geological and petrophysical interpretations, and aligning the component(s) for an optimal discriminability of lithofacies.

5.4 SHALE RESERVOIRS WITH THE PRESENCE OF CARBONATE LITHOFACIES AND WITHOUT SILICEOUS LITHOFACIES

When organic mudstone is present, it is important to separate organic-rich mudstone from nonorganic mudstone, especially when argillaceous lithofacies are abundant. In such cases, GR and Vclay are generally not highly correlated as seen in conventional resource plays. This is the case for many well-known unconventional formations, noticeably the Marcellus and Bakken. GR is generally important for lithofacies classification, in particular, separating organic mudstones from nonorganic mudstones (Ma et al., 2014a). However, a combination of several other logs can achieve a similar discriminability in some cases. When a GR histogram shows multimodality, it often suggests multiple lithofacies, but the number of lithofacies might be more numerous than number of modes in the histogram. Examples are given in a later section.

Classification of three lithofacies for a Marcellus' well is shown in Fig. 5.5. The lithofacies clustered by ANN using first three PCs from the PCA of five logs (GR, Vclay, Neutron, sonic DT, and density) are geologically sensible. The organic-rich mudstone has much higher GR, higher DT, higher neutron, and low density values. The correlation between GR and Vclay is only 0.323, but density and GR have a high correlation, at −0.701, because the organic shale has significantly lower density and higher GR (Fig. 5.5(c) and (d)). Three lithofacies similar to the above-clustered lithofacies were observed in an outcrop study (Soeder, 2011; Walker-Milani, 2011). Notice that the interval in the well shows little presence of silica and thus no siliceous lithofacies were classified. Examples with siliceous lithofacies are discussed in the next section.

It is worth comparing the two different uses of PCA as a preprocessor for lithofacies clustering; the major component, PC1, was not used in the previous clustering example, but it is used in the example above. When only one of the three major rock types exists, i.e., dolostone, limestone, or sandstone, the first PC may contain meaningful information for the clustered lithofacies, and thus should not be excluded. On the other hand, when two or three of those major lithofacies are present and have a wide range of porosity, the second and/or third PCs generally contain the essential information for separating the main lithofacies, while the first PC carries mostly porosity information (Ma, 2011). Exceptions do exist, for example, when porosity is highly correlated to lithofacies or each lithofacies has a narrow range of response to the input logs (e.g., NPHI, DT, and density).

5.5 FORMATIONS WITH A MIXTURE OF CLAYEY, SILICEOUS, CARBONATE AND ORGANIC LITHOFACIES

Unconventional resources sometimes reside in formations with a mixture of claystone, sandstone and carbonate lithofacies. For example, the Bakken and Three Forks contain all these lithofacies (LeFever et al., 1991), although only some of these lithofacies may be present in a given location. The Eocene formations in California contain oil-bearing tight rocks, and they are often a mixture of muddy, siliceous, and carbonate rocks (Peters et al., 2007). Here, we discuss a lithofacies-classification example with the presence of organic-rich mudstones using multiple wireline logs.

Figure 5.6 shows an example of clustering four lithofacies using NPHI, density and DT for a well in the Williston basin. The lithofacies clustered using ANN with those three logs are quite consistent with the benchmark charts (Fig. 5.3(a) and (b)); yet unlike what was recommended earlier, PC1 was used in the clustering. This is because PC1 can separate the organic shale from the other lithofacies (see Fig. 5.6(a) and (c)). PC3 is the main discriminator for separating siltstone/sandstone from silty dolostone and dolostone (see Fig. 5.6(a) and (d)). Organic shale was well classified even though GR was not used to separate it from the other lithofacies. In fact, PC1 from the PCA of NPHI, density, and DT is highly correlated to GR, with the correlation coefficient equal to 0.836, as a result of the high correlation between GR and density (−0.841) and between GR and NPHI (0.749). This explains why PC1 was useful in this classification and why organic shale was well clustered despite not using GR or Vclay.

When including a larger interval above the Bakken and below the Three Forks, not only all the three major lithofacies are present, but their log response ranges are wider as a result of wider porosity range. Hence, PCA will be highly useful or even necessary because the major component, PC1, will

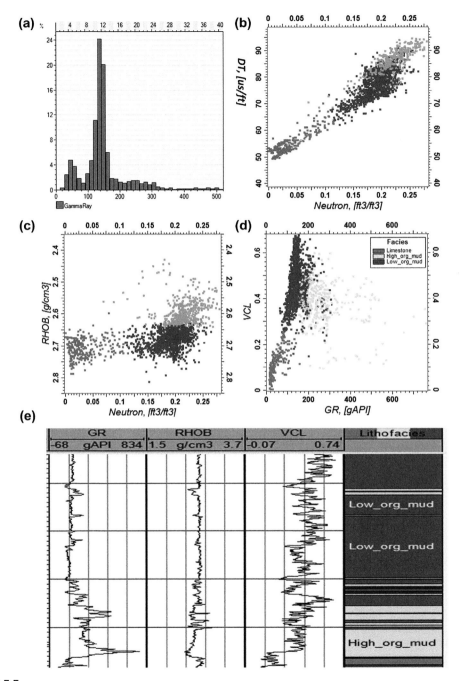

FIGURE 5.5

(a) GR Histograms from a wireline log in the Appalachian basin. (b) Neutron–DT cross-plot overlain by the lithofacies clustered using ANN with the first three PCs from five logs (GR, Vclay, Neutron, sonic DT, and density). (c) Neutron–density cross-plot overlain by the lithofacies clustered using ANN with the first three PCs. (d) GR–VCL (volume of clay or Vclay) cross-plot overlain by the lithofacies clustered using ANN with the first three PCs. Notice the weak correlation between GR and Vclay. (e) Well section showing the wireline logs and the clustered lithofacies. Gray, low organic mudstone (Low_org_mud); Green, high organic mudstone (High_org_mud); and blue, limestone.

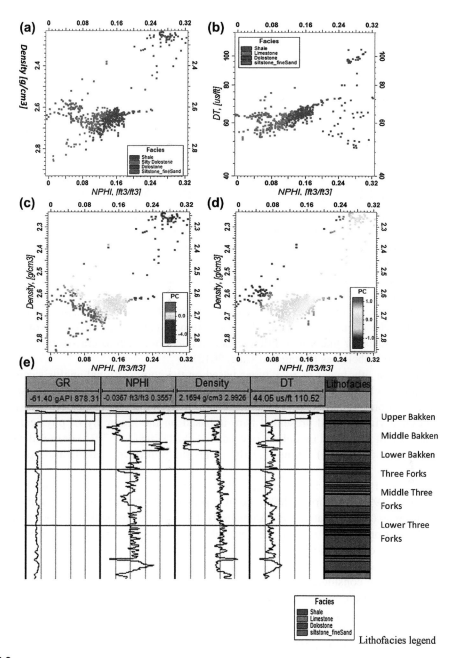

FIGURE 5.6

Examples of lithofacies clustering (single well) for the Bakken and Three Forks formations with a mixture of shale/mudstone, siltstone, fine-grain sandstone, limestone, silty dolostone, and dolostone. (a) NPHI-density cross-plot overlain by lithofacies clustered using ANN with the first three PCs of PCA from NPHI, DT, and density. (b) NPHI–DT cross-plot overlain by lithofacies clustered using ANN with the first three PCs. (c) NPHI density cross-plot overlain by PC1 of PCA from NPHI, DT, and density. (d) NPHI density cross-plot overlain by PC3 of PCA from NPHI, DT, and density. (e) Well section showing four wireline logs and clustered lithofacies (GR was not used for the clustering).

carry little information on the lithofacies, but it will mainly convey information on porosity. By using PCA, PC1 can be excluded in the clustering so that PC2 (possibly PC3) can separate sandstone and limestone from dolomite according to the benchmark charts (Fig. 5.3).

Notice also the pronounced separation between shale and the other lithofacies in these cross-plots (Fig. 5.6(a)–(d)). It is possible to analyze the data based on the mixture decomposition and refine the classification with multimodality consideration (Ma et al., 2014b).

5.6 MULTILEVEL CLUSTERING OF LITHOFACIES AND ROCK TYPES

The lithofacies classifications presented in the previous sections represent the first-order separation of rock fabrics, lacking detailed classifications. Often, subclasses of lithofacies are required for detailed reservoir characterization for unconventional resources. For example, sandy facies may contain more or less clay content. Shale can be distinguished into organic and nonorganic. These subclassifications can be based on mineral compositional analysis. The mineral composition is very important for unconventional resource evaluation because compositional analysis enables examining geochemical elements, of which some are more directly related to TOC or other reservoir quality variables, and/or completion quality. The popularity of mineral compositional analysis for unconventional resources has already led to studies for fine-scaled lithofacies classification, such as classifying the Marcellus shale into a number of subclasses (Wang and Carr, 2012).

Detailed clusters of lithofacies or rock types based on the compositional analysis can be highly useful. Shale may be classified as organic and nonorganic shale, siliceous, or muddy shale. More generally, it is often convenient to classify the lithofacies using a multilevel clustering methodology. The multilevel clustering method presented here is based on the hierarchy of physical properties under the framework of multilevel modeling, or cascaded clustering (Ma, 2011) or multilevel mixture decomposition (Ma et al., 2014b). Figure 5.7 illustrates the multilevel classification of lithofacies and rock types.

Incidentally, there is a clustering method, called hierarchical clustering analysis (HCA) in the literature (Jain et al., 1999; Kettenring, 2006), but it is performed in successive steps, either top down (divisive) or bottom up (agglomerative). HCA generally does not use physical properties to distinguish the hierarchical order of scales of heterogeneities, unless the user converts physical properties into a distance metric for the classification.

Historically, facies classified using wireline logs have been termed electrofacies, and as such, they may or may not be considered the same as lithofacies. In some cases, electrofacies coincide with lithofacies, but they are not generally the same because these clusters are based on the input logs with little or no geologic or petrophysical interpretation. In the example of clustering using neutron and density logs, the clusters (Fig. 5.3(d) and (e)) are commonly labeled electrofacies (Wolff and Pelissier-Combescure, 1982). They are not lithofacies as they are highly inconsistent with the benchmark based on the laboratory experiments. Lithofacies are closely related to geology because lithology represents compositional mineral content with a certain texture and geologic facies are commonly correlated with depositional environments. Despite the difference in definition between lithofacies and electrofacies, it is possible to model both of them in a hierarchical order (Ma, 2011). This can be carried out using the cascaded PCA–ANNs methodology discussed above.

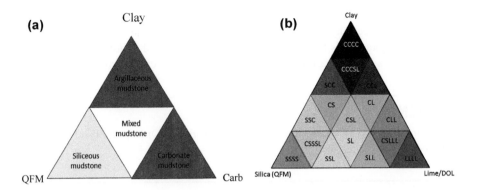

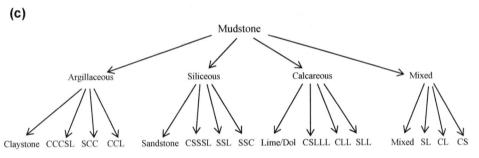

FIGURE 5.7
(a) Ternary plot showing four main lithofacies based on mineral composition. (b) Second level of the hierarchical classification of 16 lithofacies and rock types using mineral compositions. (c) Classification of shale rock types with organic consideration.

5.7 CONCLUSIONS

Some of the most important characteristics of well-log signatures of organic mudstones include high GR, high resistivity, high neutron, slow sonic, low density, and low photoelectric absorption (Pe). Lithofacies are complex for most unconventional formations, often including shale, organic-rich lithofacies, carbonates, siltstone, and fine-grained sandstone. Characterization of lithofacies is important for evaluation of unconventional resources. Classification of lithofacies from wireline logs is quite different for shale reservoirs than for conventional formations because of organic matters. The latter often causes not only different characteristics in individual wireline logs, but also their relationships. One extreme case is the reversal of the correlation between two wireline logs from conventional formations to shale reservoirs, which is a manifestation of Simpson's paradox. In this chapter, we have proposed a workflow that combines the classical petrophysical charts with statistical methods and neural networks for lithofacies clustering. In particular, PCA is used as a preproposing method, which enables the selection of the appropriate PC or PCs for the lithofacies classification. Moreover, it is possible to rotate PCs to improve the discriminability of lithofacies. These methods enable classifying lithofacies by integrating geological and petrophysical interpretation.

ACKNOWLEDGMENT

The authors thank Schlumberger Ltd for permission to publish this work, and Dr Jinjuan Zhou for reviewing and commenting the manuscript.

REFERENCES

Abdi, H., Williams, L.J., 2010. Principal Component Analysis. In: Statistics & Data Mining Series, vol. 2. John Wiley & Sons, pp. 433–459.

Ajayi, B., et al., 2013. Stimulation design for unconventional resources. Oilfield Review 25 (2), 34–46.

Al Duhailan, M.A., Cumella, S., 2014. In: Niobrara Maturity Goes Up, Resistivity Goes Down; What's Going On? Paper URTeC 1922820, Presented at the URTeC, Denver, Colorado, USA, August 25–27, 2014

Alexander, T., et al., 2011. Shale gas revolution. Oilfield Review 23 (3), 40–57.

Allix, P., et al., 2011. Coaxing oil from shale. Oilfield Review 22 (4), 4–16.

Basilevsky, A., 1994. Statistical Factor Analysis and Related Methods: Theory and Applications. In: Wiley Series in Probability and Mathematical Statistics.

Bhuyan, K., Passey, Q.R., 1994. Clay estimation from GR and neutron-density porosity logs. In: SPWLA 35th Annual Logging Symposium.

Clavier, C., Hoyle, W., Meunier, D., 1971. Quantitative interpretation of thermal neutron decay time logs: Part I. Fundamentals and techniques. Journal of Petroleum Technology 23, 743–755.

Dewan, J.T., 1983. Essentials of Modern Open-Hole Log Interpretation. PennWell Books, Tulsa.

Eslinger, E., Everett, R.V., 2012. Petrophysics in gas shales. In: Breyer, J.A. (Ed.), Shale Reservoirs-Giant Resources for the 21st Century. AAPG Memoir 97, pp. 419–451.

Everitt, B.S., Dunn, G., 2002. Applied Multivariate Data Analysis, second ed. Arnold Publisher, London.

Gamero-Diaz, H., Miller, C.K., Lewis, R., 2013. sCore: a mineralogy based classification scheme for organic mudstones. In: SPE Paper 166284, Presented at the SPE Annual Technical Conference and Exhibition, 30 September–2 October, New Orleans, Louisiana, USA.

Hickey, J.J., Henk, B., 2007. Lithofacies summary of the Mississippian barnett shale, Mitshell 2 T.P. Sims well, Wise County, Texas. AAPG Bulletin 91 (4), 437–443.

Jain, A., Narasimha, M., Flynn, P., 1999. Data clustering: a review. ACM Computing Surveys 31 (3), 264–323.

Jarvie, D.M., 2012. In: Breyer, J.A. (Ed.), Shale Resource Systems for Oil and Gas: Part 2-Shale-Oil Resource Systems. AAPG Memoir 97, pp. 89–119.

Kettenring, J.R., 2006. The practice of clustering analysis. Journal of Classification 23, 3–30.

LeFever, J.A., Martiniuk, C.D., Dancsok, D.F.R., Mahnic, P.A., 1991. Petroleum Potential of the Middle Member, Bakken Formation, Williston Basin, vol. 6. Saskatchewan Geological Society. Special Publication, pp. 74–94, Report #11.

Loucks, R.G., Ruppel, S.C., 2007. Mississipian Barnett shale: lithofacies and depositional setting of a deep-water shale-gas succession in the Fort Worth Basin, Texas. AAPG Bulletin 91 (4), 579–601.

Lu, J., Ruppel, S.C., Rowe, H.D., 2015. Organic matter pores and oil generation in the Tuscaloosa marine shale. AAPG Bulletin 99 (2), 333–357.

Ma, Y.Z., Gomez, E., 2015. Uses and abuses in applying neural networks for predicting reservoir properties. Journal of Petroleum Science and Engineering 133, 66–75.

Ma, Y.Z., 2011. Lithofacies clustering using principal component analysis and neural network: applications to wireline logs. Mathematical Geosciences 43 (4), 401–419.

Ma, Y.Z., Moore, W.R., Gomez, E., Clark, W.J., 2015. Tight gas sandstone reservoirs – Part 1: overview and lithofacies. In: Ma, Y.Z., Holditch, S., Royer, J.J. (Eds.), Handbook of Unconventional Resource. Elsevier.

Ma, Y.Z., et al., 2014a. Identifying hydrocarbon zones in unconventional formations by discerning Simpson's paradox. In: Paper SPE 169496 Presented at the SPE Western and Rocky Regional Conference, April.

Ma, Y.Z., Wang, H., Sitchler, J., et al., 2014b. Mixture decomposition and lithofacies clustering using wireline logs. Journal of Applied Geophysics 102, 10–20. http://dx.doi.org/10.1016/j.jappgeo.2013.12.011.

Ma, Y.Z., Gomez, E., Young, T.L., Cox, D.L., Luneau, B., Iwere, F., 2011. Integrated reservoir modeling of a Pinedale tight-gas reservoir in the Greater Green River Basin, Wyoming. In: Ma, Y.Z., LaPointe, P. (Eds.), Uncertainty Analysis and Reservoir Modeling. AAPG Memoir 96, Tulsa.

Ma, Y.Z., Seto, A., Gomez, E., 2009. Depositional facies analysis and modeling of Judy Creek reef complex of the Late Devonian Swan Hills, Alberta, Canada. AAPG Bulletin 93 (9), 1235–1256.

Ma, Y.Z., Seto, A., Gomez, E., 2008. Frequentist meets spatialist: a marriage made in reservoir characterization and modeling. In: SPE 115836, SPE ATCE, Denver, CO, 12 p.

Passey, Q.R., Bohacs, K.M., Esch, W.L., Klimentidis, R., Sinha, S., 2010. From oil-prone source rock to gas-producing shale reservoir – geologic and petrophysical characterization of unconventional shale-gas reservoirs. In: SPE 131350, CPS/SPE Intl. Oil & Gas Conference & Exhibition, June 8–10, 2010, Beijing, China.

Passey, Q.R., Creaney, S., Kulla, J.B., Moretti, F.J., Stroud, J.D., 1990. A practical model for organic richness from porosity and resistivity logs. AAPG Bulletin 74 (12), 1777–1794.

Pearson, K., 1901. On lines and planes of closest fit to systems of points in space. Philosophical Magazine 2 (11), 559–572.

Peters, K.E., et al., 2007. Source-rock geochemistry of the San Joaquin Basin Province, California. In: Scheirer (Ed.), Petroleum Systems and Geologic Assessment of Oil and Gas in the San Joaquin Basin Province. USGS, California.

Pitman, J.K., Price, L.C., LeFever, J.A., 2001. Diagenesis and Fracture Development in the Bakken Formation, Williston Basin. USGS professional paper 1653, 19 p.

Rogers, S.J., Fang, J.H., Karr, C.L., Stanley, D.A., 1992. Determination of lithology from well logs using a neural network. AAPG Bulletin 76 (5), 731–739.

Schlumberger, 1989. Log Interpretation Principles/Applications, 3rd print, Houston, Texas.

Soeder, D.J., 2011. Petrophysical characterization of the Marcellus and other gas shales. In: Presentation for PTTC/DOE/RSPEA Gas Shales Workshop, September 28, 2011. AAPG Eastern Section Meeting, Arlington, Virginia.

Sondergeld, et al., 2010. Petrophysical considerations in evaluating and producing shale gas resources. In: SPE 131768, presented at the 2010 SPE Unconventional Gas Conference 23–25 February, Pittsburgh, PA, USA.

Steiber, 1970. In: Pulsed Neutron Capture Log Evaluation in the Louisiana Gulf Coast (SPE 296 1): Paper Presented at SPE 45th Annual Fall.

Szabo, N.P., Dobroka, M., 2013. Extending the application of a shale volume estimation formula derived from factor analysis of wireline logging data. Mathematical Geosciences 45, 837–850.

Tang, H., White, C.D., 2008. Multivariate statistical log log-facies classification on a shallow marine reservoir. Journal of Petroleum Science and Engineering 61, 88–93.

Theloy, C., Sonnenberg, S.A., 2013. In: Integrating Geology and Engineering: Implications for Production in the Bakken Play, Williston Basin. Unconventional Resources Tech. Conference, Paper 1596247, 12 pages.

Walker-Milani, M.E., 2011. Outcrop Lithostratigraphy and Petrophysics of the Middle Devonian Marcellus Shale in West Virginia and Adjacent States (M.S. thesis). West Virginia University.

Wang, G., Carr, T.R., 2012. Marcellus shale lithofacies prediction by multiclass neural network classification in the Appalachian basin. Mathematical Geosciences 44, 975–1004.

Wolff, M., Pelissier-Combescure, J., 1982. FACIOLOG - automatic electrofacies determination. In: Proceeding of Society of Professional Well Log Analysts Annual Logging Symposium, Paper FF.

APPENDIX A: A TUTORIAL ON PRINCIPAL COMPONENT ANALYSIS

Principal component analysis (PCA), introduced by Pearson (1901), is an orthogonal transform of correlated variables into a set of linearly uncorrelated variables, i.e., principal components (PCs). Each PC is a linear combination of weighted original variables. The number of PCs is equal to the number of original variables, but the number of meaningful PCs might be fewer depending on the correlations between the original variables.

The transform is defined in such a way that the first PC represents the most variability in the data, in fact, as much as possible, under the condition of the orthogonality between any pair of components. Each succeeding component in turn has the highest variance possible not accounted for by the preceding PCs, under the orthogonality condition. Hence, PCs are uncorrelated between each other.

PCA is mathematically defined as a linear transform that converts the data to a new coordinate system such that the first PC lies on the coordinate that has the largest variance by projection of the data, the second PC lies on the coordinate with second largest variance, and so on. The procedure includes several steps:

1. Calculating the (multivariate) covariance or correlation matrix from the sample data,
2. Computing eigenvalues and eigenvectors of the covariance or correlation matrix, and
3. Generating the PCs; each PC is a linear combination of optimally weighted original variables, such as:

$$P_i = b_{i1}X_1 + b_{i2}X_2 + \cdots + b_{ik}X_k \tag{A.1}$$

where P_i is the ith PC, b_{ik} is the weight (some call regression coefficient) for the variable, X_k. It is often convenient that all the variables, X_k, are standardized to zero mean and one standard deviation.

The weights, b_{ik}, are calculated using covariance or correlation matrix. As the covariance or correlation matrix is symmetric positive definite, it yields an orthogonal basis of eigenvectors, each of which has a nonnegative eigenvalue. These eigenvectors, multiplied by the original inputs (as Eq. (A.1)), correspond to PCs and the eigenvalues are proportional to the variances explained by the PCs. For more mathematical insights of PCA, readers can refer to Basilevsky (1994), Everitt and Dunn (2002), and Abdi and Williams (2010).

PCA is a nonparametric statistical method that provides analytical solutions based on linear algebra; statistical moments, such as mean and covariance, are simply calculated from the data without any assumption. Because of its efficiency in removing redundancy and capability of extracting interpretable information, PCA has a wide range of applications, spanning over nearly all the industries, from computer vision to neuroscience, from medical data analysis to psychology, and from chemical research to seismic data analysis, among others. In fact, PCA is one of the most used multivariate statistical tools; with the explosion of data in modern society, its application is ever increasing.

A simple bivariate example with two petrophysical variables, neutron and density (RHOB), is presented here to illustrate the method. The two PCs from PCA of neutron and RHOB logs are overlain on the neutron–RHOB cross-plots (Fig. A.1(a) and (b)). The first PC (PC1) represents the major axis that describes the maximum variability of the data and the second PC (PC2) represents the minimum axis that describes the remaining variability not accounted for by the first PC. In this example, the major axis, PC1, approximately represents porosity and the minor axis, PC2, approximately represents the lithology. This explains why lithofacies clustering by ANN or statistical clustering methods using

APPENDIX A: A TUTORIAL ON PRINCIPAL COMPONENT ANALYSIS

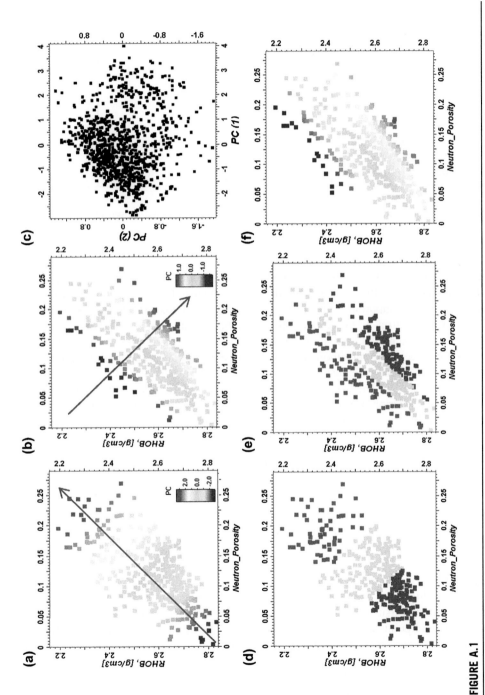

FIGURE A.1

Illustration of two principal components from PCA of neutron and density (RHOB) on neutron–RHOB or their PC1–PC2 cross-plots. (a) Overlay of PC1 on neutron–RHOB cross-plot (arrow indicates the coordinate on which PC1 is defined). (b) Overlay of PC2 on neutron–RHOB cross-plot (arrow indicates the coordinate on which PC2 is defined). (c) PC1–PC2 cross-plot (their correlation is zero). (d) Overlay of lithofacies clustered by ANN using PC1 on neutron–RHOB cross-plot (red, sandstone; green, limestone; and blue, dolostone). (e) Overlay of lithofacies clustered by ANN using PC2 on neutron–RHOB cross-plot. (f) Overlay of a rotated PC2 on neutron–RHOB cross-plot.

Table A.1 Correlation matrix between Pairs of Six Variables: Neutron (NPHI), Density (RHOB), Their PCs (PC1 and PC2), and Two Rotated Component (PC1_rotated and PC2_rotated)

	NPHI	RHOB	PC1	PC2	PC1_rotated	PC2_rotated
NPHI	1					
RHOB	−0.693	1				
PC1	0.920	−0.920	1			
PC2	0.392	0.392	0	1		
PC1_rotated	0.960	−0.867	0.993	0.119	1	
PC2_rotated	0.028	0.737	−0.414	0.910	−0.302	1

PC1 are poor (Fig. A.1(c)), but lithofacies clustered using PC2 are more consistent with the benchmark chart (Fig. A.1(d)). In other cases, major PCs, such as PC1, are important; sometimes, lithofacies classification using PC1 alone is good enough (Ma et al., 2011).

PCs can be rotated to align with a physically more meaningful variable. This can be illustrated with a bivariate example, in which the two original variables are equally weighted in the PCs before rotation. In the neutron–RHOB analysis, neutron and RHOB equally contribute to both PC1 and PC2. However, for example, if neutron is more important than RHOB for porosity determination, PC1 can be rotated to be correlated higher with neutron. Figure A.1(e) shows a rotated component that has an increased correlation to neutron, and decreased correlation to RHOB (Table A.1). Similarly, if RHOB is more important than neutron in determining lithofacies, PC2 can be rotated to reflect that. Figure A.1(f) shows a rotated component from PC2 that has an increase correlation to RHOB and a decreased correlation to neutron (Table A.1). The two rotated components do not have to be orthogonal as shown in this example. The main criterion of rotation is to make a component physically meaningful.

The original data can be reconstructed from the principal components. The general equation of reconstructing the original data can be expressed as the following matrix formulation:

$$\boldsymbol{D} = \boldsymbol{PC}^t \boldsymbol{\sigma} + \boldsymbol{uM}^t \tag{A.2}$$

where $\boldsymbol{D}$ is the reconstructed data matrix of size $k \times n$ (k being the number of variables, n being the number of samples), $\boldsymbol{P}$ is the matrix of principal components of size $q \times n$ (q is the number of PCs, equal or less than k), $\boldsymbol{C}$ is the matrix of correlation coefficients between the PCs and the variables of size $k \times q$, t denotes the matrix transpose, $\boldsymbol{\sigma}$ is the diagonal matrix that contains the standard deviations of the variables of size $k \times k$, $\boldsymbol{u}$ is a unit vector of size n, and $\boldsymbol{M}$ is vector that contains the mean values of the variables of size k.

When the data are highly correlated, a small number of PCs out of all the PCs can reconstruct the data quite well. PCA is highly efficient in removing the redundancy, which is highlighted by the following seismic amplitude versus offset (AVO) example (Fig. A.2(a)). Consider different offsets as variables and common mid points as observations or samples. The first PC (Fig. A.2(b)) from PCA represents more than 99.6% variance explained and can be used to reconstruct the original data. This is done simply by one dimensional (1D) vector multiplication of PC1 (Fig. A.2(b)) and its correlation coefficients to each offset normalized by the respective standard deviation and mean of each offset (Fig. A.2(c)). The result is very much similar to the original AVO data (compare Fig. A.2(a) and (d)).

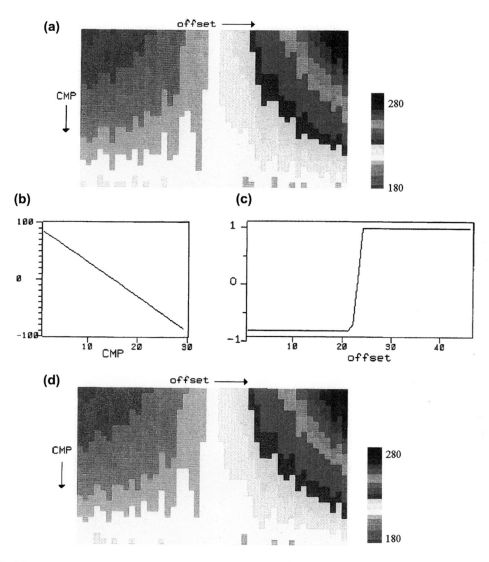

FIGURE A.2

PCA of AVO data and reconstruction of AVO data by using one PC. (a) Original AVO data. (b) PC1 (as a function of common midpoint or CMP). (c) Correlations between PC1 and each offset. (d) The reconstructed AVO data using PC1, i.e., vector multiplication of (b) and (c) normalized by the respective standard deviation and means of each offset (see Eq. (A.2)).

In the AVO example above, q is set to 1 as PC1 represents more than 99% of the information in the data. This explains the surprising reconstructed two-dimensional map (Fig. A.2(d)) simply by vector multiplication of two 1D functions of different size (Fig. A.2(b) and (c)), and normalizations by the standard deviations and means.

CHAPTER 6

THE ROLE OF PORE PROXIMITY IN GOVERNING FLUID PVT BEHAVIOR AND PRODUCED FLUIDS COMPOSITION IN LIQUIDS-RICH SHALE RESERVOIRS

Deepak Devegowda[1], Xinya Xiong[2], Faruk Civan[1], Richard Sigal[3]

Petroleum Engineering, University of Oklahoma, Norman, OK, USA[1]; Gas Technology Institute, Chicago, IL, USA[2]; Independent Consultant, Las Vegas, Nevada, USA[3]

6.1 INTRODUCTION

The significance of organic-rich shales to domestic energy policy and energy security is reflected in net reductions in oil imports and near-term forecasts of United States energy independence (US EIA, 2013). This growth in development activity in shales is largely a testament to technological advances in drilling and completions technology that enable contact with large productive reservoir volumes through horizontal well laterals and multistage hydraulic fracturing. However, because organic-rich shales are characterized by nanoscale pore throats and pore size distributions (Curtis et al., 2011) that are challenging to probe with conventional tools, the recovery mechanisms in both gas and liquids-rich shale reservoirs continue to be poorly understood.

Recent advances in modeling gas transport in shale nanopores have underscored the need for nonDarcy flow models that account for the effects of pore size on fluid transport (Michel et al., 2011; Fathi et al., 2012; Swami et al., 2012; Civan et al., 2013). Although there appears to be a lack of consensus regarding the choice of model, these advances indicate that currently available simulation tools may be inadequate in describing gas transport in shale nanopores. These complexities become even more challenging to quantify when considering nanoscale pore size distributions (Curtis et al., 2010; Michel et al., 2011), the distribution and connectivity of organics in shales, the role of adsorption (Sakhaee-Pour and Bryant, 2012; Xiong et al., 2013), the heterogeneous wettability (Sigal, 2013) and the presence of microcracks and natural fracture systems (Tinni et al., 2012).

Although our understanding of single-component gas transport in shale nanopores is limited, the related topic of gas storage mechanisms has also received considerable attention because of its relevance to quantifying reserves, understanding well productivity and estimating recovery factors.

Ambrose (2011) demonstrate the need to account for adsorbed gas volumes when estimating organic pore volume available for free gas storage and show that gas-in-place estimates may be overestimated by as much as 25% for the smallest of organic pores when the effects of adsorption is neglected. Ambrose (2011) also propose a model for multicomponent hydrocarbon adsorption and its impact on gas reserves calculations while Sigal (2013) extends these findings to develop a model for quantifying total storage in organic shale nanopores as a function of pore pressure including the effects of pore compressibility and the influence of pressure on free and adsorbed storage.

These advances in describing single-component hydrocarbon storage and recovery mechanisms have largely been a result of recognizing the relevance of pore proximity effects on confined fluids. Fluid intermolecular forces in macroscale pores such as those present in conventional reservoirs are known to be the dominant interactions governing fluid behavior (Peng and Robinson, 1975; McCain, 1990). This is because an appreciable percentage of fluid molecules are outside of the range of influence of the pore walls. Consequently, Pressure-Volume-Temperature (PVT) properties for compositional reservoir simulation may be derived from laboratory-based experiments on representative fluid samples. On the other hand, at the range of pore sizes encountered in shales, the number of fluid molecules may be somewhat limited. For example, a pore of diameter 2 nm is only five times larger than a methane molecule. This is a crucial difference between macroscale pores and shale nanopores. Within shale nanopores, a large percentage of fluid molecules may be in considerable proximity to the pore wall where pore wall interactions with the fluid may be as significant as inter-molecular interactions within the fluid thereby governing fluid densities and storage (Ambrose, 2011; Michel et al., 2011; Didar, 2012), multicomponent phase behavior (Sapmanee, 2011; Devegowda et al., 2012; Alharthy et al., 2013; Zhang et al., 2013; Travalloni et al., 2014) and multicomponent hydrocarbon transport (Xiong et al., 2013). Because there are no known approaches to apply some of these developments to the complex pore networks of shales, operators often resort to the use of conventional simulation tools that lack the necessary physics to describe pore proximity effects on fluid behavior in both organic and inorganic pore systems.

Consequently, as exploration and development activity continues to grow in liquids-rich shale plays, the examination of multicomponent and multiphase transport and recovery mechanisms in extremely low permeability nanoporous media is merited.

Our work encompasses a critical review of the relevant issues and exploratory modeling studies of phase behavior and fluid property variations affected by pore proximity for both the organic and inorganic pore systems that are commonly found in shales and their impact on well productivity and fluid transport. These studies are based on information gathered from published literature on various issues of relevance and, therefore, the results of this exploratory work should be interpreted for instructional purposes. Further refinement of these advances in describing the influence of pore wall proximity on confined fluids and the corresponding modifications to simulator design is necessary to model fluid recovery in shales with complex pore geometries and a broad distribution of pore sizes and pore throats.

This chapter is organized as follows: we first present a literature review of some of the recent advances in quantifying pore proximity effects on multicomponent fluid phase behavior. We then follow this with modeling studies to quantify differences in shale oil well performance and drainage areas when comparing confined fluid behavior with classical compositional reservoir simulation. Currently, commercial simulators do not readily allow fluid properties to vary by pore size so the class of models that can be investigated is rather limited; therefore, we then discuss recent developments to extend the modeling of confinement effects to complex pore size distributions through the use of equations-of-state and finally discuss some of the issues that need further refinement in order to provide a valid and accurate model for compositional simulation in organic-rich shales for performance predictions, reserves estimates and quantifying recovery factors.

We then develop extensions to single-component gas transport to account for multicomponent fluids including both the gas and liquid phases. In this section, we describe multicomponent adsorption and its role in controlling phase behavior and storage of reservoir fluids in organic nanopores of shales and its impact on produced fluid compositions.

6.2 PORE CONFINEMENT EFFECTS ON FLUID PROPERTIES

Pore throat diameters in typical shale formations have been shown to vary between 0.5 and 100 nm (Ambrose et al., 2010) while the chain diameter of straight-chain hydrocarbons is in the range of four to six Ångström (Mitariten, 2005) (10 Å = 1 nm) and consequently in shales, the sizes of fluid molecules and pores are comparable. However, the understanding of the impact of pore confinement effects on shale recovery mechanisms continues to be limited. Considerable knowledge gaps continue to exist although some of these effects have been known for several years. For example, several studies have documented evidence of alteration of properties, such as the reduced freezing point of clay-bound water in shales in the Arctic permafrost (Civan and Sliepcevich, 1985; Civan, 2010). Chemical engineering literature also contain several examples of significant deviations in fluid properties relative to their bulk behavior in nanoscale capillary tubes (Morishige et al., 1997; Beskok and Karniadakis, 1999 Kalluri et al., 2011, Hu et al., 2014). In general, however, the intermolecular interactions and the interactions between the molecules and the pore surface have been shown to alter the fluid properties because the pore surface available per unit volume increases as the pore size decreases. The alterations are also a consequence of the limited number of molecules present in these pores and the increasing importance of interaction between the fluid molecules (van der Waals interactions) as well as between the fluid molecules and the pore surfaces. Consequently, the thermophysical and dynamic properties of confined fluids may deviate significantly from those measured in large PVT cells and in sufficiently large pores. Some developments have focused on addressing variations in alkane critical properties, adsorption, solubility in water, viscosity, and interfacial effects in carbon nanotubes (Kanda et al., 2004; Chen et al., 2008; Campos et al., 2009; Singh et al., 2009, Singh and Singh, 2011; Moore et al., 2010 and Travalloni et al., 2010, 2014). Accurate estimation of some of these properties is crucial for accurate simulation of production from shale gas and liquids-rich shale reservoirs.

6.2.1 EFFECTS OF CONFINEMENT ON ALKANE CRITICAL PROPERTIES

Because the length scale involved to study fluid properties in nanopores prohibits the use of physical laboratory-based measurements, virtual experiments based on molecular dynamics simulations have increasingly become the preferred method to understand fluid configuration and behavior in pores with different surface characteristics. Molecular dynamics simulations rely on solving Newton's laws of motion to capture the behavior of fluid molecules under the influence of prespecified fluid–fluid and fluid–pore wall intermolecular interactions.

Initial work focused on addressing adsorption in synthetic nanotubes experimentally verified that the gas–liquid critical temperature of many chemical compounds decreases relative to the values for the bulk fluid and that these deviations showed a strong pore size dependency (Morishige et al., 1997; Morishige and Shikimi, 1998). This initial work was followed by attempts by Zarragoicoechea and Kuz (2004) to correlate these deviations of critical temperature using a generalized van der Waals equation of state theory for a Lennard–Jones fluid confined in nanopores with nonadsorbing pore walls.

Hamada et al. (2007) studied the behavior and thermodynamic properties of confined Lennard–Jones (LJ) particles in slit and cylindrical pore systems by using grand canonical Monte Carlo numerical simulations. They indicate that fluid phase behavior, thermodynamic properties, and the interfacial tension between the fluid and pore surface are affected by the pore size of nanoporous media. They attribute this to increases in the potential energy, interaction between molecules, and interaction between pore surface and molecules.

Singh et al. (2009) determined distinctive deviations in the critical properties for different hydrocarbon compounds. They investigated the behavior of methane (C1), n-butane (C4), and n-octane (C8) inside nanoscale slits with widths between 0.8 and 5 nm by means of the grand canonical transition matrix Monte Carlo numerical simulator together with a modified Buckingham exponential intermolecular potential. For grand canonical simulations volume, chemical potential (μ), and temperature are held constant and the particle number (density) and energy fluctuate. They determined that the critical temperatures exponentially decrease when the pore size is reduced and theoretically approach a value of zero as the pore diameter diminishes to zero as shown in Fig. 6.1b. However, they observed that the critical pressure of n-butane and n-octane tends to first increase to values above the corresponding value of the bulk fluid as the pore size decreases and subsequently decreases to values below the original bulk value with further reduction in the pore size as seen in Fig. 6.1a. Moreover, they also suggested that the deviations of critical properties for methane, n-butane, and n-octane are also different for different pore surfaces, such as mica and graphite. Because shales are characterized

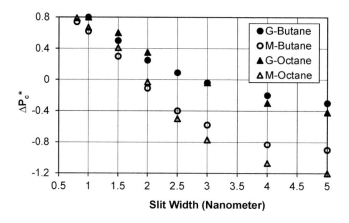

FIGURE 6.1(a)

Relative deviation of critical pressure of n-butane and n-octane in graphite (G) and mica (M) slit pores.

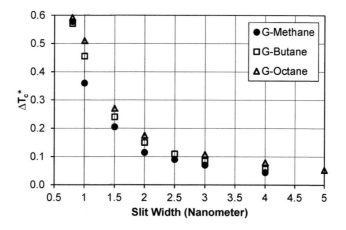

FIGURE 6.1(b)

Relative deviation of critical temperature of methane, *n*-butane and *n*-octane in graphite slit pores.

Modified from Singh et al. (2009).

by organic and inorganic pore systems for which graphite and mica form reasonable proxies in terms of surface chemistry, the work of Singh et al. (2009) is particularly significant for describing fluid behavior in shale nanopores. The relative shift of the critical temperature and pressure of Singh et al. (2009) are defined below as:

$$\Delta T_c = \frac{T_{cb} - T_{cp}}{T_{cb}} \tag{6.1}$$

$$\Delta P_c = \frac{P_{cb} - P_{cp}}{P_{cp}} \tag{6.2}$$

where P_{cp} and T_{cp} are the critical pressure and temperature in confined pores and P_{cb} and T_{cb} are the corresponding values in bulk.

Singh and Singh (2011) follow Singh et al. (2009) and also document critical temperature and pressure variations in adsorbing and nonadsorbing slit and cylindrical pores as a function of pore size. An example of these variations is shown in Fig. 6.2 for ethane. Critical temperature was found to vary monotonically with pore size while the critical pressure variations were nonmonotonic depending on whether the pore surface was adsorbing or nonadsorbing. Figure 6.2 shows these differences for hard (nonadsorbing) and attractive (adsorbing) pores.

Sapmanee (2011), Devegowda et al. (2012), Alharthy et al. (2013), and Zhang et al. (2013) document the impact of these deviations for liquids-rich shale recovery for several different hydrocarbon mixtures. These deviations will be discussed in a later section.

Consequently, with the variations described above for both the critical pressures and temperatures, several of the most commonly employed equations-of-state (EOS) such as the Peng-Robinson EOS (1975) may be employed to quantify PVT behavior of alkanes or alkane mixtures in shale nanopores of specified size. Some of these case studies are described next.

178 CHAPTER 6 THE ROLE OF PORE PROXIMITY

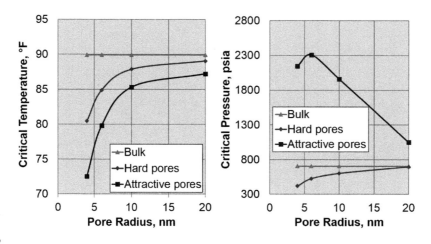

FIGURE 6.2

Variations in critical temperature and critical pressure for ethane in adsorbing (hard) and nonadsorbing (attractive) pores.

Modified from Singh and Singh (2011).

6.2.2 EFFECT OF CONFINEMENT ON PHASE BEHAVIOR OF MULTICOMPONENT MIXTURES

We demonstrate the effect of confinement on the phase behavior of confined multicomponent hydrocarbon mixtures by utilizing the critical temperature and pressure values provided in Table 6.1 for different pore radii. These values were obtained from Singh and Singh (2011) and also reported in Xiong et al. (2013) and are provided for nonadsorbing pores and adsorbing pores. In the context of shales, adsorbing pores correspond to organic kerogen pores while the nonadsorbing pores refer to the inorganic pore systems. However, we would like to reiterate that these numbers are developed for

Table 6.1 Critical Temperatures of Bulk and Confined Fluids (Zhang et al., 2013)

		Critical Temperature T_c, °F							
Component		C1	C2	C3	NC4	NC5	C6	C7+	C10
Bulk		−116.7	89.9	206.1	305.6	385.8	453.6	729.3	652.0
Nonadsorbing pore	4 nm	−104.4	80.5	184.4	273.5	345.3	405.9	652.7	583.5
	6 nm	−110.2	84.9	194.6	288.6	364.3	428.3	688.6	615.6
	10 nm	−114.0	87.9	201.3	298.6	377.0	443.2	712.6	637.1
	20 nm	−115.5	89.0	204.1	302.7	382.1	449.2	722.3	645.7
Adsorbing pore	4 nm	−94.1	72.5	166.2	246.5	311.2	365.9	588.3	525.9
	6 nm	−103.5	79.8	182.9	271.2	342.4	402.6	647.3	578.7
	10 nm	−110.7	85.3	195.5	290.0	366.0	430.4	692.0	618.6
	20 nm	−113.2	87.2	199.9	296.5	374.2	440.0	707.5	632.4

6.2 PORE CONFINEMENT EFFECTS ON FLUID PROPERTIES

cylindrical pore geometries with either strongly adsorbing or nonadsorbing pore walls. In reality, shale pore geometries and pore surface chemistry remains poorly understood and are highly complex functions of mineralogy and thermal maturity; therefore, the numbers provided here represent upper and lower bounds for the behavior of hydrocarbon fluid behavior in nanoporous rocks.

The case studies focus on two different oils with the respective compositions presented in Table 6.3. One of the oils is a synthetic mixture while the other mixture was obtained from McCain (1990). Interactions between the pore wall and the fluid molecules cause significant changes to the phase behavior of the reservoir fluids in ultra-tight formations characterized by nanoscale pore systems. This can easily be illustrated by generating the modified phase envelopes using any appropriate EOS with the critical point data for each hydrocarbon species in a commercially available PVT package. In this chapter, we input the critical properties provided in Tables 6.1 and 6.2 in to a

Table 6.2 Critical Pressures of Bulk and Confined Fluids (Zhang et al., 2013)

Component		Critical Pressure P_c, psia							
		C1	C2	C3	NC4	NC5	C6	C7+	C10
Bulk		666.4	706.5	616.4	550.6	488.6	483.0	318.4	305.2
Nonadsorbing pore	4 nm	393.7	417.4	364.2	325.3	288.7	285.4	188.1	180.3
	6 nm	495.4	525.2	458.2	409.3	363.2	359.1	236.7	226.9
	10 nm	567.1	601.2	524.6	468.6	415.8	411.0	271.0	259.7
	20 nm	657.8	697.3	608.4	543.5	482.3	476.7	314.3	301.2
Adsorbing pore	4 nm	2025.0	2146.8	1873.1	1673.1	1484.7	1467.7	967.5	927.4
	6 nm	2175.6	2306.5	2012.4	1797.5	1595.1	1576.9	1039.5	996.4
	10 nm	1849.5	1960.7	1710.7	1528.1	1356.0	1340.5	883.7	847.0
	20 nm	989.9	1049.5	915.7	817.9	725.8	717.5	473.0	453.4

Table 6.3 Fluid Composition for Synthetic Oil and Black Oil Studies

	Mole Fraction, %		
Component	Synthetic Oil	Black Oil (Modified from McCain, 1990)	Molecular Weight, lb/mole
Reservoir temperature, °F	250	150	—
C1	53.01	37.54	16.043
C2	—	9.67	30.070
C3	—	6.95	44.097
NC4	10.55	5.37	58.123
NC5	—	2.85	72.150
C6	—	4.33	86.177
C7+	—	33.29	218.000
C10	36.44	—	142.285

commercially available PVT modeling package (CMG WinPropTM, 2008). We examined the phase diagrams for synthetic oil and black oil in both nonadsorbing and adsorbing cylindrical pores as shown in Figs 6.3 and 6.4. The figures demonstrate that the phase envelopes predicted using shifted critical properties deviate significantly from the expected behavior derived using bulk properties. A noteworthy observation is that for larger pores (20 nm) the phase diagrams approach that of the bulk fluid

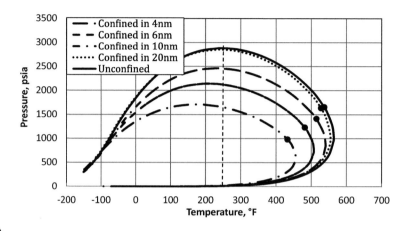

FIGURE 6.3(a)

Phase diagram for the synthetic oil of Table 6.2 in nonadsorbing pores of various diameters. The reservoir temperature is represented by a vertical dashed line.

Modified from Xiong et al. (2013).

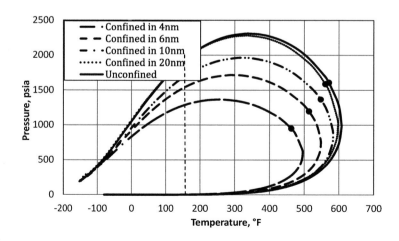

FIGURE 6.3(b)

Phase diagram for the black oil of Table 6.2 in nonadsorbing pores of various diameters. The reservoir temperature is represented by a vertical dashed line.

Modified from Xiong et al. (2013).

6.2 PORE CONFINEMENT EFFECTS ON FLUID PROPERTIES

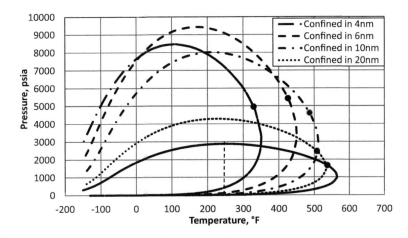

FIGURE 6.4(a)

Phase diagram for the synthetic oil of Table 6.2 in adsorbing pores of various diameters. The reservoir temperature is represented by a vertical dashed line.

Modified from Xiong et al. (2013).

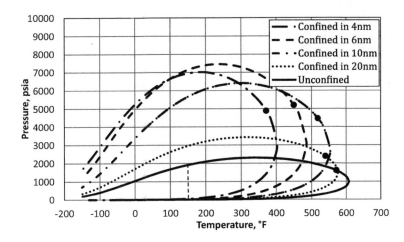

FIGURE 6.4(b)

Phase diagram for the black oil of Table 6.2 in adsorbing pores of various diameters. The reservoir temperature is represented by a vertical dashed line.

Modified from Xiong et al. (2013).

which is to be expected; however it illustrates that the phase behavior can be dramatically different if the shale is characterized by a very narrow and small pore size distribution in the sub20 nm range.

Additionally, the figures also indicate severe bubble-point suppression for the inorganic, nonadsorbing pores across a wide range of temperatures delaying the transition to the two-phase region.

This has also been reported in Sapmanee (2011), Devegowda et al. (2012), Firincioglu (2013) and Alharthy et al. (2013). This implies that fluid transport in inorganic pores may be characterized by single-phase liquid flow for substantially longer periods of time prior to the onset of two-phase oil and gas flows with the accompanying multiphase relative permeability effects that may hinder the flow of oil. In other words, if the rock is characterized by a large percentage of inorganic pores, pore proximity effects are likely to be beneficial. Examples of such plays include the Bakken complex where most of the production is from the dolomitic Middle Bakken formation characterized largely by sub100 nm pores (Ramakrishna et al., 2010).

In the organic adsorbing pores there are significant differences between the phase behaviors for a bulk fluid in comparison to the confined fluids. However in adsorbing pores, because critical pressures tend to be higher than the corresponding values in the bulk, there is an elevation of the bubble-point pressure. The elevation of the bubble-point pressure implies that at high enough pressures, some pores may be experiencing two-phase flow which is generally less efficient than single-phase transport. Phase trapping due to the capillary pressure effects and wettability of the pore surface can further reduce the ability of the organic pores to transport the heavier and intermediate hydrocarbon components. However, adsorption effects can potentially lead to slower declines in oil and gas rates because of the increasing contribution of desorbed hydrocarbon to the in situ fluid mixture as reservoir pressures decline. Shales with a significant proportion of porosity confined to the organic material may be characterized by such behavior. Examples of such plays include the Eagleford, Barnett, Woodford, and Marcellus shale plays (Sondhi, 2012; Curtis et al., 2011).

The conclusions of Fig. 6.4 are also reinforced in Fig. 6.5 where the liquid mole fraction is plotted as a function of pore pressure for different pore sizes. Fluids that exist as a liquid at a given temperature and pore pressure in nonadsorbing pores (Fig. 6.3) become two phases at the same conditions in organic adsorbing pores due to attractive nature of the wall potentials. The heavier components of the fluid mixture are likely to take the form of an adsorptive film coating the pore surface with an adsorbed phase density close to liquid densities (Singh and Singh, 2011). The remaining fluid with a lighter composition is likely to be in the gas phase and therefore reservoir fluids flow is primarily two-phase flow in organic pores.

Another consequence of the results presented above is that at any given pressure and temperature conditions, the fluid mixture may exist as a single-phase liquid in a few pores or as two phase oil and

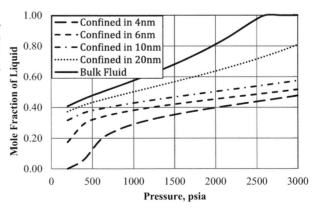

FIGURE 6.5(a)

Liquid mole fraction versus pore pressure for the synthetic oil in Table 6.2 in the bulk and confined in adsorbing, organic pores of various diameters at reservoir temperature of 250 °F.

Modified from Xiong et al. (2013).

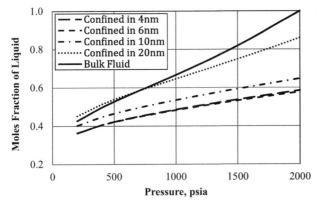

FIGURE 6.5(b)

Liquid mole fraction versus pore pressure for the black oil in Table 6.2 in the bulk and confined in adsorbing, organic pores of various diameters at reservoir temperature of 250 °F.

Modified from Xiong et al. (2013).

gas mixture with varying phase saturations in other pores depending on the pore size. This adds another layer of complexity in describing multiphase storage and transport in shale nanopores because the concept of relative permeability that is derived for a specific saturation may not be entirely appropriate. Pores rarely exist as isolated volumes and instead form complex interconnecting networks with several different pore throats and pore bodies in communication with each other. In order to describe fluid behavior in realistic porous media, an accurate description of fluid storage specific to a given pore geometry becomes necessary. The variation of the critical properties of fluids has also been investigated inside systems characterized by a distribution of pore sizes. The Monte Carlo numerical simulation of Ortiz et al. (2005) indicates that the effective deviation of critical temperature is lower for their selected distribution of nanotubes than in a uniformly sized pore system. Because shales may be characterized by distributions of pore sizes from 0.5 to 100 nm, due consideration must be given to the effects of pore proximity in such systems to accurately model fluid behavior.

However, the description of effective fluid phase behavior for relatively larger reservoir or rock volumes continues to remain challenging. This merits further research to extend formulations derived for a single pore to distributions of pore sizes that characterize shales. At this time, however, the connectivity of pores in the organics and inorganics in shales is poorly understood (Curtis et al., 2010). It is likely that the complex interplay between pore geometry and fluid behavior under confinement will continue to be an area of active research in the near future.

6.3 MULTICOMPONENT FLUID TRANSPORT IN NANOPORES

The pore proximity and pore wall interactions with fluids not only impact liquid and gas saturations depending on pore sizes and pore surface chemistry but also tend to create conditions for in situ fractionation of different hydrocarbon species. This is because of the impact of slippage at the pore walls. Because the slippage of different gas molecules occurs at different velocities in nanopores, pore proximity may also contribute to varying produced fluid compositions with time. Xiong et al. (2012) proposed an apparent permeability correction in organic and inorganic pores for single components. The apparent permeability varies with different hydrocarbon species. For the liquids-rich shale reservoirs, the apparent permeability correction for multicomponent mixtures should be made by using a

Table 6.4 Lennard–Jones Potential Parameters

Component	C1	C2	C3	NC4	NC5	C6	C7	C10
LJ parameter (Å)	4	4.8	5.5	6.1	6.7	7.3	7.9	9.6

mean free path for a gas molecule in mixture. Freeman et al. (2009) proposed the mean free path λ_i of a single gas species, i in a multicomponent gas mixture using the ideal gas law as:

$$\lambda_i = \frac{v_i}{\sum_{j=1}^{n} \pi d_i d_j \sqrt{v_i^2 + v_j^2} \frac{N_A P_i}{RT}} \tag{6.3}$$

where N_A is the Avogadro constant, R is the gas constant, T is the temperature, d_i and d_j are the Lennard–Jones potential parameters for molecules i and j, P_i is the pressure and v_i and v_j are the average velocities of the components, i and j in the gas mixture. The Lennard–Jones potential parameters are defined where the potential is zero given in Table 6.4 at the end of this section. In pores with liquid drop-out, the hydraulic radius will change with the volume available to gas flow. The effective hydraulic radius (R_{eff}) in two phase flow region is modified using the gas phase saturation S_g as:

$$R_{eff} = R\sqrt{S_g} \tag{6.4}$$

Each species has a Knudsen number Xiong et al. (2012). corresponding to its own mean free path and a corresponding permeability correction factor. Because of the difference in the permeability correction factors for different hydrocarbon species, the produced fluid composition is likely to vary with time because of the fractioning effect of different species. Thus, we define the instantaneous flowing gas composition $ý_l$ as:

$$ý_l = y_i \frac{f(Kn_i)}{\sum_{i=1}^{n} f(Kn_i)} \tag{6.5}$$

in which y_i is the static composition of component i in the vapor phase. The molar flux of the multicomponent mixture $\dot{n}$ is the summation of molar flux of each component:

$$\dot{n} = \sum_{i=1}^{n} \frac{n_{vi} K_\infty f(Kn_i)}{\mu} \nabla P \tag{6.6}$$

n_{vi} is the number of moles of component i in the vapor phase, K_∞ is the intrinsic permeability of the rock, $f(Kn_i)$ is the permeability correction or multiplier corresponding to component i in the fluid mixture, μ is the viscosity of the gas mixture and ∇P is the pressure gradient. The apparent permeability for the multicomponent gas mixture K_{amix} is therefore given in the formula below:

$$K_{amix} = K_\infty \sum_{i=1}^{n} y_i f(Kn_i) \tag{6.7}$$

Additionally we assume that the gas permeability varies linearly with gas saturation. This is a strong assumption and is generally not merited for porous media; however, the work outlined here is for instructional purposes and until a more satisfactory quantitative model of gas flow is developed, then the effective gas permeability K_g and the effective oil permeability are:

$$K_g = K_{amix} S_g$$
$$K_o = K_{amix} S_o$$
(6.8)

S_o is the oil phase saturation.

6.3.1 COMPOSITIONAL VARIATIONS IN PRODUCED FLUIDS

6.3.1.1 Synthetic Oil Case Study

The case study involves production from an oil reservoir with the composition of the synthetic oil provided in Table 6.2 with a reservoir temperature of 250 °F from an initial reservoir pressure of 3000 psia to 200 psia. The total number of moles of each species in the vapor phase is modeled using conventional PVT model for bulk fluids. For fluids confined in nanopores, the conventional approach is modified using the shifted critical temperatures and pressures appropriate to hard (inorganic) and attractive (organic) pores. The corresponding compositional ariations are shown in Figs 6.6–6.8 for methane, n-butane, and decane, respectively. Figures 6.6(a), 6.7(a) and 6.8(a) plot the static fluid composition as predicted by the PVT model appropriate for the chosen pore size in hard (inorganic) pores while Figs 6.6(c), 6.7(c) and 6.8(c) give the corresponding trends for organic pores. Figures 6.6, 6.7 and 6.8(b) plot flowing fluid composition predicted by considering slip flow effects for each component and using Eq. (6.5) for inorganic hard pores and Figs 6.6, 6.7, and 6.8(d) provide the same for organic attractive pores.

The figures illustrate that pore proximity effects tend to impact the flowing fluid composition substantially across a wide range of pressures creating a compositional fraction effect. This is because under pore proximity, different hydrocarbon species may move at different velocities as predicted by Eq. (6.3). In addition, the flowing fluid composition is significantly different from that predicted for the bulk fluid. However, the behavior of the fluid approaches that of the bulk PVT behavior as pore sizes approach 20 nm. In effect, this implies that fluid phase behavior and flowing fluid compositions may be substantially different from those predicted by conventional PVT models if the smallest of pores contribute substantially to the total pore volume and the pore connectivity.

In general, lighter components tend to have a larger mean free path and a larger Knudsen correction and therefore move more rapidly. If the predominant locations for hydrocarbon storage are the organic pore systems and if a substantial percentage of the total pore volume is present in the smallest sub20 nm pores, then the flowing vapor compositions are likely to be substantially lighter than the compositions predicted by conventional PVT models and may ultimately contribute to larger gas-oil ratios (GORs) at the wellhead.

By simply employing PVT data obtained in the lab, we may be overlooking the effect of pore confinement and fractioning effect between dry gas and liquids. This may have severe implications when estimating reserves and quantifying well productivity.

The permeability multipliers for the gas and liquid phases are shown in Fig. 6.9. The permeability multiplier for the gas phase may exceed values of one depending on the pore size and pore pressure due to slip flow effects in shale nanopores. For example, the enhancement of gas permeability may be as high as 15 times the absolute permeability in 4 nm pore at low oil saturations. As pressures drop below the saturation pressure, oil permeability may be severely restricted while the flow capacity of gas increases dramatically with increasing vapor saturations. In adsorptive pores, the strength of the wall

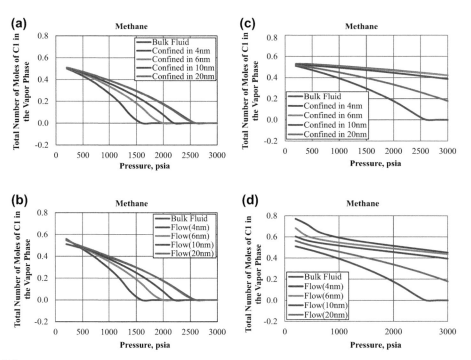

FIGURE 6.6

Total number of moles of methane in the vapor phase of the synthetic oil. (a) Static fluid in inorganic pores, (b) flowing fluid in inorganic pores, (c) static fluid in organic pores, (d) flowing fluid in organic pores.

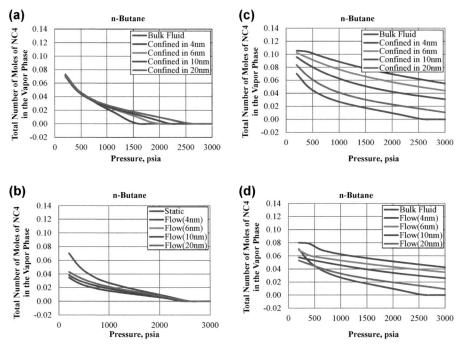

FIGURE 6.7

Total number of moles of n-butane in the vapor phase of synthetic oil. (a) Static fluid in inorganic pores, (b) flowing fluid in inorganic pores, (c) static fluid in organic pores, (d) flowing fluid in organic pores.

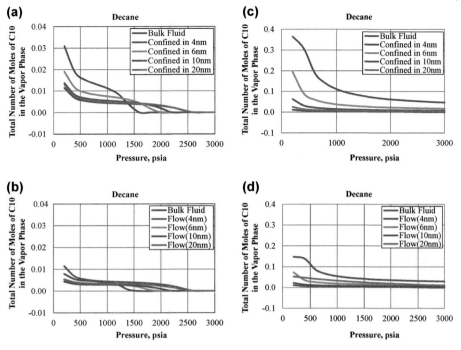

FIGURE 6.8

Total number of moles of decane in the vapor phase of synthetic oil. (a) Static fluid in inorganic pores, (b) flowing fluid in inorganic pores, (c) static fluid in organic pores, (d) flowing fluid in organic pores.

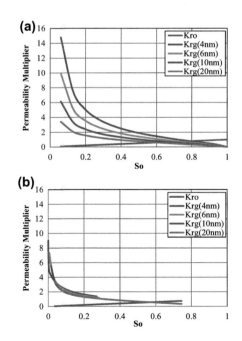

FIGURE 6.9

Permeability multipliers for gas and oil phases for the synthetic oil case study (a) in inorganic pores, (b) in organic pores.

188 CHAPTER 6 THE ROLE OF PORE PROXIMITY

interactions inhibits flow and results in lower values for the permeability multipliers across a wide range of saturations.

6.3.1.2 Black Oil Case Study

A similar numerical experiment was conducted with the black oil specified earlier in Table 6.2 with a reservoir temperature of 150 °F and reservoir pressures varying from 2000 psia to 200 psia. The results are shown in Figs 6.10–6.12 and shows similar trends seen with the synthetic fluid studied in the previous section. The permeability multipliers for the gas and liquid phases in organic and inorganic pores are shown in Fig. 6.13.

6.4 IMPLICATIONS OF PORE PROXIMITY ON WELL DRAINAGE AREAS AND PRODUCTIVITY

The previous section was devoted to describing the effects of pore proximity on fluid phase behavior using modified critical properties appropriate to the selected pore size. Although one of the key advantages of the approach adopted here is easy modification of simulation models without need for access to simulator code, we are currently restricted to assessing pore proximity effects in models with a single pore diameter. Extensions to realistic porous media with a distribution of pore sizes require

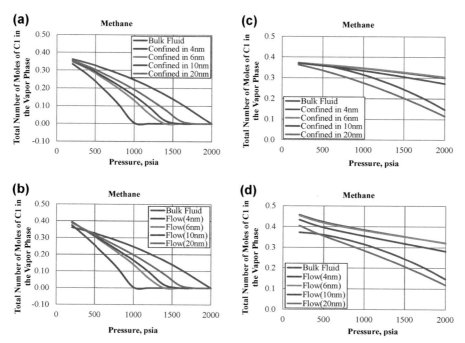

FIGURE 6.10

Total number of moles of methane in the vapor phase of black oil. (a) Static fluid in inorganic pores, (b) flowing fluid in inorganic pores, (c) static fluid in organic pores, (d) flowing fluid in organic pores.

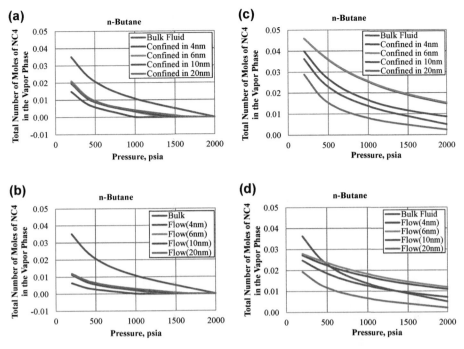

FIGURE 6.11

Total number of moles of *n*-butane in the vapor phase of black oil. (a) Static fluid in inorganic pores, (b) flowing fluid in inorganic pores, (c) static fluid in organic pores, (d) flowing fluid in organic pores.

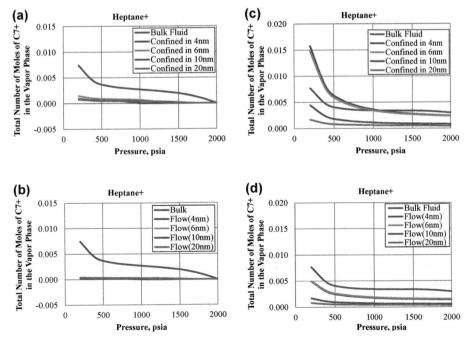

FIGURE 6.12

Total number of moles of heptane+ in the vapor phase of black oil. (a) Static fluid in inorganic pores, (b) flowing fluid in inorganic pores, (c) static fluid in organic pores, (d) flowing fluid in organic pores.

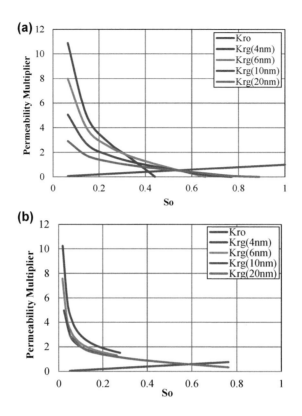

FIGURE 6.13
Permeability multipliers for gas and oil for the black oil case study (a) in inorganic pores, (b) in organic pores.

further investigation and are not the focus of this study. At the current time, there are no known approaches to address this issue.

However, in this section we demonstrate the impact of these variations on well performance for the black oil described in Table 6.2 (McCain, 1990). Our subsequent modeling results using the Peng-Robinson EOS (1975) reveal some very interesting conclusions of practical significance for liquids-rich shale reservoirs. However, again we would like to point out that these results are exploratory in nature and neglect the impact of shale pore geometries and therefore should only be utilized for instructional purposes.

6.4.1 DESCRIPTION OF THE NUMERICAL SIMULATION MODEL

The reservoir model utilized for this study comprises a vertical well intersected by a biwing hydraulic fracture in a shale reservoir of uniform thickness. The synthetic reservoir model is characterized by the following attributes (Zhang et al., 2013):

1. The reservoir is homogeneous with constant porosity, ϕ, of 6% and constant thickness of 150 ft. Initial reservoir pressure is 6000 psia and the well is operated at a constant bottom hole pressure (BHP) of 500 psia.

2. The internal flow structure of the reservoir rock is assumed to be a bundle of straight capillary tubes with a uniform pore throat diameter of specified diameter. The effect of the number of capillary tubes is neglected.
3. The reservoir model assumes that only the inorganic pores are connected while the organics volumes are restricted to hydrocarbon storage and therefore fluid properties appropriate to confinement in inorganic pores are considered.
4. The absolute permeability of the reservoir is quantified using $K(nd) = \phi r^2/8$ where the radius of the capillary tube is r. The reservoir with pore throat diameter of 4 and 8 nm are characterized by permeability ϕ of 30 nD and 120 nD.

Well performance in the synthetic reservoir with the specified pore size is compared to an identical reservoir with the same absolute permeability but using the bulk fluid properties instead. A comparison of the oil and gas rates for the wells are shown in Fig. 6.14 below. In general, they indicate that there are sufficiently large differences in well performance due to differences in the fluid properties when considering the effects of confinement.

This is also reflected in Fig. 6.15 showing the distribution of pressure transients in the reservoir at the end of four years of production. The results from Fig. 6.15 indicate that for the chosen pore radii, the wells tend to drain a larger area in comparison with the case where bulk fluid properties are used for modeling purposes.

The results shown in Figs 6.14 and 6.15 have several significant implications. It is likely that predictions of well performance based on bulk fluid properties may be compromised. Ignoring the effects of fluid confinement are also likely to impact well spacing considerations because of the different predicted drainage areas when considering fluid properties under nanoporous confinement. Therefore one of the key findings of this study is that careful consideration needs to be made

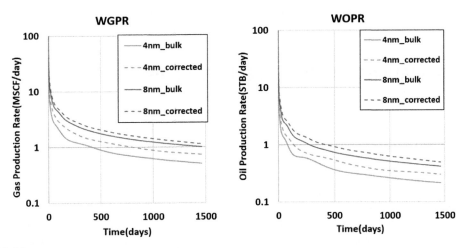

FIGURE 6.14

Gas rates on the left and oil rates on the right showing differences in well performance when using fluid critical properties appropriate to the pore size for nanoporous confinement.

Modified from Zhang et al. (2013).

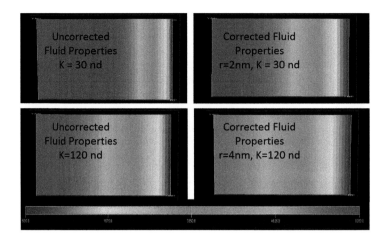

FIGURE 6.15

The pressure transient at the end of four years of production. The figures on the left show the results when using bulk fluid properties while the figures on the right utilize the fluid critical properties appropriate to the specified pore size. The hydraulic fracture is located to the right.

Modified from Zhang et al. (2013).

of pore proximity effects when quantifying reserves per well, estimating well spacing and predictions of condensate dropout for gas–condensate plays (Sapmanee, 2011). It is likely that when a significant percentage of the pore volume is contained in the smallest of pores, pore proximity effects are likely to dominate and predictions based on bulk fluid properties may be erroneous (Zhang et al., 2013).

6.5 MODIFICATIONS TO EXISTING EQUATIONS-OF-STATE

The critical property variations described above allow users to model PVT behavior using any of the commonly employed EOS such as the Peng-Robinson, PR-EOS (1975) and the Soave-Redlich-Kwong EOS (1972). One of the drawbacks of this approach is that the critical property variations presented in published literature depend strongly on the assumed values for the strength of the fluid-pore wall interactions. These are often unknown quantities and the use of critical property variations only provides a reasonable starting point to explore PVT behavior dependent on pore sizes. Consequently, the values reported for confined fluid critical properties only provide a means for exploratory analysis of the effects of confinement on fluid properties and should only be utilized for informative purposes (Devegowda et al., 2012). However, kerogen is likely to have different chemical compositions based on its maturity. It is well known that the carbon/oxygen and hydrogen/oxygen ratios in kerogen vary considerably with maturity (Vandenbroucke and Largeau, 2007) and is therefore also likely to influence pore surface chemistry (Hu et al., 2014) and thereby also influence fluid behavior and fluid–rock interactions such as wettability, adsorption, and transport (Hu et al., 2013).

A considerable body of work has therefore focused on developing new EOS or modifying existing EOS in order to overcome the drawbacks associated with the use of modified critical properties of alkanes derived for prespecified fluid–rock interaction potentials. In general, the modifications include additional terms comprising a tunable parameter that quantifies the energy of interaction between the molecules and the pore wall. Derouane (2007) discusses modifications to the van der Waals EOS for gas under microporous confinement which includes adsorption-specific terms as well as terms to adjust the interaction between the adsorbed molecules and the pore walls. Modifications to the van der Waals equation of state are also described in Travalloni et al. (2010) with examples to show the effects of confinement on the critical temperature and density of confined N_2. They determine that the pore size effects on the critical properties of nitrogen tend to be substantial when the ratio of pore diameter to the effective diameter of the molecule is lower than a value of 20.

Travalloni et al. (2014) focus on modifying the Peng-Robinson EOS (PR–EOS) and the confinement effects for different values of interaction energy between the confined molecules and the pore surface. The modified version of the PR–EOS shares similarities with the traditional form of the EOS; however as shown in Eq. (6.1), an additional third term is included to account for the molecule–wall interactions when these become significant. Under the influence of the energy of interaction between the pore wall and the fluid molecule, ε_p, and F_{pa}, respectively, that is, the fraction of the confined molecules subject to the pore wall attractive field for a random distribution of the molecules inside the pore, the modified version of the PR–EOS is given by:

$$P = \frac{RT}{V_m - b_p} - \frac{a_p}{V_m^2 + 2bV_m - b_p^2} - \theta \frac{b_p}{v^2}\left(1 - \frac{b_p}{v}\right)^{\theta-1}(1 - F_{pa})\left(RT\left(1 - \exp\left(\frac{N_{av}\varepsilon_p}{RT}\right)\right) - N_{av}\varepsilon_p\right) \quad (6.9)$$

V_m is the molar volume and N_{av} is the Avogadro number.

In Eq. (6.9), b_p is the confinement-modified volume parameter given by:

$$b_p = \frac{N_{av}}{\rho_{max}} \quad (6.10)$$

a_p is the confinement-modified energy parameter:

$$a_p = \frac{\sqrt{2}N_{av}^2 \varepsilon_p}{\rho_{max}} \alpha \quad (6.11)$$

F_{pa} is the fraction of the confined molecules subject to the pore wall attractive field for a random distribution of the molecules inside the pore:

$$F_{pa} = \frac{\left(r_p - \frac{\sigma}{2}\right)^2 - \left(r_p - \frac{\sigma}{2} - \delta_p\right)^2}{\left(r_p - \sigma/2\right)^2} \quad (6.12)$$

θ is a geometric term representing the ratio of the pore size r_p to the molecule diameter, σ, and the width of the square-well width of the molecule–wall interaction potential, δ_p:

$$\theta = \frac{r_p}{\delta_p + \sigma/2} \quad (6.13)$$

The modified version of the EOS was derived by considering three different regions within a pore (represented by a capillary tube) where the predominant forces are likely to be different. Figure 6.16 (modified from Travalloni et al., 2014) shows a cross-section of a capillary tube and illustrates these

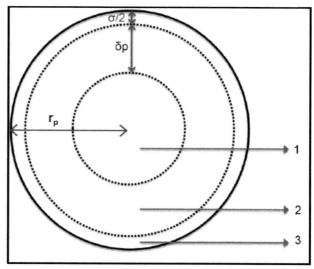

FIGURE 6.16

Cross-sectional area of a capillary tube demarcated in to three regions. Region two comprises of fluid molecules where the pore walls exerts a relatively strong influence while in Region 1, the fluid molecules largely behave independent of the pore wall.

three different regions denoted as Regions 1, 2, and 3. In Region 1, the fluid molecules are sufficiently far from the pore walls and consequently fluid behavior depends solely on intermolecular forces, while in Region 2, the fluid molecules are influenced strongly by the pore walls. The diameter of the fluid molecules restricts fluid molecules to only Regions one and two assuming the molecules are rigid spheres. In order to account for either attractive or adsorbing and nonadsorbing pore walls, the energy of interaction, ε_p needs to be appropriately modified. However, because of the limited availability of such data, Travalloni et al. (2014) have discussed PVT behavior and absorption for a diverse set of pore sizes for different levels of molecule–wall interaction.

6.6 IMPACT TO PRODUCERS

The most immediate impact of non-darcy flow in shale organic and inorganic nanopores relates to the production gas/oil ratio as well as the produced well stream compositions. In general, the behavior of the organic and inorganic pores are not similar. However, if we were to consider transport in shale nanopores restricted to only the organics, that in turn feed natural or induced fracture systems where nonDarcy flow effects vanish, then it is likely that the produced fluid composition may always be lighter than the in situ fluid composition, although as reservoir pressure declines and some of the heavier components desorb, the composition of the produced fluid progressively becomes heavier. This has an impact on the GOR values with a predicted high GOR diminishing to lower values later in the life of the well. This assumes that all flow is through the organic pores and depends on the organic pore connectivity.

Therefore, with a production GOR and produced fluid composition that is likely to evolve with time depending on the contribution of organic and inorganic pores, the net result is a substantial impact on surface facilities design and economics of the project. However there are several aspects of shale microstructures that remain poorly understood and until these are fully resolved, the analysis presented in this chapter remains an exploratory work of the effects of nonDarcy flow in liquids-rich shales.

Additionally, we also demonstrate that while the liquid relative permeability in organic pores are similar, there is a pore-size dependency on the gas relative permeabilities. This will play a significant role in production forecasting and evaluation of project economics because the relative permeability curves have a substantial effect on production GORs. In general, irrespective of the pore size, matrix flow of gas is likely to be substantially larger than Darcy flow predictions because of the effect of slip flow. This in turn is also likely to impact the design of surface facilities and the evolution of produced fluid compositions as well.

6.7 SUMMARY AND CONCLUSIONS

In this chapter, we qualitatively demonstrate how phase behavior and vapor compostion of liquid-rich reservoirs changes under pore proximity, nonDarcy flow and adsorption effects. The new compositional modeling approach is shown to predict very different flowing fluid compositions in comparison with traditional PVT models. Depending on whether the fluid is confined to inorganic or organic pores, a bubble-point suppression or elevation may be observed. At the same time, slip flow effects for the gas phase when considering two-phase flow may accelerate the production of gas and lead to higher GOR values than predicted by bulk fluid properties and Darcy's law. These conditions are likely to be exacerbated if a substantial percentage of the pore volume is contained in the smallest sub 20 nm pores.

This exploratory work also focuses on some of the challenges and issues impacting quantification of fluid properties in shale nanopores where pore wall proximity becomes significant. Our study is based on critical analyses and interpretation, and applicable extensions of previous studies reported in the field of nanosciences to describe pore proximity effects on fluid properties and therefore the results presented here should be refined when quality information becomes available for example, by investigations conducted experimentally and simulation such as by molecular dynamics studies. Simultaneously, the present article can provide guidance as to the type of data required for accurate description of fluid behavior and its consequences in extremely low permeability porous media such as shale. However, the modeling studies provided here provide key insights into the behavior of gas—condensate transport in shale gas reservoirs and has the advantage of being utilized in current numerical compositional simulation software without the need for any modifications to existing code as demonstrated by several examples in this article.

The present approach is particularly significant and of practical importance because we develop and demonstrate a simple effective methodology for convenient manipulation of the presently available commercial simulators which were originally designed for conventional reservoirs where the bulk volume fluid properties are used directly without any modification for the effect of pore size.

We can draw the following conclusions from this work:

1. The methodology proposed in this work can be used as an exploratory tool for the qualitative analysis of the modified behavior and physical properties of a gas–condensate system confined in nanopores.
2. The implications of these considerations are important for routine reservoir engineering calculations such as reserves, the productive life of a well, and well productivity indices.

3. The impact of pore proximity effects on fluid behavior across larger reservoir volumes and across longer time scales is likely to be a function of the percentage of the pore volume contained in the smallest of pores.

REFERENCES

Alharthy, N., Nguyen, T., Teklu, N., Kazemi, H., Graves, R., 2013. Multiphase Compositional Modeling in Small Scale Pores of Unconventional Shale Reservoirs. Paper SPE166306 presented at the SPE Annual Technical Conference and Exhibition, New Orleans, September 30–October 2.

Ambrose, R.J., Hartman, R.C., Campos, M.D., Akkutlu, I.Y., Sondergeld, C.H., 2010. New Pore-Scale Considerations for Shale Gas in Place Calculations. http://dx.doi.org/10.2118/77546-MS. Paper SPE 131772–MS, presented at SPE Unconventional Gas Conference, Pittsburgh, Pennsylvania, 23–25 February.

Ambrose, R.J., 2011. Micro-structure of Gas Shales and Its Effects on Gas Storage and Production Performance (Ph.D thesis). University of Oklahoma, Norman, OK.

Beskok, A., Karniadakis, G.E., 1999. A model for flows in channels, pipes, and ducts at micro and nanoscales. Microscale Thermophysical Engineering 3 (1), 43–77.

Campos, M.D., Akkutlu, I.Y., Sigal, R.F., 2009. A Molecular Dynamics Study on Natural Gas Solubility Enhancement in Water Confined to Small Pores. http://dx.doi.org/10.2118/124491-MS. Paper SPE124491–MS, presented at SPE Annual Technical Conference and Exhibition, New Orleans, Louisiana, 4–7 October.

Chen, X., Cao, G., Han, A., Punyamurtula, V.K., Liu, L., Culligan, P.J., Kim, T., Qiao, Y., 2008. Nanoscale fluid transport: size and rate effects. Nano Letter 8 (9), 2988–2992.

Civan, F., 2010. Effective correlation of apparent gas permeability in tight porous media. Transport in Porous Media 82 (2), 375–384.

Civan, F., Sliepcevich, C.M., 1985. Comparison of the thermal regimes for freezing & thawing of moist soils. Water Resources Research 21 (3), 407–410.

Civan, F., Devegowda, D., Sigal, R.F., 2013. Critical Evaluation and Improvement of Methods to Determine Matrix Permeability of Shale. Paper SPE 166473 presented at SPE Annual Technical Conference and Exhibition. New Orleans, 12–14 August.

Curtis, M.E., Ambrose, R.J., Sondergeld, C.H., Rai, C.S., 2010. Structural Characterization of Gas Shales on the Micro- and Nano-Scales. Paper SPE 137693 presented at Canadian Unconventional Resources and International Petroleum Conference, Calgary, Alberta, Canada.

Curtis, M.E., Ambrose, R.J., Sondergeld, C.H., Rai, C.S., 2011. Transmission and Scanning Electron Microscopy Investigation of Pore Connectivity of Gas Shales on the Nano-Scale. Paper SPE 144391 presented at the SPE North American Unconventional Gas Conference and Exhibition held in The Woodlands, Texas, USA.

Derouane, E.G., 2007. On the physical state of molecules in microporous solids. Microporous and Mesoporous Materials 104, 46–51.

Devegowda, D., Sapmanee, K., Civan, F., Sigal, R.F., 2012. Phase behavior of gas condensates in shale due to pore proximity effects: implications for transport, reserves and well productivity. In: SPE 160099-MS Presented at SPE Annual Technical Conference and Exhibition Held in San Antonio, Texas, USA, 8–10 October.

Didar, B., 2012. Multi-component Shale Gas-in-place Calculations (Master's thesis). University of Oklahoma, Norman, OK.

Fathi, E., Tinni, A., Akkutlu, I.Y., 2012. Shale Gas Correction to Klinkenberg Slip Theory. Paper SPE 154977 presented at SPE Americas Unconventional Resources Conference, Pittsburgh, Pennsylvania, USA.

Firincioglu, T., 2013. Bubble Point Suppression in Unconventional Liquids Rich Reservoirs and Its Impact on Oil Production (Ph.D. Dissertation). Colorado School of Mines, Golden, Colorado.

Freeman, C.M., Moridis, G.J., Ilk, D., Blasingame, T.A., 2009. A Numerical Study of Performance for Tight Gas and Shale Gas Reservoir Systems. Society of Petroleum Engineers. http://dx.doi.org/10.2118/124961-MS.

REFERENCES

Hamada, Y., Koga, K., Tanaka, H., 2007. Phase equilibria and interfacial tension of fluids confined in narrow pores. The Journal of Chemical Physics 127 (8), 084908-1–084908-9.

Hu, Y., Devegowda, D., Striolo, A., Phan, A., Ho, T.A., Civan, F., Sigal, R.F., 2014. Microscopic Dynamics of Water and Hydrocarbon in Shale-kerogen Pores of Potentially Mixed Wettability. Society of Petroleum Engineers. http://dx.doi.org/10.2118/167234-PA.

Kanda, H., Miyahara, M., Higashitani, K., 2004. Triple point of Lennard-Jones fluid in slit nanopore: solidification of critical condensate. The Journal of Chemical Physics 120 (13), 6173–6179.

Kalluri, R.K., Konatham, D., Striolo, A., 2011. Aqueous NaCl solutions within charged carbon-slit pores: partition coefficients and density distributions from molecular dynamics simulations. The Journal of Physical Chemistry 115, 13786–13795.

McCain, D.W., 1990. The Properties of Petroleum Fluids. Pennwell Books.

Michel, G.G., Civan, F., Sigal, R.F., Devegowda, D., 2011. Parametric Investigation of Shale Gas Production Considering Nano-Scale Pore Size Distribution, Formation Factor, and Non-Darcy Flow Mechanisms. Paper SPE 147438 presented at the 2011 SPE Annual Technical Conference and Exhibition, Denver, Colorado, 30 October–2 November.

Mitariten, M., 2005. Molecular Gate Adsorption System for the Removal of Carbon Dioxide and/or Nitrogen from Coalbed and Coal Mine Methane. Presented at 2005 Western States Coal Mine Methane Recovery and Use Workshop, Two Rivers Convention Center, Grand Junction, CO, April 19–20.

Moore, E.B., de la Llave, E., Welke, K., Scherlisb, D.A., Molinero, V., 2010. Freezing, melting and structure of ice in a hydrophilic nanopore. Physical Chemistry Chemical Physics 12 (16), 4124–4134.

Morishige, K., Shikimi, M., 1998. Adsorption hysteresis and pore critical temperature in a single cylindrical pore. The Journal of Chemical Physics 108 (18), 7821–7824.

Morishige, K., Fujii, H., Uga, M., Kinukawa, D., 1997. Capillary critical point of argon, nitrogen, oxygen, ethylene, and carbon dioxide in MCM-41. Langmuir 13 (13), 3494–3498.

Ortiz, V., Lópezá lvarez, Y.M., López, G.E., 2005. Phase diagrams and capillarity condensation of methane confined in single- and multi-layer nanotubes. Molecular Physics 103 (19), 2587–2592.

Peng, D., Robinson, D.B., 1975. A new two-constant equation of state. Industrial & Engineering Chemistry Fundamentals 15 (1), 59–64.

Ramakrishna, S., Balliet, R., Miller, D., Sarvotham, S., 2010. Formation Evaluation in the Bakken Complex Using Laboratory Core Data and Advanced Logging Techniques. Paper presented at the SPWLA 51st Annual Logging Symposium, Perth, Australia, June 19–23.

Sakhaee-Pour, A., Bryant, S., 2012. Gas Permeability of Shale. Society of Petroleum Engineers. http://dx.doi.org/10.2118/146944-PA.

Sapmanee, K., September 2011. Effects of Pore Proximity on Behavior and Production Prediction of Gas/Condensate (M.S. thesis). University of Oklahoma.

Sigal, R.F., 2013. The effects of gas adsorption on storage and transport of methane in organic shales. In: SPWLA D-12–00046, SPWLA 54th Annual Logging Symposium, New Orleans, Louisiana, 22–26 June.

Singh, S.K., Singh, J.K., 2011. Effect of pore morphology on vapor-liquid phase transition and crossover behavior of critical properties from 3D to 2D. Fluid Phase Equilibria 300 (1–2), 182–187.

Singh, S.K., Sinha, A., Deo, G., Singh, J.K., 2009. Vapor-liquid phase coexistence, critical properties, and surface tension of confined alkanes. The Journal Physical Chemistry 113 (17), 7170–7180.

Soave, G., 1972. Equilibrium constants from a modified Redlich-Kwong equation of state. Chemical Engineering Science 27 (6), 1197–1203.

Sondhi, N., 2011. Petrophysical Characterization of Eagleford Shale (M.S. thesis). University of Oklahoma, Norman, USA.

Swami, V., Clarkson, C., Settari, A., 2012. Non-Darcy Flow in Shale Nanopores: Do We Have a Final Answer. Paper SPE 162665 presented at the SPE Canadian Unconventional Resources Conference. Calgary. October 30–November 1.

Tinni, A., Fathi, E., Agarwal, R., Sondergeld, C., Akkutlu, Y., Rai, C., 2012. Shale Permeability Measurements on Plugs and Crushed Samples. Paper SPE-162235 presented at the SPE Canadian Unconventional Resources Conference, Calgary, October 30–November 1.

Travalloni, L., Castier, M., Tavares, F.W., 2014. Phase equilibrium of fluids confined in porous media from an extended Peng-Robinson equation of state. Fluid Phase Equilibria 362, 335–341.

Travalloni, L., Castierb, M., Tavaresa, F.W., Sandler, S.I., 2010. Critical behavior of pure confined fluids from an extension of the van der Waals EOS. Journal of Supercritical Fluids 55 (2), 455–461.

U.S. EIA, 2013. http://www.eia.gov/oil_gas/ Download on 10.12.13.

Vandenbroucke, M., Largeau, C., 2007. Kerogen origins, evolution and structure. Organic Geochemistry 38, 719–833.

WinProp Phase Behavior and Property Program, Version 2007.11 User Guide, 2008. CMG, Calgary, Alberta.

Xiong, X., Devegowda, D., Michel, G.G., Sigal, R.F., Civan, F., 2012. A Fully-Coupled Free and Adsorptive Phase Transport Model for Shale Gas Reservoirs Including Non-Darcy Flow Effects. Paper SPE 159758 presented at SPE Annual Technical Conference and Exhibition, San Antonio, Texas, USA.

Xiong, X., Devegowda, D., Michel, G.G., Sigal, R.F., Civan, F., Jamili, A., 2013. Compositional Modeling of Liquid-Rich Shales Considering Adsorption, Non-Darcy Flow Effects and Pore Proximity Effects on Phase Behavior. Paper SPE 168836-MS and URTeC 1582144, the Unconventional Resources Technology Conference, Denver, Colorado, 12–14 August.

Zarragoicoechea, G.J., Kuz, V.A., 2004. Critical shift of a confined fluid in a nanopore. Fluid Phase Equilibria 220 (1), 7–9.

Zhang, Y., Civan, F., Devegowda, D., Sigal, R.F., 2013. Improved Prediction of Multi-Component Hydrocarbon Fluid Properties in Organic Rich Shale Reservoirs. Paper SPE 166290 presented at the SPE Annual Technical Conference and Exhibition held in New Orleans, Louisiana, USA, 30 Septemper–2 October.

CHAPTER 7

GEOMECHANICS FOR UNCONVENTIONAL RESERVOIRS

Shannon Higgins-Borchardt[1], J. Sitchler[2], Tom Bratton[3]

Schlumberger, Denver, CO, USA[1]; SPE Member, Denver, CO, USA[2]; Tom Bratton LLC, Denver, CO, USA[3]

7.1 INTRODUCTION

Unconventional reservoirs are produced economically by using modern oilfield technology including horizontal wellbore drilling and stimulating wells with hydraulic fracturing. The geomechanical properties of the subsurface greatly influence both the process of drilling a wellbore and hydraulically fracturing the formation. A mechanical earth model (MEM) can be used to mathematically quantify the geomechanical behavior of the subsurface. For this purpose, the MEM is an estimate of the subsurface mechanical properties, rock strength, pore pressure, and stresses. Each of the components of the MEM is described in this chapter, with an emphasis on the realistic and practical application to operating unconventional reservoirs.

7.2 MECHANICAL EARTH MODEL

An MEM is a numerical representation of the rock properties, pore pressure and stresses in the earth (Plumb et al., 2000). This modeling tool was developed out of a practical need to understand and predict the behavior of wellbores drilled deep into the subsurface to reduce drilling risk and improve well economics (Last et al., 1995). The methods and relationships that define the MEM are well established and understood within the geomechanics community, having been applied to thousands of wells throughout the world. A MEM can be one-dimensional, following the trajectory of a wellbore, or multidimensional, such as mapping a volume of the subsurface. The models can be very simple or very complex, depending on the available data and the application. Generally, the models are nonunique and depend on the quality of the model inputs and the intended application of the model.

As the difficulty in finding and exploiting petroleum resources increases, and well costs continue to rise, unconventional reservoirs have become more attractive. These previously unconsidered reservoirs, such as hydrocarbon bearing mudrocks, often require hydraulic fracture stimulation to be economically produced. Early unconventional developments were themselves relatively expensive due to the reliance on high density, vertical drilling. A drive to lower costs and increase production has led to a shift in the industry toward smaller footprints and matrix stimulation in an attempt to improve efficiency. As a result, the main changes to well and completion design involve drilling and stimulating deviated and horizontal wellbores. The geomechanical challenges in unconventional reservoirs are wide ranging and, for example, involve well orientation, in situ stress, rock mechanical properties, layering, and well completion techniques. Because rock mechanical properties and in situ stress are primary controls on wellbore stability and hydraulic fracture propagation, the MEM is designed as a tool to be both predictive and diagnostic in these operations.

The general workflow for a simple MEM developed for engineering applications such as drilling or hydraulically fracturing is outlined. However, it is understood that this approach does not address all geomechanical scenarios. Notable challenges and exceptions to the simple MEM, which are specific to unconventional reservoirs, are included throughout this text. Gaining a deeper knowledge of these challenges and developing better models will inevitability lead to overall improvements in the recovery of hydrocarbons from unconventional reservoirs.

7.2.1 MECHANICAL PROPERTIES

An MEM can be built on the assumption that crustal rocks behave elastically and that there is a clearly defined relationship between stress and strain. This is more a matter of convenience than fact, because it allows an MEM to be constructed quickly using limited data and minimal computing power. It is understood that there are cases where this is not true, but for the majority of examples, a useful engineering answer can be derived under these simple assumptions. The following subsections describe the mechanical properties of rock that are ultimately used to calculate stress in the MEM under poroelastic assumptions.

The most common mechanical properties quantified are Young's Modulus, Poisson's Ratio, and Biot's coefficient. Quantifying these properties almost always involves acquiring acoustic data, augmented sometimes with core data. Dynamic mechanical properties are acoustic-derived properties, while static mechanical properties are obtained from laboratory tests in which core samples are physically deformed.

Slownesses measured from an acoustic tool are combined with a density measurement to determine the dynamic mechanical properties. If the formation is assumed to be isotropic, homogenous, linear, and elastic, the following equations are applied to determine the shear modulus (G), Poisson's Ratio (ν) and Young's Modulus (E):

$$G = \frac{\rho}{(\Delta t_{shear})^2}, \quad (7.1)$$

$$\nu = \frac{0.5 \times \left(\frac{\Delta t_{shear}}{\Delta t_{comp}}\right)^2 - 1}{\left(\frac{\Delta t_{shear}}{\Delta t_{comp}}\right)^2 - 1}, \quad (7.2)$$

$$E = 2G(1 + \nu), \quad (7.3)$$

where ρ = density, Δt_{shear} is shear slowness, and Δt_{comp} is compressional slowness.

Static properties are calculated from stress and strain relationships measured during deformation in the laboratory with core samples. Typically a triaxial compression test is conducted, where a lateral confining pressure, representative of in situ conditions, is applied to a core sample, and then the axial stress is increased until the sample fails. Static Young's Modulus is the ratio of the axial stress increment applied to the core divided by the corresponding change in axial strain. Graphically, when the axial stress applied to a core sample is plotted against the axial strain, the slope of the linear potion is equal to the static Young's Modulus, and the static Poisson's Ratio is the ratio of the radial strain versus axial strain of the sample.

In oilfield operations, acoustic and density measurements are typically acquired from logs across both the reservoir and bounding formations, providing data on dynamic elastic properties over continuous depth intervals. Core samples might then be collected over the same depth intervals, allowing measurements of both dynamic and static properties from samples taken at discrete depths. In the MEM process, empirical correlations are often used to relate the dynamic mechanical properties measured from logs to the static mechanical properties measured from core.

For unconventional reservoirs, the source rock and reservoir rock may often be shale or some other finely layered formation, for which the assumption of isotropy is usually not valid. Mechanically this means that elastic properties vary depending on the direction in which they are measured. It is possible to quantify mechanical properties in multiple orientations using acoustic measurements and cores taken in different directions to bedding.

For shale, elastic properties can best be described as being transversely isotropic, with an axis of symmetry perpendicular to the bedding plane. This property anisotropy is a result of textural layering. Mechanical properties measured parallel and perpendicular to bedding in some core samples have shown up to a 400% difference (Suarez-Rivera et al., 2011). Therefore, quantification of the effect of anisotropy on mechanical properties can have a significant impact on geomechanics analyses for unconventional reservoirs, because these properties are used in both wellbore stability and hydraulic fracture modeling.

To determine elastic mechanical properties in multiple orientations from acoustic measurements, consider Hooke's law, which states that strain is directly proportional to stress. For an anisotropic formation, Hooke's law can be written as (Sayers, 2010):

$$\sigma_{ij} = C_{ijkl} \times \varepsilon_{kl}, \tag{7.4}$$

where σ_{ij} denotes the second-rank stress tensor, ε_{kl} denotes the second-rank strain tensor, and C_{ijkl} denotes the fourth-rank elastic stiffness tensor. For a transversely isotropic formation with a vertical axis of symmetry, the elastic stiffness matrix, C_{ij}, can be written in the conventional two-index notation, with five independent stiffness coefficients including, $C_{11} = C_{22}$, C_{33}, $C_{12} = C_{21}$, $C_{13} = C_{31} = C_{23} = C_{32}$, $C_{44} = C_{55}$, and by symmetry $C_{66} = (C_{11}-C_{12})/2$ (Sayers, 2010; Nye, 1985):

$$C_{ij} = \begin{pmatrix} C_{11} & C_{12} & C_{13} & 0 & 0 & 0 \\ C_{21} & C_{22} & C_{23} & 0 & 0 & 0 \\ C_{31} & C_{32} & C_{33} & 0 & 0 & 0 \\ 0 & 0 & 0 & C_{44} & 0 & 0 \\ 0 & 0 & 0 & 0 & C_{55} & 0 \\ 0 & 0 & 0 & 0 & 0 & C_{66} \end{pmatrix}, \tag{7.5}$$

Acoustic measurements can be made directly on core samples to determine all five stiffness coefficients, by measuring velocities at different orientations to the bedding plane. With logging measurements, depending on wellbore orientation with respect to bedding, three of the five stiffness coefficients can be measured, while the other two must be estimated. For example, in a vertical wellbore with flat bedding planes, C_{33} represents the vertically polarized compressional wave, C_{44} represents the vertically polarized shear wave, and C_{66} represents the horizontally polarized shear wave. This is shown in Fig. 7.1. The dynamic elastic properties can be determined from the elastic stiffness coefficients by:

$$E_1 = C_{33} - 2\frac{C_{13}^2}{C_{11} + C_{12}}, \tag{7.6}$$

$$E_3 = \frac{(C_{11} - C_{12}) \times (C_{11}C_{33} - 2C_{13}^2 + C_{12}C_{33})}{C_{11}C_{33} - C_{13}^2}, \tag{7.7}$$

$$\nu_{31} = \frac{C_{13}}{C_{11} + C_{12}}, \tag{7.8}$$

$$\nu_{12} = \frac{C_{33}C_{12} - C_{13}^2}{C_{33}C_{11} - C_{13}^2}, \tag{7.9}$$

$$\nu_{13} = \frac{C_{13}(C_{11} - C_{12})}{C_{11}C_{33} - C_{13}^2}, \tag{7.10}$$

where E_1 is the dynamic Young's Modulus corresponding to the horizontal axis, E_3 is Young's Modulus corresponding to the vertical axis, ν_{31} is the dynamic Poisson's ratio where compression is applied to the vertical axis and the expansion is horizontal, ν_{12} is the Poisson's ratio where the compression is applied to the horizontal axis and the expansion is in the other horizontal axis, and ν_{13} is the Poisson's ratio where the compression is applied to a horizontal axis and the expansion is vertical (Sayers, 2010).

In addition to calculating mechanical properties at multiple orientations from acoustic measurements, the static properties can also be measured from stress-strain relationships during deformation tests on core samples plugged at various orientations to the bedding planes.

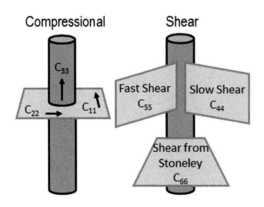

FIGURE 7.1

Stiffness coefficients for a vertical well with flat bedding.

In unconventional reservoirs, quantification of mechanical properties at various orientations has a direct impact on wellbore stability, hydraulic fracture propagation, and fracture geometry predictions.

7.2.2 ROCK STRENGTH

When a wellbore is excavated, the strength of the rock relative to the near-wellbore stresses determines whether the rock in the wellbore walls will remain mechanically stable, or will undergo yield and permanent deformation. From a strictly mechanical perspective, the elements necessary to predict stability include the rock strength parameters, the in situ stress, the borehole geometry and orientation in relation to the in situ stresses, and a shear failure criterion.

The key rock strength parameters include unconfined compressive strength (UCS), tensile strength (T_0), cohesion (C_0), and the coefficient of internal friction (ϕ). Rock strength parameters are typically measured with core, but can be estimated from log data using empirical correlations. The UCS is the ultimate or peak strength of the material when compression is applied uniaxially, under zero confining pressure, on a cylindrical core. Indirect testing procedures to determine UCS by deduction or proxy include single or multistage triaxial tests, scratch tests, and rigid indenter tests. Tensile strength is the maximum strength of the material when tension is applied, usually at the moment when the material fails completely. Tensile strength can be measured directly on core samples in the lab with a direct tensile test, or indirectly through Brazil tests or flexural tests. In a Brazil test a thick disc of rock is placed under diametral compression, along a single axis across the disk, to cause it to split diametrically in a plane parallel to the applied load. Cohesion and the coefficient of internal friction can be estimated by fitting uniaxial and/or triaxial compressive strengths to a Mohr–Coulomb failure or other criterion.

Because many unconventional reservoirs contain formations with an anisotropic rock fabric, it is also understood that the measured rock strength will depend on the orientation of the loading relative to the bedding plane. Typically, rock strengths measured parallel and perpendicular to bedding planes are stronger than orientations in between. This is significant because many unconventional reservoirs are drilled with deviated wellbores that intersect the formation at orientations that are not directly parallel or perpendicular to bedding.

7.2.3 PORE PRESSURE

Pore pressure is the pressure of the fluid in the pore spaces within the subsurface. In unconventional reservoirs pore pressure is important but challenging to quantify. It has a direct impact on the state of stress, it impacts flow rates and production, during drilling it can cause kicks from permeable formations, and it influences borehole instability in all formations.

Pore pressure is referred to as being normally-pressured if it is equal to the hydrostatic pressure of a column of water from the surface to a given depth. Likewise, it is referred to as being underpressured if the pressure is less than hydrostatic, and overpressured if the pressure is greater than hydrostatic. Under-pressuring is commonly due depletion. Overpressuring can be caused by rapid sediment burial with trapped fluid, fluid generation caused by the maturation of kerogen, uplift and exhumation, temperature changes, or diagenetic processes.

For example, in a passive margin setting such as along the coast of the Gulf of Mexico, the mechanisms for overpressuring can often be attributed to undercompaction and diagenetic processes.

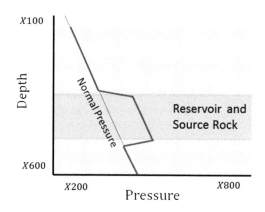

FIGURE 7.2

Typical pore pressure profile for an unconventional reservoir.

However, in unconventional reservoirs, where the reservoir rock is often the source rock, the genesis of pressure can be more difficult to define and estimate.

The most common methods for estimating pore pressure for oilfield geomechanics involve empirical methods relating velocity to pressure, through an estimate of the effective stress acting on the rock (Eaton, 1975; Bowers, 1995). For some reservoirs, a rock physics approach can be used to relate velocity and effective stress (Ciz et al., 2005; Sayers, 2006). If overpressure is caused by the maturation of kerogen, as is often the case in continuous unconventional mudrock reservoirs, the velocity may be used as a proxy for effective stress, but this relationship may not be unique. Therefore, basin modeling can be used to reconstruct the burial history, model the hydrocarbon generation and migration, and estimate the pressure generation.

In the absence of basin modeling or quality velocity versus effective stress relationships, a continuous pore pressure profile can be developed using trends that are fit to calibration data. Direct measurements of pore pressure for calibration in unconventional reservoirs are difficult, because measurements require communication between the matrix and the wellbore where gauges are located, and flow from the matrix to the gauges. The very low permeability nature of these reservoirs means that it often takes many days, weeks, or even months for the pressure to reach near-equilibrium between the formation and wellbore. Therefore, the pressure profile is often loosely constrained to indirect pore pressure calibration, such as mud weights used during drilling when connection gas is present, an influx or kick, or production test data. With this calibration data, a pore pressure profile is developed, which can be simple and continuous, but should honor valid calibration data and the known geological setting. A typical continuous pore pressure profile for an unconventional reservoir is shown in Fig. 7.2. The amount of overpressure and transition from normal pressure to overpressure, and back, is unique to each reservoir.

7.2.4 STRESSES

The in situ stress state is a fundamental component of any geomechanical model and has a significant impact for both drilling and completions of unconventional hydrocarbon reservoirs. The stress state can be represented by three perpendicular principal stresses. If one of the principal stresses is vertical, caused by the overburden stress for example, then the other two are horizontal. These are the minimum horizontal stress and the maximum horizontal stress.

7.2.4.1 Vertical Stress

The vertical stress (σ_V) referred to as the overburden stress in the earth, is the stress exerted by gravitational loading of the overlying mass of rock. The following equation is used to estimate the overburden stress by integrating the mass of the material above the depth of interest:

$$\sigma_V(Z_0) = \int_0^{Z_0} \rho_b g\, dz, \tag{7.11}$$

where ρ_b is the density of the material and g is the gravitational acceleration. If a density log is acquired from the point of interest to the surface, the calculation is very straightforward. In practice, density logs are seldom acquired to the surface and, therefore, the measured density is usually extrapolated to the surface before it is integrated to obtain the overburden or vertical stress.

7.2.4.2 Minimum and Maximum Horizontal Stress

If the vertical stress is one of the principal stresses, the other two principal stresses are the minimum horizontal stress (σ_h) and maximum horizontal stress (σ_H). The minimum horizontal stress magnitude, which is often the least principal stress, has a direct influence in the propagation of hydraulic fractures. The maximum horizontal stress magnitude is important, because depending on the wellbore orientation, it impacts wellbore stability and the likelihood of creating complex nonplanar hydraulic fractures.

The minimum horizontal stress can be determined directly from a small injection test where fluid is pumped into a formation, a small fracture is created, the well is shut in, and the pressure at which the fracture closes is measured. The closure pressure is equivalent to the minimum horizontal stress. A test can be done with wireline tools, but more commonly it is done with hydraulic fracture equipment where a small injection is created, and the resulting fracture is allowed to close prior to the main fracture treatment.

A measurement of minimum horizontal stress is often taken within the reservoir interval, but not in the intervals above or below the reservoir. It is important to quantify the least principal stress in these adjacent formations for containment, or vertical growth, of the hydraulic fracture. Therefore, a continuous minimum horizontal stress curve can be calculated, and then calibrated with the available measurements. Likewise, a continuous maximum horizontal stress can be calculated, although direct calibration is difficult as it cannot be determined directly.

Given the assumption that the formation is poroelastic and homogenous, the following equations can be used to calculate minimum and maximum horizontal stress:

$$\sigma_h - \alpha_h P_p = \frac{\nu}{1-\nu}(\sigma_V - \alpha_V P_P) + \frac{E}{1-\nu^2}\varepsilon_h + \frac{E\nu}{1-\nu^2}\varepsilon_H, \tag{7.12}$$

$$\sigma_H - \alpha_h P_p = \frac{\nu}{1-\nu}(\sigma_V - \alpha_V P_P) + \frac{E}{1-\nu^2}\varepsilon_H + \frac{E\nu}{1-\nu^2}\varepsilon_h, \tag{7.13}$$

where σ_h is minimum horizontal stress, σ_H is maximum horizontal stress, E is Young's Modulus, ν is Poisson's Ratio, σ_V is overburden stress, P_p is pore pressure, α_V is the vertical Biot's coefficient, α_H is the horizontal Biot's coefficient, ε_h is the minimum principal horizontal strain, and ε_H is the maximum principal horizontal strain.

For an anisotropic formation, given the assumption that it is transversely isotropic with a vertical axis of symmetry, the following equations can be used to calculate minimum and maximum horizontal stresses:

$$\sigma_h - \alpha_h P_p = \frac{E_{horz}}{E_{vert}} \frac{\nu_{vert}}{1 - \nu_{horz}} (\sigma_V - \alpha_V P_p) + \frac{E_{horz}}{1 - \nu_{horz}^2} \varepsilon_h + \frac{E_{horz} \nu_{horz}}{1 - \nu_{horz}^2} \varepsilon_H, \tag{7.14}$$

$$\sigma_H - \alpha_h P_p = \frac{E_{horz}}{E_{vert}} \frac{\nu_{vert}}{1 - \nu_{horz}} (\sigma_V - \alpha_V P_p) + \frac{E_{horz}}{1 - \nu_{horz}^2} \varepsilon_H + \frac{E_{horz} \nu_{horz}}{1 - \nu_{horz}^2} \varepsilon_h, \tag{7.15}$$

where E_{horz} is the horizontal Young's Modulus, E_{vert} is the vertical Young's Modulus, ν_{horz} is the horizontal Poisson's Ratio (ratio of horizontal strain, to strain induced by the other horizontal stress), and ν_{vert} is the vertical Poisson's Ratio (ratio of horizontal strain, to strain induced by a vertical stress).

Sometimes it is necessary to consider that the formation has orthorhombic anisotropy, such as when a horizontally bedded formation contains vertically parallel natural fractures. Given the assumption that the formation is orthorhombic, the following equations can be used to estimate minimum and maximum horizontal stresses:

$$\sigma_h - \alpha_h P_p = \frac{C_{13}}{C_{33}} (\sigma_V - \alpha_V P_p) + \left(C_{11} - \frac{C_{13} C_{13}}{C_{33}} \varepsilon_x \right) + \left(C_{12} - \frac{C_{13} C_{23}}{C_{33}} \varepsilon_Y \right), \tag{7.16}$$

$$\sigma_H - \alpha_H P_p = \frac{C_{23}}{C_{33}} (\sigma_V - \alpha_V P_p) + \left(C_{12} - \frac{C_{13} C_{23}}{C_{33}} \varepsilon_x \right) + \left(C_{22} - \frac{C_{23} C_{23}}{C_{33}} \varepsilon_Y \right), \tag{7.17}$$

where the C_{ij} parameters are the stiffness coefficients discussed in Section 7.2.1.

In the equations shown above, α_V is the vertical Biot's coefficient and α_H is the horizontal Biot's coefficient. Biot's coefficient is a tensor that describes the ability of the pore pressure to counteract the stresses (Biot, 1941). In theory this value may be anisotropic and can vary between 0 and 1 based on the bulk moduli of the constituents and the whole rock. However, Biot's theory applies to ideal poroelastic rocks with connected porosity, meaning that Biot's coefficient may be less representative of the actual pore pressure effectiveness in unconventional rocks, which often have unconnected porosity, microfractures, slot porosity, very low porosity, and heterogeneity. This parameter has a first-order impact on the stress result. Calculated values using Biot's equation tend to be very small in low porosity rocks, which results in unreasonably high calculated total horizontal stresses. Therefore, an argument can be made that this parameter should be generalized as an effective stress coefficient, derived empirically, rather than obeying a strict definition of Biot's coefficient.

7.2.4.3 Stress Direction

Hydraulic fractures tend to propagate in a plane perpendicular to the minimum in situ stress. Therefore, if the principal stresses are vertical and horizontal, a planar fracture will tend to propagate in the direction of the maximum horizontal stress within a normally stressed environment.

The maximum horizontal stress direction can be determined in numerous ways. First, with an image log in a vertical well, if breakout occurs it will be observed in the direction of minimum horizontal stress and if drilling induced tensile fractures are created they will occur in the direction of maximum horizontal stress. Oriented caliper logs can also be used to detect breakout direction. Also, the maximum horizontal stress direction can be obtained with acoustic logs where shear wave splitting is caused by stress. In a vertical well with flat bedding, if shear wave splitting is caused by stress then the azimuth of the fast flexural shear wave will align in the direction of maximum horizontal stress. Microseismic data acquisition can also be used to determine the direction of stress during hydraulic fracturing, by monitoring the direction of fracture propagation.

7.2.5 MODEL VALIDATION AND CALIBRATION

The MEM systematically captures the geomechanical properties of the subsurface, but to be useful for practical decision making it must be both calibrated and validated. Calibration involves comparing the computed properties to measured properties. Validation involves using the computed MEM properties in a simulator or in other engineering calculations and comparing the modeled results to observed results.

Calibration data available in unconventional reservoirs are often different from calibration data available for conventional reservoirs. Mechanical properties and rock strength are normally calibrated to measurements taken with core data in the laboratory. Pore pressure is challenging to directly measure, and therefore calibration is more often inferred from drilling events. Because hydrocarbon production from unconventional reservoirs often requires hydraulic fracturing, the minimum horizontal stress is most often calibrated to a closure measurement taken with a small fracture injection test.

To validate the computed geomechanical properties and parameters for unconventional reservoirs, wellbore stability analyses or hydraulic fracture simulations are typically used. With wellbore stability analyses, the geomechanical properties are input, along with the downhole pressures caused by the mud and drilling practices, and the computed near-well stresses are compared with shear and tensile failure criteria to predict wellbore instabilities such as breakouts or drilling induced tensile fractures. The modeled instability is then directly compared with any observed instability present on the image logs, caliper logs, or drilling report. With a hydraulic fracture simulator, the geomechanical properties, along with the pumped fluid, materials, and rates, are used to predict the hydraulic fracture pressure and geometry, which are compared with the measured pressure during the treatment. When the computed instability or pressure matches the observed instability or pressure, the model is supported and can be used to make further drilling and completion decisions.

7.3 DRILLING APPLICATIONS FOR UNCONVENTIONAL RESERVOIRS

Economic production of hydrocarbons from unconventional reservoirs often requires drilling deviated or horizontal wellbores to maximize contact with the reservoir. The geomechanical properties and parameters of the subsurface can greatly influence the optimal well construction and drilling practices.

7.3.1 WELLBORE STABILITY

One of the common challenges when drilling wells in unconventional reservoirs is to drill the well while maintaining wellbore stability. To avoid wellbore instability, a drilling engineer cannot alter the geomechanical properties of the subsurface, but can be strategic in both well construction and the fluid and mud properties used. The downhole pressure required to drill a stable wellbore without taking a kick or influx, having the hole collapse, or losing drilling fluid can be determined with the MEM and wellbore stability analyses. The ranges of downhole pressures that prevent such problems are often referred to as the safe and stable mud weight windows. The safe mud weight window is the pressure between the pressure at which a kick might occur and the pressure at which losses will commence. The stable mud weight window lies between the pressures needed to prevent breakout and wellbore damage through shear failure, and the pressure at which formation breakdown will occur. For example, see Fig. 7.3.

7.3.1.1 Kick
The kick threshold is equal to the pore pressure. To avoid taking a kick or influx, the downhole pressure must be kept above the kick threshold. However, in unconventional reservoirs the downhole pressure is often less than the pore pressure and because of the low permeability a kick does not occur. This is referred to as underbalanced drilling and is often used to decrease the drilling time.

7.3.1.2 Losses and Breakdown
Significant fluid loss while drilling occurs when the downhole pressure exceeds both the breakdown pressure and loss threshold. The loss threshold is equal to the minimum horizontal stress. The breakdown pressure, which causes tensile failure, for an isotropic formation is computed by the following equation (Jaeger et al., 2007):

$$P_B = 3\sigma_3 - \sigma_1 + T_0, \tag{7.18}$$

where P_B is the breakdown pressure, σ_1 is the maximum far-field stress, σ_3 is the minimum far-field stress, and T_0 is the tensile strength. The origins of this equation are in the calculation of near wellbore stresses and apply to fracture initiation at the wellbore wall (Kirsch, 1898). The breakdown pressure can be higher or lower than the loss threshold. This approach to estimating formation breakdown is commonly used, but for anisotropic rocks the same equation can either overestimate or underestimate the breakdown pressure. Other numerical and analytical solutions are available and are being developed to better quantify the breakdown pressure and tensile failure for anisotropic formations (Prioul et al., 2011).

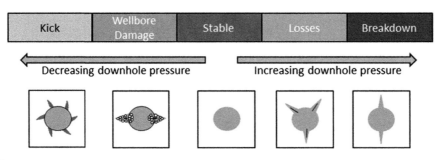

FIGURE 7.3

Mud weight window schematic.

7.3.1.3 Wellbore Damage

To prevent the initiation of hole collapse or hole enlargement due to breakout, often called wellbore damage, the downhole pressure must be kept above the pressure at which shear failure of the rock in the wellbore walls can initiate. Shear failure occurs when the induced near-well stresses exceed the shear strength of the rock. The most commonly used criterion to compute shear failure is from Coulomb (1773) which can be written in many different but equivalent forms, one being:

$$\tau = C_o + \sigma_n \tan \phi, \tag{7.19}$$

where ϕ is the coefficient of internal friction, C_o is the cohesion, σ_n is the normal stress, and τ is the shear stress. Other shear failure criteria are widely available and used, including but not limited to, Mohr–Coulomb, Mogi–Coulomb, Modified Lade, Hoek–Brown, Drucker–Prager (Jaeger et al., 2007; Mogi, 1966; Hoek and Brown, 1980; Drucker and Prager, 1952; Lade, 1984).

To compute the near wellbore stresses from the far-field in situ stresses, pore pressure, and hole orientation, the analytical equations for elastic stresses around cylindrical openings are most commonly used. For the particular solutions when the cylinder is parallel to one of the principal stresses, these elastic equations reduce to the Kirsch solution (Kirsch, 1898). The calculations assume that the formation is continuous, homogeneous, isotropic, linear, and elastic.

For most wellbore stability solutions, only isotropic mechanical properties and rock strengths are used, along with isotopic borehole failure criteria. As emphasized in this chapter, these assumptions can provide models that are useful, but are not always valid in unconventional reservoirs where there is significant textural anisotropy. At the time of this publication, anisotropic wellbore stability models are being developed and introduced into the industry (Yan et al., 2014; Suarez-Rivera et al., 2006). For example, Yan et al. (2014) introduced an anisotropic wellbore stability model with a plane of weakness failure criterion and methodologies for computing near wellbore stresses considering anisotropy (Amadei, 1983; Jaeger et al., 2007; Lekhnitskii, 1963; Yan et al., 2014).

7.3.1.4 Depth of Failure

When the downhole pressure used to drill a well falls below the shear failure limit for the surrounding rock, the rock will break. The shear failure often occurs along intersecting conjugate surfaces, leading to pieces of broken rock falling into the wellbore, often called breakout, causing borehole enlargement.

The wellbore damage threshold corresponds to the minimum downhole pressure required to keep the rock in the elastic range and prevent shear failure. If wellbore damage does occur, it does not usually mean complete hole collapse. When drilling wells in unconventional reservoirs, the downhole pressure is often maintained below the wellbore damage threshold because a lower mud pressure may result in faster drilling. Although rock in the wellbore wall may be damaged it can still remain in place, but has the potential to be released into the wellbore some time after the initial damage has occurred.

Eventually there is a point at which the damaged material, if it were to collapse into the wellbore, could not be transported away by the circulating mud, and hence a pack-off or stuck pipe incident could ensue. A risk approach may be taken in the quantification of collapse pressure, by computing pressure gradients at which different depths of failure or damage might occur (Higgins-Borchardt et al., 2013). This can be done using simple analytical solutions assuming linear-elasticity, or using more advanced formulations such as thermoporoelasticity (Frydman and da Fontoura, 2000). The result is a series of well pressures or downhole mud weights, corresponding to increasing depths of damage and hence increasing risk of hole collapse. This concept is shown in Fig. 7.4.

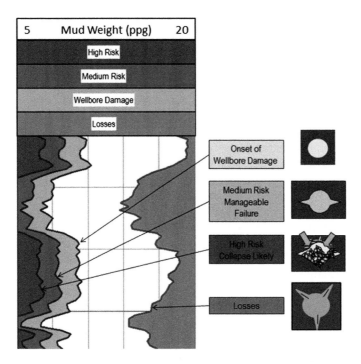

FIGURE 7.4

Depth of failure schematic.

7.3.2 DEVIATION AND AZIMUTH

Wellbore stability is highly dependent on the orientation of the well with respect to the orientations and magnitudes of the in situ stresses. By quantifying the rock strength of each formation, the formation pressures, and the in situ stresses, it becomes possible to estimate safe and stable mud weight windows for any wellbore deviations and azimuths, and use these results to aid in well construction planning and successful drilling.

In a normal stress regime, where $\sigma_h < \sigma_H < \sigma_V$, the preferred direction of horizontal drilling is usually parallel to the minimum in-situ stress so that transverse, rather than longitudinal, hydraulic fractures can be achieved. This orientation also results in a larger safe mud weight window than drilling parallel to the maximum horizontal stress, where a longitudinal hydraulic fracture would be created.

If the stress state is one of reverse-faulting, where $\sigma_V < \sigma_h < \sigma_H$, drilling a horizontal well in the direction of maximum horizontal stress σ_H, will provide a wider safe mud weight window than a horizontal well drilled in the direction of the minimum horizontal stress. However, production from hydraulic fractures will be challenging because the hydraulic fracture will tend to propagate horizontally rather than vertically.

If the stress state is strike–slip, where $\sigma_h < \sigma_V < \sigma_H$, then the horizontal direction with the largest safe window will depend on the contrasts between the magnitudes of the vertical stress and the two horizontal stresses.

7.4 COMPLETION APPLICATIONS FOR UNCONVENTIONAL RESERVOIRS

Economic hydrocarbon production from unconventional reservoirs often requires the use of hydraulic fracturing. For this method of stimulation to succeed, an adequate fracture not only needs to be created but its conductivity needs to be maintained over sufficient time. Hydraulic fracture propagation and geometry are strongly dependent on the formation geomechanics, particularly the in situ stresses. The containment or vertical growth of a hydraulic fracture is primarily controlled by the contrast in minimum horizontal stress between the producing formation and the bounding layers. However, other parameters such as the mechanical properties of the formation, fracture toughness, bedding planes, planes of weakness, rock fabric, and natural fractures can also influence hydraulic fracture propagation.

Fracture complexity and nonplanar fracture geometry are often desirable, as these can help maximize the contact area between the hydraulic fracture and producing interval. Fracture complexity is greatly influenced by the stress state, the presence and characteristics of any natural fractures, the fluids and proppant that are pumped, and by the rock fabric and any planes of weakness (Suarez-Rivera et al., 2013a,b).

As mentioned in the previous section, within a normal stress environment, where $\sigma_h < \sigma_H < \sigma_V$, hydraulic fractures will tend to be vertical and will propagate in the direction of the maximum horizontal stress. For any field development, the optimal well spacing needed to economically drain the reservoir will depend on the direction the well is drilled relative to the direction the hydraulic fracture will propagate, as well as the productive half-lengths created by the hydraulic fracture.

In unconventional reservoirs with dense horizontal well spacing and hydraulic fracturing, asymmetric hydraulic fractures often occur (Mata et al., 2014). Over time, hydrocarbon production in producing wells causes a localized decrease in pore pressure in the previously fractured stimulated volume, which reduces the total in situ stress magnitudes. For a new well, which is hydraulically fractured and located near a producing well, the hydraulic fracture will propagate preferentially into the depleted region of a producing well compared to virgin pressure areas. Therefore well completion timing related to production should be considered during field development.

7.5 CONCLUSIONS

The topic of geomechanics plays a vital role in subsurface oil and gas exploration and development. The MEM, describing mechanical properties, pore pressure, and earth stresses, has an important role in both drilling and hydraulic fracture stimulations.

The general workflow for a MEM created with the assumption that the rock behaves of poroelasticity has been discussed within this chapter. Although this assumption may be a simplification in complex and layered unconventional reservoirs, such MEMs can still be used to conduct very useful engineering estimates and designs. Incremental improvements in the model accuracy and predictivity can be achieved by exploring other more complex stress and failure models, which will help further improve the understanding of the behavior of unconventional reservoirs.

ACKNOWLEDGMENTS

The authors would like to thank Schlumberger for supporting this effort. In particular, we would like to thank John Fuller and Rob Marsden for their editorial suggestions.

REFERENCES

Amadei, B., 1983. Rock Anisotropy and the Theory of Stress Measurements. Springer-Verlag, Berlin, Germany.
Biot, M.A., 1941. General theory of three-dimensional consolidation. Journal of Applied Physics 12 (2), 155–164.
Bowers, G., 1995. Pore pressure estimation from velocity data: accounting for overpressure mechanisms besides undercompaction. SPE Drilling and Completion 10 (2), 89–95.
Coulomb, C.A., 1773. Application des regles de maxima et minima a quelques problemes de statique relatifs a l'Architecture. Acad. Roy. Sci. Mem. Math. Phys. 7, 343–382.
Ciz, R., Urosevic, M., Dodds, K., 2005. Pore pressure prediction based on seismic attributes response to overpressure. APPEA Journal 1, 1–10.
Drucker, D.C., Prager, W., 1952. Soil mechanics and plastic analysis of limit design. Quarterly of Applied Mathematics 10, 157–165.
Eaton, B., September 28–October 1, 1975. The Equation for Geopressure Prediction from Well Logs. Paper SPE 5544. 50th Annual Fall Meeting of SPE of AIME, Dallas, Texas, USA.
Frydman, M., da Fontoura, S.A.B., October 16–19, 2000. Wellbore Stability Considering Thermo-poroelastic Effects. Paper IBP 264 00. Brazilian Petroleum Institute–IBP, Rio Oil & Gas Conference, Rio de Janeiro, Brazil.
Higgins-Borchardt, S., Krepp, T., Frydman, M., Sitchler, J., August 12–14, 2013. New Approach to Geomechanics Solves Serious Horizontal Drilling Problems in Challenging Unconventional Plays. 168675-MS. Unconventional Resources Technology Conference, Denver, Colorado, USA.
Hoek, E., Brown, E.T., 1980. Underground Excavations in Rock, Institution of Mining and Metallurgy. Elsevier Applied Science, London, England.
Jaeger, J.C., Cook, N.G., Zimmerman, R.W., 2007. Fundamentals of Rock Mechanics, fourth ed. Blackwell Publishing, Malden, MA, USA.
Kirsch, E.G., 1898. Die Theorie der Elastizitat und die Bedurfnisse der Festigkeitslehre. Zeitschrift des Vereines Deutscher Ingenieure 42, 797–807.
Lade, P.V., 1984. Failure Criterion for Frictional Materials. Mechanics of Engineering Materials, Wiley, New York, p. 385.
Last, N., Plumb, R., Harkness, R., Charlez, P., Alsen, J., McLean, M., October 22–25, 1995. An Integrated Approach to Evaluating and Managing Wellbore Instability in the Cusiana Field, Colombia, South America. Paper SPE 30464. SPE Annual Technical Conference and Exhibition, Dallas, Texas, USA.
Lekhnitskii, S.G., 1963. Theory of Elasticity of an Anisotropic Body. MIR Publishers, Moscow, Russia.
Mata, D., Cherian, B., Gonzales, V., Higgins-Borchardt, S., Han, H., April 1–3, 2014. Modeling the Influence of Pressure Depletion in Fracture Propagation and Quantifying the Impact of Asymmetric Fracture Wings in Ultimate Recovery. SPE-169003-MS. SPE Unconventional Resources Conference, The Woodlands, Texas, USA.
Mogi, K., 1966. Some precise measurements of fracture stress of rocks under uniform compressive stress. Rock Mechanics and Engineering Geology 4, 41–55.
Nye, J., 1985. Physical Properties of Crystals. Oxford University Press.
Plumb, R., Edwards, S., Pidcock, G., Lee, D., Stacey, B., February 23–25, 2000. The Mechanical Earth Model Concept and Its Application to High-Risk Well Construction Problems. Paper SPE 59128. IADC/SPE Drilling Conference, New Orleans, Louisiana, USA.

Prioul, R., Karpfinger, F., Deenadayaulu, C., Suarez-Rivera, D., November 15–17, 2011. Improving Fracture Initiation Predictions on Arbitrarily Oriented Wells in Anisotropic Shales. Canadian Unconventional Resources Conference, Calgary, Alberta, Canada.

Sayers, C., 2006. An introduction to velocity-based pore-pressure estimation. The Leading Edge 25 (12), 1496–1500.

Sayers, C., 2010. Geophysics under Stress: Geomechanical Applications of Seismic and Borehole Acoustic Waves. SEG Distinguished Instructor Short Course, Series No 13.

Suarez-Rivera, R., Burghardt, J., Edelman, E., Stanchits, S., June 23–26, 2013a. Geomechanics Considerations for Hydraulic Fracture Productivity. Paper ARMA 13-666. 4th US Rock Mechanics/Geomechanics Symposium, San Francisco, California, USA.

Suarez-Rivera, R., Burghardt, J., Stanchits, S., Edelman, E., Surdi, A., March 26–28, 2013b. Understanding the Effect of Rock Fabric on Fracture Complexity for Improving Completion Design and Well Performance. Paper IPTC 17018. International Petroleum Technology Conference, Beijing, China.

Suarez-Rivera, R., Deenadayalu, C., Chertov, M., Hartanto, R., Gathogo, P., Kunjir, R., November 15–17, 2011. Improving Horizontal Completions on Heterogeneous Tight Shales. Paper CSUG/SPE 146998. Canadian Unconventional Resources Conference, Calgary, Alberta, Canada.

Suarez-Rivera, R., Green, S.J., McLennan, J., Bai, M., September 24–27, 2006. Effect of Layered Heterogeneity on Fracture Initiation in Tight Gas Shales. Paper SPE 103327. SPE Annual Technical Conference and Exhibition, San Antonio, Texas, USA.

Yan, G., Karpfinger, F., Prioul, R., Tang, H., Jiang, Y., Liu, C., December 10–12, 2014. Anisotropic Wellbore Stability Model and Its Application for Drilling through Challenging Shale Gas Wells. Paper IPTC 18143-MS. International Petroleum Technology Conference, Kuala Lumpur, Malaysia.

CHAPTER 8

HYDRAULIC FRACTURE TREATMENT, OPTIMIZATION, AND PRODUCTION MODELING

Domingo Mata[1], Wentao Zhou[2], Y. Zee Ma[3], Veronica Gonzales[1]

Technology Integration Group TIG, Schlumberger, Denver, CO, USA[1]; Production Product Champion, Schlumberger, Houston, TX, USA[2]; Schlumberger, Denver, CO, USA[3]

8.1 INTRODUCTION

Unconventional reservoirs generally require hydraulic fracturing treatment (HFT) for extracting the hydrocarbons. HFT is a process of creating or restoring fractures in a geological formation using fluids and proppants to stimulate production from oil and gas wells. The completion of hydraulically fractured wells involves many processes and variables, including completion type, stage design, lateral landing, hydraulic fracture geometry, fracturing fluid, proppant type/size/schedule, hydraulic fracture monitoring, fracture spacing, tubular selection, selection of fracturing point, simultaneous versus sequential fracturing, fracture initiation points, hydraulic fracture's interaction with natural fractures, perforation strategies, and production analysis (Chong et al., 2010; King, 2010, 2014; Allix et al., 2011; Baihly et al., 2010; Manchanda and Sharma, 2014). These variables and processes impact the fracture complexity, stimulated reservoir volume, and effectiveness of a fracturing treatment.

Some of the most basic variables and tasks in hydraulic fracture design include fracture fluid-selection, proppant selection, evaluation of reservoir quality (RQ) and completion quality (CQ), and evaluation of fracture geometries and propagations based on rock mechanics. The completion design in developing unconventional reservoirs should be based on a combination of geomechanical and reservoir properties. It is important to distinguish completion effort from CQ. While CQ describes the properties of the rocks that make completion straightforward or complicated, completion effort describes the various completion methods and tools used in the completion to access the hydrocarbon in the formation. The combination of CQ and completion effort determines the completion efficiency.

Rock properties and natural fracture distribution in tight reservoirs have significant implications for stimulation and recovery. Because of the high degree of bed laminations and variations of rock texture in organic shales and other tight formations, the hydraulic fracture geometry can be complex. Stress is a critical parameter that will govern fracture geometry. Heterogeneities of mechanical properties in both the horizontal and vertical directions result in complex stress profiles. Model calibration is required to provide a fracture design that limits fracture-height growth and promotes fracture half-length and maximizes reservoir coverage.

CHAPTER 8 HYDRAULIC FRACTURE TREATMENT, OPTIMIZATION

This chapter first presents main principles and considerations in fracturing fluid selection and proppant selection. It then discusses the optimization of hydraulic fracture design based on the mechanical earth model (MEM), and formation stress field. One of the tasks in completing a hydraulically fractured well is to estimate the fracture properties; we present the method of fracture pressure history match to estimate the fracture geometries.

The second part of this chapter covers production modeling. We discuss analytical and numerical methods for modeling hydrocarbon production from unconventional resources. The advantages and disadvantages of these methods are elaborated. We present how to develop a predictive production model by integrating geological, petrophysical, and fluid data. We also discuss applications and uncertainty analysis of production modeling.

In the last part of the chapter, we briefly discuss economic analysis of completion using an example of determining an optimal value for a production parameter. Operational and logistic considerations are briefly discussed.

8.2 FRACTURE FLUID AND PROPPANT SELECTIONS

Fracture design is a complex process, especially when data is scarce and the fracture model cannot be adequately calibrated. The important parameters, including formation permeability, stress contrast, conductivity requirements, and formation fluid/chemistry, may not be available or have large uncertainties. Thus, it is often necessary to use a multidisciplinary approach and an iterative process to narrow down the uncertainty and determine the best stimulation technique for an exploration well. This section provides some general guidelines for fracture design when data is scarce and models cannot be calibrated. Applications of fracture fluids and proppants are presented.

8.2.1 FLUID SELECTION

One of the most important processes in fracture design is the fluid selection. The fracturing fluid is a bridge between reservoir fluids and the wellbore. A good fluid minimizes complications in HFT such as screen-outs or difficulty with fracture placement. Several objectives are considered when deciding what fluid to use for providing the conduit for reservoir fluids. Several properties are examined to ensure an efficient conveyance of proppant through the well assembly into the formation.

Based on rheology fluids can be Newtonian and nonNewtonian. Newtonian fluids include water, heavy brines, and slickwater, and they are nonwall-building and have exceptionally high leak-off capabilities. On the other hand, wall building fluids with higher viscosities, such as gel-based systems, have lower leak-off capabilities and provide conveyance for heavy laden proppant schedules. Most wall building fluids are linear or cross-linked polymers that leave residual damage in the fracture network, reducing the fracture conductivity. More advanced fluids, such as viscoelastic surfactants, provide the conveyance offered by gel laden, and heavy viscosity type of fluids, while retaining the permeability of the proppant pack and maximizing fracture conductivity.

Understanding of some basic reservoir parameters is vital for the selection of fluid. The most important variables include formation permeability, stress contrast, conductivity requirements, and formation mineralogy and fluids because fracturing fluids behave differently as a function of these variables. Table 8.1 summarizes important considerations in fracture fluid selection.

8.2 FRACTURE FLUID AND PROPPANT SELECTIONS

Table 8.1 Important Variables and Considerations in Fracture Fluid Selection

Property	Scenario	Consideration	Recommended Fluid
Formation permeability	Low permeability	Leak-off to formation decreased	Newtonian or wall building fluid with low gel loading
	High permeability	Leak-off to formation increased	Crosslink fluid with higher gel loading
Stress contrast	Low contrast	Height growth expected	Low viscosity fluid
	High contrast	Lower heights/containment	Higher viscosity or gelled fluid
Conductivity requirements	Low	Low permeability/gas wells	Newtonian or low viscosity fluids
	High	High permeability/multiphase wells	Cross-link fluid, higher gel loading fluids, viscoelastic fluids, energized fluids
Formation mineralogy and fluids	Swelling clays/emulsions	Clay swelling, emulsions, scale	Use of surfactants and nonemulsifying agents
	High calcite	Calcite present	Use of acid

The formation permeability impacts the selection of fluid. In low permeability environments, leak-off from the fractures to the formation is generally low, whereas, in high permeability environments, leak-off from the fractures to the formation is high.

The stress contrast is one of the most important parameters for fluid selection in a stimulation treatment because it will determine the potential for fracture height growth. When height growth is prevalent, fluids with low viscosity are better choices, as they develop low net pressures maximizing fracture containment through low stress contrast layers. In areas with a high stress contrast, a higher viscosity gelled fluid is generally a better choice.

The dimensionless conductivity of hydraulic fractures, defined as the ratio of the ability of the fracture to carry flow over the ability of the formation to feed the fracture (Economides and Nolte, 2000), is another important variable for fluid flow and reservoir deliverability from a stimulation treatment. Notice that fracture conductivity is correlated to formation permeability.

Understanding the geochemistry of the formation and its compatibility with water and oil-based products allows a more accurate fluid selection. For example, in areas where emulsions or sludging is frequent, the use of nonemulsifying agents is recommended. Complications resulting from unintended chemistry reactions between formation and fracture fluids can result in a poor stimulation treatment.

8.2.2 PROPPANT SELECTION

The purpose of a propping agent is to maintain a conductive channel between the reservoir and the wellbore when the hydraulic force is withdrawn and the fracture closes. Selecting the proper grain size and material strength can be a convoluted process that requires good understanding of reservoir properties, fracture geometry, in situ stresses and well economics.

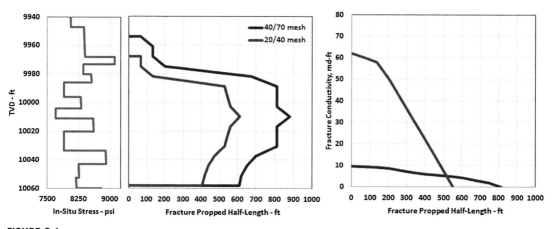

FIGURE 8.1

Impact of proppant mesh size on fracture geometry and fracture conductivity based on an experiment using a fracture simulator.

A good starting point is to determine the proppant mesh. Smaller mesh proppants such as 100 mesh or 40/70 have a diameter of 0.006–0.0011 inches and the particles will travel further inside the fracture and create a longer propped fracture lengths. The disadvantage of smaller size proppants is that the void space between the particles is small, resulting in low fracture conductivity. Figure 8.1 shows an example of the impact of proppant mesh size on fracture geometry and conductivity. In the simulation, two mesh sizes were used: 40/70 and 20/40 mesh, and the conveying fluid and stress conditions are the same. The 40/70 proppant mesh travels further into the fracture, resulting in fracture conductivities between 5 and 10 md ft throughout the whole 800 ft interval. On the other hand, the 20/40 mesh provides significantly better fracture conductivity (up to 60 md ft), but it only reaches about 500 ft of conductive length.

Another important parameter for proppant is the material strength that should be high enough to avoid fatigue under high closure stress. Grain failure can create small fines that migrate and reduce the permeability of the proppant pack. The grain strength can be reinforced using chemical coats (e.g., resin). Alternatively, synthetic proppants that have less internal porosity can perform significantly better under high stress. However, materials that are resistant to high closure stress often have higher specific gravity and settle faster as they travel in the fracture. It is important to couple these types of proppant with a carrying fluid that is capable of conveying the proppant to the designed length.

The effective stress on the proppant pack will determine the required strength of the material. Effective stress is the difference between the formation closure and bottom hole flowing pressure as shown in Eq. (8.1) (Economides and Nolte, 2000).

$$\sigma_{eff} = \sigma_{closure} - P_{wf} \tag{8.1}$$

As the effective stress increases, the fracture conductivity decreases. Table 8.2 lists the main features of several proppant types.

Table 8.2 Proppant Selection Based on the Effective Stress

Proppant Type	Specific Gravity	Effective Stress on Proppant	Cost
Conventional sand	2.65	<6000 psi	$
Resin-coated	2.55–2.60	<8000 psi and <250 °F	$$–$$$
Intermediate strength	2.60–2.90	<8000 psi and >250 °F	$$$–$$$$
High strength	>3.0	>10,000 psi	$$$$

Furthermore, it is important to ensure that the material selected can deliver a proppant pack that is capable of transporting the reservoir fluids efficiently. The dimensionless fracture conductivity is a measure that relates fracture performance and reservoir deliverability, as described by Eq. (8.2) (Cinco-Ley et al., 1978; Cinco-Ley and Samaniego, 1981).

$$C_{FD} = \frac{k_{prop} \cdot w}{k_{formation} \cdot x_f} \qquad (8.2)$$

The numerator represents the ability of the fracture to transport reservoir fluids into the wellbore and it depends on the fracture width and proppant permeability. The denominator refers to the ability of the reservoir to feed the fracture and it is the product of formation permeability and fracture half-length.

The main principle of proppant selection is that the fracture should not restrict production. The requirements for fracture conductivity will depend on the flow regime of the well. For pseudosteady state, a C_{FD} value of 2 is considered to have an infinite conductivity by some researchers (Prats et al., 1962). This means that the proppant pack is able to transport two times the rate at which the reservoir can feed and further improvements will not result in higher production. For unconventional reservoirs, because wells are in transient flow during the majority of their operating life, a dimensionless conductivity of 30 for dry-gas reservoirs and 100 for liquids-rich reservoirs is preferable.

The following example illustrates the importance of selecting a propping agent that can sustain formation closure stress and provide sufficient fracture conductivity across the effective fracture area. In Fig. 8.2, "Well 1" was completed with ceramic proppant, capable of resisting formation closure of >10,000 psi. On the other hand, "Well 2" and "Well 3" were completed with intermediate strength proppant. Finally, "Well 4" was completed mostly with conventional sand. During the first year of production, "Well 1" exhibits a performance fivefold higher than "Well-4". Economic analysis determined that a 1.5-fold increase would cover the additional cost of the ceramic proppant.

8.3 OPTIMIZING FRACTURE DESIGN AND COMPLETION STRATEGIES

When completion data is available to calibrate models the completion strategy can be better optimized. In developing unconventional resources, cost control and good engineering practices are crucial because profit margins are generally small and highly sensitive to commodity prices. The ability to integrate RQ and CQ will allow proper assessment of a given prospect. Petrophysical models provide critical input parameters to determine reservoir and acoustic properties that will

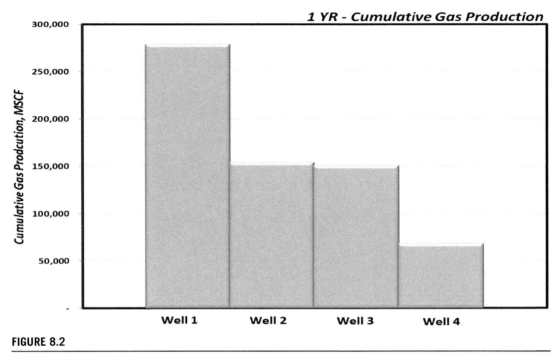

FIGURE 8.2

Comparison of first year performance for four wells in the same field. Petrophysical properties are similar but CQ changed based on different proppant selection.

serve as input to production and stimulation simulations. Although these models are built using measurements, such as core samples, and log data, and constrained with production history match, they often still carry a high level of uncertainty. It is critical to understand fracture propagation, net pressures, spacing etc., in order to design an optimum stimulation strategy tailored to the rock characteristics. Figure 8.3 shows a workflow for calibrating a hydraulic fracture model to completion and reservoir qualities.

A reliable MEM is highly important for accurate fracture estimation. Mechanical properties and stress calculations directly affect the net pressure in the simulation with a significant impact on the fracture geometry. Selecting the adequate fracture simulator is vital and the stress calculation in the MEM can provide insights on the model required for a specific stress condition. The incorporation of the log data can be valuable if the elastic properties are honored, the fracture simulation is adequately performed, and the results show repeatability as well as validity by other measurements.

8.3.1 BUILDING A CALIBRATED MECHANICAL EARTH MODEL

An MEM is a numerical representation of the rock mechanical properties and in situ stress state of the subsurface. The rock properties, including density, Poisson's Ratio, Young's Modulus, pore pressure, and Earth stresses, are the primary inputs for hydraulic fracture modeling and evaluation. The

8.3 OPTIMIZING FRACTURE DESIGN AND COMPLETION STRATEGIES

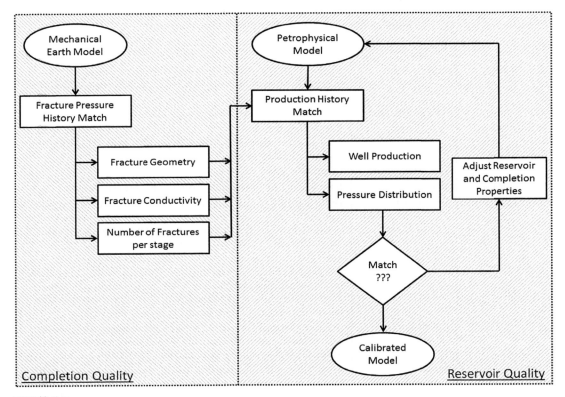

FIGURE 8.3

Illustration of calibrating hydraulic fracture model to completion quality and reservoir quality.

complexity of an MEM depends on its application to hydraulic fracturing of unconventional resources, and sometimes it requires not only an understanding of these properties along the wellbore (one-dimensional (1D) or two-dimensional), but also beyond (three-dimensional (3-D) or four-dimensional). The majority of the commercial fracture simulators rely on a 1D MEM to predict fracture propagation and overall geometry while assuming transverse isotropy about a vertical axis of symmetry (Transverse Isotropic Media or TIV) in the vicinity of the HFT and extend these properties uniformly laterally.

Here we give an overview on the workflow to build and calibrate a 1D MEM that has been applied in numerous basins since the mid 1990s. Primary inputs for the MEM are the acousto–elastic rock properties and density as derived from advanced sonic measurements and the minimum horizontal stress, which is assumed to be a function of the lithostatic stress (dependent on rock elastic properties, pore pressure, and overburden) and the tectonic loading, such as:

$$\sigma_h = \frac{E_h}{E_v} \frac{v_v}{1 - v_h} (\sigma_z - \alpha P_p) + \alpha P_p + Tectonics \tag{8.3}$$

where the first term on the right hand is the lithostatic stress component.

Table 8.3 Common Mechanical Properties and Their Data Sources

Property	Data Source	Comments
Poisson's ratio	Acoustic data	Correlations from dynamic to static condition are applicable. During the exploration phase it is recommended that values get calibrated to actual lab measurement.
	Lab measurement	Often most reliable data source
Young modulus	Acoustic data	Correlations from dynamic to static condition are applicable. During the exploration phase it is recommended that values get calibrated to actual lab measurement.
	Lab measurement	Often most reliable data source
Overburden stress	Bulk density	Integration of log measurement (subject to hole condition)
Pore pressure	Drilling data	Drilling reports can help understand transitions in the pore pressure regime throughout the entire drilling process. Key information such as mud weight, loses, and kicks must be considered and incorporated to the mechanical earth model construction.
	Testing tools	Modular testers can acquire in situ values of pore pressure with great accuracy. Additionally, measurements can be taken at different depth to understand changes in the pore pressure regime throughout the interval of interest. In fact, most testers can perform small injection fall-off tests that can help calibrate the stress value.
	Completion data	If completion data is available, it can help constrain pore pressure value. Shut-in casing pressures and instantaneous shut-in pressures represent absolute upper bounds for reservoir pressure.
Biot constant	Mineralogy	A modeling process to predict mineralogy would allow estimation of continuous mineral bulk modulus. This would provide more accurate Biot's coefficients and fluid substitution results (Havens, 2011).
Tectonic stress	Closure test	Tectonic stress depends on rock strain, a value that can't be measured with current technology. Therefore, if the influence of tectonic forces is expected, calibration of this parameter most be performed last. Closure tests allow estimating the influence of tectonic stress by subtracting the lithostatic component from the overall closure value.

The accuracy of the minimum horizontal stress calculation relies on the qualities of the inputs in Eq. (8.3). However, the heterogeneous nature of unconventional reservoirs poses a challenge to the current reservoir characterization tools and requires in most of the cases integration of various data to narrow down the uncertainty to a manageable range. Table 8.3 summarizes common data sources and techniques that can be used to estimate reservoir and rock properties with an acceptable degree of approximation.

The maximum horizontal stress is determined using a wellbore stability simulator to match predicted borehole breakouts with caliper enlargements. An example of an MEM performed in a well located in Mountrail County, North Dakota is shown in Fig. 8.4.

There are different techniques to calibrate earth models. Core measurements, injection fall-off test and Fracture Pressure History Match (when completion data is available) can be used to calibrate the initial mechanical model. Figure 8.4 corresponds to an injection fall-off in an Oil Shale formation in

8.3 OPTIMIZING FRACTURE DESIGN AND COMPLETION STRATEGIES

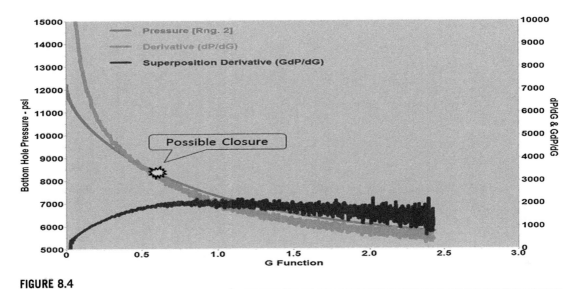

FIGURE 8.4

G-function plot for an injection fall-off test. Instantaneous shut-in pressure ~12,000 psi. Estimated closure ~8500 psi. Resultant fracture net pressure ~3500 psi. For this particular example, closure occurs within 1 log cycle which raises concerns about the validity of the test.

the Continental United States. In this example, closure seems to occur in the vicinity of 8500 psi. This value indicates that the net pressure developed during the short injection period was ~3500 psi. However, closure occurs within one log cycle in the time axis. This raises valid concerns regarding test integrity. It could be possible that the test did not develop a competent fracture. Therefore, it is always recommended to run several test to validate closure results. If reliable conclusions are obtained from the injection fall-off test, these serve as calibration points to the minimum horizontal stress in the MEM (Fig. 8.5).

8.3.2 SELECTING THE ADEQUATE FRACTURE MODEL

Modeling unconventional well completion is a challenging task. Large uncertainty exists and integration of the various domains, including geology, geomechanics, completion, and reservoir engineering, is needed to estimate the possible ranges for fracture geometry and more importantly fracture propagation. Too often, analysis is performed in a basin with idealized or synthetic data while ignoring some basic assumptions in log processing, properties calculations and fracture simulation. The lack of understanding of data limitation compromises the process and frequently results in models that are not robust and may even violate key engineering concepts.

A critical dataset that can facilitate the integration process is pilot-hole log-data. It is recommended that the information is captured above and below the target zone because the properties in the surrounding formations are important for accurate modeling and understanding of the fracture propagation process. Sometimes data acquisition programs focus exclusively in RQ evaluation without

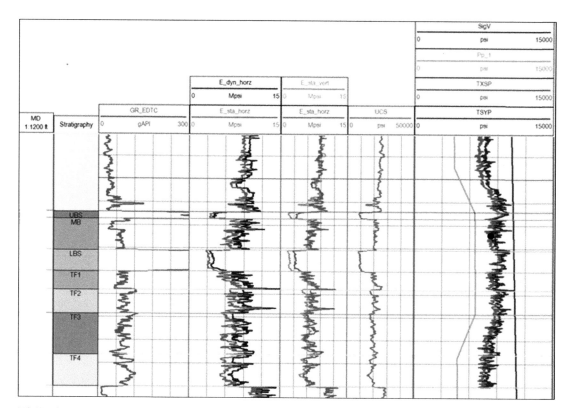

FIGURE 8.5

Mechanical earth model. Track 1: measured depth, Track 2: stratigraphy, Track 3: gamma ray, Track 4: horizontal Young's modulus, dynamic (blue, black in print versions), static (orange, lighter gray in print versions), Track 5: vertical Young's modulus, dynamic (pale blue, gray in print versions), static (orange, lighter gray in print versions), Track 6: unconfined rock strength, Track 7: overburden stress (red, darker gray in print versions), pore pressure (magenta, dark gray in print versions), minimum horizontal stress (green, light gray in print versions), and maximum horizontal stress (blue, black in print versions). The black dot in the minimum horizontal stress line corresponds to the closure value observed during the injection fall-off test.

Modified from Mata et al. (2014b).

an attempt to understand completion challenges. Ideally, a complete dataset will include not only the pilot-hole log data but also lateral measurements in case of horizontal wells to understand variability and identify areas that will pose challenges for well completions.

In most commercial fracture simulators, planar fractures are assumed, and height growth is calculated using an analytical pseudo3-D (P3-D) or a 3-D numerical model. The resultant geometry from these two models may vary dramatically. It is well known that P3-D models have limitations when barriers are not present (run-away height growth). Additionally, in these simulators, mechanical properties are averaged across the fracture height, preventing the user from incorporating important variables, such as rock laminations and high stress sections (i.e., bentonite layers) to determine the

8.3 OPTIMIZING FRACTURE DESIGN AND COMPLETION STRATEGIES

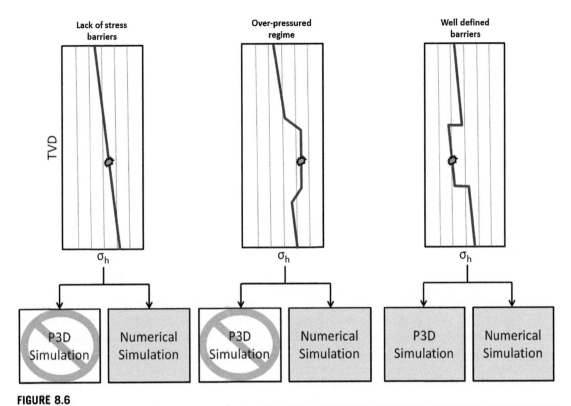

FIGURE 8.6

Geometry comparison for different stress regimes using a Pseudo-3-D analytical simulator and a numerical simulator (Left: Uintah–Piceance Basin, Middle: Bakken–Three Forks system, and Right: Niobrara/Frontier system).

effect on fracture geometry. It is recommended to use a 3-D numerical simulator when low stress contrast is expected or thin laminations exist, since these models honor the mechanical properties for every layer in the fracture grid.

Another limitation of many conventional P3-D fracture simulators is the need for upscaling. This can limit the ability of the model to recognize laminations and variability between rock layers. It is important to preserve the log resolution as much as possible to construct a valid model.

Figure 8.6 shows three different stress profiles and the gun shots correspond to the fracture initiation point. The left box indicates a stress regime without significant barriers. The middle box represents fracture initiation from an overpressured interval. The left box indicates fracture initiation point in a contained interval with well-defined barriers. The boxes below indicate the model that is applicable for each case.

Another factor in the selection of adequate simulation tool is the difference in magnitude between the principal horizontal stresses, as this will promote or limit the ability to propagate a

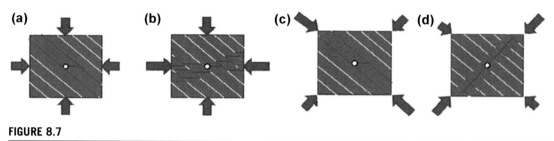

FIGURE 8.7

Various fracture propagation schemes.

Modified from Suarez-Rivera et al. (2013a).

planar fracture. When the difference between principal horizontal stresses is large enough, a planar fracture is more likely, even if planes of weakness exist (Suarez-Rivera et al., 2013a). On the other hand, if the stress field is in an isotropic state, fracture complexity is likely to occur, especially if planes of weakness, such as natural fractures, are present. Figure 8.7(a) shows fracture propagation on a medium with planes of weakness oriented oblique to the direction of maximum in-plane stress, in which the difference between the principal stresses is relatively small. Figure 8.7(b) shows the same case but assuming high stress contrast. Figures 8.7(c) and (d) illustrate the fracture propagation on a medium with planes of weakness oriented parallel and perpendicular to the direction of maximum stress. The difference in principal stress for these two cases is high.

These experiments lead to the conclusion that the assumption of planar fractures in a simulator is valid for areas with significant difference in principal horizontal stress. However, a complex model is needed to accurately estimate the surface area stimulated for formations without significant differences between principal horizontal stresses (Wu et al., 2012).

8.3.3 ESTIMATING FRACTURE PROPERTIES

One of the best techniques available to estimate fracture geometry and properties is the fracture pressure history match (FPHM). This method uses a simulator to recreate the injecting pressure recorded in the field during the actual treatment. The simulation takes into account earth stresses, fluid rheology, proppant properties, pipe friction, and other parameters and provides an estimation of the resultant geometry, number of fracture propagating (for horizontal wells), fracture conductivity, and fluid efficiency.

The first parameter to be analyzed is the number of fractures that can propagate. Completions in multistage horizontal wells are often segmented using either ball and sleeves or plug and perforation. The ball and sleeve method is more efficient because it allows continuous pumping but reservoir coverage might be limited since only one fracture propagates through a sleeve port. In contrast, the plug and perforation method requires wireline intervention for each stage and uses an explosive gun to open three to six sections of casing at the same time. The advantage of the plug and perforation technique is that multiple fractures can propagate in a single pumping stage, which gives a larger reservoir coverage.

Propagating multiple fractures depend on the interaction of several variables such as fracture net pressure, stress contrast and fluid selection. High net pressures are required to facilitate breakdown of multiple clusters. Low stress contrast results in the dominant fracture breaking through the barriers, height growth occurs and this fracture absorbs the majority of the slurry, limiting the possibility of propagation through other clusters.

A step down tests before the main treatment (prior to sand stages) can be very useful. It is recommended to repeat this exercise multiple times in the well to determine consistency of the results. The analysis of these tests helps understand near-wellbore pressures, a parameter governed by loses related to perforation friction, phase misalignment and tortuosity.

Figure 8.8 shows a closure analysis and step-down test performed in the Williston basin, North Dakota. Results indicate that the area open to flow is approximately equivalent to a third of the holes shot. It is evident (intuitively and from modeling) that if the treatment is distributed equally across each perforation cluster (neglecting fluid friction down the lateral), the injection rates per cluster would be too low and screenouts (early unplanned treatment terminations) may be more likely and offset well fracture interference may not be visible.

FPHM can be useful for this analysis. FPHM consists of using the MEM, petrophysical properties, fluid and proppant properties, and perforation parameters (number of clusters, spacing and shot density) for fracture simulation. Adjustments are made until the simulated pressure response mimics the surface treating pressure recorded in the field during the fracture treatment. Although this method does not provide a unique solution, it can help reduce the uncertainty and set boundaries around fracture geometry and fracture propagation.

An example of FPHM is shown in Fig. 8.9. The exercise determined that: (1) it is not possible to obtain a match assuming that only one fracture propagates in the stage. Even if the treatment fluid is assumed to be ideal (no friction loses in pipe) the simulated treating pressure is higher than the actual value (driven by the near wellbore loses). (2) It is possible to match treating pressure assuming both cases two or three fractures propagating per stage. In conclusion, using FPHM it was determined that it was possible to propagate at least two fractures. The implications of this finding in well productivity will be discussed in the next section, Production Modeling.

After understanding fracture propagation, the next step is to analyze the fracture geometry resulting from a successful FPHM. Figure 8.10 shows an example of a fracture geometry estimation using a numerical fracture simulator.

Although the simulation indicates propped lengths reaching up to 1400 ft, a fair amount of that interval contains very low sand concentrations (less than 1.0 ppa) which is small compared with the initial sand stages. The fracture conductivity profile confirms that behavior. Based on these results, the maximum effective fracture half-length is estimated to be approximately 500 ft (Fig. 8.11).

Finally, it is important to estimate the number of propagating fractures, fracture geometry and fracture conductivity. However, uncertainty intrinsic to the analysis requires an additional validation loop to determine the portion of propped rock that will actually contribute to production. This can be accomplished through production modeling.

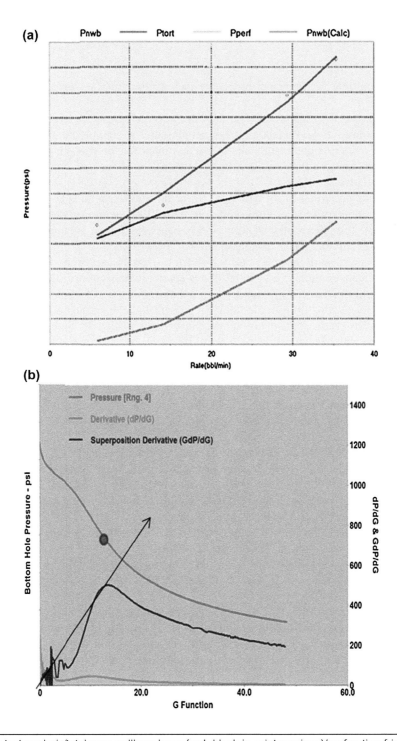

FIGURE 8.8

(a) Step down test analysis/total near wellbore loses (red, black in print versions)/perforation friction (green, dark gray in print versions)/tortuosity loses (blue, darker gray in print versions). (b) G-function closure analysis. Blue (darker gray in print versions) dot represents the estimated closure value. This parameter is used to calibrate the Mechanical Earth Model (Suarez-Rivera et al., March 26, 2013b). Understanding the Effect of Rock Fabric on Fracture Complexity for Improving Completion Design and Well Performance. International Petroleum Technology Conference http://dx.doi.org/10.2523/17018-MS.

8.3 OPTIMIZING FRACTURE DESIGN AND COMPLETION STRATEGIES

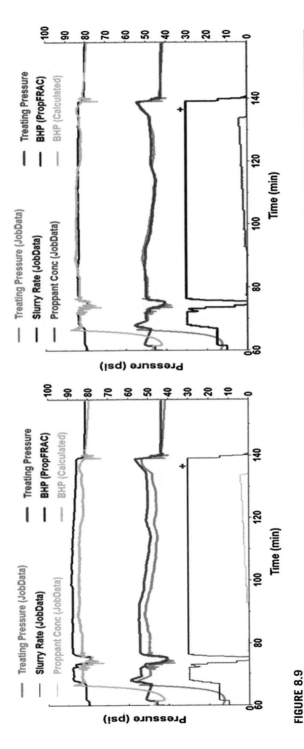

FIGURE 8.9

Fracture pressure history match. Left: match attempt assuming that only one fracture is propagating. Right: fracture Pressure history match assuming that two fractures are propagating.

Modified from Mata et al. (2014b).

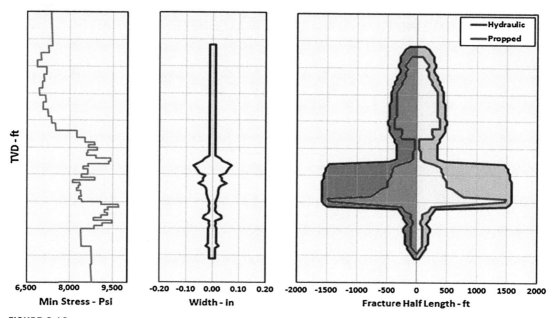

FIGURE 8.10

Left: minimum horizontal stress versus True Vertical Depth (TVD); Center: fracture width profile; Right: fracture half-length. For the first well in the unit, fracture wings are symmetrical. The yellow (light gray in print versions) area corresponds to one fracture wing while the orange (gray in print versions) area represents its counterpart.

Modified from Mata et al. (2014a).

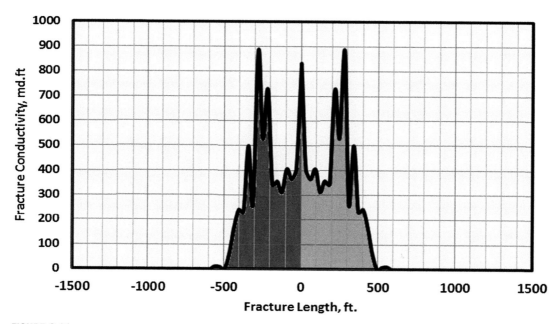

FIGURE 8.11

Fracture conductivity vs. fracture length. Conductivity is similar for both wings. The yellow (light gray in print versions) area corresponds to the wing facing the parent well while the orange (gray in print versions) area represents its counterpart. Fracture conductivity passed 500 ft is negligible due to critical low concentrations per area.

Modified from Mata et al. (2014a).

8.4 PRODUCTION MODELING

Production models are often built for two purposes in relation to completion: optimizing the prejob design, and evaluating the postjob completion effectiveness.

For prejob optimization, production models are created for a number of completion scenarios, for example, to select proppant and fluid type and volume and find the optimal design that yields the maximum recovery or net present value (NPV) under field constraints.

For postjob evaluation, production history is matched by tuning key completion parameters in the production model, for example, hydraulic fracture half-length and conductivity, to obtain an estimation of these parameters, which can be related back to the original design for completion optimization on future wells.

8.4.1 ANALYTICAL VERSUS NUMERICAL MODELS

Various production models are built for production optimization and they can be analytical or numerical, each with its own advantages and limitations. The main types of analytical and numerical models are shown in Table 8.4.

In analytical models based on inflow performance relationship (IPR), hydraulic fracture is treated as an equivalent or pseudoskin. It is a function of fracture parameters, such as half-length, conductivity, fracture damage skin, formation permeability. For example, for a fractured oil well in steady-state, well productivity index can be expressed by

$$PI = \frac{kh}{141.1 B \mu \left[\ln \frac{r_e}{r_w} + s_f\right]} = \frac{kh}{141.1 B \mu \left[\ln \frac{r_e}{r'_w}\right]} \qquad (8.4)$$

where $r'_w = r_w e^{-s_f}$ is the effective well radius. Dimensionless fracture conductivity $F_{cd} = \frac{k_f b_f}{k X_f}$ is related to pseudoskin s_f as shown in Fig. 8.12. At high conductivity, effective well radius $r'_w = 0.5 X_f$.

IPR models for multiply fractured horizontal well in conventional reservoirs are available (Raghavan and Joshi, 1993). In fact, IPR models are widely used in conventional wells, but they are rarely used for unconventional wells because they are not applicable in ultratight formations. Instead, analytical reservoir simulation is more useful.

Table 8.4 Main Types of Production Models

Type	Main Types
Analytical model	Inflow performance relationship (IPR)
	Analytical reservoir simulation
Numerical model	Implicit fracture modeling
	Explicit fracture modeling

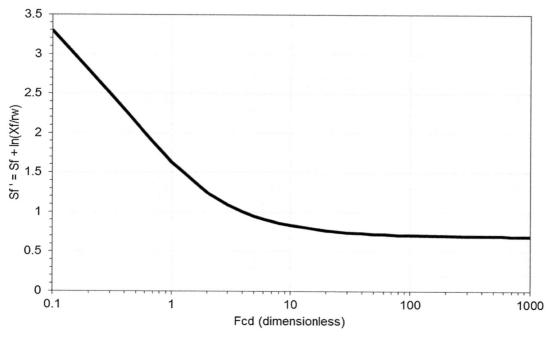

FIGURE 8.12

Pseudoskin factor for a well with a vertical fracture.

Reproduced from Cinco-Ley et al. (1978).

The fluid flow in porous medium is governed by the pressure diffusion equation

$$\frac{\partial p}{\partial t} = \eta_x \frac{\partial^2 p}{\partial x^2} + \eta_y \frac{\partial^2 p}{\partial y^2} + \eta_z \frac{\partial^2 p}{\partial z^2} \qquad (8.5)$$

subject to initial and boundary conditions. Eq. (8.5) can be solved analytically. Detailed assumptions, solution process and an exhaustive compilation of solutions are discussed in Thambynayagam (2011). Different inner boundary conditions give rise to models for single or multistage planar fractures (Cinco-Ley and Samaniego, 1981), orthogonal and unorthogonal fracture networks (Zhou et al., 2014).

While the IPR method is only applicable to steady state or pseudosteady state, analytical solution captures all the flow regimes, including the transients, which is essentially to model production in tight and shale formations.

Figure 8.13 shows the pressure distribution around a horizontal well with six transverse fractures at different times, and the flow regimes on the pressure log–log plot.

This type of models are widely used in pressure transient analysis (PTA) and rate transient analysis (RTA) for postjob evaluation in both conventional and unconventional wells. They are also widely used for prejob design in unconventional wells.

Analytical models can be created quickly and the computational cost is very small. These analytical methods are easy to use and require minimal amount of data. Transient flow regimes,

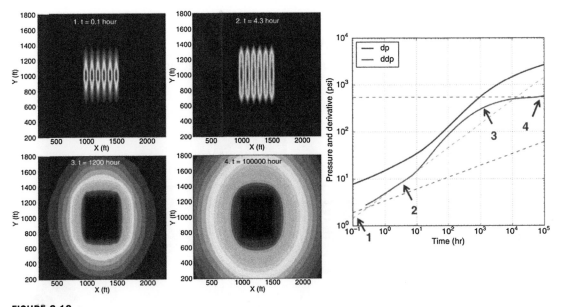

FIGURE 8.13

Flow regimes of a multistage fractured horizontal well (after Zhou et al., 2014): formation linear flow (1 and 2), stimulated reservoir volume flow (3), and pseudoradial flow (4).

essential for PTA and RTA, are easily captured while numerical models may suffer from numerical issues. Nonetheless, limitations of analytical models include:

- Reservoir model is typically single-layer homogeneous shoe box. Structural complexity and reservoir heterogeneity cannot be handled.
- Mathematics of analytical simulation is based on single-phase fluid. Dynamic multiphase behavior, such as water-coning, gas-coning, solution gas drive, cannot be modeled.
- Compositional fluid model cannot be used in analytical models.

When it comes to numerical models, hydraulic fractures may be treated implicitly, using equivalent wellbore skin similar as in the IPR model, or by mapping the fractures to a high-permeability zone and/or dual porosity model (Du et al., 2009; Fan et al., 2010; Li et al., 2011). These models give quick estimation of well performance, but the direct link between completion design and production outcome is compromised in the mapping process.

Another type of numerical models treat hydraulic fractures explicitly, as high permeability grid cells. Planar fractures and orthogonal fracture networks (Xu et al., 2010) are usually modeled with a structured grid system. Unorthogonal fracture networks, however, poses serious challenges in terms of gridding and numerical solutions. The combination of unstructured grid and other advances in reservoir simulation technology has made this possible (Cipolla et al., 2011; Mirzaei and Cipolla, 2012). Numerical models can handle the complexity of hydraulic fractures, but they require complex inputs and take more time and expertise to build and run.

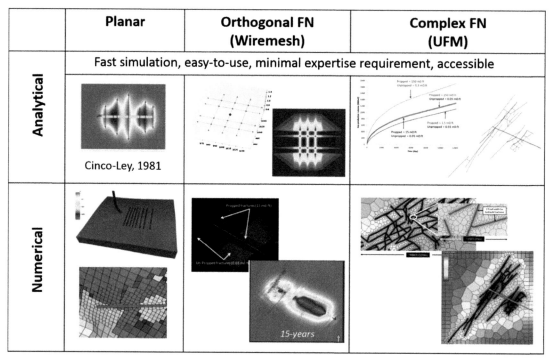

FIGURE 8.14

Analytical and numerical simulation for hydraulically fractured wells.

Synthesized from a number of sources: Zhou et al. (2013), Zhou et al. (2014), Cipolla et al. (2011), and Li et al. (2011).

Figure 8.14 summarizes the different methods to model hydraulic fractures, depending on the type of fractures, planar, orthogonal, or complex fracture networks (Zhou et al., 2013).

The use of either analytical or numerical models depends on many aspects, for example:

- If rich amount of data is available, such as, well logs, fluid Pressure, Volume and Temperature (PVT), geological information, numerical models are more appropriate because they can readily integrate the data. On the other hand, if data is limited, analytical models can be used.
- Analytical models are preferable when fast turnaround is required, or reservoir modeling and simulation expertise is limited.

8.4.2 CONSOLIDATING A PREDICTIVE MODEL

Production model is created by consolidating the petrophysical, geological, and fluid inputs with the result from hydraulic fracturing modeling.

The petrophysical, geological and fluid inputs include porosity, permeability and water saturation distribution, rock compressibility, reservoir depth and structure, fluid PVT, pressure, capillary pressure, and relative permeability curves.

Hydraulic fracture model includes the 3-D geometry of the fractures and properties, including width, proppant concentration, conductivity, etc.

For planar fractures, analytical model takes simple inputs. No gridding is required. The parameters required for hydraulic fractures include:

- Number of fractures (or number of stages and number of perforation clusters per stage).
- Fracture location in terms of measured depth along the well.
- Fracture orientation (longitudinal or transverse).
- Fracture half-length.
- Fracture conductivity.

Numerical model handles more complexity. Besides the parameters in the list above, it can also accommodate complex 3-D fracture geometry to better model fracture lateral and height growth, and varying fracture conductivity along the fracture and vertically due to proppant settlement. Several types of gridding methods may be used:

- Cells to represent hydraulic fractures honor fracture width, as shown in Fig. 8.15 (left). These cells are assigned real fracture permeability and porosity. Cell size may grow logarithmically into the formation. This is the closet to reality but may cause numerical issues because of the size difference and huge permeability contrast between these tiny fractures cells and the larger formation cells. It also generates large number of cells.
- Linear grid is used so cell size for hydraulic fractures and the formation are similar, as shown in Fig. 8.15 (right). Fracture cell permeability and porosity are then assigned honoring conductivity:

$$\overline{k} = \frac{k_f b_f}{\Delta x}, \overline{\phi} = \frac{\phi_f b_f}{\Delta x}$$

where k_f is the fracture permeability, ϕ_f is the fracture porosity, b_f is the fracture width, Δx is the fracture cell width, $\overline{k}$ and $\overline{\phi}$ are the permeability and porosity to be assigned to the fracture cells.

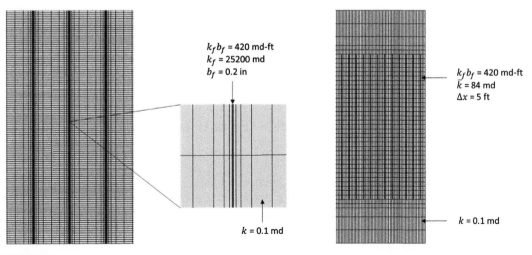

FIGURE 8.15

Representing hydraulic fractures with logarithmic cells and honoring fracture width (left), representing hydraulic fractures with linear cells (right).

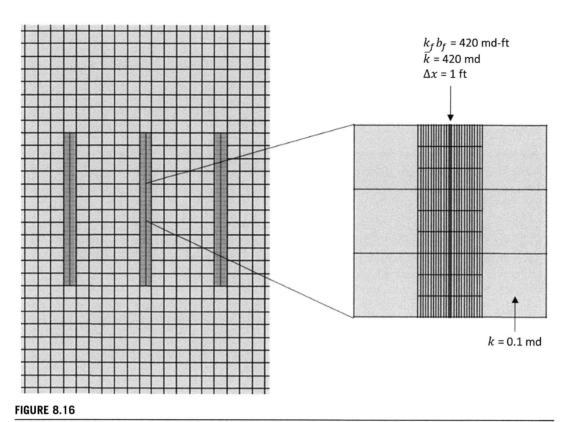

FIGURE 8.16

Representing hydraulic fractures with local grid refinement.

- Local grid refinement (LGR) is used on those cells that contain the hydraulic fracture, as shown in Fig. 8.16. LGR is built on top of a reservoir simulation model with no need of changing the global grid system, which is its key advantage.

When it comes to complex fracture networks, the analytical model requires the fracture information as below.

- Fracture network description (location of fracture panels, how the panels are connected to each other and to the wellbore).
- Conductivity of fracture panels.

An analytical fracture network model is shown in the middle of Fig. 8.17.

For numerical models, advanced nonstructured gridding is required to model the complex network as shown on the right of Fig. 8.17. Fractures may be represented by small cells. Similar to planar fractures, numerical models can better model fracture height growth and vertically-varying fracture conductivities.

8.4.3 MANAGING UNCERTAINTY

Compared with conventional vertical wells, production models for fractured horizontal wells in unconventional reservoirs contain more uncertainty because of the complex physics and input parameters.

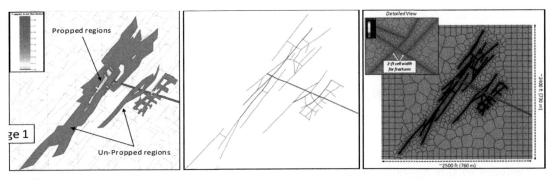

FIGURE 8.17

Hydraulic fracturing model (left), analytical production model, pink (dark gray in print versions) panels are propped fractures with high conductivity and cyan (gray in print versions) ones are unpropped fractures with low conductivity (middle), numerical model (right).

Taken from Cipolla et al. (2011) and Zhou et al. (2014).

Complexity of the production mechanism in unconventional reservoirs comes from multiple aspects:

- Slip flow, Knudsen effect for shale gas due to small pore sizes.
- Source rock physics, such as the presence of kerogen and gas desorption.
- Hydraulic fracturing and flow back process.
- Presence of natural fractures and their interaction with hydraulic fractures.
- RQ and CQ heterogeneity across the long horizontal lateral.
- Hydraulic fracture degradation.
- Wellbore flow dynamics in the lateral section and the vertical section. The low production rate and large wellbore size (to accommodate stimulation operation) tend to make the flow unstable.

The parameters required for the production model in unconventional resources, as summarized in the section above, are more uncertain, because:

- Rich suite wireline logs and core measurements are often reserved only for high-tier data wells, leaving many wells without accurate measurements.
- The measurements are often conducted in the vertical section. A laterally homogeneous reservoir model (layer-cake model) is assumed to cover the long lateral section.
- Interpreting shale logs and measuring shale cores are more challenging due to the ultra-low permeability.
- Due to the ultra-low permeability, there is a lack of accurate measurements for certain parameters, such as pressure, permeability, saturation functions including relative permeability curves, and capillary pressure curves, etc.
- The hydraulic fracture properties are generated from hydraulic fracturing modeling, which carry their own uncertainty.

All effort should be made to obtain maximum set of data for the well and offset wells, from multiple domains. For example, the geometry of hydraulic fractures can be estimated from pumping pressure matching, microseismic observation, and production history matching. The estimations should be cross-checked with each other to obtain a realistic estimation of effective geometry.

Production history-matching, with either numerical model or analytical model (rate transient analysis), is a way to reduce uncertainty too. Parameters that are influential on production and with high uncertainty should be tuned to reproduce the history, such as formation permeability, fracture half-length, stimulated reservoir volume, and end points for relative permeability curves, etc.

8.4.4 MODEL APPLICATIONS

A calibrated completion model can be used to determine:

- Optimum lateral length.
- Optimum number of stages per lateral.
- Optimum fracture design.
- Optimum number of laterals per drilling unit.
- Optimum field development plan.

8.5 ECONOMIC AND OPERATIONAL CONSIDERATIONS

As a function of economics, variables such as volume, logistics, and operations pose either constraints or inhibitors for the fracture design and fluid/proppant selection. Examination into the NPV for the well must be examined in order to determine the ideal amount of return that will supplement the cost invested into the well. The adage "the more you pump the more the production" has proven to be more of an anomaly, expensive, and unpredictable. Depending on economic constraints, attaining the return on investment will dominantly dictate much of the design parameters and what is feasible within a controlled budget. Ultimately, through science or execution, production and reservoir drivers should be established in order to estimate the NPV and the overall point of diminishing returns. Through the life of treatments in unconventional plays, the point of investment and seeing a positive return that is invested hits a peak and then begins going down. The objective of implementation of stimulation treatments coupled with the cost for drilling and planning a well is to find the optimum payout period, which subsequently is the peak on the NPV curve versus the driver (Fig. 8.18). Production drivers can

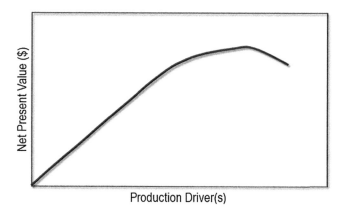

FIGURE 8.18

Illustration of finding the optimal net present value as a function of a production driver.

take months or years to establish through modeling or statistical analyses. Understanding these drivers proves to be very valuable in order to create operations through efficiency as well as reducing overall capital invested in wells to make economic successes.

8.5.1 OPERATIONAL AND LOGISTIC ANALYSES

Completion design optimization and NPV calculations are essential for understanding economics of the well, but execution of the design and handling material transport for placement becomes an equally hard challenge. Understanding placement of sand transporting, truck placement, and the horsepower needed on location is ideal for bringing the appropriate team and equipment for execution. Although designed through technical engineers, a series of questions arise as to check whether the appropriate operational equipment and materials are available. This includes the availability of the sand/proppant for placement within a reasonable location, whether the treatment can be pumped in compliance with safety standards, and whether we have the equipment available for pumping. With many more questions in the forefront, these are some common guidelines in order to provide a series of checks and balances from the design engineer to the operation engineer. Having both agreed with the timing and feasibility of the materials and equipment help execution to transition smoothly and safely.

It is also necessary to review how theory approaches actual implementation. For example, it may be easy to execute through modeling the application of three or more types of proppant, but during execution this makes placement and the hauling of proppant to location very difficult. Operators have to keep each type separated and have to regulate placement within the sand master on location. If the same result through modeling can be attained with only having two proppant types being placed, the risks associated with an incident reduce and make execution much more feasible.

8.6 CONCLUSION AND DISCUSSIONS

Production of oil and gas from unconventional resources commonly requires stimulation. Advanced fracturing and completion can maximize the reservoir contact through effective reservoir stimulation. A variety of proven reservoir stimulation technologies exists. Two basic processes in HFT design are fluid selection and proppant selection. These selections are based on a number of physical variables.

Given a certain formation with adequate hydrocarbon resources, the formation permeability is a fundamental variable that impacts fracture fluid selection and fracture design; low-permeability formations are more accommodative to less viscous fracture fluids as the leak-off of the fluids to the formation tends to be low; higher-permeability formations, on the other hand, may require more viscous fracture fluids as the potential for the leak-off to the formation can be high. Stress contrast is critical for fracture containment and adequate fracture length. Moreover, formation fluid, mineralogy, and conductivity also impact the fracture fluid selection. For example, a significant presence of swelling clays will reduce the conductivity and may not be accommodative to certain fluids geochemistry. Furthermore, closure stress, conductivity and bottom-hole flowing pressure impact the selection of proppant; low-stress formations typically are less demanding for proppant selection.

Typically, without knowledge of formation characteristics, a uniform design for perforation clusters is used, ignoring the variability in rock properties. Completion effectiveness can be improved when hydraulic fracturing and perforating operations take into account the anisotropies in stress and other reservoir parameters. Geometric spacing in a heterogeneous zone may not be applicable as the lateral penetrates a heterogeneous reservoir. Sequenced fracturing in combination with a tailored design can increase the fracture coverage (Kraemer et al., 2014). Understanding lateral heterogeneity can help place the perforation clusters in optimal locations to maximize reservoir coverage and boost hydrocarbon production (d'Huteau et al., 2011).

Production models integrate information from the rock, the fluid, and the hydraulic fracturing design to estimate well production performance, for prejob treatment design and postjob production analysis. There are two types of models, analytical and numerical. Analytical models based on IPR treats fractures as negative skin. Analytical reservoir solution treats the fractures explicitly and captures the flow regimes, making it suitable for pressure transient analysis and rate transient analysis. Some numerical models treat hydraulic fractures implicitly, as negative skin or by mapping the fractures to enhanced permeability areas. Hydraulic fractures may also be handled explicitly in numerical models as high-permeability cells. Unstructured grid and advanced reservoir simulation techniques are required to model complex fracture networks. Production models carry higher level of uncertainty when compared to conventional models, because of the complexity of production mechanism and hydraulic fracturing and the lack of high-quality data measurement. Effort should be made to integrate all available data and match the production history to reduce uncertainty and produce a reliable predictive model.

NOMENCLATURE

b_f Fracture width, ft
F_{cd} Fracture conductivity, dimensionless
h Pay zone thickness, ft
k Formation permeability, md
k_f Fracture permeability, md
p Pressure, psi
q Production rate, B/D or Mscf/D
r_e Drainage radius, ft
r_w Well radius, ft
r'_w Effective well radius, ft
s'_f $s_f + \ln X_f/r_w$, pseudo skin factor for hydraulic fracture
X_f Fracture half-length, ft
μ Viscosity, cp
ϕ Porosity, fraction

REFERENCES

Allix, P., Burnham, A., Fowler, T., Herron, M., Kleinberg, R., Symington, B., 2011. Coaxing oil from shale. Oilfield Review 22 (4), 4–16.
Baihly, et al., 2010. Unlocking the Shale mystery: How Lateral Measurements and Well Placement Impact Completions and Resultant Production. Paper SPE 138427 presented at the SPE Annual Technical Conference and Exhibition.

REFERENCES

Chong, K.K., Grieser, B., Jaripatke, O., Passman, A., June 8–10, 2010. A Completion Roadmap to Shale-Play Development. SPE 130369, presented at the CPS/SPE International Oil & Gas Conference and Exhibition in China, Beijing, China.

Cinco-Ley, H., Samaniego, V.F., Dominguez, A.N., 1978. Transient pressure behavior for a well with a finite-conductivity vertical fracture. SPE Journal 18 (4), 253–264.

Cinco-Ley, H., Samaniego, V.F., 1981. Transient-pressure analysis for fractured wells. SPE Journal 33 (9), 1749–1766.

Cipolla, C.L., Fitzpatrick, T., Williams, M.J., Ganguly, U.K., October 9–11, 2011. Seismic-to-Simulation for Unconventional Reservoir Development. Paper SPE 146876 presented at the SPE Reservoir Characterisation and Simulation Conference and Exhibition, Abu Dhabi, UAE.

D'Huteau, E., et al., 2011. Open-channel fracturing—a fast track to production. Oilfield Review 23 (3), 4–17.

Du, C., Zhang, X., Melton, B., Fullilove, D., Suliman, B., et al., May 31–June 3, 2009. A Workflow for Integrated Barnett Shale Gas Reservoir Modeling and Simulation. Paper SPE 122934 presented at the Latin American and Caribbean Petroleum Engineering Conference, Cartagena de Indias, Colombia.

Economides, M.J., Nolte, K.G., 2000. Reservoir Stimulation, third ed. John Wiley and Sons.

Fan, L., Thompson, J.W., Robinson, J.R., et al., October 19–21, 2010. Understanding Gas production Mechanism and Effectiveness of Well Stimulation in the Haynesville Shale through Reservoir Simulation. Paper SPE 136696 presented at the Canadian Unconventional Resources and International Petroleum Conference, Calgary, Alberta, Canada.

Havens, J., 2011. Mechanical Properties of the Bakken Formation (Thesis, Faculty and the Board of Trustees of the Colorado School of Mines).

King, G.E., September 19–22, 2010. Thirty Years of Gas Shale Fracturing. Paper SPE 133456 presented at the SPE Annual Technique and Exhibition, Florence, Italy.

King, G.E., 2014. 60 Years of Multi-Fractured Vertical, Deviated and Horizontal Wells: What Have We Learned? Paper SPE 17095 presented at the SPE Annual Technical Conference and Exhibition, 27–29 October, Amsterdam, The Netherlands.

Kraemer, C., et al., April 1–3, 2014. A Novel Completion Method for Sequenced Fracturing in the Eagle Ford Shale. Paper SPE 169010 presented at the SPE Unconventional Resources Conference, Held in The Woodlands, TX, USA.

Li, J., Du, C.M., Zhang, X., January 31–February 2, 2011. Critical Evaluations of Shale Gas Reservoir Simulation Approaches: Single Porosity and Dual Porosity Modeling. Paper SPE 141756 presented at the SPE Middle East Unconventional Gas Conference and Exhibition, Muscat, Oman.

Manchanda, R., Sharma, M.M., 2014. Impact of completion design on fracture complexity in horizontal shale wells. Society of Petroleum Engineers. http://dx.doi.org/10.2118/159899-PA.

Mata, D., Cherian, B., Gonzales, V., Higgins-Borchardt, S., Han, H., April 1, 2014a. Modeling the influence of pressure depletion in fracture propagation and quantifying the impact of asymmetric fracture wings in ultimate recovery. Society of Petroleum Engineers. http://dx.doi.org/10.2118/169003-MS.

Mata, D., Dharwadkar, P., Gonzales, V., Sitchler, J., Cherian, B., April 17, 2014b. Understanding the implications of multiple fracture propagation in well productivity and completion strategy. Society of Petroleum Engineers. http://dx.doi.org/10.2118/169557-MS.

Mirzaei, M., Cipolla, C.L., January 23–25, 2012. A Workflow for Modeling and Simulation of Hydraulic Fractures in Unconventional Gas Reservoirs. Paper SPE 153022 presented at the SPE Middle East Unconventional Gas Conference and Exhibition, Abu Dhabi, UAE.

Prats, M., Hazebrock, P., Stricker, W.R., June 1962. Effect of vertical fractures on reservoir behavior–compressible fluid case. SPE Journal. http://dx.doi.org/10.2118/98-PA.

Raghavan, R., Joshi, S.D., 1993. Productivity of multiple drainholes or fractured horizontal wells. SPE Formation Evaluation 8 (1), 11–16.

Suarez-Rivera, et al., 2013a. The role of stresses versus rock fabric on hydraulic fractures. In: AAPG Geoscience Technology Workshop, Geomechanics and Reservoir Characterization of Shales and Carbonate.

Suarez-Rivera, R., Burghardt, J., Stanchits, S., Edelman, E., Surdi, A., March 26, 2013b. Understanding the effect of rock fabric on fracture complexity for improving completion design and well performance. In: International Petroleum Technology Conference. http://dx.doi.org/10.2523/17018-MS.

Thambynayagam, R.M.K., 2011. The Diffusion Handbook: Applied Solutions for Engineers. McGraw-Hill Professional, New York.

Wu, R., et al., 2012. Modeling of Interaction of Hydraulic Fractures in Complex Fracture Networks. SPE 152052 presented at the Hydraulic Fracturing Technology Conference.

Xu, W., Thiercelin, M., Ganguly, U., Weng, X., et al., June 8–10, 2010. Wiremesh: A Novel Shale Fracturing Simulator. Paper SPE 132218 presented at the International Oil and Gas Conference and Exhibition in China, Beijing, China.

Zhou, W., Banerjee, R., Poe, B., Spath, J., Thambynayagam, M., 2014. Semianalytical production simulation of complex hydraulic-fracture networks. SPE Journal 19 (1), 6–18.

Zhou, W., Gupta, S., Banerjee, R., Poe, B., Spath, J., Thambynayagam, M., March 26–28, 2013. Production Forecasting and Analysis for Unconventional Resources. Paper IPTC 17176-MS presented at the 6th International Petroleum Technology Conference, Beijing, China.

CHAPTER 9

THE APPLICATION OF MICROSEISMIC MONITORING IN UNCONVENTIONAL RESERVOIRS

Yinghui Wu[1,2], X.P. Zhao[3], R.J. Zinno[2], H.Y. Wu[1], V.P. Vaidya[2], Mei Yang[2], J.S. Qin[2]

China University of Geosciences, Beijing, China[1]; Weatherford International, Houston, TX, USA[2]; ExGeo, Toronto, ON, Canada[3]

9.1 INTRODUCTION

Unconventional reservoirs are an attractive target for exploration due to abundant hydrocarbon resources. The hydraulic fracturing and horizontal drilling methods have made recovering gas from unconventional reservoirs commercially viable. In August 2014, driven by the shale revolution, United States crude oil and condensate production exceed 8.6 million barrels per day, the highest volume since 1986 (U.S. Energy Information Administration (EIA) Website). Subsequently, between Q4 2014 and Q1 2015 the price of oil plunged more than half. As a result, the industry has turned the focus to improve efficiency and productivity. The petroleum industry needs tools to determine how successfully hydraulic fractures optimize well production and field development. These tools should provide information about drainage volumes, formation isolation, hydraulic fracture conductivity, geometry (height, width, length, and azimuth), and the complexity associated with natural fracture interference. Microseismic monitoring has proved to be a unique tool to monitor the evolution of fracturing around the treated rock reservoir (Zinno, 1999; Zhao, 2010) as it provides cost-effective data that is used to mitigate uncertainty for resource evaluation (Ma, 2011). By reducing uncertainty microseismic data, and its many insights, enable decision makers to improve drilling and completion decisions and increase the amount of hydraulically fractured wells that produce hydrocarbons in economic quantities (Zinno, 1999).

The goal of microseismic monitoring of hydraulic fracturing in unconventional reservoirs is to provide data that can be used to analyze stimulated regions within the reservoir (Maxwell, 2014). Microseismic monitoring uses acoustic sensors placed as close to a treatment well as possible to record acoustic emissions created as either liquids or gases lever apart rocks at depth. Passive acoustic techniques are currently the only way to image the permeable flow paths connecting pay zones with the wellbore during hydraulic fracturing.

Since the turn of the century microseismic monitoring has been applied to unconventional plays globally. Microseismic monitoring is common and widespread in the unconventional plays of North America and it has been also successfully applied in the West Siberia Basin, Sichuan Basin, and Northwest Saudi Arabia (Zinno, 2013). The insights gained from maps of microseismicity have caused major paradigm shifts in unconventional reservoir characterization and completion engineers rely upon the information from microseismic monitoring to improve the economic viability of sizable oil and gas assets. The availability of microseismic data has driven the current evolution of completion methods and has resulted in improvement of the economics (Zinno, 2011).

Microseismic data can be integrated with existing datasets to provide a more detailed understanding of the reservoir. A holistic approach adds value to the existing datasets and maximizes research investments. Cross-functional relationships developed at the wellsite merge microseismic technology with other disciplines: geomechanics, moment tensors, source characterization, b-values, intervention between stages and wells (e.g., zipper frac), engineering analysis and interpretation, natural fracture and fault identification, fluid and proppant program optimization, completion method, nitrogen and CO_2 fracture, and conductivity studies (Britt et al., 2001).

In the interest of efficiency and productivity oil field service companies now provide fracture mapping as part of the hydraulic fracturing operations. This means that microseismic data is now routinely acquired during hydraulic fracturing operations. Development of both software and hardware enables us to monitor microseismic events in real time. At the wellsite microseismic monitoring is used to not only monitor the reservoir behavior but also to adjust fracturing parameters for optimum results (Zinno, 2011). At present operators choose to acquire microseismic data because it provides essential information; in future it is likely to be a regulatory requirement, particularly in sensitive regions under zero-discharge requirements (Moschovidis et al., 2000). Microseismic monitoring provides cost-effective answers and solutions in the ongoing drive for efficiency and productivity, and its application is likely to become even more pervasive in the near future.

9.2 MICROSEISMIC MONITORING BASICS
9.2.1 CONCEPTS AND BACKGROUND
9.2.1.1 What Is Microseismicity?

Microseisms are feeble earth tremors due to natural causes or human activities. Microseismic events are defined as a small-scale earthquake with a magnitude usually less than 0 and frequencies in the 0.1–10 kHz band (Young and Baker, 2001). Thus, the term microseismicity is used interchangeably with seismicity recorded during hydraulic fracturing operations. Microseismic events can either be triggered or induced (McGarr and Simpson, 1997). Triggered microseismicity is caused by a small stress change (i.e., tectonic loading plays a primary role), whereas induced seismicity occurs when most of the stress change or energy required to produce the seismicity is generated by the causative activity. Artificially-induced (stimulated) fractures are created by virtually all subterranean engineering.

Hydraulic fracturing, by injecting a mix of water and proppants at high pressure, imposes stress changes in the treated reservoir that can result in stimulated microseismicity with a wide range of magnitudes associated with the opening of new cracks or the mobilization of preexisting fractures. Accurate three-dimensional (3D) locations of microseismic events provide crucial information about the geometry and propagation of hydraulic fractures (Griffin et al., 2003). Processing and locating the source by calculation of the monitored waves is a critical part of evaluating the engineering objectives using microseismic data. A more detailed discussion of the full development of microseismic technology is provided in Appendix A.1.

9.2.1.2 Microseismic Applications

A single microseismic event produces seismic energy that can be recorded and processed to provide a location and origin time along with other more advanced features such as moment magnitude, S/P ratio, signal to noise ratio, root-mean-square (RMS) noise, location uncertainty and residual, corner frequency, apparent stress, static stress drop, etc. (Zinno, 1999). Time series of microseismic events and event clusters provide information about the propagation of fractures within a stimulated rock volume (SRV). Advanced techniques have also been used to link microseismic events with other geoscience information in hydraulic fracturing operations (Huang et al., 2014).

The microseismic monitoring of stimulated wells provides the oil industry with a method to analyze the many factors that influence the effectiveness of well completions. Prior to collecting microseismic data, the commercial design of well completions was limited to well centric electric logging, well centric in-situ stress measurements, pumping pressure curve analysis, and comparative production results (Walker et al., 1998). No information was collected about the actual fracturing and drainage path outside the wellbores, except for the occasional research consortia, which would investigate the SRV with mine-back or drill cores (Moschovidis et al., 2000; Warpinski and Branagan, 1989).

The microseismic maps (Fig. 9.1) produced during hydraulic fracture stimulations provide key geometrical statistics about the created fracture networks: fracture lengths, heights, widths, connectivity, complexity, time lapse, and fracture propagation. This is time lapse and three dimensional information, or four dimensional mapping (x, y, z, time), which is unavailable by any other methodology (Walker et al., 1998).

The earliest interpretations of microseismic maps found strong evidence that hydraulic fracture stimulations and production depletion activity were strongly influenced by natural fractures and faults

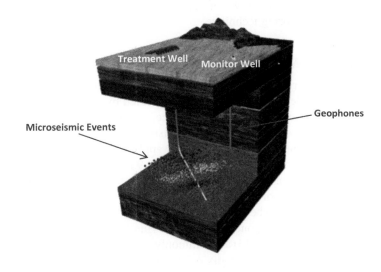

FIGURE 9.1

Microseismic downhole monitoring schematic. A set of downhole geophones are deployed in a vertical offset well, to monitor the horizontal well multi stages fracturing (events colored by stage).

(Phillips et al., 1998; Urbancic et al., 2002). Usually, operators did not have information about small-scale fractures at the reservoir level, nor did they have any way of predicting the fluid conductive properties of those small structural features. Now, it is common to use microseismic interpretation to reveal those interactions, and modify completion designs to accommodate the influences of geological structures (Zinno and Mutlu, 2015). Additionally, multiple wellbores and perforation zones are typically included in the four dimensional fracture maps, which allow operators to observe the stimulation interaction between wells, stages and formations.

Geoscientists and engineers have collaborated on the use of microseismic data by integrating the results with stimulation pumping statistics and the measured in situ stress tensors (Britt et al., 2001). These integrated time lapse plots and microseismic maps have allowed operators to observe, in real time, the changes in fracture network geometries and their rates of propagation that correspond with changes in pumping or producing activities, e.g., pressure, slurry rate, proppant, frac fluid and chemicals, production flow rates, etc. This integrated approach enables operators to improve quality control on completion activity, such as packer and plug isolation; perforation, completion sleeve, or diversion efficiency. Other direct spatial measurements from the microseismic maps are then used to evaluate the effectiveness of a well completion program, and predict the ultimate drainage area about the well. These spatial measurements include the total created fracture surface area, and total conductivity network volume (Zinno and Mutlu, 2015).

Optimizing the treatment is vital for economic and sustainable development of hydrocarbon reservoir and for minimization of potential environmental impacts. Microseismic events provide an estimate in real time of the effect of treatment and the extent of the changes in the rock reservoir properties affecting fluid conductivity. This gives critical feedback for the optimization of the treatment and development plan including well azimuth, well spacing, stage spacing, landing depth (formation), fracturing program design, completion design, refracturing programing, calibration fractures and reservoir modeling. Section 9.3 of this chapter provides case studies concerning application of the microseismic monitoring technique.

9.2.2 MICROSEISMIC MONITORING AND PROCESSING

Geophones or accelerometers are used to continuously record ground motion from induced microseismicity as rocks fracture at depth. An acoustic signal created by a fracture can be visually recognized as a seismic waveform "event", by the coherence, onset time delays, and polarities of the waveforms across multiple receivers in the recording seismic array.

There are two sensor deployment methods commonly applied in the petroleum industry: downhole monitoring and surface monitoring, as illustrated in Fig. 9.2. Downhole monitoring is the most common deployment (Zinno et al., 1998; Maxwell et al., 2010), and it commonly uses three component geophones for each level of tools. The total number of tools varies from different providers. Multiple monitor downhole arrays can be deployed when multiple monitor wells are provided. Surface monitoring normally uses vertical component geophones. Hundreds or thousands of seismometers or similar reflection sensors (Eaton et al., 2013) are commonly required. The sensors are nailed, buried, or deployed in a shallow well (refer to the Appendix A.3 for more information about the tool deployment for these two methods).

The seismic records of these microseismic events are very rich in information, which includes: location (map coordinates, depth, time), waveform information (apparent energy, frequency,

9.2 MICROSEIMIC MONITORING BASICS

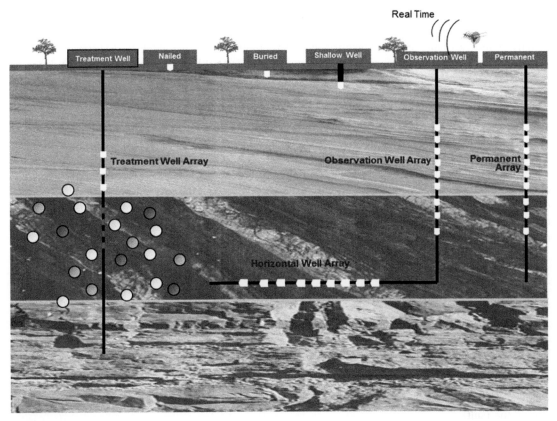

FIGURE 9.2

Tool deployment schematic. The colored balls represent microseismic events; white pentagons are the monitor tools.

background noise, signal to noise ratio), and geomechanical information (local stress state, anisotropy, seismic source mechanism), etc. Microseismic data is acquired, processed, and mapped during a hydraulic fracturing well stimulation, in real time. Results are streamed to an engineer for evaluating the development of fractures in the reservoir. Monitoring the reservoir stimulation provides direct feedback on the progress and possible success of the well stimulation (Zinno, 1999). Further advanced processing, interpretation and evaluation are applied to the data after acquisition. Refer to Appendix A.2 for more information about the microseismic processing methods.

9.2.3 IMPORTANT PARAMETERS

9.2.3.1 Velocity

The physical properties of rocks contribute to their elastic parameters and density. These properties combined with secondary properties such as fluid saturation create variations in the density and therefore acoustic impedance. First order information from microseismic relies on the accurate identification of the locations of as many induced microseismic events as possible. The accuracy of event locations is dependent on the final results of the inversion of measured travel times that rely heavily on the use of an accurate transmission velocity model. Velocity models in stimulated reservoirs are typically based on dipole sonic logs or Vertical Seismic Profile (VSP) measurements at different production and exploration boreholes and are later calibrated using active shots or early induced microseismic events (Pettitt and Young, 2007). Many formations have strongly anisotropic seismic velocity characteristics. Simple anisotropic correction factors can resolve the issue, for simple arrays (Walker et al., 1998). However, a priori anisotropy corrections are rarely accurate and velocity models have to be modified during data processing, as more data from different spatial offsets and formations can be tested against the velocity model. There are also time variant changes to the seismic velocities, which have to be addressed. The stimulation of the reservoir involves the introduction of fluids and changes in the reservoir fracture network and the rock physical properties, in turn, resulting in further changes in the velocity field. Therefore, further calibration is needed in order to retain location accuracy across the different stages of the reservoir treatment (Zhao et al., 2013) and makes velocity modeling an iterative process throughout the processing.

9.2.3.2 Moment Magnitude

The moment magnitude (M_w) is a dimensionless number used to measure the relative size of microseismic events in terms of the energy released. It is defined by (Hanks and Kanamori, 1979):

$$M_w = 2/3 \ \log \ M_0 - 6.1 \qquad (9.1)$$

where M_0 is the seismic moment which is calculated through the following relationship:

$$M_0 = \mu A \Delta u \qquad (9.2)$$

where μ is the shear modulus, A is the area covered by the slipping contacts, and Δu is the amount of slip on the fault.

9.2.3.3 Signal to Noise Ratio (SNR)

The quality and quantity of microseismic events that can be acquired is dependent on the method used for acquisition. In the field data is collected using best practices but the nature of fieldwork means that restrictions can result lower than optimal methods during some of the acquisition process. For example, using receivers in a treatment well increases the noise levels during acquisition and thus reduces the SNR. Correspondingly, the SNR of a microseismic dataset is typically low meaning that signals may be recorded but are not always resolvable about the noise. Optimal acquisition achieves the highest SNR possible. If noise cannot be eliminated then filters are available, or can be designed, to improve the SNR (Liang et al., 2009). Not all noise is random and where it exists noise can typically be identified and dealt with during processing (St-Onge and Eaton, 2011).

9.2.3.4 b-Value

The b-value is a parameter (b) in the Gutenberg–Richter power–law relationship that describes the frequency of earthquake events and their magnitudes (Eq. (9.3)) (Gutenberg and Richter, 1944). b-values have been observed in laboratory studies of rock deformation (Scholtz, 1968).

$$\log 10 N = a - bM \qquad (9.3)$$

Where *M* is magnitude, *N* is the cumulative number of earthquakes with magnitude equal to or greater than *M*, *a* and *b* are constants where *a* is the intercept, and *b* the slope on a bivariate graph of number of events versus event magnitude.

The slope and intercept of the generated bivariate (event magnitude vs number of events) graph are assumed to be constant within play boundaries. b-values are used to see if the majority of microseismic events are coming from within a reservoir. A higher b-value implies more small magnitude events occur relative to larger magnitude events, and vice versa. Statistical studies of microseismic data have shown that b-values of ~1 usually are associated with seismically active regions whereas a b-value of 2+ are associated with natural fractures and joints within a reservoir.

9.2.3.5 D-Value

Another fractal dimension, the D-value, is a statistical coefficient which reflects the spatial distribution and clustering of events (Grassberger and Procaccia, 1983). The D-value is equal to 0 if the events map onto a point, 1 for a linear distribution, 2 for a planar distribution, and 3 for uniform distribution.,

The spatial distribution of event hypocenters can be analyzed through the correlation integral which computes the number of pairs of events $N(R < r)$ separated by a distance R smaller than r (Grassberger and Procaccia, 1983):

$$C(r) = \frac{2}{N(N-1)} N(R < r) \tag{9.4}$$

where N is the total number of events. When plotting the integral value $C(r)$ versus a range of distance r on a log–log plot, part of the distribution appears linear. The gradient of this linear portion is the D-value:

$$C(r) \propto r^D \tag{9.5}$$

9.2.3.6 S/P Ratio

The S/P ratio is the ratio of shear to compressional wave energy received at a sensor. Differences in P and S wave energy received at a sensor can be exploited to provide information about the mechanics of deformation. For example, Roff et al. (1996) used the S/P ratio and clustering of events to infer that event populations were generated by the same focal mechanism. Furthermore, the S/P ratio values are predicted to be altered by changes in fracture orientation.

9.2.3.7 Focal Mechanisms

Seismic waves radiated from an earthquake reflect the geometry of the fault and the motion it experiences during rupture. As such, they can be used to obtain the fault kinematics. Focal mechanisms use the first arrival patterns of radiated seismic waves between seismic stations at different directions from an earthquake. Focal mechanisms may be based on more sophisticated optimal waveform fitting between real and synthesized records (Song and Toksöz, 2011), on considering azimuthal variations in amplitude ratios between P and S waves (Julian and Foulger, 1996; Hardebeck and Shearer, 2003), or on simple "classical" P-wave first-motion polarity readings only. The latter is still the most common procedure for masses of data, especially when recorded in the local distance and low magnitude range. Refer to Appendix A.4 for more information about the theory and method of focal mechanism inversion.

9.3 MICROSEISMIC APPLICATION TO UNCONVENTIONAL RESOURCE DEVELOPMENT

The following case studies and analyses show some of the typical applications of microseismic monitoring. The objectives of each microseismic project could be variable but the application in most unconventional reservoirs can be understood by using analogies. Any application or analysis should be based on the data processing accuracy.

9.3.1 MICROSEISMIC EVENT PARAMETERS

Normally, information from one microseismic event includes location, origin time, moment magnitude, S/P ratio, SNR, RMS noise, location uncertainty and residual, corner frequency, apparent stress, static stress drop, etc. Table 9.1 shows a typical list of event parameters.

9.3.2 APPLICATIONS AND CASE STUDIES

Microseismic monitoring in unconventional reservoirs could be applied to optimize development plan, including well azimuth, well spacing, stage spacing, landing depth (formation), fracturing program design, completion design, refracture programing, calibration of fracture, and reservoir model, etc. Prior to operations fracture modeling provides propped fracture geometry and subsequently microseismic data can be used to confirm the geometry and understand fracture propagation (Yang et al., 2014). Real-time microseismic data is used as part of the decision making process during fracturing programs.

Table 9.1 Typical Microseismic Event Parameters

Type	Parameter Description
Event identifier	Event number and type (noise, shot, earthquake)
Time information	Date and time of triggered events
Event location	Location in northing, easting and depth; event location origin time
Location error	Location RMS error in northing, easting and depth
Location residual	Location RMS residual in travel times, and angular residual from source vectors
Arrival times	Number of P and S-waves picked, and number of P and S-waves used in location
Source parameters	Seismic moment, moment magnitude, event radiated energy, S/P energy ratio, apparent stress, static stress drop, source radius, P and S-wave corner frequency
Signal quality	SNR, RMS noise level
Source solution*	Moment tensor decomposition (isotropic, double-couple, and compensated linear-vector dipole components; T-value, K-value), fault plane solutions (azimuth and plunge for two potential fault planes)

Parameters with * are only provided by source mechanism processing, which requires multiple monitor wells, parameters without * could be provided by either single monitor well or multiples.

9.3.2.1 Fracture Azimuth

One of the key factors for horizontal well drilling, azimuth planning, is directly related to the azimuth of the encountered fractures. As the tensile hydraulic fracture opens orthogonal to the minimum stress, horizontal wells are typically drilled perpendicular to the expected hydraulic fracture azimuth so that transverse fractures are created to maximize reservoir contact. More reservoir contact generally means more SRV and drain area that are directly related to the production of hydrocarbons. Figures 9.3 and 9.4 show unsuccessful and successful drilling and completions designs, respectively.

The horizontal well in Fig. 9.3 was drilled parallel to the fracture azimuth. Fractures from each of the stages along the wellbore follow the same azimuth as the wellbore and create an overlapping drainage area, as indicated by microseismic monitoring. As a result, the network of connected fractures does not extend far from the wellbore, i.e., relatively low SRV, and the well productivity is expected to be lower than an optimally configured completions design. On the other hand, Fig. 9.4

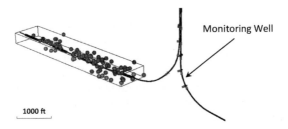

FIGURE 9.3

Fracture azimuth parallel to drilling azimuth. The horizontal well drilling direction is parallel to the orientation of natural fractures. Fractures from each stage follow the same azimuth and overlap in the same drainage area. This well produced relatively low volumes of hydrocarbons. Different colors represent the event-depth difference. In multiple stages the fractures overlap.

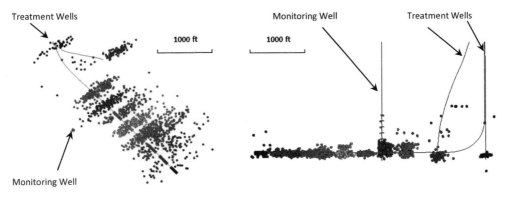

FIGURE 9.4

Map (left) and side (right) views of microseismic results for a successful multiwell and multistage stimulation. Different colors indicate different stages.

Modified from Zinno (2010), and Brooks et al. (2010).

shows a successful fracturing operation where a horizontal well was drilled perpendicular to natural fractures, thereby creating a relatively large nonoverlapping drainage volume and a relatively higher volume of produced gas. Optimization of the fracture azimuths was verified by microseismic mapping that defined the transverse propagation of microseismic events during each stage.

9.3.2.2 Natural Fractures

Microseismic measurements and other evidence suggest that expansion and development of preexisting complex fracture networks during fracturing treatments may be a common occurrence in many unconventional reservoirs (Maxwell et al., 2002; Fisher et al., 2002; Warpinski et al., 2005). The created complexity is strongly influenced by the preexisting natural fractures or mechanically weak planes relative to the rock matrix and in situ stresses in the formation.

Figure 9.5 shows the microseismic fracture network in the first microseismic monitoring job completed in the Barnett Shale. As indicated by the microseismic monitoring results, in situ stress and orientations of preexisting fractures, the stimulated fracture pattern was controlled by preexisting joints. Wells were found to be most productive when the stress directions were not oriented similarly to joint set orientations. The microseismic images enabled several test horizontal wells to be drilled in order to recreate the successes achieved using complex fracture patterns in the most productive vertical wells. As a result, vast tracts of the Barnett Shale became commercial.

Interaction between the hydraulic fracture and natural fractures could cause fluid loss into the natural fractures, dilation of the natural fractures either due to shear or in tension, or even branching or alteration of the hydraulic fracture path, leading to complex fractures (Zhao and Young, 2011; Weng, 2014).

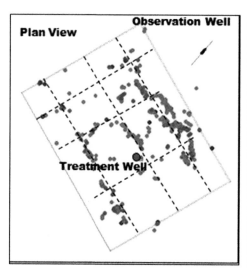

FIGURE 9.5

The microseismic fracture network in the first microseismic monitoring job completed in the Barnett Shale. (Left) Principal and secondary fracture sets in the Barnett Shale. Dots are hypocenters of microseismic events. (Right) Principal and secondary fracture sets seen in shale outcrop.

Modified from Zinno (2010), and Brooks et al. (2010).

As shown in Fig. 9.6, due to the preexisting fault, the fracture growth was asymmetrical and arrested in the Bonner layer, but the Bonner treatment was observed to have communicated upward into the Moore and Bossier Marker sands through a fault (Sharma et al., 2004), resulting in a significant amount of out-of-zone fracture height growth as indicated by the migration of seismicity in space and time. Furthermore, propped or effective fracture half-length, derived from pressure buildup analysis and history matching, were significantly shorter than the created fracture half-length deduced from microseismic locations (Sharma et al., 2004).

Figure 9.7 shows stimulated fractures following the natural fracture azimuth parallel to the wellbore. The well was drilled along the azimuth of the minimum horizontal stress. Microseismic events in stage 2 started from the perforated region in the wellbore and followed fractures opened in a previous stage to the west of the wellbore (events in a light green color, left hand side of wellbore). The microseismic events trend parallel to the natural fracture azimuth, corroborated by imaging logs. As a result of the incorrect azimuthal orientation production of relatively low volumes of hydrocarbons ensued. Figure 9.8 shows another example where microseismic events propagated along a fault.

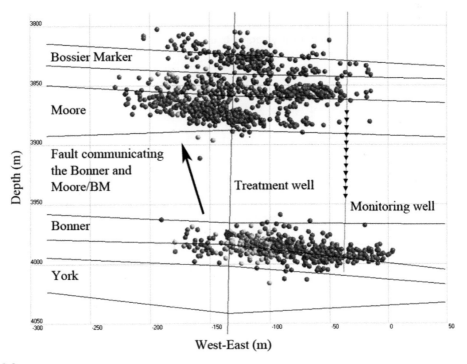

FIGURE 9.6

Microseismic data (color-filled circles) recorded through the Bonner treatment. The color corresponds to the timing of the microseismic events (green (gray in print versions) = early, red (dark gray in print versions) = late). A perforation shot (blue (darkest gray in print versions) triangle) is located at the treatment well within the Bonner layer and a linear receiver array is located at the monitoring well.

Modified from Zhao and Young (2011).

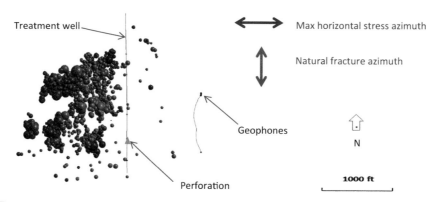

FIGURE 9.7

Map view of stage 2 events showing fractures perpendicular to the maximum horizontal stress azimuth. The color corresponds to the occurring time of microseismic events (green (gray in print versions) = early, red (dark gray in print versions) = late), and the size of events corresponds to the moment magnitude (bigger ball = bigger moment magnitude). Over 90% of events propagated to the left hand side of the illustrated wellbore showing linear fracture azimuth mainly parallel to the wellbore.

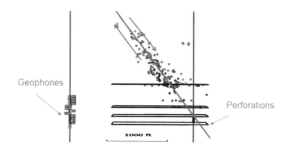

FIGURE 9.8

Side view of vertical well identified fracture following a fault. The color corresponds to the occurring time of microseismic events (blue (darker gray in print versions) = early, red (dark gray in print versions) = late). The microseismic map shows the fracture propagating upwards into the overlying stratigraphy during the first stage treatment.

Zinno (2010).

9.3.2.3 Real Time Processing and Analysis

The development of microseismic technology has led to real-time processing and analysis that have been used to image subsurface fractures propagation during treatment. These microseismic maps are used to optimize the pumping schedule in the field on the fly. Figure 9.9 shows a real-time example where fractures started propagating from upper shale formation into lower limestone at stage 1. Originally all events were confined in the shallow shale formation, while later in the stage events with a bigger moment magnitude propagated down into the deeper limestone formation.

9.3 MICROSEISMICS IN UNCONVENTIONAL RESOURCE DEVELOPMENT

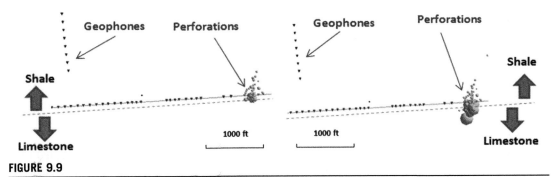

FIGURE 9.9

Side view showing fractures from upper shale formation to lower limestone. The color corresponds to the occurring time of microseismic events (light green (light gray in print versions)/dark green (gray in print versions) = early/later), and the size of events corresponds to the moment magnitude (bigger ball = bigger moment magnitude). All bigger events with a large moment magnitude located in the harder limestone formation, while the smaller events located in the shallower shale formation.

Figure 9.10 demonstrates the screenout process recorded by real-time microseismic monitoring. All large events lined up indicate a minifault (Fig. 9.10(a)). During stages 3–4, the fracture continued propagating to the west leading to the minifault (identified by red events). After that, the minifault was keeping taking pumping fluid as soon as the channels (fractures) to perforations were built up. As significant screenout occurred during stages 3 and 4, pumping was suspended.

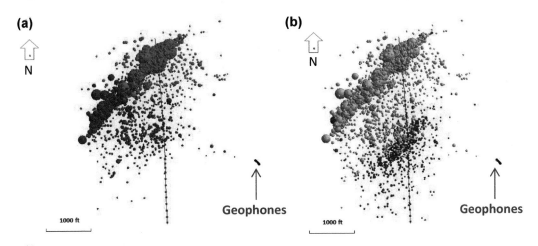

FIGURE 9.10

Map view of stage 1–4 (a) and stage 1–5 (b). The size of events corresponds to the moment magnitude (bigger ball = bigger moment magnitude). All bigger events lined up in the minifault were located in the limestone as showing in Fig. 9.9. (a) Events are colored by time (green (gray in print versions)/red (dark gray in print versions) = early/later), roughly speaking green (gray in print versions) = stage 1–2 and red (dark gray in print versions) = stage 3–4 due to long pause between stage 2 and 3 caused by screenout. (b) Events are colored by stage (gray (light gray in print versions) = stage 1–4, red (dark gray in print versions) = stage 5).

Pumping crew took the advice based on the microseismic results to apply less pad and more fine sand, and change perforation program on stage 5 and later stages. After the resumption, stage 5 was successfully isolated from the minifault without screenout. When several events were located at the minifault (a few red events), real-time microsesimic data was again used to warn the pumping crew and change fracture program on the fly. After that, the main fracture trend stopped propagating to the min-fault, and more SRV was created for stage 5 (Fig. 9.10(b)) and the following stages.

9.3.2.4 Fracture Encounter
Figure 9.11 provides an example showing a new offset well fractured into the minifault created previously. The new fracturing was monitored after the previous job two years ago. As indicated the combined results of both microseismic jobs (Fig. 9.11(b)), the sharp new fractures just followed the path of the previous fractures. As a result, two fractures were encountered in the minifault.

9.3.2.5 Refracturing and Diversion
Refracturing is the repeat application of hydraulic fracture stimulations into a zone after a period of time of well production. It has been applied for many years in unconventional reservoirs. Refracturing is considered if unconventional wells suffer from low production because of ineffective initial completions, or after the hydrocarbon flow has declined. The current low oil price has become a driver for companies to reinvest the refracturing of their old or mature oil wells because refracturing becomes economically attractive when the additional stimulation costs can be offset by revenues from improved well performance. During a refracturing job, diversion agents, e.g., diverter balls or biodegradable

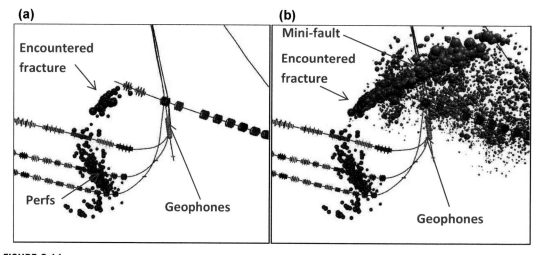

FIGURE 9.11

3D view showing two fractures encountered. View azimuth from bottom to top. The size of events corresponds to the moment magnitude (bigger ball = bigger moment magnitude). (a) Microseismic results from the new treatment. (b) Microseismic results from the new treatment superposed by events from the previous job as shown in Fig. 9.10.

materials, are pumped downhole to temporarily block open perforations, thus diverting the refracturing pressure into new target regions (Potapenko et al., 2009). Real-time microseismic imaging and pressure/rate responses can be used to evaluate the diverter effectiveness and allow engineers to take actions (such as adding more diverters) as needed.

As shown in Fig. 9.12, microseismic image from the initial treatment was first used to identify the missing portion of horizontal well and understimulated reservoir zones for the refracturing candidates. After the initial stage of the refracturing, the microseismic activity occurred in similar regions as the initial treatment. After the first diversion attempt, the microseismicity shows only some extension of previous fractures. Furthermore, with a more aggressive diversion, microseismic shows successful fracturing in new portions of the well.

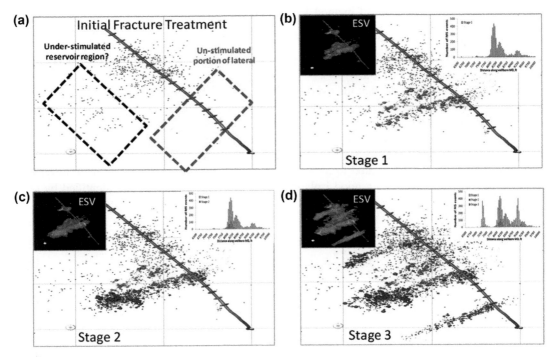

FIGURE 9.12

Example of utilizing microseismic image to investigate the effectiveness of a fracture diversion method, attempting to initiate fracturing along different parts of a horizontal well. (a) Map view of the microseismicity recorded during the original stimulation. Remaining panels include microseismicity recorded during various stages between diversion attempts. They also include stimulated volume contours and a histogram of events along the well. (b) Events for the initial stage of the refracturing. (c) Events after the first diversion attempt. (d) Events after a more aggressive diversion.

After Cipolla et al. (2012). Reproduced with permission of Society of Petroleum Engineers Inc. Further reproduction prohibited without permission.

9.3.2.6 Isolation and Overlapping

Microseismic can monitor the degree of overlap of created fracture networks from the hydraulic stimulation of each perforation point, also called a "frac stage," during a multiperforation well completion, also called a "multistage frac." Generally we refer to this as "stage isolation and stage overlap," which are related to stimulation efficiency, as measured by the relative productivity of the completed well. Figure 9.13 shows an example with a bad isolation. Stage 2 was 98% overlapping the previous stage 1.

The SRV is calculated by creating an event density volume enveloping the main cloud of events. Volumes are calculated for each stage separately (Method 1) and for all events (method 2). Excess overlap (Eq. (9.6)) is used to calculate the overlap of the combined stages, and it is a factor of overall stimulation. Stage overlap (Eq. (9.7)) is used to calculate the efficiency for each stage.

$$\text{Excess Overlap} = (\text{Vsum} - \text{Vcom})/\text{Vsum} \tag{9.6}$$

Where Vsum is sum of all stages of method 1, and Vcom is combined stages.

$$\text{Stage } i \text{ Overlap} = (\text{Vsum} - \text{Vcom})/\text{V}i \tag{9.7}$$

Where i is the stage number. With the example of Fig. 9.13, Table 9.2 summarizes the overlapping results calculated based on Eqs (9.6)–(9.7).

It does not necessarily mean that the overlap must be "0" or such a number. However, 98% stage overlapping would not be acceptable in most case studies, although different formation characterizes differently, and a different operator has its own study, understanding, and experiences.

Figure 9.14 shows stage overlapping in a vertical well. Fractures from stages 2 and 3 propagated into stage 1, and imaging logs suggest numerous vertical fractures were present in this region.

As productivity is directly related to SRV, the production of above example is much lower than another offset well which shows a good isolation. Stage isolation could also be identified by other

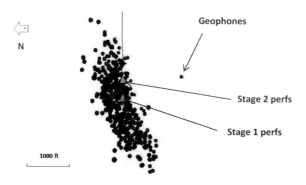

FIGURE 9.13

Map view showing stages overlapping. Events are colored by stage: stage 1 (blue (darkest gray in print versions)) and stage 2 (red (dark gray in print versions)), and sized by moment magnitude (bigger ball = bigger moment magnitude). There was no isolation between these two stages. Stage 2 was overlapping frac into stage 1 by 98%.

9.3 MICROSEISMICS IN UNCONVENTIONAL RESOURCE DEVELOPMENT

Table 9.2 Overlapping Calculation Example

Stage	Cubic ft	Acre-ft
# Well Example		
No. of Stages in Calculation	2	
Method 1: individual stages		
Stage 1	9.51E + 08	21,833.00
Stage 2	7.06E + 08	16,208.00
Sum of stages	**1.66E + 09**	**38,041.00**
Method 2: combined stages		
Combined stages	**9.68E + 08**	**22,223.00**
Difference method 1 and 2	6.92E + 08	15,818.00
Excess overlap	71%	
Stage to stage overlap	98%	
SRV overlap/stage	3.46E + 08	7909.00

Excess overlap 71% is the overlap percentage of the two stages total volume, while 98% of Stage 2 is overlapping the previous stage 1, only 2% of stage 2 was creating new SRV.

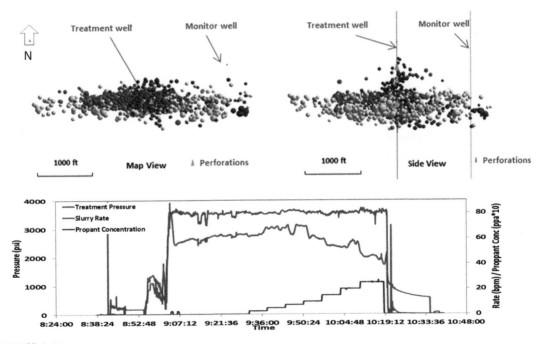

FIGURE 9.14

Map (top left) and side (top right) view of an overlapping example. Event colored by stage (gray (light gray in print versions) = stage 1, light red (dark gray in print versions) = stage 2, green (gray in print versions) = stage 3), and sized by moment magnitude. Both stages 2 and 3 started from the perforations but propagated downwards to the first stage during the stimulation. (Bottom) stage 3 engineering curves (red (dark gray in print versions) = pressure, blue (darkest gray in print versions) = pump rate, green (gray in print versions) = proppant concentration) showing the pressure was dropping down during the second half of the treatment.

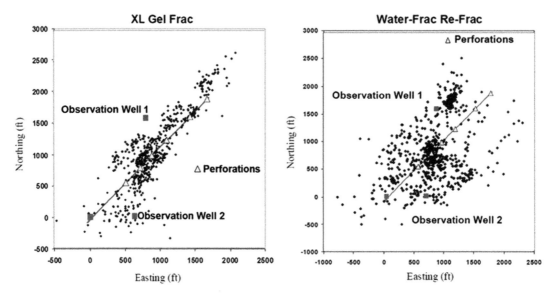

FIGURE 9.15

Comparison of XL Gel frac and water–frac. Map View, Barnett horizontal Wells. The water–frac refrac (right) is wider and shorter comparing to the fracture pattern of XL Gel frac (left).

Modified from Warpinski et al. (2005). Reproduced with permission of Society of Petroleum Engineers Inc. Further reproduction prohibited without permission.

measurements, or an integrated solution, e.g., treatment engineering records, radial or chemical frac fluid and proppant, distributed temperature/pressure/sonic sensors, gravity, resistivity, tilt, etc.

9.3.2.7 Different Fracture Fluid

Fractures act differently with different fracture fluid. Figure 9.15 shows the difference between XL Gel frac and water–frac refrac, using a typical example in Barnett shale. The fracture pattern of XL Gel frac is narrower and longer comparing to the water–frac refrac, while the water–frac refrac is more complex than the XL Gel frac.

Figure 9.16 shows the comparison of slick water (well #1) and cross-link (well #2) performed in two different formations of Mississippian shale. A two-stage treatment was applied in two vertical wells. The fracture pattern of cross-link is narrower and shorter compared with the slick water. The second stage of cross-link treatment (well #2) was isolated to the first stage, while the slick water treatment (well #1) fracture was overlapped with its second stage treatment: Fig. 9.16(c) shows the stage 2 events of well #1 with time sequence, indicating that the fracture started from its perforations, then broke down into the previously stimulated formation of stage 1.

9.3.2.8 Different Completions

Figure 9.17 demonstrated a test using microseismic to map and compare the results of three different horizontal well completion methods (jetted port, perf and plug, and openhole completion) in a clastic formation with high clay content. Production results were compared with the size and complexity of

9.3 MICROSEISMICS IN UNCONVENTIONAL RESOURCE DEVELOPMENT

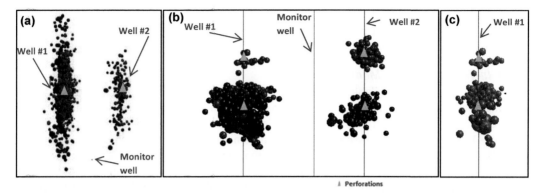

FIGURE 9.16

Slick water and cross-link treatment in two vertical wells. Map view (a) and depth view (b) of all stages, and depth view (c) of Well #1 Stage 2. Both depth views (b and c) are viewing from the fracture azimuth. (a) and (b) are colored by stage: blue (darkest gray in print versions) = stage 1, red (dark gray in print versions) = stage 2; (c) is colored by time: green (gray in print versions)/red (dark gray in print versions) = earlier/later. Event sized by moment magnitude.

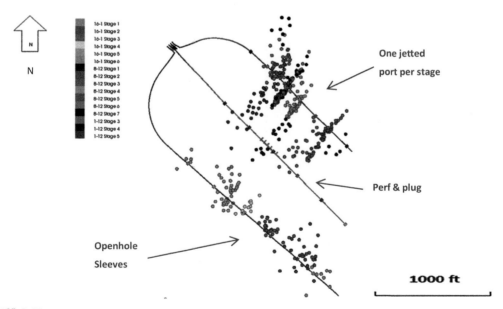

FIGURE 9.17

Comparison of jetted port, perf–plug, and openhole completion in horizontal well completions, colored by stage. Events located by combined horizontal and vertical arrays. The single jetted port (top right) created the most activity of microseismic events, the open hole completion created medium events, while the perf–plug (middle) had the least.

Zinno (2010).

the created fracture networks. Methods to optimize each type of completion procedure were studied. The procedure that had the best prospects for increased production was selected for use (with recommended modifications) on numerous other multiwell pads.

9.3.2.9 In-Treatment Well Monitoring

When no monitor well is available, the in-treatment well monitoring applies. It generally starts monitoring immediately after the cessation of injecting and shutting-in of the wellhead under pressure (Mahrer et al., 2007). Figure 9.18 shows an example of a 3D map of microseismicity.

Figure 9.19 provides another case from a field where no horizontal wells had been drilled and the Triaxial Borehole Seismic (TABS) treatment well tool was deployed to analyze frac height and orientation. The results were surprising in that two orientation directions were measured in different zones and formations. This would be largely invisible to an observation well array due to error uncertainty and events superimposing in the total data set to give a combined frac orientation. We also observe symmetrical development of a penny shaped frac that is bounded below by a formation and fades out with vertical distance above the perforated zone.

9.3.2.10 Permanent Monitoring

The ability to have seismic sensors collecting data in operational wells opens many applications derived from permanently listening to the reservoir (Hornby and Birch, 2008). There is an increasing

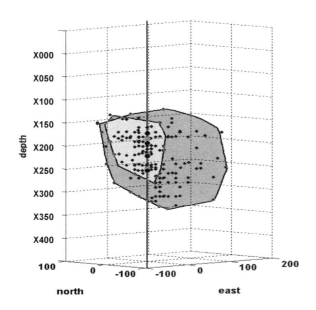

FIGURE 9.18

Sample of microseismicity (small dots) and overlain fracture (shaded) zones. Green zone is a small gel injection which was followed by the main, proppant-laden treatment (red zone). Vertical line of large dots is TABS array in well. Depth is in feet (Mahrer et al., 2007).

Reproduced with permission of the Society of Petroleum Engineers Inc. Further reproduction prohibited without permission.

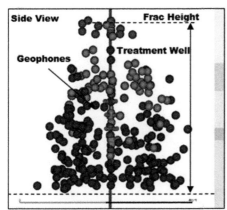 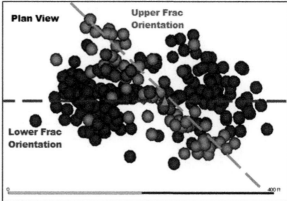

FIGURE 9.19

The advantage of being able to move the tool within the frac zone in the treatment well includes increased accuracy of frac height interpretation. Subtle differences in individual formation response can be seen.

Brooks et al. (2010). Reproduced with permission of the Society of Petroleum Engineers Inc. Further reproduction prohibited without permission.

use of three component and wide azimuth surface seismic data, sometimes in time lapse mode, to image the subsurface and show the effects of production on the reservoir. Having permanent seismic arrays in operational wells enables each new seismic survey to be calibrated at the reservoir level in both velocity and signature terms and to optimize the chances of obtaining the true four-dimensional (4D) signal. Combinations of monitoring sensors are now available. The established single point pressure/temperature gauges can now be combined with distributed temperature giving a calibrated temperature profile for the whole well. With seismic monitoring on an occasional imaging basis and continuous microseismic monitoring, the system can now be used together with full bore downhole flow meters for use in determining flow fractions from multizone production. Implementation of this complete monitoring strategy is becoming known as the "smart field" approach.

Figure 9.20 shows processed microseismic events before, during, and after injection of the sulfur back into the oil-bearing formation. This permanent monitoring was to assess the capability of the system to detect events and to help determine the reasons for those events. The near real time analysis extended for many months helping to solve some of the questions from a reservoir engineering perspective. The located microseismic events cover the entire field and within the reservoir depths. There is significant clustering of events within the field along what may appear to be natural fracture drainage paths leading to production wells. These events are consistent with production from the field as they seem to cease when the field is shut in. The monitoring also shows other benefits of microseismic monitoring such as the delineation of fractures and fracture networks. Other applications using surface seismic sources are now being considered for this permanent installation.

9.3.2.11 Geomechanics

Geomechanical models have been introduced to qualify the impact of key parameters that control the extent and complexity of productive stimulated rock volume (Huang et al., 2014). Microseismic data is

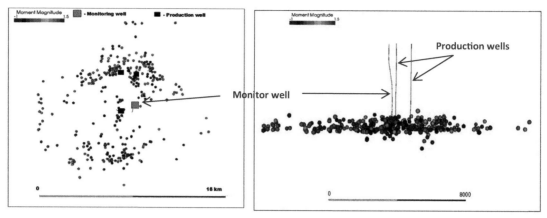

FIGURE 9.20

Production related microseismic events recorded from permanent monitoring installation, with the Map View on the left and Side View on the right. Event colored by moment magnitude, three production wells colored on black box on the left. Lineation and clustering seen are likely fractures and faulting in response to drainage that might reveal reservoir compartmentalization.

Brooks et al. (2010). Reproduced with permission of Society of Petroleum Engineers Inc.
Further reproduction prohibited without permission.

used to calibrate the geomechanical model. Figure 9.21 shows the complex fracture network geometry coupling with microseismic data and synthetic microseismic events. The model was calibrated and showed a similar half lengths and coverage area. This study was based on a set of calibrated Discrete Fracture Network (DFN) properties derived from a sensitivity analysis, and could be applied for stimulation design optimization (Huang et al., 2014).

9.3.2.12 Source Mechanism

The microseismic record has a wealth of information beyond event locations that can provide information about the fracturing. One of the most informative observations from the seismic event record is provided by interpretation of the event fracturing mechanism. The source mechanism during hydraulic fracturing could provide the direct information about the in situ stress and strain, and the relationship between the fracture propagation and the fracturing operation, such as the injections of fluid and proppant (Zhao and Young, 2011).

Typical surface and near surface arrays have adequate focal-sphere coverage to create beach balls for microseismic events recorded with sufficient SNR on individual sensors to pick first motions. Downhole single linear arrays are particularly limited in azimuthal coverage of the source. Although certain constraints can be imposed to determine focal mechanisms, more robust solutions can be produced by grouping events and performing a composite solution. Selections of events with similar waveforms can be grouped together (Rutledge et al., 2004) and jointly analyzed for a common mechanism (Maxwell, 2014).

Figure 9.22 shows the locations and moment tensor solutions for selected microseismic events recorded from three borehole arrays during a fracturing operation. The color of the events is scaled by the proportion of isotropic (ISO), double couple (DC), and compensated linear-vector dipole (CLVD)

9.3 MICROSEISMICS IN UNCONVENTIONAL RESOURCE DEVELOPMENT 265

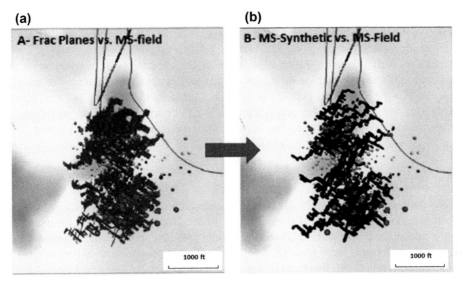

FIGURE 9.21

Coupling microseismic events with geomechanical modeling. Map view (a) on the left shows microseismic field data coupling with geomechanics frac planes modeling; Map view (b) on the right shows the matching of synthetic modeling.

After Huang et al. (2014).

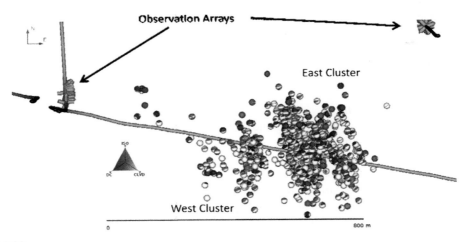

FIGURE 9.22

Map view of the event distributions and moment tensors from three downhole monitoring wells. The events seemed to separate into a number of linear of sublinear clusters, not aligned with the direction of regional Shmax, but showing a large amount of growth away from the treatment well. Two of the three observation arrays are depicted.

After Baig et al. (2010). Reproduced with permission of Society of Petroleum Engineers Inc. Further reproduction prohibited without permission.

components to the full source mechanism as prescribed with the color triangle. The events themselves reveal a variety of mechanisms, suggesting that the events cannot be considered as simple shear failures but include volumetric components of failure (Baig et al., 2010).

More studies on both field and synthetic data showing the uncertainty of moment tensor should be taken care, because of the potential location errors (and other errors). The errors are relatively smaller for events located in the middle portion of the depth straddled by the sensors, but are larger as event locations extend farther vertically (Du and Warpinski, 2011). Synthetic data analysis shows the event location error could be reduced by applying multimonitor wells, but the moment tensor results (first motion and its arrival) are influenced by random noise.

9.3.3 STATISTICAL ANALYSIS

Microseismic data provides not only geometric information but also statistical information. The analysis of statistical information will provide more understanding of the tool, formation, and geomechanics, and helps to understand detection limitation, fracture depth contribution, fracture complexity, fracture length, well spacing, and geomechanics analysis of S/P ratio, b-value, and D-value, and moment tensor. Furthermore, the geomechanics analysis helps understanding of natural fracture or preexisting fractures which most likely to break, in-situ stress which relates to the magnitude and azimuth, the local rock properties defined by the Young's modulus and Poisson's ratio, and the source mechanism, etc.

9.3.3.1 Moment Magnitude Versus Distance

Figure 9.23 provides the plot of the moment magnitude with respect to the source-receiver distance. The furthest detected event is located at 2140 m from the tool center, with moment magnitude of −3.0. A fairly sensitive slim tool array was deployed in a hard formation. On the contrary, Fig. 9.24 shows different tool array deployed in a softer formation with a furthest event at 1250 m (4100 ft) with moment magnitude of −2.2. The possible detection limitation shows less detection capability for Fig. 9.24 comparing to Fig. 9.23.

9.3.3.2 Depth Contribution

Figure 9.25 shows the relationship between the moment magnitude distribution and depth. The max and min M_w of both wells show consistency with sonic logs (brittleness).

For well 1H (left red events), target formation is softer with less and smaller M_w events compared to the harder formations above and below, but more events observed with bigger M_w compared with blue events (on the right) when fracture was initiated from the target formation. Furthermore, for well 2H (right blue events), fracture preferred propagated upwards and it initiated from lower target formation through 1H target formation to the same fracture top of 1H fracture. Similar to 1H, larger M_w presents in harder formation.

In addition, scattered events above formation mark "A" and below "B" were leading by preexisting fracture or offset well production activity.

Figure 9.26 shows the development of fracture height growth by Stages 1–7 of well 2H.

9.3.3.3 Fracture Complexity

Fracture complexity index (FCI) is the ratio between the total width of the fracture network (microseismic cloud width) to the total length of the fracture (Fig. 9.27). Low FCI numbers indicate longer fracture networks and larger FCI values indicate wider fracture networks relative to the extension of the fracture network (fracture complexity).

9.3 MICROSEISMICS IN UNCONVENTIONAL RESOURCE DEVELOPMENT

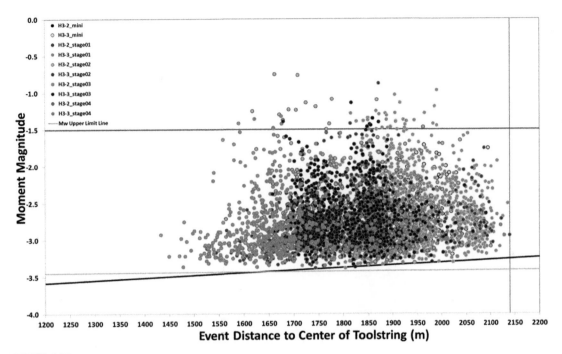

FIGURE 9.23

Moment magnitude vs. distance plot. Colored by stage, the distance is from event location to the center of tool string. The majority of the moment magnitude range is between −3.4 and −1.5, and the majority distance is within 2100 m (7000 ft).

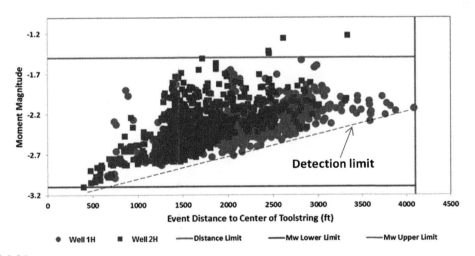

FIGURE 9.24

Moment magnitude vs. distance plot. Colored by well, the distance is from event location to the center of tool string. The majority of the moment magnitude range is between −3.0 and −1.6, and the majority distance is within 3000 ft (900 m).

268 CHAPTER 9 THE APPLICATION OF MICROSEISMIC MONITORING

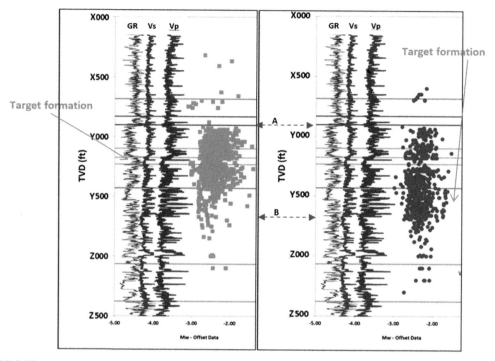

FIGURE 9.25

Moment magnitude distribution vs. depth. The x-axis is moment magnitude (M_w) distribution; y-axis is event depth. Red dots on the left presents horizontal well 1H, and blue dots on the right presents horizontal well 2H.

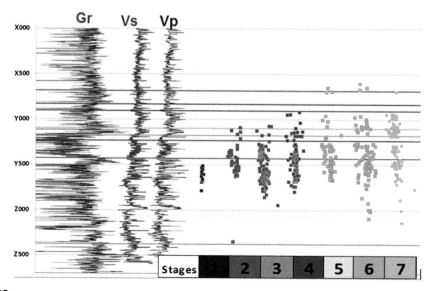

FIGURE 9.26

Depth contribution by stage. Event colored by stage. The X axis is divided by stage with different color. Y axis is the depth (ft).

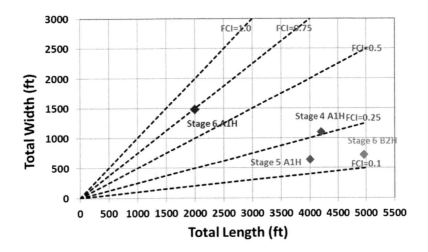

FIGURE 9.27

Fracture complexity index plot.

FCI values are defined as follows:

<0.1: Very low complexity: Long and narrow fractures
>0.1 and <0.25: Low complexity
>0.25 and <0.5: Moderate complexity
>0.5 and <0.75: Moderate-to-high
>0.75 and <1: High complexity
>1: Very high complexity: Short and wide fractures Complexity

9.3.3.4 Fracture Length and Well Spacing

Well costs are enormous, so economically optimized well spacing can have a significant impact on reservoir viability. Determining the optimal well spacing is a key field development decision with significant economic implications. Spacing the wells too far will result in bypassed reserves. Spacing wells too close increases the well density and hence cost, and may result in reduced production due to well interference between neighboring wells draining overlapping intervals (Maxwell et al., 2011).

Microseismic can be used to image if fractures are growing into or just approaching neighboring wells. Well spacing can be decided by analyzing hydraulic fracture microseismic half lengths and locating neighboring wells so that each fracture overlaps or is spaced out depending on the expected matrix drainage. Fracture extent can be interpreted directly from the microseismic locations. The cumulative number of events within distance to wellbore studies has been introduced to understand the relationship between fractures and well spacing. As shown in Fig. 9.28, about 98% of all events locate within 900 ft of the wellbore and the curve levels off close to 900 ft. The dot line with 90% event located within 600 ft could be advised as highly communicated fractures, i.e., fracture half-length. The 300 ft difference could be caused by natural fracture communication or previously opened stages activities. Fracture half-lengths calculated using this method yield more conservative measurements and are suggested to be used for well spacing design.

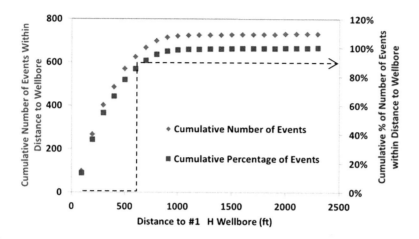

FIGURE 9.28

Cumulative number and percentage of events with respect to the distance to treatment wellbore for a fracturing job.

9.3.3.5 S/P Ratio

The ratio of S/P in the seismic wave can be used to obtain information of the relative contribution of shear or tensile component in the fracture mechanics. Tensile fracturing can be associated to opening of new fractures, while shear component indicates reactivation or growth of previous fractures. The analysis has been used to obtain further information on the behavior of different geomechanical domains and model validation.

Downhole single-well arrays are particularly limited in azimuthal coverage of the source. The azimuthal variations in amplitude ratios between P and S waves can be imposed to constrain focal mechanisms (Julian and Foulger, 1996; Hardebeck and Shearer, 2003). More robust solutions can be produced by grouping events and performing a composite solution. Selections of events with similar waveforms can be grouped together and jointly analyzed for a common mechanism (Rutledge et al., 2004). This can be achieved by plotting amplitude ratios as a function of source angle and fitting a radiation-pattern model. This method also can be used to estimate a mechanism (Fig. 9.29), by fitting a single-source radiation pattern to groupings of events (Maxwell and Cipolla, 2011). Different source types also can be investigated by grouping events into common mechanisms. Various mechanisms can be modeled, which is useful for single-well monitoring, where mechanisms can be ambiguous for a single event because of limited radiation-pattern sampling.

9.3.3.6 b-Value and D-Value

Many natural phenomena show statically fractal properties. Two common fractal dimensions can be statistically characterized from data: the distribution of magnitude sizes (b-value) and spatial event hypocenters (D-value). A combined analysis of these two dimensions can indicate the variations of local stress regimes in a reservoir (Grob and van der Baan, 2011). In addition, the variation and systematic deviation of b-value from the natural average of 1.0 gives an insight into the geomechanical behavior of the reservoir.

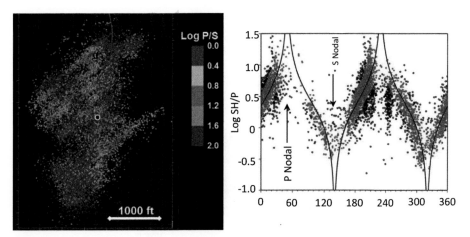

FIGURE 9.29

(left) Map view of a multistage stimulation of a Barnett Shale horizontal well (purple (gray in print versions)), with a microseismic monitoring well in the center (black dot). Events are color coded by the log of the P/SH amplitude ratio. (right) Composite display of SH/P amplitude ratio as a function of azimuth and overlain with a theoretical curve representing the radiation pattern of a vertical dip-slip mechanism striking at 50°. Orange symbols are within 15° of the theoretical mechanism.

After Rutledge et al. (2013). © 2013, Reproduced with permission of Society of Petroleum Engineers Inc. Further reproduction prohibited without permission.

The b-value represents the frequency-magnitude distribution in a catalog of seismic events (Gutenberg and Richter, 1944). A larger b-value implies more small magnitude events occurring relative to larger magnitude ones. Conversely, a small b-value implies that more large magnitude events occurring relative to smaller magnitude ones. The b-value is believed to be an indicator of the stress regime as this will influence the size of rupture, and thus the magnitude of an event. Schorlemmer and Wiemer (2005) found a b-value above one for normal (extensional) faulting, b-value around one for the strike-slip regime and b-value below one for reverse (compressive) types of stress regime.

Although within a single stage, the temporal variation of b-value is associated to the combination of the treatment parameters and the reservoir mechanical properties, the b-value usually presents a positive correlation with the D-value if the new fractures are extended within the stage interval unless there is a local stress transfer, which can be inferred from the temporal variations of both b-values and D-values. However, a negative relation between these two fractal dimensions exists if the reactivation of preexisting fractures from the previous stage happens.

Figure 9.30 shows microseismic locations of the two selected stages for a horizontal fracking treatment. Clearly, events from these two stages are overlapping. Figure 9.31 examines the temporal evolution of b- and D-values for two stages. As both fractal dimensions are calculated with 20 events overlapping, we could investigate both the fracture interstage interaction between stages from the first few sample points and the fracture propagation within a stage using the samples from the rest of the stage. For the interaction between adjacent stages, if microseismic events are constrained within the stage interval, we would expect an increasing D-value between stages, and a decreasing D-value when

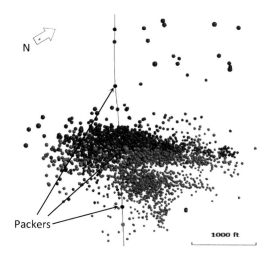

FIGURE 9.30

Map view of the stimulation results of the two selected stages for a horizontal treatment. The microseismic locations are colored by stage, red for stage "a" and green for stage "b." The arrow indicates the packer position for selected stages.

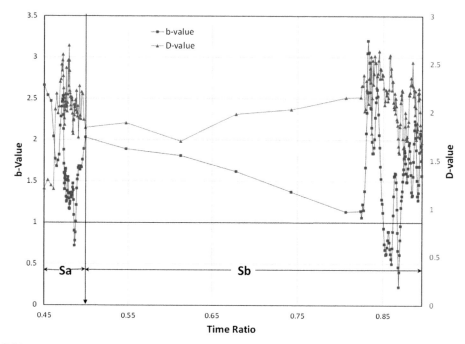

FIGURE 9.31

Temporal variation of b and D-values from Fig. 9.30. The time scale is normalized to the total monitoring time for the entire treatment. The stage "a" and stage "b" are divided by the black line along the vertical axis.

events locations from the later stage continue dominating on the former stage. For the latter case, a different variation of the b-value may happen depending on the relative proportion of reactivation of existing fractures and extension of new fractures. As shown in Fig. 9.31, an initially decreasing D-value on the stage "b" is consistent with the actual location distribution (Fig. 9.30). Noticeably, the b-value is decreasing close to one for the first part of the stage b, which suggests reactivation of existing planar fractures as indicated by the D-value around 2. Potentially, the relationship between D and b-values across interstages can be used as a guide to adjust the treatment operation accordingly.

Next, from the event locations including only a single stage we could estimate the local mechanics during the operation if we have enough sampling data. In the Stage "a," as revealed by D-values, the hydraulic fractures form a linear ($D < 1.5$) to a planar plane ($D = 2$) while b-values vary inversely. The whole process might infer local stress transfers from the normal to reverse stress regime (Huang and Turcotte, 1988; Henderson et al., 1994; Helmstetter et al., 2005). On the other scenario for the Stage "b," the b-value shows a positive relation with D-value. The b-value over two and the D-value between 1.5 and 2.5 suggest an extensional stress regime with the newly created fractures propagating in a planar to a uniform pattern.

This analysis would benefit for its application to a larger number of treatments. Analysis on multiple treatment projects can provide a first order guidance on selecting optimal treatment parameters.

9.4 CONCLUSIONS

Since the late 1990s microseismic technology and applications have been developed at exciting fast pace. Both laboratory and field based studies have proven the utility of microseismic monitoring as applied to hydraulic fracturing operations. Furthermore, new methods for microseismic data are enabling more detailed interpretation of the data acquired and helping to reduce uncertainty. Microseismic monitoring is a key technology that can be used to unlock the full efficiency and productivity of unconventional reservoirs.

This chapter gives an overview for microseismic monitoring including the basic theory, development, data acquisition and tool deployments, processing methods, with a focus on applications for improving efficiency and productivity. Microseismic data mitigates uncertainty during completions evaluation in unconventional reservoirs which is critically important as the economic challenges of development increase.

APPENDIX
A.1 MICROSEISMIC TECHNOLOGY DEVELOPMENT HISTORY

Results of laboratory tests of brittle rock cores concluded that fracture initiation is associated with the onset of microseismic emissions (Thill, 1972). Subsequent studies of a deep South African gold mine, and earthquakes in Denver and Matsushiro in Japan by McGarr (1976) proposed a direct relationship between cumulative seismic magnitudes and volumes of material removed. The presence of microseismic was further supported by work done by the Gas Research Institute at the Multi Well Experimental Site near Rifle, Colorado (Warpinski et al., 1994; Peterson et al., 1996).

Most well stimulations induce detectable seismic events caused by the opening or slipping of fractures within the reservoir. The ability to detect induced seismicity depends upon the sensitivity of the recording system, the range from hypocenter to geophone, and the magnitude of the event (Niitsuma, 1997). Furthermore, based on laboratory studies and theoretical models (Sondergeld and Estey, 1981; Chouet, 1986; Madariaga, 1976) the likelihood of a hydraulic fracture treatment to induce microseismicity depends on localized variables that include: formation porosity and permeability, the elastic parameters of the rock, and the local stress state. Tight and brittle rocks loaded with stress are likely to produce microseismic events during hydraulic stimulation. Combined, these variables can be summarized as the geomechanical stratigraphy of a given location.

In shallow treatment zones where the regional horizontal stresses approach the magnitude of the vertical stress component it is difficult for fractures to open (Pine and Batchelor, 1984) and in this case rock failure would be dominated by tensile failure mode fractures that radiate less seismic energy than shearing mode. The orientation of lines of weakness within the rocks such as joints and pre-existing fracture networks also control the amount of radiated seismic energy (Niitsuma, 1997).

Earthquake seismologists developed a number of analysis techniques (Brune, 1970) employed by the mining, geothermal, gas storage, waste injection, coal bed methane, and petroleum industries to map fault and fracture planes, and to determine rock failure modes (Hanks and Wyss, 1972; House, 1987; Fix et al., 1989). In the Fenton Hill Hot Dry Rock experiment, Albright and Pearson (1982) recorded microseismic emissions induced by fluid injection and mapped the fractures. In the late 1990s real time microseismic monitoring was underway by ARCO Alaska Inc. in waste disposal wells (Keck and Withers, 1994), ESG Canada Ltd. for the mining industry (Trifu and Urbancic, 1996), and Valhalla in the North Sea (Dyer et al., 1999). Further experimental work in the Applied Seismology Lab at Imperial College London provided quantification of the micromechanical processes of failure associated with dilation and shear (Pettitt et al., 1998).

In May 1994, Union Pacific Resources conducted a pilot study to confirm the existence of detectable hydraulic fracture induced seismicity in the Carthage Cotton Valley Formation (Zhu et al., 1996; Zinno et al., 1998; Zinno, 1999). A commercially available wireline was used to deploy a single three component clamped slimhole tool in one monitor well and numerous seismic events were subsequently recorded that were associated temporally and spatially with either the stimulation or production in the surrounding wells. Microseismicity including both P and S-waves were successfully recorded before, during, and after pumping fluid and proppant. Imaging of the recorded events was performed using hodogram analysis and P to S- arrival time differences. The pilot study had two conclusions that led to the development of modern day microseismic monitoring: (1) that recording and mapping microseismic events is possible, and (2) the signal quality could infer rock failure mechanisms (Urbancic et al., 1993). Further work during the ARCO/Vastar Resources project detailed in Truby et al. (1994) and Keck and Withers (1994) and the GRI/DOE M-Site project described in Peterson et al. (1996) were particularly important in addressing both the design and operational issues of fracture imaging. Built on these early successes recognition of fracture azimuths, lengths, and heights was achieved (Rutledge et al., 2004) and subsequently led to commercialization.

The first commercial microseismic monitoring job was successfully completed in the East Texas, Carthage Gas Field (Zinno, 2011); and the first Shale Gas well stimulation was recorded in the Barnett Shale formation (Urbancic et al., 2002) which delineated a complex fracture network and resulting fluid flow along preexisting fractures (Maxwell et al., 2002). Development of the acquisition and processing methods have continued to increase the accuracy of data processing and have enabled more detailed interpretation of data from the unconventional reservoirs that have ultimately increased efficiency and productivity (Zinno and Mutlu, 2015).

Following these first commercial microseismic projects, and largely, coincident with the dramatic increase in unconventional reservoir development; microseismic monitoring technology has enjoyed a great deal of commercial interest and technical innovation (Zinno, 2011).

A.2 MICROSEISMIC PROCESSING METHODS

An accurate velocity model for the depth interval of interest based on a detailed understanding of the local geology is a critical component of microseismic processing. The velocity model provides travel time so that a solution for distance can be calculated. Usually P (compressional) and S (shear) wave velocities acquired using a dipole sonic logging tool are integrated with what is known about the local geology to create a fit for purpose velocity model.

Passive seismic events in terms of acoustic waveforms are recorded by three component geophones (Fig. A.1). Two main methods are typically used to solve the location of a microseismic event: the first is the traditional Geiger method inspired, "arrival time" based method; and the second is the "migration" based method. The Geiger method comes from classic earthquake seismology and has been the mainstay of earthquake seismology for generations, and is a proven method for locating earthquake hypocenter (Geiger, 1912; Zinno, 1999). This is the approach that has been used in many industries, when microseismic emissions are analyzed (Zinno, 1999). The advantages of this robust approach are that it has been well developed to address earthquake sources, calculates all known earthquake source parameters, is generalizable across many acquisition geometries, works very well

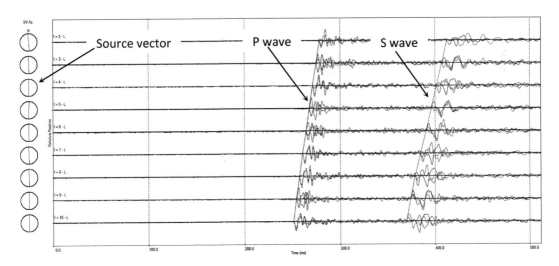

FIGURE A.1

A typical recorded microseismic event. The x-axis is time, y-axis includes a total 10-level tools waveforms, and each tool shows three component waveforms in different colors, X red (gray in print versions), Y blue (dark gray in print versions), Z black. Sampling rate is 0.25 ms. Both the compressional (P-wave) and share (S wave) arrivals are picked, and the source vector on the left shows P wave source vector for each tool.

with limited numbers of geophones, is computationally efficient, and provides ample opportunity for quality control and/or refinement of processing parameters, including complex velocity anomalies. There are several liabilities with using this method of source point location. In cases where the data are recorded from a single array of geophones placed near reservoir level in a monitor well, which is usually the case for oil field applications; the algorithms require clear waveforms, or high signal to noise ratios. When multiple monitoring wells can be employed, or when the geophone arrays are, spatially, more three dimensional; the signal quality standards can be relaxed. Another problematic aspect of this traditional method is that there is a high degree of processor interaction with software; which, while allowing ample quality control and adjustments to processing parameters, will also become time-consuming and very demanding upon the skills of the processor. Different processors will usually produce slightly different results; or a very different result, in the case of inexperienced users. However, the Geiger method has distinct advantages in versatility, completeness of analytic output, and works well with minimal acquisition expense (Zinno, 2011).

The Geiger method relies on the P and/or S wave arrival times. As shown in Fig. A.1, P and S wave arrivals could be identified and picked. The distance (D) from the geophones to the event is given by Eqs (A.1) and (A.2).

$$\Delta t = t_s - t_p \tag{A.1}$$

$$D = \frac{\Delta t \cdot V_p \cdot V_s}{V_p - V_s} \tag{A.2}$$

Where t is time, t_s is S wave arrival, t_p is P wave arrival, V_p is P velocity, and V_s is S velocity.

The azimuth of the event is determined using the particle motion of components through time, or polarization analysis. The same analysis using hodogram is used to calibrate the sensor orientations since the directions of horizontal components are not known in the borehole. As a hodogram can be used to determine the arrival direction of the impinging wave, the azimuth of horizontal components could be calibrated by an event with known location, such as perforation shot (Fig. A.2), string shot, etc.

The depth of the event can be found by examining the arrival time delays of P and S waves at different receiver heights. Together with the hodogram information, the resulting separation between receiver and source, termed moveout, provides the location of an event.

An interactive search is applied to minimize the difference between predicted and observed P and S arrival times using a least-squares method (Urbancic and Rutledge, 2000) in what can be considered a variation to the Geiger method.

Different "migration methods" have been exported from reflection seismic processing to the continuous seismic data records by passive seismic arrays, to identify and locate possible microseismic events. All these migration techniques carry assumptions regarding the consistency of forward modeled, earthquake source characteristics, which are, in reality, quite complex and azimuthally variable. Therefore, the accuracy of event source locations, derived by these migration techniques, is subject to the applicability of those assumed characteristics. Migration algorithms are fundamentally, a spatial summation technique. As such, they depend on a volume of evenly distributed, spatial, data bins, and further require that the apparent recorded signal amplitudes that populate each bin have been normalized to correct for variations in travel path attenuation. To meet these processing needs, field acquisition geometries are made much more elaborate, with many more receiver stations, than the geometries used for traditional Geiger earthquake location processing.

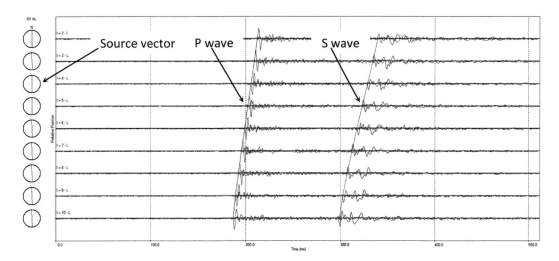

FIGURE A.2

A typical recorded perforation shot waveform. The x-axis is time, y-axis shows a total 10-level tools waveforms, and each tool is showing three component waveforms in different colors: X, red (gray in print versions); Y, blue (dark gray in print versions); Z, black. Sampling rate is 0.25 ms. All tools are oriented and rotated to the perforation.

Several of the variations of these migration methods are popularly known as: reverse time migration, beam steering, Kirchhoff migration, semblance, or source imaging (Maxwell, 2014). All these methods share the process of summing, into discrete 3D spatial bins, coherent arrivals from each time sample; using data from a short, trailing, window in time, long enough to capture the longest travel time from the furthest bin to the further geophone receiver station. The summation captures signal amplitudes that are consistent with forward-modeled arrival times and polarizations, predicted by the algorithm. Then these processes will continue, scanning all those spatial bins at each sample time, to find trial source locations, in space and time; which have strong concentrations of summed coherent arrivals (Maxwell, 2014). Reverse time migration by wave equation finds a maximum energy focus point, by migrating the entire data set summing all possible diffraction paths of (assumed) radially, consistent waveforms; as opposed to single radiation pathways taken from a ray trace model, as in Kirchhoff migration. Gaussian-beam search methods locate source locations by back shooting a ray, through a velocity model, starting along the incident angle, calculated though Hodogram analysis. That ray trace drives a spatially limited, migration of a semblance weighted stack of the raw recorded data. The semblance weighting attempts to reduce the effect of the polarity variations of an earthquake source (Rentsch et al., 2007; Zhao et al., 2010). A search engine technique, then, finds an area of maximum energy within the migrated volume, as a possible source location (Zhang et al., 2013). Semblance-weighted deconvolution, used prior to a full waveform, vector, migration, tries to address earthquake radiation patterns and source parameter variables; and is representative of "source imaging techniques" (Haldorsen et al., 1994; Chen et al., 2010). This method assumes that microseismic events located close to one another in 3D space would produce a similar pattern of different waveforms across a passive seismic array. The waveforms of those multiple colocated events, along a particular vector from the receivers, can be used to build a deconvolution operator, to precondition the raw seismic records prior to full-waveform migration; thereby attempting to satisfying the migration assumption of, radially and temporally, homogeneous source waveform characteristics. These migration methods could be applied for either fully-automated or semiautomated processing; reducing the impact of varying skill levels of individual processors, but retaining systematic errors in the output data, and eliminating most quality control opportunities afforded by traditional Geiger earthquake location techniques.

Some new "hybrid methods" attempt to retain the quality control features of traditional Geiger techniques; but add some of the efficiencies of "migration" methods. Two such hybrid methods are: first, full-waveform inversion, which matches forward modeled full synthetic seismic waveforms with the recorded seismograms using a grid search technique (Song and Toksöz, 2011); and second, the relative location method by investigating the time differences between strong, high amplitude, master events against weaker events, with similar moveout characteristics (Waldhauser and Ellsworth, 2000; Reyes-Montes and Young, 2006).

A.3 TOOL DEPLOYMENT
A.3.1 Downhole Monitoring

Three different methods are used to collect borehole microseismic data, depending on the locations of the deployed tools. The monitoring tools can be situated in: (1) observation well, the observation well could be vertical, deviated, or horizontal with a tractor deployment, wireline or fiber optics cable is required for data transit and power supply between downhole tools and surface recording system; (2) treatment well; and (3) permanent arrays (Brooks et al., 2010; Zinno, 2011).

Offset Well Monitoring

Downhole microseismic monitoring is a mature method that has been used to understand and increase the efficiency and productivity of hydraulic fracturing (Baig and Urbancic, 2010) in tight-gas completions, fault mapping, reservoir imaging, waterflood conformance, and drilling-waste disposal and injection, and therefore offers the assurance that only a proven, mature technology can give. Microseismic acquisition tools are typically deployed on conventional wirelines or fiber optic cable, a tractor can be applied for a horizontal deployment.

In-Treatment Well Monitoring

For microseismic monitoring in the treatment well the Triaxial Borehole Seismic (TABS or SPEAR) system can be used (Fig. A.3). A TABS or SPEAR system is a real time monitoring tool built to the same standards as logging tools and it is able to be used in the borehole via wellhead lubricators. The in-treatment well monitoring technology was developed for remote areas without offset monitor well options (Bailey and Sorem, 2005; Mahrer et al., 2007). Comparisons of treatment well and observation well microseismic monitoring have been conducted and suggest that under the right conditions treatment well sensors can record more data than sensors in an observation well (Mahrer et al., 2007). The advantage of this in-treatment well monitoring method is its accuracy of the fracture depth and azimuth as the fracture is very close to the tools, the challenge is the borehole fluid noise.

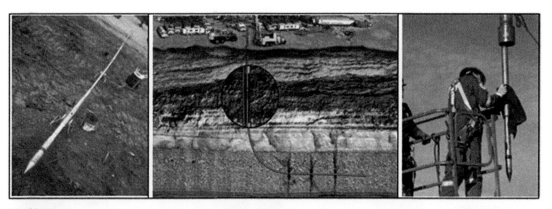

FIGURE A.3

In-Treatment Well Array Deployment. The TABS or SPEAR array is deployed via a lubricator, the tool itself is 72 ft long, descends during pumping deployed with wireline or eCoil, a gyro built in the tool identified the geophones orientation. The tool could be deployed on eCoil, which uses a seven conductor cable threaded through coiled tubing, with this deployment, tools can be pushed down to high deviated or horizontal wells, the coiled tubing will allow for safer deployment and retrieval.

Zinno, 2010; Brooks et al., 2010. Reproduced with permission of Society of Petroleum Engineers Inc. Further reproduction prohibited without permission.

Permanent Monitoring

Technological advances have resulted in permanently instrumented smart wells and 4D monitoring (Wu et al., 2011). Permanently installed sensors are attractive because data can be collected from nearby seismic sources. In offshore wells permanently installed sensors also have allowed opportunistic data acquisition from passing seismic sources (Hornby and Burch, 2008). The permanent instruments could also be applied to the treatment or injection wells, as well as production well, by deployment with casing or production tubing (Fig. A.4).

A.3.2 Surface Monitoring

Surface monitoring of microseismic events uses arrays of geophones placed at or near the Earth's surface (buried geophones). The disadvantage of this method is added noise from surface sources and increased distance from the microseismic sources. The greater distance from the source and the additional complexity of the velocity model, leads to more uncertainty in the hypocenter location. This is especially true for the depth dimension of the event locations (Mohammad and Miskimins, 2010). The small magnitude makes precise picking of arrival times and azimuths problematic at the surface, and data processing is usually carried out by one of the interferometry or migration methods. For the surface or near surface data acquisition, sufficient aperture coverage of the target zone is required for the full-waveform migration (Duncan, 2012).

An advantage of the surface or near surface method is the relative ease of acquiring failure mechanism and magnitude, and it suggests the moment tensors that can be used to effectively map local stress orientations (Wessels et al., 2011).

A.3.3 Combination of Downhole and Surface Monitoring

A monitoring method combining surface or near surface with downhole, or multi downhole monitor wells has been introduced to maximize the monitor coverage, detecting sensitivity, source mechanism,

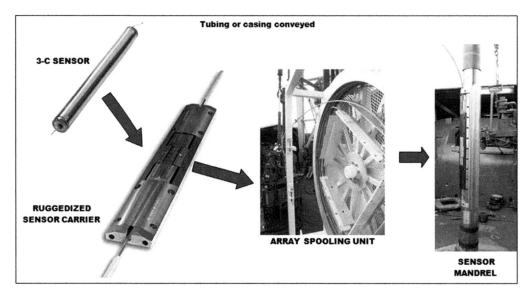

FIGURE A.4

Deployment of Permanent Monitor System. Permanent seismic monitor system can be deployed either behind casing or in the annulus between tubing and casing; sensors are cemented or released from mandrel downhole. No downhole electronics is required for the fiber sensors and cable, 4D monitoring can be applied without interference to the wellbore operations such as fracturing, production, or shut in. Microseismic data can be recorded continuously such as fracture, refracture, production fluid movement, and fluid type exchange etc.

Zinno, 2010; Brooks et al., 2010. Reproduced with permission of Society of Petroleum Engineers Inc. Further reproduction prohibited without permission.

information variation, and accuracy, etc. The difficulty with merging surface and downhole recordings is that all anisotropic velocity anomalies are accentuated, and it is very difficult to solve those issues in a way that satisfies both receiver placements. Although each monitoring method has its advantage and negative, the combination could maximize all the positives of microseismic technology which has being improved during each of the applications.

A.4 SOURCE MECHANISM

The classic source mechanism is shearing along a preexisting fault. Figure A.5 illustrates the first motion concept for a strike-slip earthquake on a vertical fault. The first motion is either compressional (material initially displaced towards the station) or dilatational (material initially displaced away from the station). The first motions define four quadrants, two compressional and two dilatational. The division between quadrants occurs along the fault plane and a plane perpendicular to it, i.e., auxiliary plane. The orthogonal planes are called nodal planes, and if their orientations can be found, then the fault geometry is known. However, first motions alone cannot distinguish between the fault plane and the auxiliary plane. Additional information, such as geology, aftershock, etc., is needed to identify which is the actual fault plane.

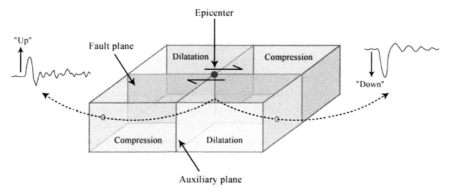

FIGURE A.5

Fault example of first motions from a pure strike-slip earthquake (MIT OpenCourseWare, 2008).

Focal mechanisms and the associated radiation pattern of different wave types are traditionally displayed on a lower-hemisphere stereographic projection representing a focal sphere around the source, i.e., the "beach ball" diagram (Fig. A.6). Beach balls can be interpreted in terms of the orientation of the fault plane, along with the direction of shearing. Shear-wave first motions and S and P-wave amplitude ratios also can be used to help constrain the mechanism.

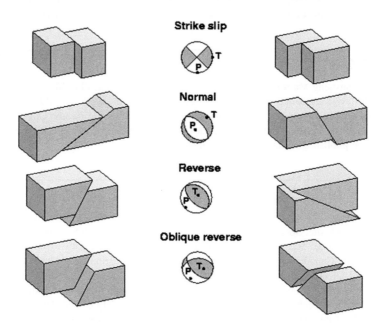

FIGURE A.6

Schematic diagram of focal mechanisms (USGS, 1996). The gray quadrants contain the tension axis (T), which reflects the minimum compressive stress direction, and the white quadrants contain the pressure axis (P), which reflects the maximum compressive stress direction.

As shown in Fig. A.6, the block diagrams adjacent to each focal mechanism illustrate the two possible types of fault motion that the focal mechanism could represent. As mentioned above, the ambiguity may sometimes be resolved by comparing the two fault plane orientations to the alignment of small earthquakes and aftershocks. The first three examples describe fault motion that is purely horizontal (strike-slip) or vertical (normal or reverse). The oblique-reverse mechanism illustrates that slip may also have components of horizontal and vertical motion.

Moment tensor inversion (MTI) is a more general solution that can describe a variety of source types including shearing, tensile opening, explosions, or any combination of these source types. Moment tensors completely describe in a first order approximation the equivalent forces of general seismic point sources. The equivalent forces can be correlated to physical source. The moment tensor inverted from waveform amplitude can be interpreted as different type of source mechanism, such as explosion/implosion, model I opening crack, model II in-plane crack, or model III antiplane crack. Interpretation of the second-order moment tensor is nontrivial, and various methods have been developed to decompose MTI into specific, end-member source types. A common decomposition is ISO, DC, and CLVD components.

Downhole geophone strings and narrow aperture surface arrays present a challenge when trying to obtain source mechanisms for microseismic events due to limited sampling of the focal sphere. To determine the complete moment tensor (six independent components) in an isotropic medium, we have to use the amplitudes of P-waves from at least three boreholes, or the amplitudes of P and S-waves from two boreholes (Vavrycuk, 2007; Eaton, 2010; Zhao et al., 2014). Furthermore, since rocks are in general anisotropic, particularly shales, it is important to consider what impact anisotropy may have on microseismic sources and their inversion (Vavrycuk, 2005; Leaney et al., 2011).

ACKNOWLEDGMENTS

Thanks to Professor Yongsheng Ma for his instruction; also thanks Xianhuai Zhu, Mark Milam, Rob Hull, Vladimir Grechka, Carlos Bejarano for their kindly help and encouragement. We would also like to thank Jon Musselman and Jim Rangel for their support. Special thanks to Alex Barnard, Yassine Oukaci, Dawei Fang, and Haidong Chen for supplying images, text and data. Last but not least, thanks to Weatherford International for giving us permission to publish the data contained in this chapter.

REFERENCES

Albright, J.N., Pearson, C.F., 1982. Acoustic emissions as a tool for hydraulic fracture location: experience at the Fenton Hill hot dry rock site. Society of Petroleum Engineers Journal 22, 523–530.

Baig, A., Urbancic, T., Prince, M., 2010. Microseismic moment tensors: a path to understanding growth of hydraulic fractures. In: Canadian Unconventional Resources & International Petroleum Conference, Calgary, Alberta, Canada, 19–21 October. Society of Petroleum Engineers (SPE). CSUG/SPE 137771.

Bailey, J.R., Sorem, W., 2005. ExxonMobil logging tool enables fracture characterization for enhanced field recovery. In: SPE Annual Technical Conference and Exhibition. Society of Petroleum Engineers.

Britt, L.K., Smith, M.B., Cunningham, L.E., Hellman, T.J., Zinno, R.J., Urbancic, T.I., 2001. Fracture optimization and design via integration of hydraulic fracture imaging and fracture modeling: East Texas Cotton Valley. In: Paper SPE 67205, SPE Production and Operations Symposium, Oklahoma City, Oklahoma, USA, 24–27 March.

REFERENCES

Brooks, N., Gaston, G., Rangel, J., 2010. Three different methods of mapping the characteristics of induced fractures related to both hydraulic frac and production as measured with microseismic array technology from observation wells treatment wells and in a permanent setting. In: SPE Deep Gas Conference and Exhibition. Society of Petroleum Engineers. SPE 131123.

Brune, J.N., 1970. Tectonic stress and the spectra of seismic shear waves from earthquakes. Journal of Geophysical Research 75 (26), 4997–5009.

Chen, C.W., Miller, D.E., Djikpesse, H.A., Haldorsen, J.B.U., Rondenay, S., 2010. Array conditioned deconvolution of multiple component teleseismic recording. Geophysical Journal International 182, 967–976.

Chouet, B., 1986. Dynamics of a fluid-driven crack in three dimensions by the finite difference method. Journal of Geophysical Research: Solid Earth (1978–2012) 91 (B14), 13967–13992.

Cipolla, C., Maxwell, S., Mack, M., 2012. Engineering guide to the application of microseismic interpretations. In: Hydraulic Fracturing Technology Conference, SPE 152165.

Du, J., Warpinski, N., 2011. Uncertainty in fault plane solutions from moment tensor inversion due to uncertainty in event location. In: SEG San Antonior 2011 Annual Meeting.

Duncan, P.M., January 2012/26. Microseismic monitoring for unconventional resource development. Geohorizons.

Dyer, B.C., Barkved, O., Jones, R.H., Folstad, P.G., Rodriguez, S., June 1999. Microseismic monitoring of the Valhall reservoir. In: 61st EAGE Conference and Exhibition.

Eaton, D.W., van der Baan, M., Tary, J.B., Birkelo, B., Cutten, S., 2013. Low-frequency tremor signals from a hydraulic fracture treatment in Northeast British Columbia, Canada. In: 4th EAGE Passive Seismic Workshop.

Eaton, D.W., 2010. Resolution of microseismic moment tensors. SEG Expanded Abstracts 28, 2789–2793.

Fisher, M.K., Davidson, B.M., Goodwin, A.K., Fielder, E.O., Buckler, W.S., Steinberger, N.P., 2002. Integrating fracture mapping technologies to optimize stimulations in the barnett shale. In: Presented at the SPE Annual Technical Conference and Exhibition, San Antonio, Texas, 29 September–2 October. SPE-77441-MS.

Fix, J.E., Adair, R.G., Fisher, T., Mahrer, K., Mulcahy, C., Myers, B., Woerpel, J.C., 1989. Development of Microseismic Methods to Determine Hydraulic Fracture Dimensions. Gas Research Institute. Technical Report No. 89-0116.

Geiger, L., 1912. Probability method for the determination of earthquake epicenters from the arrival time only (translated from Geiger's 1910 German article). Bulletin of St. Louis University 8, 56–71.

Grassberger, P., Procaccia, I., 1983. Measuring the strangeness of strange attractors. Physica 9D, 189–208.

Griffin, L.G., Sullivan, R.B., Wolhart, S.L., Waltman, C.K., Wright, C.A., Weijers, L., Warpinski, N.R., 2003. Hydraulic fracture mapping of the high-temperature, high-pressure Bossier Sands in East Texas. In: Paper Presented at the SPE Annual Technical Conference and Exhibition, Denver, Colorado, 5–8 October.

Grob, M., van der Baan, M., 2011. Inferring in-situ stress changes by statistical analysis of microseismic event characteristics. The Leading Edge 30 (11), 1296–1302.

Gutenberg, B., Richter, C.F., 1944. Frequency of earthquakes in California. Bulletin of the Seismological Society of America 34 (4), 185–188.

Hanks, T.C., Kanamori, H., 1979. A moment magnitude scale. Journal of Geophysical Research 84 (B5), 2348–2350.

Hanks, T.C., Wyss, M., 1972. The use of body-wave spectra in the determination of seismic-source parameters. Bulletin of the Seismological Society of America 62 (2), 561–589.

Haldorsen, J.B.U., Miller, D.E., Walsh, J.J., 1994. Multichannel Wiener deconvolution of vertical seismic profiles. Geophysics 59, 1500–1511.

Hardebeck, J.L., Shearer, P.M., 2003. Using S/P amplitude ratios to constrain the focal mechanisms of small earthquakes. Bulletin of the Seismological Society of America 93 (6), 2434–2444.

Helmstetter, A., Kagan, Y.Y., Jackson, D.D., 2005. Importance of small earthquakes for stress transfers and earthquake triggering. Journal of Geophysical Research 110, B05S08. http://dx.doi.org/10.1029/2004JB003286.

Henderson, J., Main, I., Pearce, R., Takeya, M., 1994. Seismicity in north-eastern Brazil: fractal clustering and the evolution of the b value. Geophysical Journal International 116, 217–226.
Hornby, B.E., Burch, T., 2008. Passive "drive by" imaging in a deep water production well using permanent borehole seismic sensors. In: 2008 SEG Annual Meeting. Society of Exploration Geophysicists.
House, L., 1987. Locating microearthquakes induced by hydraulic fracturing in crystalline rock. Geophysical Research Letters 14 (9), 919–921.
Huang, J., Turcotte, D.L., 1988. Fractal distributions of stress and strength and variations of b-value. Earth and Planetary Science Letters 91, 223–230.
Huang, J., Safari, R., Mutlu, U., Burns, K., Geldmacher, I., McClure, M., Jackson, S., 2014. Natural-hydraulic fracture interaction: microseismic observations and geomechanical predictions. In: Unconventional Resources Technology Conference, Denver, Colorado, USA.
Julian, B.R., Foulger, G.R., 1996. Earthquake mechanisms from linear-programming inversion of seismic-wave amplitude ratios. Bulletin of the Seismological Society of America 86 (4), 972–980.
Keck, R.G., Withers, R.J., 1994. A field demonstration of hydraulic fracturing for solids waste injection with real-time passive seismic monitoring. In: SPE Annual Technical Conference and Exhibition. Society of Petroleum Engineers.
Leaney, S., Chapman, C., Ulrych, T., 2011. Microseismic Source Inversion in Anisotropic Media. Recovery–2011 CSPG CSEG CWLS Convention.
Liang, C., Thornton, M.P., Morton, P., Hulsey, B.J., Hill, A., Rawlins, P., January 2009. Improving signal-to-noise ratio of passsive seismic data with an adaptive FK filter. In: 2009 SEG Annual Meeting. Society of Exploration Geophysicists.
Ma, Y.Z., 2011. Uncertainty Analysis in Reservoir Characterization and Management: How Much Should We Know About What We Don't Know? AAPG Memoir 96.
Madariaga, R., 1976. Dynamics of an expanding circular fault. Bulletin of the Seismological Society of America 66 (3), 639–666.
Mahrer, K.D., Zinno, R.J., Bailey, J.R., DiPippo, M., 2007. Simultaneous recording of hydraulic-fracture-induced microseismics in the treatment well and in a remote well. In: SPE Hydraulic Fracturing Technology Conference. SPE 106025.
Maxwell, S.C., Cipolla, C., 2011. What does microseismicity tell us about hydraulic fracturing? In: SPE Annual Technical Conference and Exhibition SPE 146932.
Maxwell, S.C., Urbancic, T.I., Steinsberger, N., Zinno, R., 2002. Microseismic imaging of hydraulic fracture complexity in the Barnett shale. In: SPE Annual Technical Conference and Exhibition. Society of Petroleum Engineers.
Maxwell, S.C., Rutledge, J., Jones, R., Fehler, M., 2010. Petroleum reservoir characterization using downhole microseismic monitoring. Geophysics 75 (5), 75A129–75A137.
Maxwell, S.C., Cho, D., Pope, T., Jones, M., Cipolla, C., Mack, M., Henery, F., Norton, M., Leonard, J., 2011. Enhanced reservoir characterization using hydraulic fracture microseismicity. In: Hydraulic Fracturing Technology Conference. SPE 140449.
Maxwell, S., 2014. Microseismic Imaging of Hydraulic Fracturing: Improving Engineering of Unconventional Shale Reservoirs. Distinguished Instructor Short Course (No.17). Society of Exploration Geophysicists, Tulsa.
McGarr, A., Simpson, D., 1997. Keynote lecture: a broad look at induced and triggered seismicity, "Rockbursts and seismicity in mines". In: Gibowicz, S.J., Lasocki, S. (Eds.), Proceedings of the 4th International Symposium on Rockbursts and Seismicity in Mines, Poland, 11–14 August, 1997. A.A. Balkema, Rotterdam, pp. 385–396.
McGarr, A., 1976. Seismic moments and volume changes. Journal of Geophysical Research 81 (8), 1487–1494.
MIT OpenCourseWare, Spring 2008. 12.510 Introduction to Seismoloty. http://ocw.mit.edu/courses/earth-atmospheric-and-planetary-sciences/12-510-introduction-to-seismology-spring-2010/lecture-notes/lec18.pdf.

Mohammad, N.A., Miskimins, J.L., 2010. A comparison of hydraulic fracture modeling with downhole and surface microseismic data in a stacked fluvial pay system. In: Paper: SPE 134490, SPE Annual Conference and Technology Exhibition, Florence, Italy.

Moschovidis, Z., Steiger, R., Peterson, R., Warpinski, N., Wright, C., Chesney, E., Hagan, J., Abou-Sayed, A., Keck, R., Frankl, M., Fleming, C., Wolhart, S., McDaniel, R., Sinor, A., Ottesen, S., Miller, L., Beecher, R., Dudley, J., Zinno, R., Akhmedov, O., 2000. The Mounds drill-cuttings injection field experiment: final results and conclusions. In: IADC/SPE Drilling Conference. Society of Petroleum Engineers. SPE 59115.

Niitsuma, H., 1997. Integrated Interpretation of Microseismic Clusters and Fracture System in a Hot Dry Rock Artificial Reservoir.

Peterson, R.E., Wolhart, S.L., Frohne, K.H., 1996. Fracture Diagnostics Research at the GR/DOE Multi-site Project: Overview of the Concept and Results (No. CONF-961003–). Society of Petroleum Engineers (SPE), Inc., Richardson, TX (United States).

Pettitt, W.S., Young, R.P., 2007. InSite Seismic Processor: User Operations Manual Version 2.14. Applied Seismology Consultants Ltd, Shrewsbury, UK.

Pettitt, W.S., Young, R.P., Marsden, J.R., 1998. Investigating the mechanics of microcrack damage induced under true-triaxial unloading. In: EUROCK 98. Symposium.

Phillips, W.S., Rutledge, J.T., Fairbanks, T.D., Gardner, T.L., Miller, M.E., 1998. Reservoir fracture mapping using microearthquakes: two Oilfield case studies. In: Paper: SPE 36651.

Pine, R.J., Batchelor, A.S., 1984. Downward migration of shearing in jointed rock during hydraulic injections. International Journal of Rock Mechanics and Mining 21 (5), 249–263.

Potapenko, D.I., Tinkham, S.K., Lecerf, B., Fredd, C.N., Samuelson, M.L., Gillard, M.R., Le Calvez, J.H., Daniels, J.L., 2009. Barnett Shale refracture stimulations using a novel diversion technique. In: Hydraulic Fracturing Technology Conference. SPE 119636.

Reyes-Montes, J.M., Young, R.P., 2006. Interpretation of fracture geometry from excavation induced microseismic events. In: Proceedings of the European Regional ISRM Symposium, Eurock06. Liege, Belgium.

Rentsch, S., Buske, S., Gutjahr, S., Kummerow, J., Shapiro, S.A., 2007. Migration-based location of the SAFOD target-earthquakes. In: EAGE, 69th Meeting, London.

Roff, A., Phillips, W.S., Brown, D.W., September 1996. Joint structures determined by clustering microearthquakes using waveform amplitude ratios. International Journal of Rock Mechanics and Mining Sciences and Geomechanics Abstracts 33 (6), 627–639.

Rutledge, J.T., Phillips, W.S., Mayerhofer, M.J., 2004. Faulting induced by forced fluid injection and fluid flow forced by faulting: an interpretation of hydraulic-fracture microseismicity, Carthage Cotton Valley gas field, Texas. Bulletin of the Seismological Society of America 94 (5), 1817–1830.

Rutledge, J.T., Downie, R.C., Maxwell, S.C., Drew, J.E., 2013. Geomechanics of Hydraulic Fracturing Inferred from Composite Radiation Patterns of Microseismicicty. SPE 166370.

Scholtz, C.H., 1968. The frequency-magnitude relation of microfracturing in rock and its relation to earthquakes. Bulletin of the Seismological Society of America 58 (1), 399–415.

Schorlemmer, D., Wiemer, S., 2005. Earth science: microseismicity data forecast rupture area. Nature 434 (7037), 1086.

Sharma, M.M., Gadde, P.B., Sullivan, R., Sigal, R., Fielder, R., Copeland, D., Griffin, L., Weijers, L., 2004. Slick water and hybrid fracs in the Bossier: some lessons learnt. In: Presented at the SPE Annual Technical Conference and Exhibition.

Sondergeld, C.H., Estey, L.H., 1981. Acoustic emission study of microfracturing during the cyclic loading of Westerly granite. Journal of Geophysical Research: Solid Earth (1978–2012) 86 (B4), 2915–2924.

Song, F., Toksöz, M.N., 2011. Full-waveform based complete moment tensor inversion and source parameter estimation from downhole microseismic data for hydro- fracture monitoring. Geophysics 76 (6), WC103–WC116. http://dx.doi.org/10.1190/geo2011–0027.1.

St-Onge, A., Eaton, D., 2011. Noise examples from two microseismic datasets. CSEG Record 36 (8), 46–49.

Thill, R.E., 1972. Acoustic methods for monitoring failure in rock. In: The 14th US Symposium on Rock Mechanics (USRMS). American Rock Mechanics Association.

Trifu, C.I., Urbancic, T.I., 1996. Fracture coalescence as a mechanism for earthquakes: observations based on mining induced microseismicity. Tectonophysics 261 (1), 193–207.

Truby, L.S., Keck, R.G., Withers, R.J., 1994. Data Gathering for a Comprehensive Hydraulic Fracturing Diagnostic Project: A Case Study. IADC/SPE 27506. In: IADC/SPE Drilling Conference, Dallas, Texas.

Urbancic, T.I., Rutledge, J., 2000. Using microseismicity to map Cotton Valley hydraulic fractures. SEG-2000-144. In: SEG Annual Meeting, Calgary, Alberta.

Urbancic, T.I., Trifu, C.I., Young, R.P., 1993. Microseismicity derived fault-Planes and their relationship to focal mechanism, stress inversion, and geologic data. Geophysical Research Letters 20 (22), 2475–2478.

Urbancic, T.I., Maxwell, S.C., Zinno, R.J., 2002. Assessing the effectiveness of hydraulic fractures with microseismicity. In: Expanded Abstracts: SEG Annual Meeting, Salt Lake City, Utah, USA.

USGS, 1996. Focal Mechanisms. http://earthquake.usgs.gov/learn/topics/beachball.php.

Vavrycuk, V., 2005. Focal mechanisms in anisotropic media. Geophysical Journal International 161, 334–346. http://dx.doi.org/10.1111/j.1365-246X.2005.02585.x.

Vavrycuk, V., 2007. On the retrieval of moment tensors from borehole data. Geophysical Prospecting 55 (3), 381–391.

Waldhauser, F., Ellsworth, W.L., 2000. A double-difference earthquake location algorithm: method and application to the northern Hayward fault, California. Bulletin of the Seismological Society of America 90, 1353–1368.

Walker Jr., R.N., Zinno, R.J., Gibson, J.B., Urbancic, T.I., Rutledge, J.T., 1998. Cotton Valley hydraulic fracture imaging project-imaging methodology and implications. In: Paper: SPE 49194, SPE Annual Technical Conference and Exhibition, New Orleans, Louisiana.

Warpinski, N.R., Branagan, P.T., 1989. Altered stress fracturing. JPT 41 (9), 990–997. http://dx.doi.org/10.2118/17533-PA. SPE-17533-PA.

Warpinski, N.R., Moschovidis, Z.A., Parker, C.D., Abou-Sayed, I.S., 1994. Comparison study of hydraulic fracturing models—test case: GRI staged field experiment no. 3 (includes associated paper 28158). SPE Production and Facilities 9 (01), 7–16.

Warpinski, N.R., Kramm, R.C., Heinze, J.R., Waltman, C.K., 2005. Comparison of single- and dual-array microseismic mapping techniques in the Barnett shale. In: Paper SPE 95568, SPE Annual Technical Conference and Exhibition, Dallas, Texas, 9–12 October.

Weng, X., 2014. Modeling of complex hydraulic fractures in naturally fractured formation. Journal of Unconventional Oil Gas Resourc. http://dx.doi.org/10.1016/j.juogr.2014.07.001.

Wessels, S.A., De La Pena, A., Kratz, M., Williams-Stroud, S., Jbeili, T., 2011. Identifying faults and fractures in unconventional reservoirs through microseismic monitoring. First Break 29 (7), 99–104.

Wu, H., Kiyashchenko, D., Lopez, J., 2011. Time-lapse 3D VSP using permanent receivers in a flowing well in the deepwater Gulf of Mexico. The Leading Edge 30 (9), 1052–1058.

Yang, M., Araque-Martinez, A., Abolo, N., 2014. Constrained Hydraulic Fracture Optimization Framework. IAPG, 165.

Young, R.P., Baker, C., 2001. Microseismic investigation of rock fracture and its application in rock and petroleum engineering. International Society for Rock Mechanics News Journal 7, 19–27.

Zhang, W., Zhang, J., 2013. Microseismic migration by semblance-weighted stacking and interferometry. SEG-2013-0970.

Zhao, X.P., Young, R.P., 2011. Numerical modeling of seismicity induced by fluid injection in naturally fractured reservoirs. Geophysics 76, 169–184.

Zhao, X.P., Collins, D., Young, R.P., 2010. Gaussian-Beam polarization-based location method using S-wave for hydraulic fracturing induced seismicity. CSEG Recorder 35, 28–33.

Zhao, X.P., Reyes-Montes, J.M., Paul Young, R., 2013. Time-lapse velocities for locations of microseismic events–a numerical example. In: The 75th EAGE Annual Meeting, London, UK.

Zhao, X.P., Reyes-Montes, J.M., Paul Young, R., 2014. Analysis of the stability of source mechanism solutions for microseismic events from different receiver configurations. In: Proceedings of EAGE Annual Meeting, Amsterdam, Netherlands.

Zhao, X.P., 2010. Imaging the Mechanics of Hydraulic Fracturing in Naturally-fractured Reservoirs Using Induced Seismicity and Numerical Modeling (Ph.D. Thesis). University of Toronto.

Zhu, X., Gibson, J., Ravindran, N., Zinno, R., Sixta, D., 1996. Seismic imaging of hydraulic fractures in Carthage tight sands: a pilot study. The Leading Edge 15 (3), 218–224.

Zinno, R.J., Mutlu, U., 2015. Microseismic data analysis, interpretation compared with geomechanical modeling. In: EAGE Workshop on Borehole Geophysics: Unlocking the Potential, Athens, Greece, 19–22 April.

Zinno, R.J., Gibson, J., Walker Jr., R.N., Withers, R.J., 1998. Overview: Cotton Valley hydraulic fracture imaging project. In: Annual Meeting Abstracts. Society of Exploration Geophysicists, pp. 926–929.

Zinno, R.J., 1999. The Cotton Valley Hydraulic Fracture Imaging Consortium: Implications for Hydraulic Stimulation Design and Commercial Passive Seismic Monitoring (MSc Thesis). Dedman College, Southern Methodist University.

Zinno, R.J., 2010. Microseismic monitoring to image hydraulic fracture growth. In: AAPG Geosciences Technology Workshop, June 28–30, Rome, Italy.

Zinno, R.J., 2011. A brief history of microseismic mapping in unconventional reservoirs. In: EAGE Borehole Geophysics Workshop - Emphasis on 3D VSP, Istanbul, Turkey, 16–19 January.

Zinno, R.J., 2013. Microseismic mapping in eastern hemisphere, results and general criteria for successful application, comparisons to North American experience. In: SPE Workshop: Addressing the Petrophysical Challenges Relevant to Middle East Reservoirs, Dubai, UAE, 28–30 October.

IMPACT OF PREEXISTING NATURAL FRACTURES ON HYDRAULIC FRACTURE SIMULATION

10

Xiaowei Weng[1], Charles-Edouard Cohen[2], Olga Kresse[1]

Production Operations Software Technology, Schlumberger, Sugar Land, Texas, USA[1]; Production Operations Software Technology, Schlumberger, Rio de Janeiro, Brazil[2]

10.1 INTRODUCTION

In many field-scale hydraulic fracturing experiments in which the fractured formations are mined back or cored through to directly observe the created fracture geometry it has been observed that the interaction of hydraulic fractures with natural fractures can result in branching and offset at the natural fractures and consequently lead to complex fractures (Warpinski and Teufel, 1987; Jeffrey et al., 1994; Jeffrey et al., 2009; Warpinski et al., 1993). Figure 10.1 shows an example of complex parallel fractures and offsets created as a hydraulic fracture propagates through natural fractures and zone boundaries, observed by Warpinski and Teufel (1987) in a mineback experiment.

However, it is not always clear if these complexities are only small-scale features relative to an otherwise planar fracture at large scale and whether they also occur in formations at greater depth. Limited direct observations available at depth by coring through the hydraulically fractured interval at the Gas Research Institute/Department of Energy-sponsored Multiwell Experiment Site (Warpinski et al., 1993) revealed multiple closely spaced hydraulically induced fractures filled with the residue of the fracturing fluid. In spite of the general awareness of potential complexity in the hydraulically induced fractures, based on the observations like these, over the years, hydraulic fracturing treatment design continued to be simulated based on the models that assume a planar fracture.

Since 2005, following the success of horizontal drilling and multistage fracturing in the Barnett Shale, exploration and drilling activities in shale gas and shale oil reservoirs have skyrocketed in the United States and abroad. Economic production from these reservoirs depends greatly on the effectiveness of hydraulic fracturing stimulation treatment. Microseismic measurements and other evidence suggest that creation of complex fracture networks during fracturing treatments may be a common occurrence in many unconventional reservoirs (Maxwell et al., 2002; Fisher et al., 2002; Warpinski et al., 2005). The created complexity is strongly influenced by the preexisting natural fractures and in situ stresses in the formation. Fracture simulation can provide information such as induced overall fracture length and height, propped versus unpropped fracture surface areas, and proppant distribution

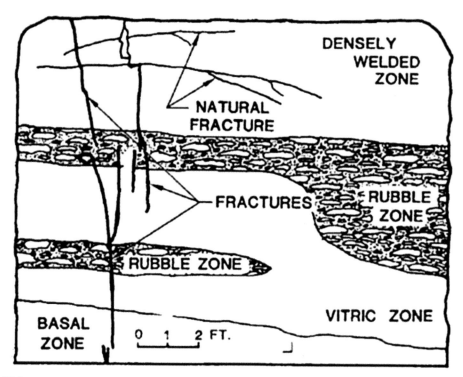

FIGURE 10.1

Complex fractures observed in mineback experiments (Warpinski and Teufel, 1987).

and its conductivity, all of which influence the short and long-term production from the unconventional reservoir and cannot be obtained from microseismic measurement alone (Cipolla et al., 2011).

However, modeling the process of hydraulic fracture network creation and interaction between hydraulically induced fractures and natural fractures presents many technical challenges. Significant progress has been made in recent years in the development of complex fracture models to address the needs for more suitable design tools for the unconventional reservoirs than the conventional planar fracture models. Some aspects of this complex fracturing process are still not fully understood in terms of their impact or importance to the overall fracture geometry creation.

One of the difficulties in fracturing design is the lack of clear understanding of the nature of fracture complexity created during fracture treatment. Microseismic monitoring does not provide sufficient resolution to delineate the exact hydraulic fracture planes. Microseismic events are mostly attributed to shear failures along natural fractures or faults surrounding a hydraulic fracture (Rutledge et al., 2004; Williams-Stroud et al., 2012). The events cloud forms a "halo" surrounding the hydraulic fracture. In conventional sandstone formations, the observed events cloud has a relatively narrow width, whereas in unconventional reservoirs, a much wider events cloud is often observed (Fisher et al., 2002). A wide microseismic cloud may possibly be explained by either deep fluid penetration into natural fractures in the shale while the induced hydraulic fracture remains planar or simple (Savitski et al., 2013), or by the creation of complex hydraulic (tensile) fracture network. Although deep fluid penetration into a highly permeable and initially well-connected natural fractures network is certainly possible (Zhang et al.,

2013), many unconventional plays have very low effective permeability, and observation of cores shows that most natural fractures in these shales are mineralized (Gale et al., 2007; Gale and Holder, 2008; Han, 2011; Williams-Stroud et al., 2012). Therefore, in very low permeability shale, fluid penetration in the natural fracture network is limited. Fluid penetration into natural fractures can also occur due to dilation of natural fractures as a result of shear, but this typically occurs under the condition of large stress anisotropy and for natural fractures oriented 30–60° from the principal stress directions (Murphy and Fehler, 1986). For many shale reservoirs where the tectonic environment is relaxed and the difference between the horizontal stresses is low, a wide microseismic cloud is a strong indication that complex, tensile-opening hydraulic fracture networks are created, although the hydraulic fractures may follow the paths of the natural fractures. A field case in the Barnett Shale presented by Fisher et al. (2002), in which fracturing fluid unexpectedly connected to and brought down the production of several adjacent wells not on the expected fracture plane, provided the supporting evidence of complex hydraulic fracture networks. In the analysis of another Barnett case, Cipolla et al. (2010) showed that the predicted fracture length from a planar fracture model far exceeded the fracture length indicated by the microseismic data, unless a very low fluid efficiency (less than 10%) is assumed in the simulation in order for a planar fracture to accommodate the large volume of fluid injected. Such low efficiency is not consistent with the very slow pressure decline typically observed during the shut-in following the pumping period in most shale formations. In contrast, complex hydraulic fracture networks can explain the much larger fluid volume stored in fracture networks for the same overall fracture network length and yet still high fluid efficiency due to low leakoff. However, in a geological setting with high tectonic stress, shear fracturing may be an important mechanism accounting for the observed wide microseismic cloud and potential permeability enhancement induced by shear. Properly constructed complex fracture models and/or geomechanics models can help answer these questions and provide the tools for optimizing the fracture design and completion strategy.

For a hydraulic fracture propagating in a formation that contains preexisting natural fractures, or mechanically weak planes relative to the rock matrix, the interaction between the hydraulic fracture (HF) and natural fractures (NF) could cause fluid loss into the NF, dilation of the NF either due to shear or in tension, or even branching or alteration of the HF path, leading to complex fractures. Figure 10.2 depicts a possible hydraulic fracture network created when pumping into a perforation cluster in a horizontal well (Weng, 2014) and shows some of the scenarios of HF interaction with NF that can lead to fracture branching and complexity:

1. Direct crossing. When an NF has strong mechanical bonding and/or is subjected to high normal stress, the tensile stress concentration at the tip of the approaching HF is readily transmitted across the NF interface to the rock on the opposite side of the hydraulic fracture, causing the rock to fail in tension and allowing the hydraulic fracture to directly propagate through the NF without change of direction. Consequently, the HF propagates through the formation as a planar fracture. However, if the fluid pressure can exceed the closure stress acting on the NF, it will open in tension and become a part of a now nonplanar hydraulic fracture network.
2. HF arrested by NF. This scenario occurs when the NF interface is weaker than the rock matrix and the stress condition is such that the interface fails in shear and slips. Consequently, the tensile stress at the tip of the approaching HF is not sufficiently transmitted to the opposite side of the NF interface to cause the rock to fail in tension. And the HF growth is hence arrested by the NF. If the fluid pressure in the HF continues to increase, it can exceed the closure stress acting on the NF and cause the NF to be opened in tension and become a part of the hydraulic fracture network.

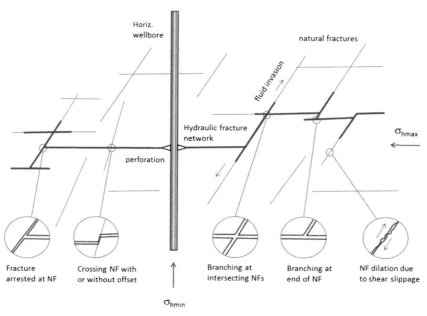

FIGURE 10.2

Map view of induced HF network depicting various scenarios of interaction with NF that can lead to fracture branching and complexity (Weng, 2014).

3. Crossing with an offset. It is often observed in laboratory and mineback experiments that when an HF crosses an NF, it can do so with a small offset at the interface, as shown in Fig. 10.2. The offset is typically on the order of one to a few inches (Jeffrey et al., 2009). The offset is created due to localized interface separation and shear slip at the point where the HF intersects the NF, which shifts the stress concentration away from the intersection point to the tip of opening/shear slip region (Thiercelin and Makkhyu, 2007).
4. Intersecting natural fractures. Once fluid pressure exceeds the closure stress on the NF, the NF opens up in tension and becomes a part of the HF network. If the NF intersects another NF, when the fluid front reaches the intersection, the HF may branch again at the intersection as long as fluid pressure exceeds the closure stress on the NF.
5. Branching or turning of fracture at end of the NF. For an HF following the path of an NF to its end, there is no longer a weak plane for fluid to preferentially open. Consequently, the fracture either turns itself to align with the preferred fracture direction or creates a T-shaped branch.
6. Shear slip along NF. If the fluid pressure in the NF stays below its closure stress, the fracture interface will not separate in tension. However, it can fail in shear. The shear-induced interfacial slip causes dilation and enhances the permeability of the NF, which can potentially enhance production. The occurrence of shear failure depends on the normal and shear stresses applied on the NF, which, in turn, depend on the in situ principal stresses, angle of the NF relative to the in situ stresses, the fluid pressure (which depends on pressure diffusion in the NF), and interfacial frictional properties.

A major challenge in numerical modeling of complex hydraulic fractures is due to the different scales the problem presents. The mechanics that controls the various behaviors at the fracture intersection point is a very localized phenomenon. For a comprehensive complex fracture simulator, both the large-scale network and localized crossing processes need to be properly modeled. In the large-scale modeling of a fracture network, the mechanical interaction among propagating hydraulic fractures also needs to be taken into account. In the most common horizontal well completion in unconventional reservoirs, a completion stage consists of multiple perforation clusters from which multiple fractures are initiated. Even in the case of noncomplex, planar fractures, the interaction among the fractures can lead to uneven fracture growth. Most conventional planar fracture models do not account for the fracture interaction, or the so-called stress shadow effect. In the case of a complex fracture network, the fracture interaction can be even stronger due to the typically higher fracture density. Therefore, stress shadowing must be considered in the fracture simulator.

The main goal of this chapter is to investigate and demonstrate how the properties of preexisting natural fractures can affect the hydraulic fracture network footprint when pumping into a formation with preexisting natural fractures, and consequently the propped fracture surface area and production.

10.2 HYDRAULIC FRACTURE AND NATURAL FRACTURE INTERACTION

As mentioned above, the interaction between HF and NF plays a critical role in creating fracture complexity during hydraulic fracturing treatments in formations with preexisting NFs. Understanding and proper modeling of the mechanisms governing HF–NF interaction are keys to explain fracture complexity and microseismic events observed during hydraulic fracturing treatments, and, ultimately, to be able to properly predict fracture geometry and reservoir production.

When an HF intersects an NF it can cross or be arrested by the NF, and the HF can subsequently open (dilate) the NF (Fig. 10.2). If the HF crosses the NF, it remains planar, with the possibility to open the intersected NF if the fluid pressure at the intersection exceeds the normal stress acting on the NF. If the HF does not cross the NF, it can dilate and eventually propagate into the NF, which leads to a more complex fracture network. Therefore, the crossing criterion, in general, strongly influences the complexity of the resulting fracture network.

The interaction between the HF and the NF depends on the in situ stresses, mechanical properties of the rock, properties of natural fractures, and the hydraulic fracture treatment parameters, including fracturing fluid properties and injection rate. During the last decades, extensive theoretical and experimental work has been done to investigate, explain, and develop the rules controlling HF–NF interaction (Warpinski and Teufel, 1987; Blanton, 1982, 1986; Renshaw and Pollard, 1995; Hanson et al., 1982; Leguillon et al., 2000; Beugelsdijk et al., 2000; Potluri et al., 2005; Zhao et al., 2008; Gu and Weng, 2010; Gu et al., 2011) as well as numerical simulations to model fracture behaviors at the intersection (Heuze et al., 1990; Zhang and Jeffrey, 2006b, 2008; Thiercelin and Makkhyu, 2007; Zhang et al., 2007a,b, 2009; Zhao and Young, 2009; Chuprakov et al., 2010; Meng and de Pater, 2010; Dahi-Taleghani and Olson, 2011; Sesetty and Ghassemi, 2012; Chuprakov et al., 2013). A short overview is provided below based on a previous detailed review of numerical and analytical approaches to model HF–NF interaction (Weng, 2014).

Based on relatively simple analytical considerations of the stress field at the propagating fracture tip, several analytical crossing criteria have been developed (Warpinski and Teufel, 1987; Blanton,

1982, 1986; Renshaw and Pollard, 1995; Gu and Weng, 2010). The Blanton (1982, 1986) criterion is based on linear elastic fracture mechanics without considering the mechanical interaction between the hydraulic and the natural fracture, and relies on differential stress and angle of interaction between the hydraulic and the natural fracture. The Warpinski and Teufel criterion (1987) defines conditions for shear slippage on the interface based on Coulomb frictional law and prescribes a simple criterion for intersection of the permeable NF by a hydraulic fracture and/or dilation of NF. Renshaw and Pollard (1995) developed a simple criterion for predicting if a fracture will propagate across a frictional interface orthogonal to the fracture based on the linear elasticity fracture mechanics solution for the stresses near the fracture tip. It determines the stresses required to prevent slip along the interface at the moment when the stress on the opposite side of the interface is sufficient to reinitiate a fracture.

The Renshaw and Pollard crossing criterion is given as

$$\frac{-\sigma_H}{T_0 - \sigma_h} > \frac{0.35 + \frac{0.35}{\lambda}}{1.06}. \tag{10.1}$$

where σ_H and σ_h are maximum and minimum horizontal stresses (positive for tension), respectively; T_0 is the rock tensile strength; and λ is the coefficient of friction of the NF interface. The criterion was validated by laboratory experiments for dry cracks. The conditions for NF fracture opening and crossing and HF reinitiation from the NF fracture tip or flaw are summarized also in Polturi et al. (2005).

Natural fractures are often not aligned with the contemporary principal in situ stress directions in the rock formation, in which case the intersection angle of an HF approaching an NF is between 0° and 90°. Because the intersection angle has a significant effect on crossing, the Renshaw and Pollard criterion has been extended to fracture intersection at nonorthogonal angles by Gu and Weng (2010). Their work showed that it becomes more difficult for a hydraulic fracture to cross an interface when the intersection angle decreases from 90°. This crossing criterion (referred as extended Renshaw and Pollard, or eRP criterion, in this chapter) was compared to the laboratory results from block experiments and good agreement was obtained (Fig. 10.5; also see Gu et al., 2011).

With their relative simplicity, these analytical criteria do not take into account the detailed evolution of the stress field and rock deformation after the initial contact by the tip. These criteria capture the first-order effects of fracture intersection angle, NF friction coefficient, and in situ stresses, but they are insensitive to the parameters of fluid injection affecting the hydraulic fracture geometry and the fluid infiltration into the natural fracture after the contact. However, field and laboratory observations showed that fluid properties are important and should also be accounted for.

The experimental studies by Beugelsdijk et al. (2000) and de Pater and Beugelsdijk (2005) showed that flow rate and fracturing fluid viscosity strongly influence hydraulic fracture complexity in a prefractured block. With a low value of the product of the injection rate Q and fracturing fluid viscosity μ ($Q\mu$), fluid tends to leak into the preexisting discontinuities in the rock and creates tortuous fracture paths following the discontinuities. With a large $Q\mu$ value, the hydraulic fracture tends to cross most discontinuities, and the overall fracture path is nearly straight (Fig. 10.3).

A similar observation is made based on the microseismic monitoring of a treatment using gel fracturing fluid and a treatment using slickwater in the same well in Barnett Shale (Warpinski et al., 2005). Figure 10.4 shows the microseismic events observed in the same well first treated with a cross-linked gel and then refractured with slickwater. Cross-linked gel was pumped at 70 bbl/min for about 3 h

10.2 HYDRAULIC FRACTURE AND NATURAL FRACTURE INTERACTION

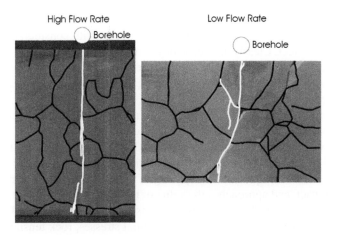

FIGURE 10.3

Cross-sections of blocks with shrinkage cracks in black and hydraulic fractures in white. Left is a base case with high flow rate and right is a low flow rate test. The high flow rate induced a new hydraulic fracture whereas the low flow rate resulted in fluid flow into the shrinkage cracks (de Pater and Beugelsdijk, 2005).

with sand concentration ramped up to 3 ppg. Most of the microseismic activity suggests longitudinal fracturing with only modest activation of natural fractures, resulting in a narrow stimulated network (less than 500 ft from the wellbore in many sections of the lateral), as seen in Fig. 10.4(a), with resulting stimulated reservoir volume (SRV) equal to 430 million cubic feet. During the full refracturing conducted several months later, 60,000 bbl of slickwater and 285,000 lbm of sand were

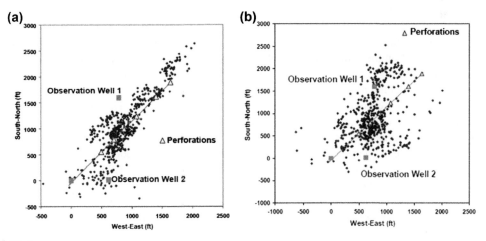

FIGURE 10.4

Single-well microseismic event locations for (a) cross-linked gel stimulation and (b) slickwater refracturing treatment in a horizontal Barnett Shale well (Warpinski et al., 2005).

pumped at 125 to 130 bbl/min for most of the treatment, which lasted 6.5 h. The stimulated network was approximately 1500 ft wide and 3000 ft long (Fig. 10.4(b)) with considerable height growth and SRV of 1450 million cubic feet. Clearly, the refracturing treatment stimulated a much larger volume of rock than the initial gel treatment (1450 million cubic feet versus 430 million cubic feet), and showed the patterns of development that suggested the opening of both north-east and north-west-trending fractures (Warpinski et al., 2005).

To improve the description of HF–NF interaction, a much more sophisticated analytical crossing model that takes into account the mechanical influence of the HF opening and the hydraulic permeability of the NF was developed by Chuprakov et al. (2013). The model solves the problem of elastic perturbation of the NF at the contact with a blunted HF tip, which is represented by a uniformly open slot (thereby, giving it the name OpenT). The opening of the HF at the junction point w_T (blunted tip) develops soon after contact, and approaches the value of the average opening of the hydraulic fracture $\overline{w}$, defined by the injection rate Q and the fluid viscosity μ. The new OpenT crossing model incorporates the influence of rock properties (local horizontal stresses, rock tensile strength, toughness, pore pressure, Young's modulus, Poisson's ratio), natural fracture properties (friction coefficient, toughness, cohesion, permeability), intersection angle between hydraulic and natural fractures, fracturing fluid properties (viscosity, tip pressure), and injection rate to define crossing rules.

The model computes the elastic stress field along the interface and in the vicinity of the activated NF and determines the size of the open and shear sliding zones. It examines the generated stress field to determine the sites of tensile stress. A combined stress and energy criterion determines if the fracture can be initiated. If the fracture initiation criterion is satisfied, a secondary fracture is initiated on the opposite side of the interface, leading to either direct crossing (if the initiation site is right at the intersection point) or crossing with an offset (if initiation site is located away from the intersection point).

The solution shows that the spatial extent of the open and sliding zones strongly depends on the fluid pressure inside the activated part of the NF. The larger the inner fluid pressure, the larger the open and sliding zones at the NF. Consequently, it is expected that after the HF contacts the NF, the injected fracturing fluid will gradually penetrate the NF with finite hydraulic permeability κ and thus enhance the fluid pressure within the NF. The average pressure inside the NF is computed approximately based on the fluid diffusion into the fracture. As a result, the fluid pressure builds up much more easily in a highly permeable NF or with a low-viscosity fracturing fluid, leading to greater likelihood of the HF being arrested. Conversely, it is more difficult to penetrate the NF with a high-viscosity fluid, leading to greater likelihood of the HF crossing the NF. The OpenT model has been validated against laboratory experiments and against rigorous numerical models (Thiercelin and Makkhyu, 2007; Chuprakov et al., 2013).

For more accurate modeling of HF–NF interaction, numerical models are used. Zhang et al. (2007a) and Zhang et al. (2009) applied a generalized two-dimensional (2D) elasticity equation for arbitrary fracture surfaces Ω in an infinite elastic medium:

$$\sigma_n(\mathbf{x}) - \sigma_n^\infty(\mathbf{x}) = \int_\Omega [G_{11}(\mathbf{x},s)w(s) + G_{12}(\mathbf{x},s)v(s)]ds$$
$$\tau(\mathbf{x}) - \tau^\infty = \int_\Omega [G_{21}(\mathbf{x},s)w(s) + G_{22}(\mathbf{x},s)v(s)]ds \qquad (10.2)$$

where $\mathbf{x} = (x, y)$ is a point on the fracture surface Ω, ds is an infinitesimal length increment along the fracture, $\sigma_n(\mathbf{x})$ and $\tau(\mathbf{x})$ are the normal and shear stresses acting on the fracture face, σ_n^∞ and τ^∞ are the normal and shear stress induced by the remote in situ stresses on the fracture at point $\mathbf{x}$, w and v are opening and shear displacements along the fracture, and G_{ij} are hypersingular Green's functions given by Zhang et al. (2005, 2007a). Equation (10.2) contains the normal components of the stress and displacement, as well as shear components, and it applies not only to the open hydraulic fracture, where $\sigma_n(\mathbf{x}) = p(\mathbf{x})$ and $\tau(\mathbf{x}) = 0$, but also to the closed fracture where shear stress $\tau(\mathbf{x})$ may be nonzero. The corresponding numerical code (MineHF2D) accounts for the mass conservation equation and the equation for the flow of Newtonian or power-law fluid in the fracture, with the aperture of the closed fracture due to surface roughness ϖ taken into account. For the part of the fracture system that is closed, the Coulomb frictional law applies, and the aperture of a closed fracture is a function of both effective normal stress and shear displacement. When the shear stress acting on the fracture interface reaches the maximum shear dictated by Coulomb frictional law, the interface slips. This shear slip causes dilation of the interface, i.e., an increase in $\overline{w}$ and, consequently, its conductivity to fluid flow. In the Zhang et al. (2007a) formulation, the mechanical aperture that gives rise to an increase in porosity and the hydraulic aperture that dictates fracture conductivity are considered the same for simplicity. These two quantities can be significantly different, especially when mechanical aperture is small. Refer to Zhang et al. (2009) and Yew and Weng (2015) for further detailed discussions on this subject.

Dividing the fractures into small, equal-size elements, and discretizing the governing equations by using the displacement discontinuity method (DDM) (Crouch and Starfield, 1983), the resulting coupled nonlinear system of equations can be solved for the pressure, opening width, and stresses along the fractures. Using this model, Zhang and Jeffrey (2006a,b, 2008), Zhang et al. (2007a,b, 2009), and Jeffrey et al. (2009) investigated various problems related to a hydraulic fracture crossing a natural fracture or a bedding plane, effect of fracture offset, and interaction of multiple fractures reinitiated from the interface. In Zhang et al. (2007a), the effect of fracture arrest at a formation interface on the pressure response was studied, for both situations of stiff-to-soft and soft-to-stiff interfaces. The results of the MineHF2D model, widely recognized in industry, have been compared with experimental observations in Fig. 10.5, and the model has been used to validate the analytical OpenT model (Fig. 10.6).

Among other numerical crossing models, we will mention the models of Cooke and Underwood (2001), Chuprakov et al. (2010), Sesetty and Ghassemi (2012), and Dahi-Taleghani and Olson (2011), the short description of which can be found in Weng (2014).

Whereas the numerical models have been successfully used to simulate the detailed stress field and obtain insights to the complex process of HF–NF interaction, these simulations are computationally intensive, even though the problem is only 2D, because of the fine numerical elements needed to produce accurate results and the complexity of the coupled solid–fluid problem. It is difficult to apply these models to simulate the reservoir-scale problems that have a much larger length scale, with three-dimensional (3D) geometry (having finite fracture height), and possibly involving a large number of fractures. In contrast, the analytical crossing models discussed earlier can be computed very quickly, which makes them much more suited for integration in the large-scale complex fracture models.

Figure 10.5 shows the comparison between different analytical models by Blanton (1986), Gu and Weng (2010, eRP in Fig. 10.4), OpenT model (Chuprakov et al., 2013), and the experimental results from Gu et al. (2011), as well as numerical simulation using the MineHF2D code (Zhang and Jeffrey,

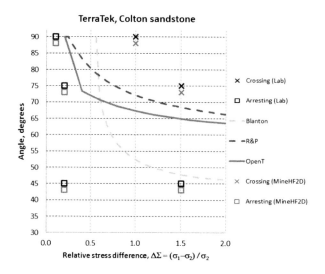

FIGURE 10.5

Comparison of laboratory "crossing-arresting" data with previous analytical models (Blanton, eRP), new analytical model OpenT, and MineHF2D simulations (Chuprakov et al., 2013).

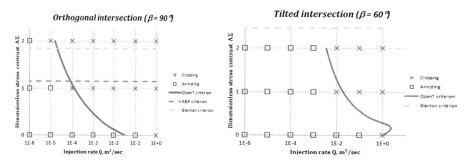

FIGURE 10.6

Comparison of HF–NF crossing-arresting behavior between analytical models and the numerical model using MineHF2D code. The red (gray in print versions) crosses and squares indicate, respectively, crossing and arresting behavior from MineHF2D code, solid green (dark gray in print versions) curves correspond to analytical predictions using the OpenT criterion, dashed yellow (light gray in print versions) curve corresponds to the Blanton criterion, and the eRP criterion is given by dashed blue lines. The interaction is studied for various injection rates and relative stress differences for two different HF–NF contact angles, $\beta = 90°$ (left) and $\beta = 60°$ (right) (Chuprakov et al., 2013).

2006a,b, 2008; Zhang et al., 2007b). It shows that both the analytical and the numerical models agree with the experiments, suggesting they capture the first-order crossing-arresting behavior.

To compensate for the lack of laboratory data for full validation, numerical experiments were conducted using MineHF2D code to assess the sensitivity of the injection rate on fracture crossing. The results for one set of parametric runs are shown in Fig. 10.6. The OpenT model agrees well with the numerical results in the sense that it captures the crossing-arresting transition as flow rate changes (Kresse et al., 2013). This model will be adopted here to investigate the effect of NF properties on the generated complex HF network.

10.3 MODELING OF COMPLEX FRACTURE NETWORK

Multiple approaches have been developed to model the propagation of a hydraulic fracture in a formation with a complex preexisting natural fracture network (Weng, 2014).

Here we will use the unconventional fracture model (UFM), incorporating the OpenT crossing criterion described above, to investigate the influence of the natural fractures' properties on the hydraulic fracture propagation pattern in naturally fractured formations (Weng et al., 2011, 2014; Kresse et al., 2012).

The UFM model is a complex fracture model capable of simulating fracture propagation, rock deformation, and fluid flow in a formation with a preexisting network of natural fractures. The model solves the fully coupled problem of fluid flow in the induced hydraulic fracture network and the elastic deformation of the fractures and has similar assumptions and governing equations as found in conventional pseudo-3D (P3D) fracture models. But, instead of solving the problem for a single planar fracture, the UFM model solves the equations for the complex fracture network. Fracture height growth is modeled in the same manner as in a conventional P3D model. A three-layer proppant transport model, consisting of a proppant bank at the bottom, a slurry layer in the middle, and clean fluid at the top, is adopted for simulating proppant transport in the fracture network. A key difference between the UFM model and the conventional planar fracture model is being able to simulate the interaction of hydraulic fractures with preexisting natural fractures, using the analytical OpenT crossing model discussed earlier in this chapter. Additionally, the UFM model also considers the interaction among hydraulic fracture branches by computing the stress shadow effect on each fracture exerted by the adjacent fractures.

The basic equations implemented in the UFM model include the equations governing fluid flow in the fracture network, mass conservation, fracture deformation, and the fracture propagation/interaction criteria. The mass conservation (continuity) equation along any branch of the fracture network, is given as

$$\frac{\partial q}{\partial s} + \frac{\partial (H_{fl}\overline{w})}{\partial t} + q_L = 0, \quad q_L = 2h_L u_L \tag{10.3}$$

where q is the local flow rate inside the hydraulic fracture along the length, $\overline{w}$ is an average opening width at the cross-section of the fracture at position $s = s(x, y)$, $H_{fl}(s, t)$ is the local height of the fracture occupied by fluid, and q_L is the leakoff volume rate through the wall of the hydraulic fracture into the rock matrix per unit length (leakoff height h_L times leakoff velocity u_L), which is expressed through Carter's leakoff model. The fracture tips propagate as a sharp front, and the total length of the entire hydraulic fracture networks at any given time t is defined as $L(t)$.

The rheological behavior of the injected fluid is characterized as a power-law fluid with power-law index n' and consistency index K'. The fluid flow in the fracture could be laminar, turbulent, or Darcy flow through the proppant pack and is described correspondingly by different laws. For the case of laminar flow along any given fracture branch, the fluid flow equation is

CHAPTER 10 IMPACT OF PREEXISTING NATURAL FRACTURES

$$\frac{\partial p}{\partial s} = -\alpha_0 \frac{1}{\overline{w}^{2n'+1}} \frac{q}{H_{fl}} \left|\frac{q}{H_{fl}}\right|^{n'-1} \tag{10.4}$$

with

$$\alpha_0 = \frac{2K'}{\phi(n')^{n'}} \cdot \left(\frac{4n'+2}{n'}\right)^{n'}; \quad \phi(n') = \frac{1}{H_{fl}} \int_{H_{fl}} \left(\frac{w(z)}{\overline{w}}\right)^{\frac{2n'+1}{n'}} dz \tag{10.5}$$

and for turbulent flow

$$\frac{\partial p}{\partial s} = -\frac{f\rho}{\overline{w}^3} \frac{q}{H_{fl}} \left|\frac{q}{H_{fl}}\right| \tag{10.6}$$

Here $w(z)$ represents fracture width as a function of depth z at the current position $s(x, y)$, and f is the Fanning friction factor for turbulent flow. Fracture width is related to fluid pressure through the elasticity equation (Eq. (10.7)). The elastic properties of the rock (assumed to be homogeneous, isotropic, liner elastic material) are described by Young's modulus E and Poisson's ratio ν.

In a multilayered formation, the fracture width profile in a cross-section and fracture height depend on fluid pressure, the in situ stresses, fracture toughness, layer thickness, and elastic modulus of each layer covered by the fracture height. Neglecting vertical flow, the stress intensity factors at fracture top and bottom tips K_{Iu} and K_{Il}, and fracture width profile w can be directly computed analytically for a layered medium with piecewise constant stress σ_I in the ith layer, as given below (Fung et al., 1987; Mack and Warpinski, 2000):

$$w(z) = \frac{4}{E'}\left[p_{cp} - \sigma_n + \rho_f g\left(h_{cp} - \frac{h}{4} - \frac{z}{2}\right)\right]\sqrt{z(h-z)} + \frac{4}{\pi E'}\sum_{i=1}^{n-1}(\sigma_{i+1} - \sigma_i)$$

$$\times \left[(h_i - z)\cosh^{-1}\frac{z\left(\frac{h-2h_i}{h}\right) + h_i}{|z - h_i|} + \sqrt{z(h-z)}\arccos\left(\frac{h-2h_i}{h}\right) \right] \tag{10.7}$$

$$K_{Iu} = \sqrt{\frac{\pi h}{2}}\left[p_{cp} - \sigma_n + \rho_f g\left(h_{cp} - \frac{3}{4}h\right)\right] + \sqrt{\frac{2}{\pi h}}\sum_{i=1}^{n-1}(\sigma_{i+1} - \sigma_i)\left[\frac{h}{2}\arccos\left(\frac{h-2h_i}{h}\right) - \sqrt{h_i(h-h_i)}\right]$$

$$K_{Il} = \sqrt{\frac{\pi h}{2}}\left[p_{cp} - \sigma_n + \rho_f g\left(h_{cp} - \frac{h}{4}\right)\right] + \sqrt{\frac{2}{\pi h}}\sum_{i=1}^{n-1}(\sigma_{i+1} - \sigma_i)\left[\frac{h}{2}\arccos\left(\frac{h-2h_i}{h}\right) + \sqrt{h_i(h-h_i)}\right]$$

$$\tag{10.8}$$

where h_i is the distance from top of the ith layer to fracture bottom tip, p_{cp} is the fluid pressure at a reference (entry) depth h_{cp} measured from the bottom tip, and ρ_f is fluid density.

The fracture height can be determined at each position of the fracture by matching K_{Iu} and K_{Il}, given by Eq. (10.8), to the fracture toughness K_{Ic} of the corresponding layer containing the tips.

The fracture height computed directly using Eq. (10.8) is referred to as equilibrium height. This could be extended to nonequilibrium height growth calculations by taking into account the pressure gradient due to the fluid flow in the tip regions in the vertical direction by adding apparent toughness proportional to the fracture's top and bottom velocities (Mack and Warpinski, 2000).

In addition to the equations described above, the global volume balance condition must be satisfied:

$$\int_0^t Q(t)dt = \int_0^{L(t)} h(s,t)\overline{w}(s,t)ds + \int_{H_L}^t \int_0^{L(t)} \int_0^t 2u_L ds dt dh_L \qquad (10.9)$$

That is, the total volume of fluid pumped is equal to volume of fluid in fracture network and volume leaked from the fracture into the matrix up to time t. The boundary conditions require the flow rate, net pressure, and fracture width to be zero at all fracture tips. The total fracture network system consists of not only fractures, but also the perforations and wellbore. The fracture networks communicate through injection elements to account for perforation friction, and perforation clusters are connected through wellbore elements to account for the friction in the casing.

The system of equations (Eqs (10.3)–(10.9)), together with the initial and boundary conditions, plus the equations governing fluid flow in the wellbore and through the perforations, represent a complete set of governing equations. Combining these equations and discretizing the fracture network into small elements leads to a nonlinear system of equations in terms of fluid pressure p in each element, simplified as $f(p) = 0$. At each time step, each propagating fracture tip extends an incremental distance, according to the propagation criterion based on the local fluid velocity and fracture tip stress intensity factor, in the direction of local maximum horizontal principal stress (accounting for the stress shadow effect). The intersection of fracture tip with any natural fracture is checked, and if intersection with a natural fracture occurs, the OpenT crossing model (Chuprakov et al., 2013; Kresse et al., 2013) is applied to determine whether the tip crosses the natural fracture or is arrested by it, and an adjustment in the fracture grid is applied accordingly. The system of equations $f(p) = 0$ is then solved by using the damped Newton–Raphson method to obtain the new pressure and flow distribution in the fracture networks. Proppant transport equations are solved to update the proppant movement and settling in the fractures, and fracture height and stress shadow are also updated. A more detailed description of the model can be found in Weng et al. (2011), Kresse et al. (2012), and Weng et al. (2014).

Fracture network growth pattern is affected by the mechanical interaction between the adjacent fractures. Generally known as the stress shadow effect, the stress field acting on each fracture is perturbed by the opening and shearing displacements of other nearby fractures. In a 2D plane-strain displacement discontinuity solution, Crouch and Starfield (1983) described the normal and shear stresses (σ_n and σ_s) acting on one fracture element induced by the opening and shearing displacement discontinuities (D_n and D_s) from all fracture elements as following (Fig. 10.7):

$$\sigma_n^i = \sum_{j=1}^N A^{ij} C_{ns}^{ij} D_s^j + \sum_{j=1}^N A^{ij} C_{nn}^{ij} D_n^j$$

$$\sigma_s^i = \sum_{j=1}^N A^{ij} C_{ss}^{ij} D_s^j + \sum_{j=1}^N A^{ij} C_{sn}^{ij} D_n^j \qquad (10.10)$$

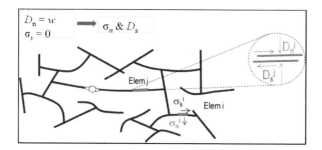

FIGURE 10.7

Stress shadow effect (Wu et al., 2012).

where C^{ij} are the 2D, plane-strain elastic influence coefficients. This method, referred to as the 2D displacement discontinuity method (2D DDM), is used in the model to compute the additional stresses induced on each fracture element from the displacements of adjacent elements. In addition, a 3D correction factor A^{ij} suggested by Olson (2004, 2008) (referred to as enhanced 2D DDM) was further introduced to modify the influence coefficients C^{ij} in the above equation to account for the 3D effect due to finite fracture height that leads to diminishing interaction between any two fracture elements when the distance increases. This additional normal stress due to the stress shadow is computed at each time step in the UFM model and is then added to the initial in situ stress field on each fracture element during pressure and width iteration. The effect of stress shadow, including the shear stress, on the directional change of propagating fracture tips is essential for proper modeling of fracture network propagation pattern (Wu et al., 2012).

The UFM model that incorporates the enhanced 2D DDM approach is validated against full 2D DDM simulator MineHF2D incorporating a full solution of coupled elasticity and fluid flow equations by Zhang et al. (2007a) in the limiting case of very large fracture height (because the 2D DDM approach does not consider the 3D effect of finite fracture height). The comparison of influence of two closely propagating fractures on each other's propagation paths has been provided in Wu et al. (2012) and is shown in Fig. 10.8.

The effect of stress shadow on fracture geometry is highly influenced by many parameters. Figure 10.9, for example, shows the fracture geometry predicted by the UFM model for five fractures propagating from a horizontal well with perforation cluster spacing of 10, 20, and 40 m, respectively. When fracture spacing is large (larger than fracture height), the effect of stress shadow dissipates, and fractures have approximately the same dimensions; as the distance between fractures is reduced, the effect of stress shadow becomes greater and is noticeably observed in the reduced width and length for the inner fractures. These observations are consistent with those from 3D simulations such as those by Castonguay et al. (2013).

The propagation of hydraulic fracture network in formation with preexisting natural fractures is highly influenced by the interaction between hydraulic fracture branches (stress shadow effect) and by the crossing rule when hydraulic fractures intersects natural fractures.

To demonstrate the validity of OpenT crossing criterion, which accounts for the pumping fluid properties and is implemented in UFM model, the field case from Warpinski et al. (2005) presented earlier in Fig. 10.4 was simulated using UFM model. Using a DFN model based on the dominant set of

10.3 MODELING OF COMPLEX FRACTURE NETWORK

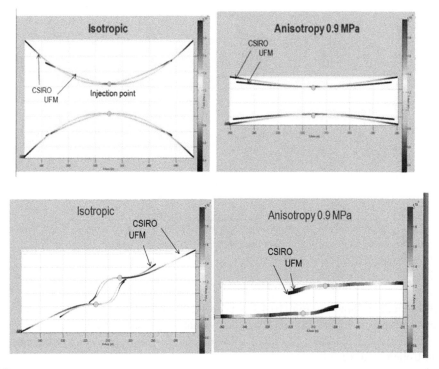

FIGURE 10.8

Comparison of propagation paths for two initially parallel fractures, with and without offset, in isotropic and anisotropic stress fields, by Wu et al. (2012).

fractures in the north 70° west direction by Gale et al. (2007), the UFM simulated hydraulic fracture network in the two treatments closely agree with the microseismic footprint and SRV, as shown in Fig. 10.10 (Kresse et al., 2013). The difference in the predicted induced fracture network from two treatments with different types of fluid matches the microseismic cloud trend observed in Warpinski et al. (2005). This confirms that the UFM model with the OpenT crossing model is able to correctly replicate the field generated hydraulic fracture patterns.

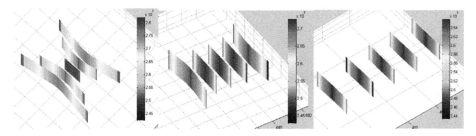

FIGURE 10.9

Fracture geometry and fluid pressure for five fractures propagating from a horizontal well when distance between perforation clusters is equal to 10, 20, and 40 m; by Kresse et al. (2012).

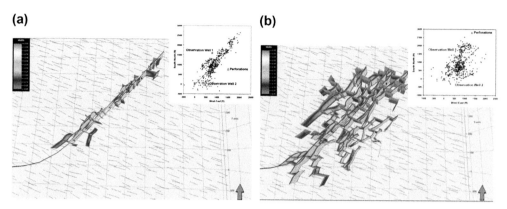

FIGURE 10.10

Hydraulic fracture network simulated by the UFM model for a Barnett case by Kresse et al. (2013); the thin blue lines (dark gray in print versions) indicate the traces of the natural fractures on a horizontal plane (a) Cross-linked gel treatment (b) Slickwater refracturing. The insets of the microseismic map are from Warpinski et al. (2005).

The UFM model as shown in Fig. 10.4 has been integrated with a reservoir simulator. The integrated fracture–reservoir model can be used to evaluate production performance from hydraulic fracturing stimulation of naturally fractured formation and to optimize completion parameters, as presented in Cohen et al. (2014).

The UFM model has been widely used to predict hydraulic fracture networks in unconventional reservoirs (Cipolla et al., 2011; Kennaganti et al., 2013; Liu et al., 2013). This model will be applied below for analysis of the impact of natural fractures on the resulting hydraulic fracture network (HFN).

10.4 IMPACT OF NATURAL FRACTURES ON INDUCED FRACTURE NETWORK

As was shown above, the hydraulic fracture propagation pattern in a naturally fractured formation could be influenced by multiple factors, including rock properties, pumping fluid properties, pumping schedule/rates, stress anisotropy, stress shadow due to interaction of different branches of hydraulic fractures, and the interaction of hydraulic fractures with preexisting natural fractures.

Below, the impact of various natural fracture configurations on the induced hydraulic fracture network are examined through a parametric study using the UFM complex fracture model discussed earlier. The results demonstrate how the natural fracture orientation, density, and length may impact the resulting fracture network footprint.

The parametric study presented here uses a synthetic case based on Cohen et al. (2014) and represents the Marcellus shale play. The reservoir description is presented in Table 10.1. The stress profile and the perforation depth are described in Fig. 10.11. The reservoir permeability is 300 nD, the porosity is 5%, the average reservoir pressure is about 2800 psi, the horizontal stress anisotropy is 200 psi, and the initial reservoir temperature (or bottomhole static temperature) is 175 °F. The unpropped hydraulic fracture conductivity is assumed to be negligible (0.001 mD ft) in this study.

10.4 IMPACT OF NATURAL FRACTURES ON INDUCED FRACTURE NETWORK

Table 10.1 Base Case Reservoir Zone Properties

Zone Name	Top TVD (ft)	Reservoir Pressure (psi)	Min. Horizontal Stress (psi)	Max. Horizontal Stress (psi)	Intrinsic Permeability (mD)	Porosity (%)	Young's Modulus (psi)	Poisson's Ratio
1	5559.86	2787	4620	4820	0.0003	5	3,750,000	0.23
2	5600.00	2803	3950	4150	0.0003	5	3,750,000	0.23
3	5653.41	2813	4150	4350	0.0003	5	3,750,000	0.23
4	5683.41	2833	4220	4420	0.0003	5	3,750,000	0.23
5	5715.00	2844	4540	4740	0.0003	5	3,750,000	0.23
6	5750.00	2864	4050	4250	0.0003	5	3,750,000	0.23
7	5775.00	2884	4400	4600	0.0003	5	3,750,000	0.23
8	5895.00	2922	4100	4300	0.0003	5	3,750,000	0.23
9	5847.75	2932	4900	5100	0.0003	5	3,750,000	0.23

TVD, true vertical depth.

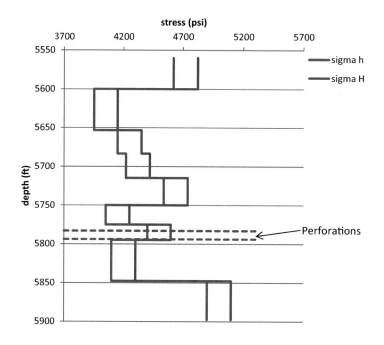

FIGURE 10.11

Minimum (blue; light gray in print versions) and maximum (red; dark gray in print versions) horizontal stress profiles and perforated interval (green; gray in print versions).

As the base case for our analysis, we consider the pumping schedule with one fluid (slickwater) pumped at 60 bbl/min and one proppant (40/70) pumped through four perforated intervals in a formation with natural fractures oriented at 90° with respect to maximum horizontal stress direction, having average length of 100 ft and average spacing of 100 ft (Table 10.2).

We will concentrate our analysis on the following cases (Table 10.3):

- Influence of pumped fluid viscosity: slickwater (SW), linear gel (LG), cross-linked gel (XL)
- Influence of friction coefficient of natural fracture: 0.1, 0.5, and 0.9
- Influence of DFN orientation: angle of 0°–90° with respect to direction of σ_H
- Influence of DFN spacing
- Influence of DFN length
- Influence of multiple DFN sets.

Table 10.2 Base Case Pumping Schedule

Step Number	Pump Rate (bbl/min)	Fluid Volume (gal)	Fluid Type	Proppant Mesh Size	Proppant Concentration (ppa)
1	60	20,000	Slick water	None	0
2	60	60,000	Slick water	40/70	0.5
3	60	60,000	Slick water	40/70	1
4	60	30,000	Slick water	40/70	1.5
5	60	30,000	Slick water	40/70	2

10.4 IMPACT OF NATURAL FRACTURES ON INDUCED FRACTURE NETWORK

Table 10.3 Input (SW: SlickWater. LG: Linear Gel. XL: Cross-linked gel)

Friction	Fluid Type	Length (ft)	Angle (degree)	Fracture Spacing(ft)	Case Study
0.1	SW	100	90	100	Friction Coefficient and fluid Viscosity
0.1	LG	100	90	100	
0.1	XL	100	90	100	
0.5	SW	100	90	100	
0.5	LG	100	90	100	
0.5	XL	100	90	100	
0.9	SW	100	90	100	
0.9	LG	100	90	100	
0.9	XL	100	90	100	
0.5	SW	100	10	100	Orientation of DFN
0.5	SW	100	30	100	
0.5	SW	100	45	100	
0.5	SW	100	60	100	
0.5	SW	100	75	100	
0.5	SW	100	90	100	
0.5	SW	50	90	100	Length of NF
0.5	SW	100	90	100	
0.5	SW	200	90	100	
0.5	SW	400	90	100	
0.5	SW	100	90	25	DFN spacing
0.5	SW	100	90	50	
0.5	SW	100	90	100	
0.5	SW	100	90	200	
0.5	SW	100	0 and 100	50 and 100	Two sets orientation and spacing
0.5	SW	100	0 and 100	100 and 100	
0.5	SW	100	0 and 100	200 and 100	
0.5	SW	100	0 and 100	400 and 100	
0.5	SW	100	45 and 135	50 and 100	
0.5	SW	100	45 and 135	100 and 100	
0.5	SW	100	45 and 135	200 and 100	
0.5	SW	100	45 and 135	400 and 100	

10.4.1 IMPACT OF NF FRICTION COEFFICIENT AND FLUID VISCOSITY

This section investigates the relation between the HFN's geometry, the fracturing fluid viscosity and the NF's friction coefficient. Figure 10.12 shows the HFN generated from the treatments using slickwater, linear gel, and cross-linked gel for an NF friction coefficient of 0.1, 0.5, and 0.9.

The first observation is that fluid viscosity has a large impact on hydraulic fracture footprint in unconventional formations (see also Kresse et al., 2013 and Cohen et al., 2013). For higher-viscosity fluids, the HFs have greater tendency to cross the orthogonal NFs and develop a less complex fracture network. Contrarily, the HFs generated by low-viscosity fluid, such as slick water, tend to be more easily arrested by the NFs, leading to greater fracture complexity. A higher-viscosity fluid also creates more fracture

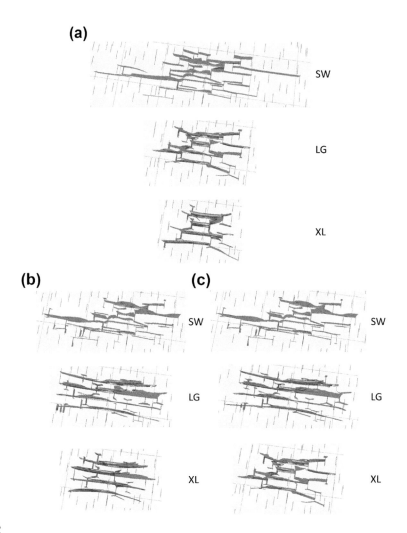

FIGURE 10.12

Hydraulic fracture network footprint for NF with friction coefficient = (a) 0.1, (b) 0.5, and (c) 0.9 for pumped slickwater (SW), linear gel (LG), and cross-linked gel (XL). The color contour shows the predicted proppant distribution in the fracture network. Pink color (gray in print versions) corresponds to the induced fracture area with zero proppant concentration; cooler colors (blue; dark gray in print versions) correspond to lower proppant concentration, and warmer colors (red (light gray in print versions) and yellow (very light gray in print versions)) correspond to greater proppant concentration.

width. Consequently, it generates a smaller fracture surface area as compared to a lower-viscosity fluid, for the same amount of fluid pumped. Figure 10.12 also shows the proppant distribution in the fractures, indicated by the color scale. For cross-linked gel, the proppant is well suspended vertically in the fracture, whereas for the slick water, the proppant mostly settles to the bottom of the fractures.

We can also observe that friction coefficient also influences the propagation pattern. Most variations of the HFN are observed when comparing cases with a friction coefficient of 0.1 and 0.5, whereas the HFNs for friction coefficients of 0.5 and 0.9 are almost the same. This is because in the simulated cross-linked and linear gel treatment, for the given orthogonal NFs and reservoir conditions, HFs

10.4 IMPACT OF NATURAL FRACTURES ON INDUCED FRACTURE NETWORK

mostly cross NFs at friction coefficient of 0.5 and 0.9, but mostly get arrested by the NFs for friction coefficient of 0.1. Note that for friction coefficients of 0.5 and 0.9, even though the HFs cross the NFs, some fracture complexity still develops due to the opening of the NFs after the crossing as a result of fluid pressure exceeding the closure stress acting on the NFs. With slick water, the HFs are mostly arrested by the NFs. There are some variations in the predicted geometry at different friction coefficients due to greater sensitivity of fracture geometry in the case of low-viscosity fluid.

10.4.2 IMPACT OF DFN ORIENTATION

The orientation of natural fractures has a great impact on the induced HFN footprints. This is clearly shown in Fig. 10.13 for the case of slick water and NF friction coefficient equal to 0.5. When natural fractures are almost parallel to the direction of the maximum horizontal stress (direction of hydraulic fractures propagation), hydraulic fractures may not be significantly affected by the natural fractures, resulting in longer planar fractures (case of NF angle = 10° in Fig. 10.13). The larger the intersection angle between hydraulic and natural fractures, the more complicated network we observe due to the greater deviation of the HFs from their original path and greater chance of their intersecting more NFs, as a direct result of the fracture arrest by the NFs and penetration of fracturing fluid into the NFs.

Figures 10.14 and 10.15 show that orientation of the NFs has a drastic impact on fracture extension and the total and propped surface areas of the induced hydraulic network. Figure 10.14 shows the total fracture surface area decreases as the NF orientation increases toward 90°. Most interestingly, the propped surface area remains almost unchanged by the variation of the NF angle. This is due to the low viscosity of the carrying fluid for which the proppant placement is dominated by the relatively high settling velocity in slickwater.

The fracture surface area is affected more than the area of propped surface mainly due to the low-viscosity fluid pumped for this case. Even though the fracture network becomes more complex, qualitatively characterized by a greater number of branch points and secondary fractures, as the NF angle increases, the total surface area decreases, primarily due to rapid decrease of the extension of the

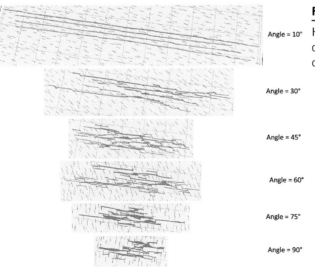

FIGURE 10.13

Hydraulic fracture network footprint for different orientations of natural fractures for NF friction coefficient = 0.5 and slickwater fluid.

FIGURE 10.14

Dependence of the total surface area and propped area of stimulated hydraulic fracture network on the orientation of natural fractures relative to direction of maximum horizontal stress (for friction coefficient = 0.5 and slickwater).

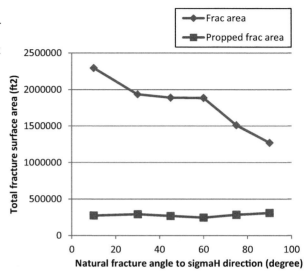

FIGURE 10.15

Dependence of the final extension of stimulated hydraulic fracture network in the direction of maximum and minimum horizontal stresses on the orientation of natural fractures relative to direction of maximum horizontal stress (for friction coefficient = 0.5 and slickwater).

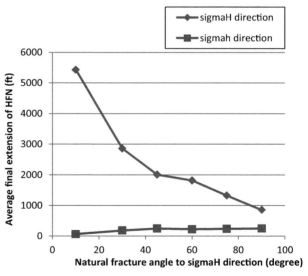

network along the primary fracture direction, as shown in Fig. 10.15 (to be further discussed below). The total surface area is relatively constant between 30° and 60°, as the increase of surface area created from greater number of secondary fractures is offset by the reduction in the overall extension of the network. However, if one considers the surface area per unit reservoir volume, it increases with the NF angle. This means more surface area can be created in a given reservoir volume when NF is at a larger angle with HF, though it would require drilling more wells (i.e., at smaller well spacing) to cover the same reservoir volume compared with more planar fractures.

10.4 IMPACT OF NATURAL FRACTURES ON INDUCED FRACTURE NETWORK

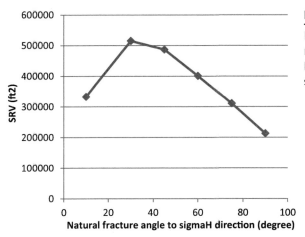

FIGURE 10.16

Dependence of the SRV on the orientation of natural fractures relative to direction of maximum horizontal stress (for friction coefficient = 0.5 and slickwater).

Figure 10.15 shows that the averaged final extension of the HFN in both the σ_h and σ_H directions as a function of the NF orientation. In this study, the HFN extension is calculated for each perforation cluster, so the averaged HFN extension is defined as the average from all four clusters. In the rest of the section, we consider the HFN extension as this averaged value. The results in Fig. 10.15 show that the extension of the HFN in the σ_H direction is significantly reduced as the angle increases whereas the extension in the σ_h direction increases only slightly. This is because of the combined effect of more intersections of the HFs with NFs as the angle increases on the one hand, and the more elevated net pressure on the other. Since the fluid pressure needs to exceed the closure stress on the NFs in order to open them, the net pressure increases as the NF angle increases, resulting in wider width in the primary fractures and, consequently, shorter fracture network length.

In Fig. 10.16, the estimated SRV is plotted as a function of the NF orientation. The SRV here is given in surface area (square feet) because the height component is neglected, since the main focus is on the extension of the HFN in the horizontal plane, and fracture height is well contained in this case. The SRV is estimated by multiplying the average HFN extensions in the σ_h direction by the average HFN extensions in the σ_H direction. The results show that for this case the optimum SRV occurs at NF angle of 30°. It should be mentioned that the SRV is used here as a crude proxy for production. The actual well production is also influenced by the density of the fractures, i.e., the fracture surface area, within the SRV, and the fracture conductivity.

10.4.3 IMPACT OF DFN LENGTH

The impact of the length of NFs on the HFN footprint is shown in Figs 10.17 and 10.18. The length of NFs was increased from 50 to 400 ft to assess the sensitivity of the HFN footprint to the NF length. Figure 10.17 shows that for small NF length, the HFN extends primarily along the maximum horizontal stress direction. This is due to the lower probability of HF intersecting an NF and the shorter secondary fractures that propagate along the NFs. For long NFs, the network extends toward the orientation of the NFs, due to the growth of fractures along the NFs.

312 CHAPTER 10 IMPACT OF PREEXISTING NATURAL FRACTURES

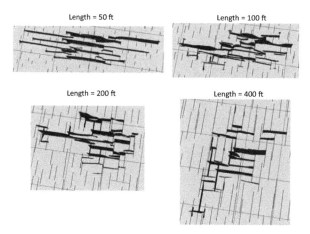

FIGURE 10.17

Dependence of the final extension of stimulated hydraulic fracture network on the length of natural fractures (for friction coefficient = 0.5 and slickwater).

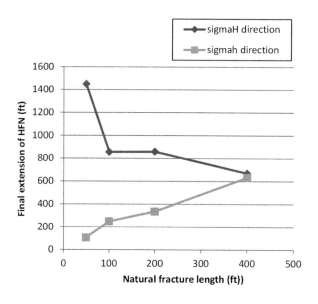

FIGURE 10.18

Dependence of the final extension of stimulated hydraulic fracture network on the length of natural fractures (for friction coefficient = 0.5 and slickwater).

This is further shown in Fig. 10.18, which compares the HFN extension in the maximum and minimum horizontal stress directions, as a function of the NF length. One observation is that the NFs start to play a dominant role in controlling the orientation of the HFN when the NF length exceeds the average fracture spacing. In that case, an HF tip cannot travel in the rock matrix a distance much larger than the fracture spacing before intersecting an NF, and when it is arrested by the NF, it is forced to

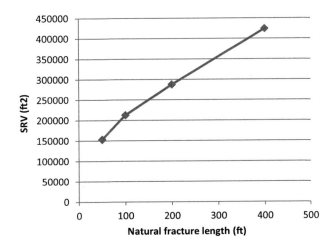

FIGURE 10.19

Dependence of the SRV on the length of natural fractures (for friction coefficient = 0.5 and slickwater).

propagate along the long NF. In conclusion, the length of natural fractures, together with orientation of natural fractures, can considerably change the hydraulic fracture propagation path.

Figure 10.19 shows that SRV increases significantly and almost linearly as the NF length increases, suggesting that greater NF length should help increase the production. However, Fig. 10.18 indicates that as the ratio of NF length to NF spacing increases, the SRV reaches a volume for which HFN extension in both horizontal directions are approximately equal, optimizing its surface area. It is likely that further increasing the length of NFs may reduce the SRV by giving a preferential extension in the minimum stress direction, and therefore potentially reduce the production.

10.4.4 IMPACT OF DFN SPACING

The NF spacing is also an important parameter that affects the induced HFN, since it controls the NF density. Figure 10.20 shows the HFN for a spacing of NFs varying from 25 to 200 ft.

As we see from these simulations (Figs 10.20 and 10.21), by increasing the fracture spacing, the extension of the hydraulic fracture network in the direction of maximum horizontal stress also increases, whereas the extension in the direction of minimum horizontal stress decreases due to less intersection between HFs and NFs. For low NF spacing, we see that the HFN mainly propagates in the direction of the NF orientation. The fracture propagation path is altered by the closely located natural fractures, resulting in smaller SRV, but more complex network geometry. Because the NF spacing is directly related to the NF density, the trends in both Figs 10.20 and 10.21 suggest that the NFs gain control of the HFN orientation as the NF density increases.

This result is consistent with the observation in Fig. 10.18. To clarify the relation between the NF's density and the shape of the HFN, Fig. 10.22 plot the ratio of the average NF's length by the average NF's spacing, versus the ratio of the HFN extension in the σ_H direction by the HFN extension in the σ_h direction, based on results from both Figs 10.18 and 10.21. It shows that whether we fix the length of NFs or their spacing, the proportions of the HFN evolve similarly as a function of the NF's length-to-spacing ratio.

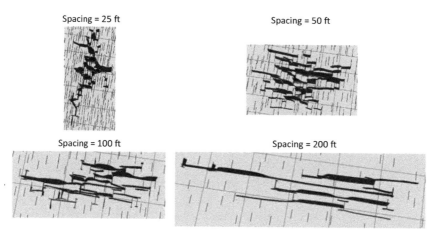

FIGURE 10.20

Dependence of the final extension of stimulated hydraulic fracture network on the spacing of natural fractures (for friction coefficient = 0.5 and slickwater).

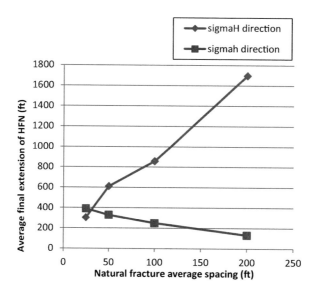

FIGURE 10.21

Dependence of the final extension of stimulated hydraulic fracture network on the spacing of natural fractures (for friction coefficient = 0.5 and slickwater).

Figure 10.23 shows that the SRV increases with the NF spacing, with the rapid increase occurring in the range of small fracture spacing that is less than the fracture length, where the high NF density inhibits the HFN propagation in the σ_H direction.

10.4 IMPACT OF NATURAL FRACTURES ON INDUCED FRACTURE NETWORK

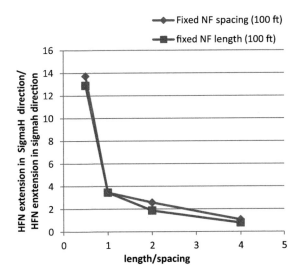

FIGURE 10.22

Ratio of the HFN extension in σ_H direction by the σ_h direction as a function of the NF's length-to-spacing ratio.

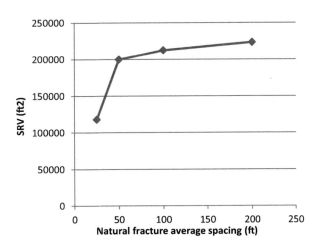

FIGURE 10.23

Dependence of the SRV on the spacing of natural fractures (for friction coefficient = 0.5 and slickwater).

10.4.5 IMPACT OF MULTIPLE SETS OF NATURAL FRACTURES

It is common for a fractured reservoir to have multiple sets of NFs, with different properties and orientations. This section illustrates how, under different conditions, the impact of a set of NFs on the HFN footprint can change from significant to negligible.

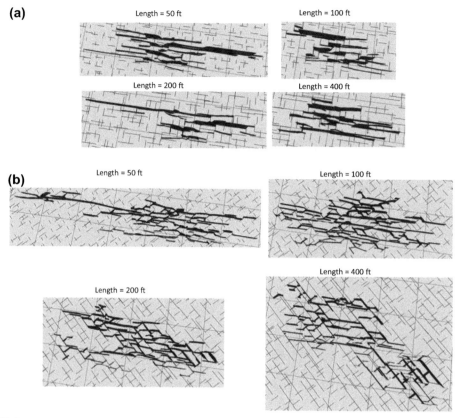

FIGURE 10.24

Dependence of the final extension of stimulated hydraulic fracture network on the DFN orientation and length of first set of fractures (for friction coefficient = 0.5 and slickwater). (a) First fracture set with orientation angle 0° and second set with orientation angle of 90° (b) First fracture set with orientation angle 45° and second set with orientation angle of 135°

Figure 10.24 compares HFNs for a case with two sets of NFs that are orthogonal to each other. The two sets have a spacing of 100 ft. The first set has an NF length varying between 50 and 400 ft whereas the second set has a fixed NF length of 100 ft. Figure 10.24(a) shows the resulting HFN in the case of an angle to the maximum horizontal stress orientation of 0° for the first set and 90° for the second set.

The results do not show any particular correlation between the HFN footprint and the length of fractures of the first set. This indicates that the HFN geometry is dominated only by the fracture set with the orientation orthogonal to the maximum horizontal stress. Figure 10.24(b) shows similar simulations, except that the first set of NFs has an orientation of 45° and the second set has an orientation of 135°. This means that both sets have an angle of 45° to the maximum horizontal stress orientation.

We can see that the geometry of the HFN can become dominated by the NF orientation, if both the length of the NFs and the angle of the NFs to the maximum horizontal stress are sufficient. These results

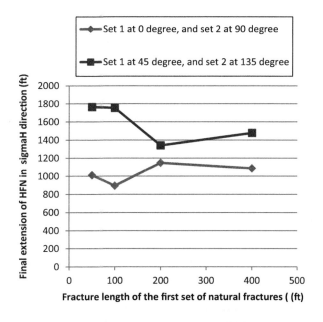

FIGURE 10.25

Dependence of the final extension of stimulated hydraulic fracture network on the length and orientation of two sets of natural fractures (for friction coefficient = 0.5 and slickwater).

are illustrated in Fig. 10.25, which compares the extension of the HFN in the maximum horizontal stress direction as a function of the NF length of the first set, for the cases with NF orientations of 0°–90° and 45°–135°. It shows that if the NF orientation is aligned with the maximum horizontal principal stress, there is little influence of the natural fracture length on the HFN, whereas at an angle of 45°, final extension of the HFN in the maximum stress direction decreases significantly as the length of the first set increases. A final observation is that the set with a 45° angle offers a much greater SRV and complexity than the set orthogonal to the horizontal principal stresses, independent of the length of the NFs.

We can conclude from the study that the simulated HFN footprint, which is generally associated with a microseismic events cloud, is greatly affected by the properties of natural fractures. Natural fracture orientation can alter the preferred hydraulic fracture propagation direction, and the length of natural fractures can greatly affect the extension of HFN in the direction of minimum horizontal stress. When multiple fracture sets are present, the fracture set that is at a greater angle to the hydraulic fracture orientation has a greater influence in generating fracture complexity.

10.5 IMPACT OF UNCERTAINTY OF DFN ON HFN SIMULATION

In the last section, we examined the influence of the natural fracture parameters on the geometry of the induced hydraulic fracture network. One important characteristic of natural fractures that needs to be considered is their statistical nature. Since the average parameters and the distribution of fractures are obtained through statistical means, for example inferred from the population of fractures observed in the borehole image logs, there is no precise determination of the exact location and geometry of the natural fractures in the formation. For a given statistical parameters that describe a natural fracture system, there are many possible realizations of the fractures according to a probability distribution. Each realization of the discrete fractures may result in a slightly different induced hydraulic fracture system. Therefore, the outcome of a hydraulic fracture treatment bears a significant degree of uncertainty. These uncertainties present additional challenges in fracture and completion design optimization, for example, through parametric analysis, due to the inherent "noise" in the simulation results. Good understanding of the inherent uncertainty and quantification of the uncertainty would help the engineers carrying out more appropriate parametric study and meaningful interpretation of the simulation results. This section presents a statistics-based uncertainty analysis on the impact of natural fractures to quantify the uncertainty in the stimulation outcome.

Several prior studies used statistical tools to analyze the uncertainty on production in shale gas reservoirs from parameters such as permeability, hydraulic fracture half-length, and skin effect (Chaudhri, 2012). For example, Hatzignatiou and McKoy (2000) presented an uncertainty analysis on the effect of natural fractures on well performance using a Monte Carlo approach.

We present here a parametric study on the relationship between parameters such as natural fracture length, density, and orientation on both the production and the uncertainty on production. The methodology begins with the generation of multiple realizations of the natural fracture system for a given set of statistical parameters that define the system. The complex hydraulic fracture network generated from a fixed treatment design for any given realization is simulated using the UFM model and the corresponding production performance is predicted using an unconventional production model (UPM) (Cohen et al., 2012), as illustrated in Fig. 10.26.

To generate an extensive number of simulations, an automated workflow runs all the simulation cases, archives the results, and generates visualization outputs and reports. The cumulative production predicted for a large number of simulations provides a statistical distribution described by a mean and a standard deviation. This workflow is illustrated in Fig. 10.27. The parametric study uncovers the relation between these two statistical parameters with the natural fracture parameters. This section describes the algorithm to generate the DFN realizations, the parametric study, and the results.

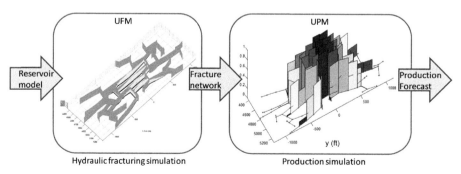

FIGURE 10.26

Illustration of the UFM–UPM workflow.

10.5 IMPACT OF UNCERTAINTY OF DFN ON HFN SIMULATION

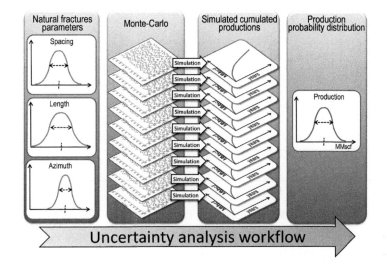

FIGURE 10.27

Illustration of the uncertainty analysis workflow.

10.5.1 STOCHASTIC GENERATION OF THE NATURAL FRACTURE NETWORK

The UFM model requires a predefined 2D DFN in a horizontal plane as input. In this study, we use an algorithm to generate a realization of the 2D DFN using the specified statistical parameters. These parameters are the fracture spacing, azimuth angle, and length. The main challenge is to distribute the fractures in 2D space. The method used here starts by predistributing the "seed points" according to the specified mean spacing along the direction normal to the mean fracture azimuth. The spacing along the fracture azimuth direction (defined as the distance between the centers of the adjacent fractures) is assumed to be mean spacing plus mean fracture length. Once the seed points are defined, each seed point is displaced by dx and dy determined randomly based on the specified standard deviation of the fracture spacing. This displaced seed point becomes the center of a natural fracture, whose actual azimuth and length are then generated randomly based on their respective mean values and standard deviations. This process is repeated for all seed points.

10.5.2 BASE CASE

The reservoir description is presented in Table 10.4. The permeability is 200 nD and the horizontal stress anisotropy is 1%. The conductivity of unpropped fractures is fixed at 0.001 mD ft.

The completion is a horizontal well, and the base case is a single stage of pumping through four perforation clusters that are 100 ft from each other, at a true vertical depth (TVD) between 5794 ft and 5784 ft. The treatment is made up of 224,576 gal of fracturing fluid and 183,700 lbm of proppant pumped at 80 bbl/min, as described in Table 10.5. The schedule begins with 18% of pad followed by slurry with a proppant concentration of 1 ppg. For simplicity of the parametric study, the fracturing fluid is assumed to have Newtonian rheological behavior.

Table 10.4 Base Case Reservoir Zone Properties

Zone	Top Depth TVD (ft)	Height (ft)	Reservoir Pressure (psi)	σ_h (psi)	σ_H (psi)	Young's Modulus (Mpsi)	Poisson's Ratio	Permeability (mD)	Porosity (%)
1	5653	60	2832	4137	4178	2	0.23	0.0002	8
2	5713	114	2863	4117	4158	2	0.23	0.0002	8
3	5827	40	2930	4124	4165	2	0.23	0.0002	8

Table 10.5 Base Case Pumping Schedule

Step	Rate (bbl/min)	Fluid Volume (gal)	Proppant (ppa)	Fluid Type	Proppant Size
1	80	40,832	0	Slick water	40/70
2	80	183,744	1	Slick water	40/70

The proppant type considered in this study is 40/70-mesh sand with an average diameter of 0.01106 in and a specific gravity of 2.65. The production is simulated at a constant bottomhole flowing pressure of 1000 psi.

The natural fracture network is made of fractures of 200 ft average length, 200 ft average spacing, and 0° from north (parallel to the wellbore and 90° from the fracture orientation) on average. The standard deviation of the length is 200 ft and the standard deviation of the fracture angle is 10°. The natural fractures are considered vertical and as extending through all three zones.

To investigate the distribution of the cumulative production, we generated 98 realizations of the natural fracture network with the same value for the mean and the standard deviation for the length. The fracturing treatment and subsequent production simulations were run for each realization. Figure 10.28 illustrates the distribution of the cumulative production for three different times of production (6 months, 1 year, and 3 years). The y-axis of the figure is the "frequency" of cases, which is the number of cases that have cumulative production falling into a certain range of ±7.5 MMscf.

Figure 10.28 shows that the cumulative production seems to follow a normal distribution. It also shows that both the mean and the standard deviation increase with time. One explanation for this result is that at early times the production comes mostly from the reservoir volume around the wellbore where only a limited number of natural fractures are contributing to the flow into the hydraulic fracture networks. At longer time, a larger part of the production comes from deeper into the hydraulic fracture network, increasing the number of interactions with natural fractures in the producing area of the network. Because each interaction with a natural fracture modifies the production behavior of the reservoir, the possibilities of a different behavior increases with time.

Because each simulation case can be relatively time-consuming, the next question is: what is the minimum number of cases needed to have an acceptable description of the production distribution? Figures 10.29 and 10.30 show the evolution of the mean and relative standard deviation of the production as a function of the number of cases considered. The relative standard deviation is calculated as the standard deviation divided by the mean. Figures 10.29 and 10.30 show that a smaller number of

10.5 IMPACT OF UNCERTAINTY OF DFN ON HFN SIMULATION

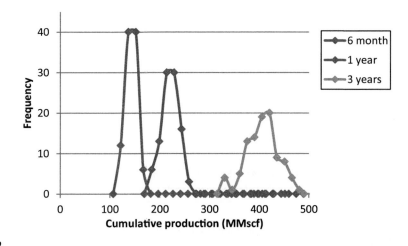

FIGURE 10.28
Distribution of cumulative production for three different times of production (6 months, 1 year and 3 years).

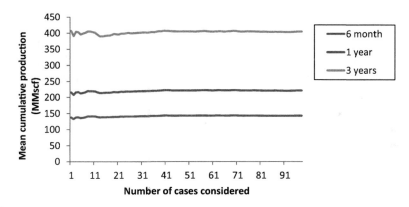

FIGURE 10.29
Mean cumulative production as a function of the number of cases considered for 6 months, 1 year, and 3 years.

cases is required for the mean to converge than for the relative standard deviation. In the results presented in the following discussion, the relative standard deviation and the mean are calculated based on 30 natural fracture network realizations and simulations.

10.5.3 NATURAL FRACTURE LENGTH

The first parameter to be investigated is the length of natural fractures. Figure 10.31 shows that the mean production increases with the length of natural fractures. This can be explained by the results in Section 10.4.3, which show that when increasing the NF length, the HFN becomes more extended in

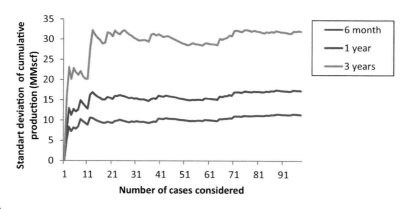

FIGURE 10.30

Relative standard deviation of the cumulative production as a function of the number of cases considered for 6 months, 1 year, and 3 years.

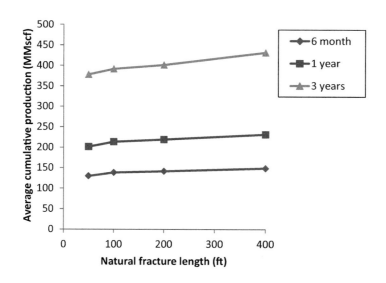

FIGURE 10.31

Mean cumulative production as a function of the mean natural fracture length, for 6 months, 1 year, and 3 years.

the NF orientation, which is along the direction of σ_h, leading to a greater SRV, as illustrated in Figs 10.18 and 10.19.

Figure 10.32 shows that the relative standard deviation seems insensitive to the length of NFs. It also shows that relative standard deviation stays almost constant (around 8%), independent of the natural fracture length.

10.5 IMPACT OF UNCERTAINTY OF DFN ON HFN SIMULATION

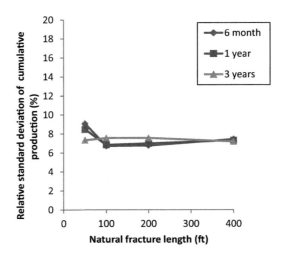

FIGURE 10.32

Relative standard deviation as a function the length of natural fracture.

10.5.4 NATURAL FRACTURE SPACING

Figure 10.33 shows the mean cumulative production as a function of the natural fracture spacing after 6 months, 1 year, and 3 years of production. It is interesting to notice that after 6 months, the production seems to decline with the spacing whereas for 3 years of production, there is a peak of production for spacing of 200 ft. This means that there is an optimum natural fracture spacing that is increasing with the time of production.

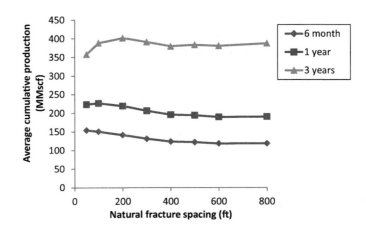

FIGURE 10.33

Mean cumulative production as a function of the mean natural fracture spacing, for 6 months, 1 year, and 3 years.

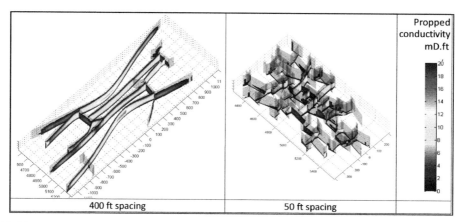

FIGURE 10.34

Comparison of conductivity profiles for treatment done in a reservoir with a natural fracture spacing of 400 ft (left) and 50 ft spacing on the (right). Pictures are at the same scale.

Small natural fracture spacing translates into a dense hydraulic fracture network within a limited SRV. In that case, at early times, the area from which the gas is produced is still within the network boundary of the SRV and, locally, the high fracture density increases the fracture surface to be produced from and maximizes the production. At longer time, the SRV has been significantly depleted because of its limited size, and most of the production comes from the boundary of the SRV, limiting the production rate. The difference of SRV and the density of hydraulic fracture networks between fracture spacing of 50 and 400 ft is illustrated in Fig. 10.34. In addition, this description is consistent with the results in Fig. 10.23. The long-term production should be related to the SRV, and the comparison of Fig. 10.33 with Fig. 10.23 shows a similar trend between the cumulative production after 3 years and the SRV for the same range of NF spacing (25–200 ft).

Figure 10.35 is the evolution of the standard deviation of the cumulative production as a function of the natural fracture spacing. It shows that the standard deviation rapidly declines as the spacing increases, from 18% for 50 ft to about 4% and less at 800 ft. Another observation, similar to that for the influence of natural fracture length (Fig. 10.32), is that the relative standard deviation does not depend on the time of production.

Hatzignatiou and McKoy (2000) found in their study a relative standard deviation of 25%, which is greater than our maximum of 18%. One explanation for this difference is that their natural fracture density seems greater than that in our study.

10.5.5 NATURAL FRACTURE ANGLE

This section investigates the relation between the natural fracture angle and both the mean cumulative production and the relative standard deviation. Figure 10.36 shows that the angle has little

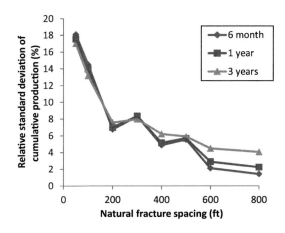

FIGURE 10.35

Relative standard deviation of the production as a function of the fracture spacing.

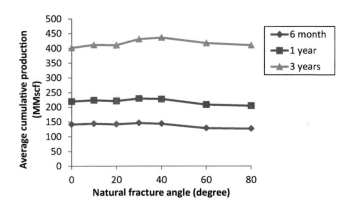

FIGURE 10.36

Mean cumulated production as a function of the natural fracture angle.

effect on the mean production. This result can partly be explained by the low fracture density of the base case (200-ft spacing). Another explanation comes from the previous study in Fig. 10.16, in which the total fracture area varies significantly with the NF orientation, but the actual propped surface area is almost unchanged due to the high settling rate. Another observation, the long-term cumulative production (3 years) is connected to the SRV, and in Fig. 10.36, we see an optimum production for an NF angle of 40°, whereas Fig. 10.16 shows a maximum SRV for an NF angle of 30°.

Figure 10.37 shows that the relative standard deviation varies between 6% and 13% as a function of the natural fracture angle, without a clearly identifiable trend.

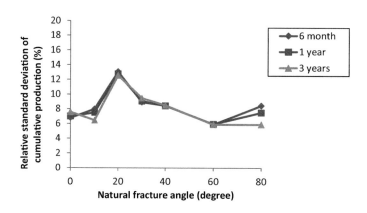

FIGURE 10.37

Evolution of the standard deviation of the cumulative production as a function of the natural fracture angle.

10.6 CONCLUSION

Even though there has been significant progress made in understanding the mechanisms influencing the interaction between hydraulic fractures and natural fractures, precise prediction of detailed fracture geometry is still very challenging. This chapter presented some observations about the influence of a natural fracture network on the hydraulic fracture footprint based on the UFM model.

The impact of natural fractures on the stimulation outcome presents new challenges to the operators and engineers who design completions and evaluate fracturing treatments. Due to the high uncertainty and stochastic nature of the natural fractures as the input for fracture simulation, there is an inherent large uncertainty in the model predictions. Obtaining good understanding of the reservoir and properly characterizing the natural fracture system provide a good baseline for model simulation and an important first step toward reducing uncertainty. Additional measurements, such as microseismic monitoring and treating pressure data, will greatly help calibrate the model and uncertain parameters to further reduce the uncertainty. Even so, the uncertainties in the predicted results need to be considered in well performance forecasts and engineering decision making.

The fracture simulations using UFM model carried out in this study showed some general observations how the natural fracture geometric attributes affect the induced hydraulic fracture network which are summarized below.

- The sensitivity of the HFN pattern to fracturing fluid viscosity increases for NFs with low friction coefficient.
- The propped surface area is less sensitive to the uncertainties on the NFs attributes for low-viscosity fracturing fluids, since the proppant placement will be controlled more by the settling velocity and less by the extension of the HFN.
- The orientation of the natural fractures plays a critical role in affecting the complexity of the induced fractures. Generally speaking, the greater the angle of the natural fractures with respect to the direction of the maximum horizontal stress, the more complex the induced fracture network is. However, when the intersection angle approaches 90°, the complexity depends to a large

degree the crossing behavior of the hydraulic–natural fracture interaction. The parametric study presented in this paper showed that maximizing the SRV requires a balance between the HFN extension in both σ_h and σ_H directions, which seems to be better achieved with NFs oriented halfway between this two stress directions.
- For natural fractures at an angle with respect to the maximum horizontal stress, longer fracture length and smaller fracture spacing result in greater extension of the induced hydraulic fracture network in the direction of the natural fractures. Smaller length and greater spacing result in greater extension of the induced fracture network in the maximum stress direction. The ratio of the HFN extension in the σ_H and to its extension in the σ_h direction seems to be simply a function of the NF's length-to-spacing ratio.
- When the angle of natural fractures relative to the maximum horizontal stress increases, the lateral extension of the fracture network in maximum stress direction decreases due to a greater number of intersections of the hydraulic fractures with the natural fractures. The total surface area of the induced fracture network generally decreases due to the increased net pressure to overcome the higher closure stress acting on the natural fractures, which results in the greater fracture width in the primary fracture branches.
- When more than one set of natural fractures are present, the fracture set at a greater angle with respect to the maximum horizontal stress has a greater influence on the extension of fracture network in the direction normal to the maximum horizontal stress, augmented by the spacing of the fracture set parallel to the maximum stress.
- The spacing of NFs that maximizes the cumulative production increases with time, because early production responds mostly to the density of the HFN close to the wellbore, while long-term production depends mostly to the size of the SRV.

While the characteristics of the natural fractures are native to the formation and are not controllable, recognizing their potential impact on the induced hydraulic fracture network will allow the engineer better plan the well spacing and completion to optimize the reservoir coverage and production. Different types of treatment designs can be used in formations with different natural fracture characteristics to achieve the best deliverability of the induced hydraulic fracture system. For a formation with abundance of large natural fractures at an angle to the maximum stress direction, treatment design with slick water and small size proppant would help connecting natural fractures and generating a large surface area of a complex fracture network favorable for an ultralow permeability formation. Contrarily, if the formation has very limited natural fractures, or the natural fractures are aligned with the maximum horizontal stress direction, the fractures will be mostly planar, and the treatment design can be focused on achieving the optimal propped length and conductivity.

REFERENCES

Beugelsdijk, L.J.L., de Pater, C.J., Sato, K., 2000. Experimental hydraulic fracture propagation in a multi-fractured medium. In: SPE 59419, SPE Asia Pacific Conference in Integrated Modeling for Asset Management, Yokohama, Japan, 25–26 April.

Blanton, T.L., 1982. An experimental study of interaction between hydraulically induced and pre-existing fractures. In: SPE 10847, SPE/DOE Unconventional Gas Recovery Symposium, Pittsburgh, Pennsylvania, USA, 16–18 May.

Blanton, T.L., 1986. Propagation of hydraulically and Dynamically induced fractures in naturally fractured reservoirs. In: SPE 15261, SPE Unconventional Gas Technology Symposium, Louisville, Kentucky, USA, 18–21 May.

Castonguay, S.T., Mear, M.E., Dean, R.H., Schmidt, J.H., 2013. Prediction of the growth of multiple interacting hydraulic fractures in three dimensions. In: Paper SPE 166259, SPE Annual Technical Conference and Exhibition, New Orleans, Louisiana, USA, 30 September–2 October.

Chaudhri, M.M., 2012. Numerical modeling of multi-fracture horizontal well for uncertainty analysis and history matching: case studies from Oklahoma and Texas shale Gas Wells. In: SPE 153888, SPE Western Regional Meeting, Bakersfield, California, USA, 21–23 March.

Chuprakov, D.S., Akulich, A.V., Siebrits, E., Thiercelin, M., 2010. Hydraulic fracture propagation in a naturally fractured reservoir. In: SPE 128715, SPE Oil and Gas India Conference and Exhibition, Mumbai, India, 20–22 January.

Chuprakov, D., Melchaeva, O., Prioul, R., 2013. Injection-sensitive mechanics of hydraulic fracture interaction with discontinuities. In: ARMA 47th US Rocks Mechanics/Geomechanics Symposium, San Francisco, California, USA, 23–26 June.

Cipolla, C.L., Williams, M.J., Weng, X., Mack, M., Maxwell, S., 2010. Hydraulic fracture monitoring to reservoir simulation: maximizing value. In: SPE 133877, SPE Annual Technical Conference and Exhibition, Florence, Italy, 19–22 September.

Cipolla, C.L., Weng, X., Mack, M., Ganguly, U., Gu, H., Kresse, O., Cohen, C., 2011. Integrating microseismic mapping and complex fracture modeling to characterize fracture complexity. In: SPE 140185, SPE Hydraulic Fracturing Technology Conference and Exhibition, The Woodlands, Texas, USA, 24–26 January.

Cohen, C.E., Xu, W., Weng, X., Tardy, P., 2012. Production forecast after hydraulic fracturing in naturally fractured reservoir: coupling a complex fracturing simulator and a semi-analytical production model. In: SPE 152541, SPE Hydraulic Fracturing Technology Conference and Exhibition, The Woodlands, Texas, USA, 6–8 February.

Cohen, C.E., Abad, C., Weng, X., England, K., Phatak, A., Kresse, O., Nevvonen, O., Lafitte, V., Abivin, P., 2013. Analysis on the impact of fracturing treatment design and reservoir properties on production from shale gas reservoirs. In: IPTC 16400, International Petroleum Technology Conference, Beijing, China, 26–28 March.

Cohen, C.E., Kamat, S., Itibrout, T., Onda, H., Weng, X., Kresse, O., 2014. Parametric study on completion design in shale reservoirs based on fracturing-to-production simulations. In: IPTC 17462, International Petroleum Technology Conference, Doha, Qatar, 20–22 January.

Cooke, M.L., Underwood, C.A., 2001. Fracture termination and step-over at bedding interfaces due to frictional slip and interface opening. Journal of Structural Geology 23, 223–238.

Crouch, S.L., Starfield, A.M., 1983. Boundary Element Methods in Solid Mechanics, first ed. George Allen & Unwin Ltd, London.

Dahi-Taleghani, A., Olson, J.E., 2011. Numerical modeling of multistrand hydraulic-fracture propagation: accounting for the interaction between induced and natural fractures. SPE Journal 16 (3), 575–581. SPE 124884-PA.

de Pater, C.J., Beugelsdijk, L.J.L., 2005. Experiments and Numerical Simulation of Hydraulic Fracturing in Naturally Fractured Rock. ARMA/USEMS 05–780.

Fisher, M.K., Davidson, B.M., Goodwin, A.K., Fielder, E.O., Buckler, W.S., Steinberger, N.P., 2002. Integrating fracture mapping technologies to optimize stimulations in the barnett shale. In: SPE 77411, SPE Annual Technical Conference and Exhibition, San Antonio, Texas, USA, 29 September–2 October.

Fung, R.L., Viayakumar, S., Cormack, D.E., 1987. Calculation of vertical fracture containment in layered formations. SPE Formation Evaluation 2 (4), 518–522.

Gale, J.F.W., Reed, R.M., Holder, J., 2007. Natural fractures in the barnett shale and their importance for hydraulic fracture treatment. AAPG Bulletin 91 (4), 603–622.

REFERENCES

Gale, J.F.W., Holder, J., 2008. Natural fractures in the barnett shale: constraints on spatial organization and tensile strength with implications for hydraulic fracture treatment in shale-gas reservoirs. In: ARMA 08-096, 42nd US Rock Mechanics Symposium and 2nd Canada Rock Mechanics Symposium, San Francisco, California, USA, 29 June and 2 July.

Gu, H., Weng, X., 2010. Criterion for fractures crossing frictional interfaces at non-orthogonal angles. In: 44th US Rock Mechanics Symposium and 5th US-Canada Rock Mechanics Symposium, Salt Lake City, Utah, USA, 27–30 June.

Gu, H., Weng, X., Lund, J.B., Mack, M., Ganguly, U., Suarez-Rivera, R., 2011. Hydraulic fracture crossing natural fracture at non-orthogonal angles, a criterion, its validation and applications. In: SPE 139984, Hydraulic Fracturing Conference and Exhibition, The Woodlands, Texas, USA, 24–26 January.

Han, G., 2011. Natural fractures in unconventional reservoir rocks: identification, characterization, and its impact to engineering design. In: ARMA 11-509, 45th US Rock Mechanics/Geomechanics Symposium, San Francisco, California, USA, 26–29 June.

Hanson, M.E., Anderson, G.D., Shaffer, R., Thorson, L.D., June 1982. Some effects of stress, friction, and fluid flow on hydraulic fracturing. SPE Journal 321–332.

Hatzignatiou, D.G., McKoy, M.L., 2000. Probabilistic evaluation of horizontal wells in stochastic naturally fractured gas reservoirs. In: CIM 65459, SPE/Petroleum Society of CIM International Conference on Horizontal Well Technolog, Calgary, Alberta, Canada, 6–8 November.

Heuze, F.E., Shaffer, R.J., Ingraffea, A.R., Nilson, R.H., 1990. Propagation of fluid-driven fractures in jointed rock. Part 1 – development and validation of methods of analysis. International Journal of Rock Mechanics and Mining Sciences and Geomechanics Abstracts 27 (4), 243–254.

Jeffrey, R.G., Bunger, A., Lecampion, B., Zhang, X., Chen, Z., As, A., Allison, D.P., de Beer, W., Dudley, J.W., Siebrits, E., Thiercelin, M., Mainguy, M., 2009. Measuring hydraulic fracture growth in naturally fractured rock. In: SPE 124919, SPE Annual Technical Conference and Exhibition, New Orleans, Louisiana, USA, 4–7 October.

Jeffrey, R.G., Weber, C.R., Vlahovic, W., Enever, J.R., 1994. Hydraulic fracturing experiments in the great Northern Coal Seam. In: SPE 28779, SPE Asia Pacific Oil & Gas Conference, Melbourne, Australia, 7–10 November.

Jeffrey, R.G., Zhang, X., Thiercelin, M., 2009. Hydraulic fracture offsetting in naturally fractured reservoirs: quantifying a long-recognized process. In: SPE 119351, SPE Hydraulic Fracturing Technology Conference, The Woodlands, Texas, USA, 19–21 January.

Kennaganti, K.T., Grant, D., Oussoltsev, D., Ball, N., Offenberger, R.M., 2013. Application of stress shadow effect in completion optimization using a reservoir-centric stimulation design tool. In: SPE 164526, SPE Unconventional Resources Conference USA, The Woodlands, Texas, USA, 10–12 April.

Kresse, O., Weng, X., Wu, R., Gu, H., 2012. Numerical modeling of hydraulic fractures interaction in complex naturally fractured formations. In: ARMA-292, 46th US Rock Mechanics/Geomechanics Symposium, Chicago, Illinois, USA, 24–27 June.

Kresse, O., Weng, X., Chuprakov, D., Prioul, R., Cohen, C., 2013. Effect of flow rate and viscosity on complex fracture development in UFM model. In: International Conference for Effective and Sustainable Hydraulic Fracturing, Brisbane, Australia, 20–22 May.

Leguillon, D., Lacroix, C., Martin, E., 2000. Interface debounding ahead of primary crack. Journal of the Mechanics and Physics of Solids 48, 2137–2161.

Liu, H., Luo, Y., Zhang, N., Yang, D., Dong, W., Qi, D., Gao, Y., 2013. Unlock shale oil reserves using advanced fracturing techniques: a case study in China. In: IPTC 16522, International Petroleum Technology Conference, Beijing, China, 26–28 March.

Mack, M.G., Warpinski, N.R., 2000. Mechanics of hydraulic fracturing. In: Economides, Nolte (Eds.), Reservoir Stimulation, third ed. John Wiley & Sons (Chapter 6).

Maxwell, S.C., Urbancic, T.I., Steinsberger, N.P., Zinno, R., 2002. Microseismic imaging of hydraulic fracture complexity in the barnett shale. In: SPE 77440, SPE Annual Technical Conference and Exhibition, San Antonio, Texas, USA, 29 September–2 October.

Meng, C., de Pater, C.J., 2010. Hydraulic fracture propagation in pre-fractured natural rocks. In: ARMA 10-318, 44th US Rock Mechanics Symposium and 5th US-Canada Rock Mechanics Symposium, Salt Lake City, Utah, USA, 27–30 June.

Murphy, H.D., Fehler, M.C., 1986. Hydraulic fracturing of jointed formations. In: SPE 14088, SPE 1986 International Meeting on Petroleum Engineering, Beijing, China, 17–20 March.

Olson, J.E., 2004. Predicting fracture swarms – the influence of subcritical crack growth and crack-tip process zone on joint spacing in rock. In: Cosgrove, J.W., Engelder, T. (Eds.), The Initiation, Propagation, and Arrest of Joints and Other Fractures. Geological Society of London Special Publication 231, pp. 73–87.

Olson, J.E., 2008. Multi-fracture propagation modeling: applications to hydraulic fracturing in shales and tight sands. In: 42nd US Rock Mechanics Symposium and 2nd US-Canada Rock Mechanics Symposium, San Francisco, California, 29 June–2 July.

Potluri, N., Zhu, D., Hill, A.D., 2005. Effect of natural fractures on hydraulic fracture propagation. In: SPE 94568, SPE European Formation Damage Conference, Scheveningen, The Netherlands, 25–27 May.

Renshaw, C.E., Pollard, D.D., 1995. An experimentally verified criterion for propagation across unbounded frictional interfaces in brittle, linear elastic-materials. International Journal of Rock Mechanics and Mining Sciences and Geomechanics Abstracts 32 (3), 237–249.

Rutledge, J.T., Phillips, W.S., Meyerhofer, M.J., 2004. Faulting induced by forced fluid injection and fluid flow forced by faulting: an interpretation of hydraulic-fracture microseismicity, Carthage Cotton Valley Gas field, Texas. Bulletin of the Seismological Society of America 94, 1817–1830.

Savitski, A.A., Lin, M., Riahi, A., Damjanac, B., Nagel, N.B., 2013. Explicit modeling of hydraulic fracture propagation in fractured shales. In: IPTC 17073, International Petroleum Technology Conference, Beijing, China, 26–28 March.

Sesetty, V., Ghassemi, A., 2012. Simulation of hydraulic fractures and their interactions with natural fractures. In: ARMA 12-331, 46th US Rock Mechanics/Geomechanics Symposium, Chicago, Illinois, 24–27 June.

Thiercelin, M., Makkhyu, E., 2007. Stress field in the vicinity of a natural fault activated by the propagation of an induced hydraulic fracture. In: Proceedings of the 1st Canada-US Rock Mechanics Symposium, vol. 2, pp. 1617–1624.

Warpinski, N.R., Teufel, L.W., 1987. Influence of geologic discontinuities on hydraulic fracture propagation (includes associated papers 17011 and 17074). SPE Journal of Petroleum Technology 39 (2), 209–220.

Warpinski, N.R., Lorenz, J.C., Branagan, P.T., Myal, F.R., Gall, B.L., August 1993. Examination of a cored hydraulic fracture in a deep gas well. SPE Production and Facilities 150–164.

Warpinski, N.R., Kramm, R.C., Heinze, J.R., Waltman, C.K., 2005. Comparison of single- and dual-array microseismic mapping techniques in the Barnett shale. In: SPE 95568, SPE Annual Technical Conference and Exhibition, Dallas, Texas, USA, 9–12 October.

Weng, X., Kresse, O., Cohen, C., Wu, R., Gu, H., 2011. Modeling of hydraulic fracture network propagation in a naturally fractured formation. In: SPE 140253, SPE Hydraulic Fracturing Conference and Exhibition, The Woodlands, Texas, USA, 24–26 January.

Weng, X., Kresse, O., Chuprakov, D., Cohen, C.E., Prioul, R., Ganguly, U., December 2014. Applying complex fracture model and integrated workflow in unconventional reservoirs. Journal of Petroleum Science and Engineering 124, 468–483.

Weng, X., 2014. Modeling of complex hydraulic fractures in naturally fractured formation. Journal of Unconventional Oil and Gas Resources 9 (2015), 114–135. http://dx.doi.org/10.1016/j.juogr.2014.07.001.

Williams-Stroud, S.C., Barker, W.B., Smith, K.L., 2012. Induced hydraulic fractures or reactivated natural fractures? Modeling the response of natural fracture networks to stimulation treatments. In: ARMA 12-667, 46th US Rock Mechanics/Geomechanics Symp., Chicago, Illinois, 24–27 June.

Wu, R., Kresse, O., Weng, X., Cohen, C.E., Gu, H., 2012. Modeling of interaction of hydraulic fractures in complex fracture networks. In: SPE 152052, SPE Hydraulic Fracturing Technology Conference, The Woodlands, Texas, USA, 6–8 February.

Yew, C.H., Weng, X., 2015. Mechanics of Hydraulic Fracturing, second ed. Gulf Professional Publishing.

Zhang, X., Jeffrey, R.G., Detournay, E., 2005. Propagation of a fluid driven fracture parallel to the free surface of an elastic half plane. International Journal of Numerical and Analytical Methods in Geomechanics 29, 1317–1340.

Zhang, X., Jeffrey, R.G., 2006a. Numerical studies on fracture problems in three-layered elastic media using an image method. International Journal of Fracture 139, 477–493.

Zhang, X., Jeffrey, R.G., 2006b. The role of friction and secondary flaws on deflection and re-initiation of hydraulic fractures at orthogonal pre-existing fractures. Geophysics Journal International 166 (3), 1454–1465.

Zhang, X., Jeffrey, R.G., 2008. Reinitiation or termination of fluid-driven fractures at frictional bedding interfaces. Journal of Geophysical Research-Solid Earth 113 (B8), B08416.

Zhang, X., Jeffrey, R.G., Thiercelin, M., 2007a. Effects of frictional geological discontinuities on hydraulic fracture propagation. In: SPE 106111, SPE Hydraulic Fracturing Technology Conference, College Station, Texas, 29–31 January.

Zhang, X., Jeffrey, R.G., Thiercelin, M., 2007b. Deflection and propagation of fluid-driven fractures as frictional bedding interfaces: a numerical investigation. Journal of Structural Geology 29, 390–410.

Zhang, X., Jeffrey, R.G., Thiercelin, M., 2009. Mechanics of fluid-driven fracture growth in naturally fractured reservoirs with simple network geometries. Journal of Geophysical Research 114, B12406.

Zhang, F., Nagel, N., Lee, B., Sanchez-Nagel, M., 2013. The influence of fracture network connectivity on hydraulic fracture effectiveness and microseismicity generation. In: Paper ARMA 13-199, 47th American Rock Mechanics Symposium, San Francisco, California, USA, 23–26 June.

Zhang, Y., Sayers, C.M., Adachi, J., 2009. The use of effective medium theories for seismic wave propagation and fluid flow in fractured reservoirs under applied stress. Geophysical Journal International 177, 205–221.

Zhao, J., Chen, M., Jin, Y., Zhang, G., 2008. Analysis of fracture propagation behavior and fracture geometry using tri-axial fracturing system in naturally fractured reservoirs. International Journal of Rock Mechanics and Mining Sciences and Geomechanics Abstracts 45, 1143–1152.

Zhao, X.P., Young, R.P., 2009. Numerical simulation of seismicity induced by hydraulic fracturing in naturally fractured reservoirs. Paper SPE 124690, SPE Annual Technical Conference and Exhibition, New Orleans, Louisiana, 4–7 October.

PART 2

SPECIAL TOPICS

CHAPTER 11

EFFECTIVE CORE SAMPLING FOR IMPROVED CALIBRATION OF LOGS AND SEISMIC DATA

David Handwerger[1], Y. Zee Ma[2], Tim Sodergren[3]

Schlumberger, Salt Lake City, UT, USA[1]; *Schlumberger, Denver, CO, USA*[2];
Alta Petrophysical LLC, Salt Lake City, UT, USA[3]

11.1 INTRODUCTION

Two main goals of a core analysis program for evaluation of unconventional reservoirs are to collect data not easily derived from wireline logs, and gather "ground-truth" data to calibrate the logs. Such data are typically used to construct or quality-control relational models developed between a reservoir parameter of interest and logs, such that the log data can be used to propagate the reservoir properties over the zone of interest or into the three-dimensional (3D) reservoir model. However, such relational calibration models are often suboptimal, either because they are built without incorporating some basic tenets of statistical modeling, or are built with biased samples.

Linear regression is a common statistical technique used to predict reservoir properties by cores, logs and seismic attributes (e.g., Woodhouse, 2002; Zhu et al., 2012). Regressions are easy to use. For example, one has input data from cores, logs and/or seismic attributes, such as a collection of discretely sampled core and log measurements, and then a model is established by relating the data to each other in a cross plot or by the mean of the mathematical correlation, and by minimizing an estimation–error term. Least squares fits, for example, are found by finding a linear descriptor that minimizes the mean square error between the predicted values and the measured data. Although the model does not have to be linear, we will focus on linear models here as they are the most commonly used.

There are several considerations regarding linear regression that are often overlooked. First, the differences between prediction and observation are approximately normally distributed for each input; second, the mean residual is approximately zero; third, the residuals about any given input are of equal variance (often termed homoscedacity, Weiss and Hassett, 1982). These assumptions imply that regressions are good at capturing the mean of the dependent variable, but are not as good at capturing the extremes (Hook et al., 1994). In other words, regressions "collapse to the mean". Furthermore, in this context, one needs to consider what exactly the correlation coefficient (or R^2) is telling us; is the correlation "good" or "bad" (as opposed to "valid" or "invalid") (Fisher and Yates, 1963)? Appendix 1 discusses, in more detail, the pitfalls related to multimodality and nonGaussian distributions of inputs using linear regression.

While the examples in Appendix 1 were constructed to illustrate the issues with nonGaussian distributions of inputs in general, one is left to consider what to do with wireline log and core data, when the trends or biases may not be so apparent. In conventional reservoirs, with "end member" log responses that map out sand, carbonate and shale facies, the above warnings are well accepted (if not explicitly acknowledged), and that is why many log interpretation charts separate the logs into facies (e.g., Schlumberger, 1991). However, the trends in unconventional reservoirs are not so apparent as mineralogy has a controlling influence on log response, and it also, together with organic matter, can control porosity and saturations (Sondergeld et al., 2010; Passey et al., 2010).

Moreover, scale differences in core, log, and seismic data often cause changes in the meaning of statistical representations of reservoir variables, including representation of reservoir heterogeneity and correlation between reservoir parameters. These problems impact the quality of calibration and thus effective use of various geoscience data for unconventional resource evaluation.

The problems in using regression or other statistical modeling techniques can be largely dealt with if one takes into account the heterogeneity inherent in unconventional reservoirs, and the need to optimize the integration of various data at different scales. Heterogeneity is a common source of nonGaussian data distributions, and isolating the reservoir into zones with approximately Gaussian input distributions can improve the results of regression, so that separate models can be developed for each zone. Multivariate classification of the inputs can be done on log data, and leveraged to pick core samples in a manner that will optimize the modeling results.

Acknowledging the assumptions inherent in linear regression modeling and formation heterogeneity, the model construction does not necessarily require more sampling, just more efficient sampling, or more fit-for-purpose sampling. To achieve this, it is useful to deconstruct the log response into zones of expected similar bulk material properties, and then sample each zone in a statistically relevant manner. What "statistically relevant" means can be a function of practical concerns such as the complexity and expense of particular types of tests, but a basic level of statistically-based sampling can improve models relative to sampling based on some preordained bias (e.g., the rock is homogeneous, but sampling is performed only in a perceived sweet-spot zone).

In this chapter, we will illustrate the problems associated with the assumption that all data within a reservoir section are appropriate for creating a single model, and show the differences between using properly conditioned data sets and improperly conditioned data sets. We then show how this applies to property prediction in unconventional reservoirs and how one can take advantage of the log data up front to maximize the value of core sampling and improve models for propagating core properties through the logs.

11.2 PATTERN RECOGNITION IN LOG DATA

To improve the core-log calibration, the multimodal distribution of inputs, such as log responses to the unconventional formations, must be carefully dealt with. One way to isolate approximately Gaussian distributions from a heterogeneous data set is through an unsupervised cluster analysis of the input data (Ma et al., 2014a). An unsupervised cluster analysis can produce clusters which are approximately more Gaussian than the whole data set.

Figure 11.1(a) shows a collection of logs from an unconventional gas field in the United States, and an associated unsupervised clustering of the log data. In an unsupervised clustering, the inputs are the

11.2 PATTERN RECOGNITION IN LOG DATA

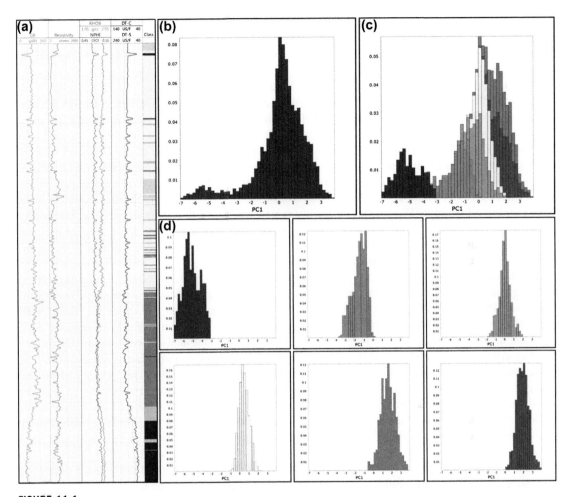

FIGURE 11.1

(a) Log data and associated clustering from a United States gas well. (b) Histogram of the first principal component resulting from a PCA of all the inputs. This represents 57.6% of the input data structure. (c) The same data as in (b), colored by cluster. (d) Individual class distributions. The X-axis range in each subplot is the same, but the Y-axes are normalized individually.

collection of logs over a zone of interest, a user-defined number of clusters, and possibly a noise rejection threshold for lower-order principal components (PCs) from principal component analysis (PCA) performed on the data prior to clustering. The chosen number of clusters is one that balances the competing influences of maximizing the uniqueness between the clusters and minimizing the distribution of the inputs that define each class of rock. Many algorithms tend to classify the data into classes of approximately equal variance. The condition of maximum uniqueness between the classes facilitates the use of the results in subsequent supervised classification. This particular aspect is beyond the scope of this chapter, and has been discussed elsewhere (Handwerger et al., 2012, 2014; Suarez-Rivera et al., 2013). The condition of minimum data distribution per class is based on the assumption that the distribution of the core data per class roughly mimics the distribution of the logs per class, as will be discussed in Section 11.3.

An additional benefit of PCA is the ability to display the data structure more completely by reducing the dimensionality of the inputs. The first PC resulting from PCA is shown in Fig. 11.1(b), as this represents the dimension of maximum correlation in the inputs. PCA is a technique that consists of identifying a set of orthogonal principal axes that orient along the direction of maximum variability in multivariate data. The components themselves then comprise the projection of the data onto each of these axes, such that the first principal component represents the maximum variability in the data, followed by projections on orthogonal axes representing decreasing amounts of the variability. Mathematically, PCA is the eigenvalue decomposition of the correlation coefficient matrix of each input to every other input. The percent variability represented by each principal axis is the ratio of its eigenvalue to the sum of all eigenvalues, and the projection of the input data to the new axes is the product of the input data and each associated eigenvector. A more detailed description of PCA can be found in Ma et al., 2016 (this volume).

In this example (Fig. 11.1), the first PC comprises 57.6% of the data structure of the six input logs, meaning that 57.6% of the correlation among the multiple input logs is reduced to the single dimension of the first PC. All the PCs (there is an equal number of PCs to inputs) add up to 100%, but with decreasing reflection of total covariance of the inputs. Note that the data in Fig. 11.1(b) are not quite Gaussian distributed as a whole. However, if one performs a cluster analysis on the data, the modes make up different classes (Fig. 11.1(c) and (d)). The dark blue class can immediately be isolated as one of the modes, which leaves behind the majority of the input data as a larger second mode. Additional classes partition that mode into smaller pieces, each of which forms a tighter approximately Gaussian distribution than the whole (Fig. 11.1(c)) and leads to increased predictive power than if that larger mode is modeled as one. Regression modeling at this more granular level should provide tighter predictions, with smaller residuals[1] per class than modeling against the whole, *even if the R^2 for each class model is lower than a single model created for all the data as one unit*. (For an explanation, see Appendix 1 while noting that R^2 not only refers to the spread of data about the regression trend, but also the significance of the trend itself, given the amount of data that define it).

Each class resulting from the cluster analysis is defined by a unique pattern of inputs and they are of approximately equal variance. Figure 11.2 shows two views of a plot of the first three PCs (representing 90.0% of the overall variability of the six input logs). Each point is colored by its assigned class from the clustering. Recall that the dark blue class comprises the data from the smaller mode of these bimodally distributed data (Fig. 11.1(b) and (c)), while the remaining classes partition the main mode of the input data.

11.3 SAMPLE SELECTION FOR CALIBRATION

If an appropriate set of logs sensitive to variability in the texture and composition of the rocks are chosen (e.g., Triple Combo), the clustered classes should represent zones of similar bulk material

---

[1] Residuals are the difference between observation and prediction, in this case between the core data and the calculation of the same parameter from the regression at the same value of the independent variable.

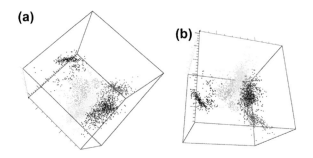

FIGURE 11.2

Two views of the 3D cross plot of the first three principal components from a PCA of the log data shown in Fig. 11.1, colored by class assignment resulting from a cluster analysis of the logs. Note that each class is unique in its log pattern. The three principal components shown here represent 90% of the overall data structure. Note that each of the colored classes is of approximately equal variance. The dark blue class represents one mode of the bimodally distributed inputs (Fig. 11.1(b) and (c)), while the other classes generally subdivide the other mode.

properties, and they also provide a better-conditioned set of inputs for regression than the whole. A basic principle of materials science states that the bulk material properties of a substance are a function of its texture and composition (Carter and Paul, 1991). Therefore, if one can isolate zones of constrained texture and composition reflected in the logs, one can consider the classes as representing zones of statistically similar bulk material properties. The material properties need not be known a priori to be useful in selecting sample locations, provided that similar zones have similar properties (once measured) and different zones have different properties (Handwerger et al., 2012). If the test results show that this link is invalid for the reservoir, then either the inputs are not adequate discriminators of the rock texture and composition, or there is an external influence on the baseline property that the logs are not sensitive to (e.g., pore pressure effects on gas adsorption).

Typically, one does not know quantifiably what the classes represent in material properties (barring some other petrophysical modeling) when logs are partitioned by an unsupervised clustering algorithm. However, the resulting classes can be used to make decisions on how to sample the reservoir section in order to determine the class properties. For example, if one assumes that a collection of samples taken in any class should represent similar bulk material properties (within the statistical bounds of the class), this assumption can be used to make intelligent decisions on where to optimize sample placement with respect to the intersection of the classes and available rock. For example, samples can be distributed across each class to best test the variability in its material properties. This can improve the ability to define appropriate regression models to predict each parameter either at the higher sampling rate of the logs within the cored well, or for use in subsequent wells and seismic data.

Given the broad range of available core tests (from simple to complicated), what is meant by "optimized sampling" is a function of the level of detail that is practical. In the simplest case, one could just choose a single sample per class, as might be the case for expensive and complicated tests, and use the result to describe the average behavior for the class. In this manner, each quantifiably unique package of log responses is characterized in an average sense, with a level of precision implied by the variance of logs within the class, and the overall heterogeneity of the section is represented.

For the above scenario, the "average" log response per class is the most efficient way to sample. It will provide the closest to an average property response that can be estimated from the input data for each class and should further map out how the properties broadly vary at the scale of heterogeneity that can be isolated from the logs through the classification. Furthermore, if one sample per class provides enough data for at least a broad global regression model (assuming the input data are suitable for this task, see Appendix 1), the training data will be the most representative of the bulk data distribution over the zone of interest. For example, a single regression could be built on the entire population of core data (at one sample per class), provided that enough classes exist to yield a relevant number of data points to justify any derived linear fits, and the inputs satisfy the assumptions of regression models discussed earlier. For the data shown in Fig. 11.1, this could mean excluding data from the dark blue class, because this represents a separate mode to the majority of the overall data distribution. Of course, this leaves five out of six classes, which would only have five data points (at only one sample per class), and the magnitude of R^2 necessary from any resulting linear regression model would need to exceed 0.77 to be a statistically valid expression of a trend in the data to the 95% confidence level, and exceed 0.91 for 99% confidence (see discussion on the Fisher and Yates criteria in Appendix 1).

Where a larger population of core samples is warranted, a two-step approach to sampling should be considered. First, the clustering of the input log data to map the heterogeneity of the formation provides a baseline, as above. Second, if one wishes to build regression models at the class level, each class needs to be sampled appropriately. This can be accomplished by analyzing the distribution of logs per class, and sampling across the distribution frequently enough to maximize the likelihood of a statistically relevant model (Appendix 1).

PCA is not only useful for the reduction of input dimensions to an easier-to-analyze number, but also collapses the correlations within the inputs to single components, allowing for easier identification of "average" bulk log responses. Figure 11.3 illustrates a scenario where the average log response for a particular yellow class is identified. The average yellow class response is most easily identified by looking for the location closest to the average of the first three PCs, which account for almost all of the variability in the larger number of raw input logs. This represents a better average of the bulk responses than just the peak of the first PC, the histogram of which is plotted in Fig. 11.3. In fact, one can see that the proposed sample location, shown on the histogram, is not exactly at the peak of the first PC distribution. The average response in each log and associated PC below the histogram is represented by a black dot and one can see that the proposed sample location is very close to the average in the first three PCs.

For increased numbers of samples, sampling can be spread out to map the distribution of the logs within each class. Instead of one sample per class, samples can be located at intervals along the distribution to make sure that it is as completely sampled as is warranted. A breakdown of each class, as shown for the hypothetical yellow class of Fig. 11.3 can be constructed, and the depths within the log section (with core) can be highlighted for their closeness to given portions of the log distributions. Again, if this is done in PC space, the appropriateness of any samples is more easily evaluated, as the number of available dimensions is usefully reduced without undue loss of variance information.

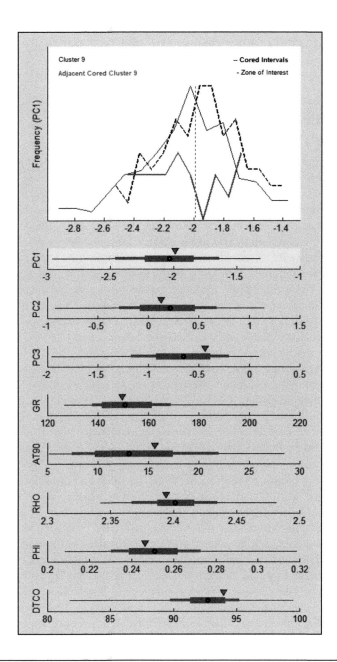

FIGURE 11.3

Example sample selection strategy. The solid black line in the top panel shows the distribution of the first PC of the log data for all occurrences of an olive-green (dark gray in print versions) class, isolated from a cluster analysis of a suite of log data over a zone of interest. The black stippled line in the panel shows the distribution of the first PC over the cored interval, and the solid olive-green line shows the distribution over the continuous block of that class being queried. The vertical blue (gray in print versions) dotted line shows the value of the first PC for a particular depth being investigated. Below the top panel are horizontal bars representing the spread of the first three PCs of the log data, and the spread of the logs themselves, for all occurrences of the olive-green class. The black dot on each bar represents the average response, and the inverted triangle above each represents the same single depth point and where its log responses lie with respect to the overall log distributions within the class. The example shown here is of a sample at the most "average" log response for the olive-green class. The definition of "average" in this case is where the chosen location is close to the average of the first three PCs of the log data simultaneously, and not just the peak of the distribution of the first PC shown in the top panel.

11.4 CASE STUDIES
11.4.1 STUDY 1 WITH TWO EXAMPLES

Figure 11.4 shows an example of a well with two cores, one from an upper zone and one from a lower zone. Unsupervised clustering of the log data shows that the two zones have different bulk log responses, as each section is represented by a different collection of classes. As a result, one should not expect that data in one zone are strictly linearly related to data from the other. The core points for each zone were selected to honor the heterogeneity shown by the logs, within the constraint of where core was actually collected (i.e., each class also exists outside the cored intervals). The prediction of gas-filled porosity (PhiG), however, is based on building a multiple linear regression model using the stepwise method (Draper and Smith, 1998) only to the samples from the lower core, and applying the fit up through the upper-cored zone. The core data from the upper zone are then plotted for comparison purposes only. What is apparent is that training a model to only the lower core severely over predicts the actual gas-filled porosities in the upper-cored interval. This is an example of nonrepresentative sampling (which may be unavoidable in Exploration and Production (E&P)).

Furthermore, the far right-hand tracks in Fig. 11.4(a) and (b) shows the Mahalanobis Distance (D_m) (Mahalanobis, 1936) between the logs associated to training data within each class, and the log responses used throughout the vertical section for prediction; basically resulting from a quasisupervised classification of the whole section to the portion with training data. D_m is a variance normalized Euclidean distance between each data point and the class whose model is being applied, and is a measure of the nearness of one set of logs to those that define its class. It is a useful metric in supervised classification, for example, where one wants to know if the chosen class—the closest possible in the model set—is actually close enough to be appropriate. Where D_m is too large, even the closest class (the one chosen) is not "close enough", and this is a sign that there is a class not represented in the model set (Handwerger et al., 2012, 2014; Suarez-Rivera et al., 2013). In this instance of applying a predictive model, where D_m is sufficiently high, the data point being predicted is well outside the data cloud that provides the regression model, and thus the prediction is flagged as suspicious.

Regression analysis should only be applied to zones equivalent to those in which the model was created (Hook et al., 1994). One can use supervised classification to determine where in a prediction well or zone a particular class model is appropriate. As a result, the D_m should also be considered because most supervised classification algorithms will pick the closest match in its training population, regardless of how close that match actually is in the model space. Therefore, a separate metric is needed to determine what constitutes "close enough" for the purposes of applying a given model.

As seen from Fig. 11.4(a), the D_m is generally elevated in the upper core, based on training data purely from the lower core; a warning that the model from lower core classes may be inappropriate for the upper core. Where D_m is lower within the upper core the predictions tend to be closer to the core measurements. For the predictions across the lower core, where the training data exist, D_m is generally very low, as would be expected since the training model is essentially being applied back on itself, though extrapolating to intervening log data.

Figure 11.4(b) shows the same prediction based on training to both cored intervals. The result is a much improved prediction, where the predicted values of gas-filled porosity are much closer to the measured values in both cored zones. Furthermore, the D_m within each cored interval is much lower. Beyond the cored intervals, however, there are regions with flagged high D_m, suggesting that a truly representative class model does not exist, based on the availability of core.

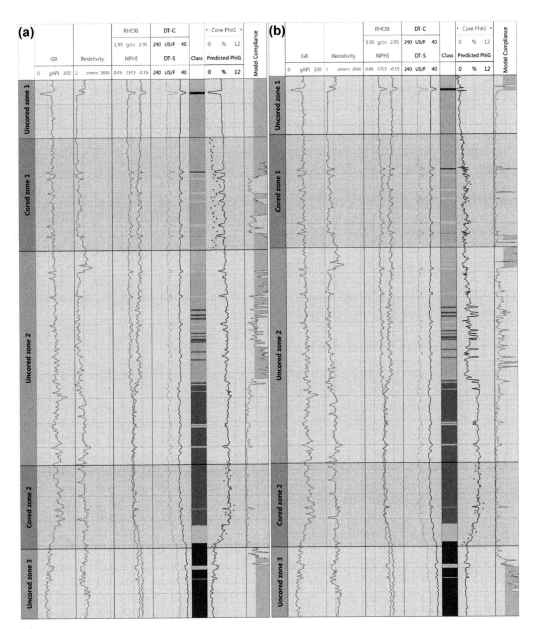

FIGURE 11.4

(a) prediction of gas-filled porosity from a regression model trained to the core data from cored zone 2, but applied everywhere. Note the lack of agreement between prediction and core in cored zone 1, and also the fact that the classes are different. The right-hand column shows the degree to which the logs used to model beyond the training interval and classes are related to the logs used to build the regression. The metric is the Mahalanobis distance. (b) The results of class-specific regression models built with all the core data from both zones. Note that the Mahalanobis distance is much lower over most of the interval, suggesting better representation of the training data for the whole interval.

In these regions of high D_m, one is left with a question of what to do. There are two main possibilities. The first possibility is to pre-define a single "global" relationship that ignores the heterogeneity from the perspective of adding granularity to the predictions, assuming that the data are appropriately conditioned; the second possibility is to apply the class model anyhow, while understanding that it may not be appropriate. Regardless, the high D_m should act as a flag for where either choice was implemented. Additionally, the flag can indicate where the aforementioned concern raised by Hook et al. (1994) is valid, and that the chosen regression model is being applied beyond the range of training inputs. In the example (Fig. 11.4), a global model was superimposed where either there was high D_m to the training data, or where the training data failed to provide a statistically relevant trend (see the discussion on the Fisher and Yates criteria in Appendix 1).

Another example of the same prediction is shown in Fig. 11.5. We have two separate training scenarios for predicting gas-filled porosity based on multiple linear regression. In each case, 14 of the original core points were chosen, but under different paradigms. In the first case, the 14 samples were chosen at average responses of each class. In the second case, the 14 samples were chosen at average responses over the cored zones without regard to class. This second scenario introduces a bias in its own right because the majority of samples come from cored zone 1, with only a few points in cored zone 2. However, the training data of the second scenario are much worse at predicting the true gas-filled porosity. This is because the sampling does not honor the variability of the log data, even to first order, but is clustered about a global mean log response.

11.4.2 STUDY 2

Figure 11.6(a) shows another example of the effects of biased sampling. The core data were filtered for those samples associated to deep resistivities from log greater than 10 Ω m (Fig. 11.6(b)). Higher resistivity is often associated to increased hydrocarbon volume, and it is not unexpected that sampling could be biased by selecting samples "only in the reservoir". When the gas-filled porosity data of these selected samples are provided for training, the predicted GFP results are quite close to the measured data with only small residuals at the training points, but elsewhere, the prediction is quite poor, especially in the upper core, where the prediction is almost systematically higher than the measurement (though matches the measurement where training data are collected). What happens in the uncored regions where we do not have ground-truth can only be speculated on, but the bias should call into question any conclusions that the prediction is adequate. Given that the upper half of Uncored Zone 2 is dominantly a yellow class, and upper Cored Zone 1 has core data in the yellow class demonstrating that the prediction overestimates GFP, we can assume a similar situation for the yellow class in Uncored Zone 2. Similarly, we can assume for the red class in the lower half of Uncored Zone 2 that the prediction is reasonable, given the training data are dominantly in the red class over Cored Zone 2. Note also that the samples associated to highest resistivities do not systematically have the highest GFP. The bias, therefore, is in the assumption surrounding the relationship between resistivity and GFP, not the GFP values themselves. Similar GFP values can have higher resistivity or lower resistivity.

Figure 11.6(c) shows a comparison of the predicted and observed core gas-filled porosities for the entire population of samples, not just those selected for training in the above scenario. Samples are colored based on which cored zone they come from. As most of the training data come from Cored Zone 2, the best-fit relationship shown is only for that section. It is visually apparent that there is

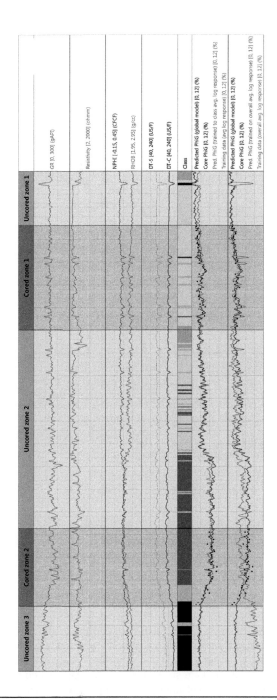

FIGURE 11.5

Predictions of gas-filled porosity based on two different sampling regimes. The one on the left (immediately to the right of the classification results) is based on filtering the core training data to 14 samples most proximal to the average log response within each class. The one to its right is a filtering of the core data to the 14 samples located closest to the average of all log responses, ignoring the classes (red curve; gray in print versions). Each is compared to the best solution from training to all the data (black curve, same prediction as in Fig. 11.4). The training core points are red dots in either channel. The black dots are the remaining core data not used for training, but displayed for comparison to the prediction results.

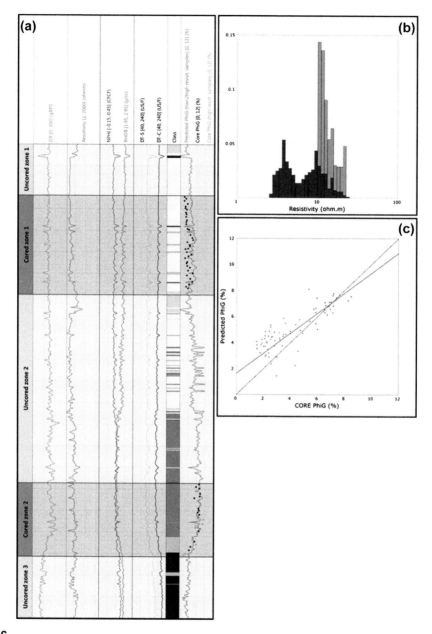

FIGURE 11.6

(a) Prediction of gas-filled porosity based on training to core data collected opposite deep resistivity log values >10 Ω m. This represents 23 out of 76 samples. Those core samples are colored green. The black dots represent the balance of the core data and are shown for QC. The black dotted samples were not used for building the regression model. (b) Histogram of input data. The blue boxes (black in print versions) are the log deep resistivity data in the two cored zones only, the green boxes (green in print versions) represent the log deep resistivity values for the samples chosen to build the regression (samples with deep resistivity >10 Ω m). (c) Correlation of predicted and observed gas-filled porosities. Data for Cored Zone 1 are in blue (gray in print versions), and for Cored Zone 2 are in red (dark gray in print versions). Black line is 1:1 line, red line (dark gray in print versions) is the best-fit line (major axis technique) to the data from Cored Zone 2 only, as most of the training data (and therefore the expected best prediction) is from there. The correlation has an R^2 of 0.71. The correlation of a best-fit line to the predictions over Cored Zone 1 (not shown) is 0.02, which is insignificant. Despite this, the prediction underestimates GFP at low values, and overestimates at high values.

11.4 CASE STUDIES

prediction bias in Cored Zone 1, as commented on above, but it is not as clear that there is bias in Cored Zone 2 as well.

While no best-fit line is plotted for Cored Zone 1, based on this prediction scenario it is apparent that there is little correlation between predicted and observed data, and in fact the R^2 of such an RMA[2] best fit (not plotted) calculates to 0.02, which signifies a statistically insignificant trend (not to mention a poor accounting of what is a reasonable amount of overall sample variance). The RMA correlation for Cored Zone 2, which is shown, has an R^2 of 0.71, which is significant. However, the slope of the best-fit line is >1, suggesting that at low GFP the prediction is an overestimation, while at higher GFP it is an underestimation. In fact, at low GFP, the overestimation is ~1 p.u., which is a ~50% error for measured values of ~2% GFP.

The above discussion shows the value in representative sampling of a heterogeneous formation. Sampling only what is perceived as "reservoir rock" (Fig. 11.6) can introduce bias when trying to use those data to model overall variability through regression models, in addition to the implied bias from assuming what is and is not reservoir based solely on the logs. This bias can occur in sampling the overall variability (see, e.g., Fig. 11.5). The effects of such bias would be more pronounced in more heterogeneous formations such as shales than in many conventional reservoirs.

11.4.3 STUDY 3

The advantage of sampling to honor a classification model of the multivariate inputs versus a one-size-fits-all model (a real-world example of Fig. 11.13 vs Fig. 11.15 in Appendix 1) is discussed here. Figure 11.7 shows the comparison of a prediction of the core effective porosity (EPOR) using a global model, where the log inputs are treated together, and a prediction from compiling models built per class. Several observations can be made. First, the class-based model provides an overall better match to the core than the global model. This is further illustrated in Fig. 11.8, which cross plots the predicted effective porosity values versus the core data for each scenario. In the global prediction (Fig. 11.8(a)), we see a decent comparison, with an R^2 of 0.778 (for 75 data points) and a slope of 0.87 to an RMA linear fit. However, for the class-based model (Fig. 11.8(b)), the R^2 improves to 0.887, and the slope is closer to 1, at 0.95. Additionally, the intercept is closer to zero for the class-based model.

What can be seen for the global model (Fig. 11.8(a)) is that the core porosities are moderately bimodally distributed (the histogram along the top of each cross plot), but this bimodality is less apparent in the predicted values (histogram along the right-hand axis). For the class-based model (Fig. 11.8(b)), the bimodality in the prediction is much more apparent, with even a break in predicted values at middle EPOR values. This suggests that the class-based model is more effective at capturing the data structure of the core data than the global model. Notice, however, that the individual class models are based on approximately Gaussian distributed data. When the entire prediction is reconstituted from the individual models, the class-based model more closely matches the overall data structure.

---

[2] A Reduced Major Axis regression fit is one that minimizes the sum-squared error (SSE) of both dependent (usually the Y-axis) and independent (usually the X-axis) axes, instead of just the dependent axis. Minimization of the SSE of just the dependent axis is what is typically done in linear regression, likely because it is the default assumption in most plotting software when regressions are chosen to fit plotted data. However, such regressions are only valid if: (a) there is no uncertainty in the independent variable; or (b) the error in the independent variable is the same in any data set to which the regression model will subsequently be applied.

348 CHAPTER 11 EFFECTIVE CORE SAMPLING FOR IMPROVED CALIBRATION

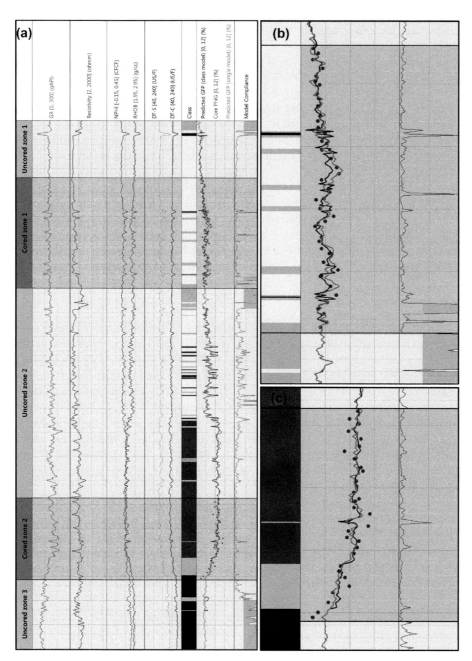

FIGURE 11.7

(a) Comparison of a global regression model prediction (green; light gray in print versions), and a class-based regression model prediction (black) of effective porosity (EPOR). (d) Zoom to the upper core zone. (c) Zoom to the lower core zone. Note the class-based model has overall better match to the core data. This is shown more diagnostically in Fig. 11.8.

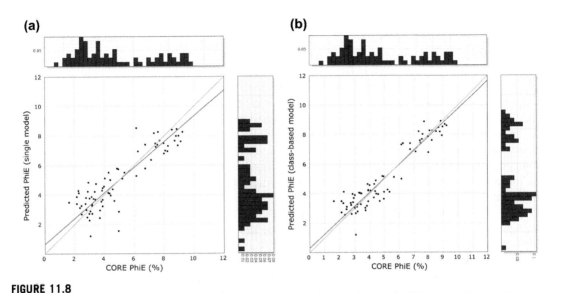

FIGURE 11.8

Comparison of predicted versus measured effective porosity (EPOR) for (a) the global regression model, and (b) the class-based model. Note from the histograms along each plot's axes that the bimodality of the core data is better represented by the class-based model. The bimodality is not so apparent in the global model, but then should also suggest the inappropriateness of a global model in the first place.

11.5 DISCUSSION

Once a prediction is made from regression, a common next step is to collate the data for use in simulations. Such simulations often require layer-dependent inputs of average properties per layer (Wei and Holditch, 2009; Waters et al., 2011). Additionally, rock classes can be propagated through seismic volumes using inverted attributes (Suarez-Rivera et al., 2013; Borgos et al., 2013; Handwerger et al., 2014), and its results can also be fed into simulations (Rodriguez-Herrera et al., 2013). In either case, layers are output, and statistics of relevant parameters for each layer are provided. These properties can either come from the descriptive statistics of the collection of core data or from the predictions of the core data from the logs. In either case, classification can be useful to define the layering in the first place because it separates the reservoir into zones of expected similar material properties rather than assuming such correlations result from a purely geologic discrimination.

Given that shales are frequently heterogeneous, potentially leading to multimodal distributions, multivariate classification can be a useful way to impose a degree of normality on the data by separating the modes and producing approximately Gaussian distributions of the inputs per class (e.g., Fig. 11.1). As one ultimately wants distributions of the core properties (whether directly from the core data, or the regression-based predictions of them), it has to be established that the core data are also approximately Gaussian distributed, or are decomposed into a number of populations of properties that are. The latter is easier to establish a priori when there are lots of core data available, but practical considerations often preclude this. For example, the Central Limit Theorem (see Appendix 2) states that a sufficiently large population of randomly chosen variables will have a mean $\bar{x}$ of a normal distribution when $N \geq 30$ (Weiss and Hassett, 1982). While one often does not have more than 30 core samples per class, there are usually more than 30 log measurements per class. Therefore, if an approximate normality can be established for the log data per class, then a similar normality in the core properties associated to them can be assumed, subject to the full range of log responses that define a class being available for sampling. Consequently, samples collected within that Gaussian framework should be approximately consistent with the Central Limit Theorem. To further increase the population of core properties, one can use the predictions instead, as the sampling of logs is typically much higher than for core. Additionally, the regression can be used to map the core results into a yet more approximately Gaussian space.

Figure 11.9 expands on the discussion around Fig. 11.1 by showing the results of one-sample Kolmogorov–Smirnov (K-S) tests (Massey, 1951) of the normality of the distributions of all the input logs in the well used for the previous examples versus the normality of the distributions of the logs per class. When one considers the major variability in all the logs together, through the first PC, the K-S test fails, showing that the logs as a unit over the reservoir section do not follow an approximately normal distribution. The cumulative distribution function of the first PC is shown in Fig. 11.9(a) alongside a "perfect" normal distribution. The K-S test calculates the probability of the data representing a normal distribution to be on the order of 10^{-63}, which basically means "zero". If one looks at the same display for the two major classes within the cored intervals (Fig. 11.9(b) and (c)), each distribution tests to be normal within the 95% confidence level.

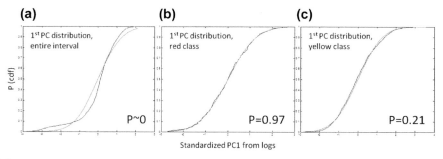

FIGURE 11.9

Comparison of the cumulative distribution function (cdf) of the logs to a standard normal distribution. The logs themselves are represented by their first principal component from a principal components analysis. The first PC is standardized so that its distribution can be compared to a normal distribution using a Kolmogorov–Smirnov test. The P-values of the test result are reported in each plot. Values >0.05 suggest that the test distribution (the first PC of the logs, in this case) are normally distributed to a 95% confidence. (a) All the log data. The results suggest that, as a whole, the logs are not normally distributed. (b) The logs that make up the red class, and (c) the logs that make up the yellow class. These are the two most represented classes within the cored intervals, and both are normally distributed to within 95% confidence.

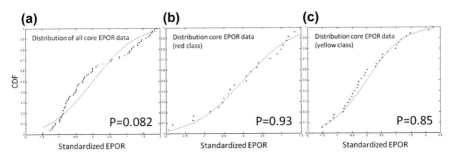

FIGURE 11.10
Comparison of the distributions of core data (black dots) and a standardized normal distribution (red curve; gray in print versions). The EPOR data are standardized so that they can be statistically tested against the standard normal cdf using a Kolmogorov–Smirnov test, the results of which are reported as the P-values in each plot. P-values >0.05 suggest the test distribution (EPOR) are normally distributed with a confidence of >95%. The higher the number, the more probable the tested distribution matches a standard normal distribution. Only the red and yellow classes contain enough core data to make a meaningful comparison. (a) All the core data, which are only weakly normally distributed; (b) The core data in the red class, which are strongly normally distributed; (c) The core data in the yellow class, which are also strongly normally distributed.

Extending this to the core data, we see in Fig. 11.10 that the distribution the entire population of effective porosity core data weakly approximates a Gaussian distribution, according to the Kolmogorov–Smirnov test, showing an only 0.08% confidence level (while it may not "look" that good, there is a degrees-of-freedom component to the K-S test, and the number of core data are sufficiently few to produce a better statistic than it "looks"). However, if we look at the class-specific K-S results, we see vast improvement. Figures 11.10(b) and (c) shows the results for effective porosity in the two most sampled classes. The yellow class shows an 85% probability of being a normal distribution, while the red class shows a 93% probability. This implies that parametric statistics calculated on a class-based model are more appropriate than the same for the global model, where the assumption of a Gaussian distribution is much weaker.

Consequently, in subsequent models that rely on layered inputs, the statistics from a class-based model should be more robust. In fact, we can see in Fig. 11.11 that the prediction of effective porosity that results from the class-based model is also more approximately Gaussian than from the global model when considered over the entire prediction range.

These concerns also come into play when one wants to upscale the data to seismic scale. As previously mentioned, the classes can be upscaled and propagated through a seismic section using supervised classification of inverted seismic attributes. The methods employed by Suarez-Rivera et al. (2013) and Handwerger et al. (2014) use a maximum likelihood estimation method based on fitting Gaussian distributions to the training data (Borgos et al., 2013). This method can be made more robust if the upscaled classes are also Gaussian distributions. Figure 11.12 shows the propagation of an upscaled classification through a seismic section. Since the classes in the seismic volume are related to classes in the logs, the statistical distributions of the log-based classes can be passed on to the interpretation of the classes in 3D through the seismic data.

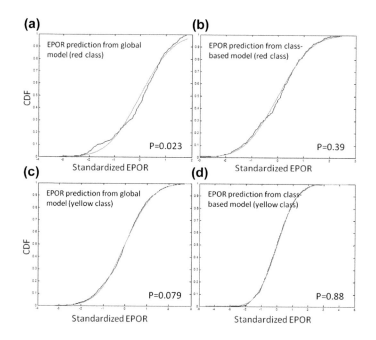

FIGURE 11.11

Comparison of the distributions of the predicted effective porosity (EPOR) to a standard normal distribution. The EPOR data (black) are standardized so that they can be statistically tested against the standard normal cdf (red; gray in print versions) using a Kolmogorov–Smirnov test, the results of which are reported as the P-values in each plot. P-values >0.05 suggest the test distribution (EPOR) are normally distributed with a confidence of >95%. (a) Global-model prediction for the red class; (b) Class-based model prediction over the red class; (c) Global-model prediction over the yellow class; and (d) Class-based model over the yellow class. These classes are the only two with enough samples to train a reliable regression model. Note that in each case, the global-model predictions are at best weakly Gaussian distributed, while the class-based predictions are strongly Gaussian (much higher P-values).

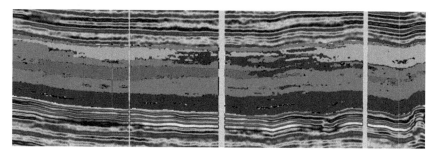

FIGURE 11.12

Propagation of a rock classification scheme defined at log scale and upscaled to seismic scale, across a seismic section. The supervised classification method used here is predicated on a maximum likelihood estimation, which itself is based on a Gaussian model of the distribution of seismic attributes trained to recognize each class.

Modified from Suarez-Rivera et al. (2013).

11.6 CONCLUSIONS

In formations such as shales, clays, and organic matter are dominant controls on the variability of logs. As a result, they make log interpretation of reservoir properties more challenging than for conventional reservoirs with their generally "clean" formations. Therefore, it is necessary to deal with a much higher degree of heterogeneity in evaluating shale reservoirs. Core data are often collected to calibrate the logs and provide more detailed information on formation properties. To do this, the core data are frequently used to build relational models with the logs so that the logs can be used as a predictive tool. In conventional reservoirs, once a basic breakdown of sand, carbonate, and/or shale is done, further degrees of heterogeneity are often not needed or considered. However, in unconventional reservoirs, rock heterogeneity is both more pronounced and subtler, and their effects on reservoir properties are broader. It is more subtle in that it is often harder to quantify, as the heterogeneity effects on the logs are harder to characterize. Moreover, the rock effect on log response in unconventional reservoirs, other than being more dominant because of the low fluid and pore volumes, is also more "shades of gray" than "end member".

Therefore, the use of regression to relate core data to log data needs to account for this heterogeneity and its effect on the basic assumptions for regression. Due to the relative homogeneity of most conventional reservoirs, this is frequently not a concern, but assumptions of homogeneity in unconventional reservoirs can lead to degradation of regression models as a useful predictive tool and erroneous estimations of reservoir properties. Based on the studies discussed in this chapter, if one can quantifiably define the heterogeneity, and do so in a manner that satisfies the conditions of successful regression modeling, these models can be powerful predictors of reservoir properties. However, the following pitfalls must be taken into consideration.

Regression models at a fundamental level assume that the input data are approximately Gaussian distributed, and consequently unimodal. Several conceptual examples shown in this chapter outline the pitfalls of assuming data so distributed when in fact they are not. Such an assumption is tenuous when it comes to heterogeneous formations such as shales. However, a variance-partitioning unsupervised multivariate classification of the input logs can isolate individual modes within the formation's log distributions, and provide classes that more closely satisfy the implicit assumptions of linear regression models. Subsequently, samples can be strategically chosen to optimize the regression(s) that can be built at the class level, and such models should be more robust than treating the formation as a monolith when it comes to sampling and model building.

These pitfalls related to building a calibration model in a core analysis program can be mitigated in the sample selection process. When such model building is part of a core analysis program, these pitfalls can be mitigated if they are considered in the sample selection process. For example, it is advantageous to choose samples with limited bias, and in a manner that accounts for the heterogeneity inherent in unconventional reservoirs. Additionally, one should sample in a manner free from assumptions that a particular log or log response is universally diagnostic of the variability of the parameter of interest (e.g., higher resistivity equals increased hydrocarbon content).

REFERENCES

Bertin, E., Clusel, M., 2006. Generalized extreme value statistics and sum of correlated variables. Journal of Physics A: Mathematical and General 39, 7607–7619.

Blyth, C.R., 1972. On Simpson's paradox and the sure-thing principal. Journal of the American Statistical Association 67 (338), 364–366.

Borgos, H.G., Dahl, G.V., Lima, A.L., Sonneland, L., Handwerger, D., Suarez-Rivera, R., 2013. Shale reservoir characterization using 3D Markov field classification. Extended Abstract. In: 75th EAGE Conference & Exhibition, London, UK, June 10–13, 2013.

Carter, G.F., Paul, D.E., 1991. Materials Science and Engineering. ASM International, Materials Park, OH, USA.

Draper, N.R., Smith, H., 1998. Applied Regression Analysis. Wiley-Interscience, Hoboken, NJ USA.

Fisher, R.A., Yates, F., 1963. Statistical Tables for Biological, Agricultural and Medical Research, sixth ed. Oliver and Boyd, Edinburgh. p. 63 (Table VII).

Handwerger, D.A., Sodergren, T., Suarez-Rivera, R., 2012. Scaling in tight gas shales. In: Monograph of the 2011 Warsaw Symposium "The Evolution of the Mental Picture of Tight Shales", pp. 191–210.

Handwerger, D.A., Martinez, C., Castañeda-Aguilar, R., Dahl, G.V., Borgos, H., Ekart, D., Raggio, M.F., Lanusse, I., Di Benedetto, M., Suarez-Rivera, R., November 3–7, 2014. Improved characterization of Vaca Muerta formation reservoir quality in the Neuquén Basin, Argentina, by integrating core, log and seismic data via a combination of unsupervised and supervised multivariate classification. In: Gas IX Congreso de Exploración y Desarrollo de Hidrocarburos, Simposio de Recursos No Convencionales. Instituto Argentino del Petrolero y del, Mendoza, Argentina, pp. 479–497.

Hook, J.R., Nieto, J.A., Kalkomey, C.T., Ellis, D., 1994. Facies and permeability prediction from wireline logs and core – a North Sea case study. In: SPWLA 35th Annual Logging Symposium, June 19–22, 1994.

Huber, P.J., 2011. Data Analysis: What Can Be Learned from the Past 50 Years. Wiley, New Jersey.

Kaminski, M., 2007. Central limit theorem for certain classes of dependent random variables. Theory of Probability and Its Applications 51 (2), 335–342.

Louhichi, S., 2002. Rates of convergence in the CLT for some weakly dependent random variables. Theory of Probability and Its Applications 46 (2), 297–315.

Ma, Y.Z., Wang, H., Stichler, J., Gurpinar, O., Gomez, E., Wang, Y., 2014a. Mixture decompositions and lithofacies clustering from wireline logs. Journal of Applied Geophysics 102, 10–20.

Ma, Y.Z., Moore, W.R., Gomez, E., Luneau, B., Handwerger, D., 2016. Wireline Log Signatures of Organic Matters and Lithofacies Classifications for Shale and Tight Carbonate Reservoirs. In: Ma, Y.Z., Holditch, S.A., Royer, J.-J. (Eds.), Unconventional Oil and Gas Resources Handbook: Evaluation and Development, pp. 151–172.

Ma, Y.Z., et al., April 2014b. Identifying Hydrocarbon Zones in Unconventional Formations by Discerning Simpson's Paradox. Paper SPE 169496 presented at the SPE Western and Rocky Regional Conference.

Mahalanobis, P.C., 1936. On the generalized distance in statistics. Proceedings of the National Institute of Sciences of India 2 (1), 49–55.

Massey, F.J., 1951. The Kolmogorov-Smirnov test for goodness of fit. Journal of the American Statistical Association 46 (253), 68–78.

Passey, Q.R., Bohacs, K.M., Esch, W.L., Klimentidis, R., Sinha, S., 2010. From oil-prone source rock to gas-producing shale reservoir – geologic and petrophysical characterization of unconventional shale-gas reservoir. In: SPE 131350, CPS/SPE International Oil & Gas Conference and Exhibition in China, Beijing, China, June 8–10, 2010.

Rodriguez-Herrera, A.E., Suarez-Rivera, R., Handwerger, D., Herring, S., Stevens, K., Marino, S., Paddock, D., Sonneland, L., Haege, M., 2013. Field-scale geomechanical characterization of the Haynesville shale. In: ARMA 13–678, American Rock Mechanics Association 47th US Rock Mechanics/Geomechanics Symposium, June 23–26, 2013.

Schlumberger, 1991. Log Interpretation Principles/Applications 1989. Schlumberger Wireline & Testing, Sugar Land, TX, USA.

Simpson, E.H., 1951. The interpretation of interaction in contingency tables. Journal of Royal Statistical Society Series B 13 (2), 238–241.

Sondergeld, C.H., Newsham, K.E., Comisky, J.T., Rice, M.C., Rai, C.S., 2010. Petrophysical considerations in evaluating and producing shale gas resources. In: SPE 131768, SPE Unconventional Gas Conference, Pittsburgh, PA USA, February 23–25, 2010.

Suarez-Rivera, R., Handwerger, D., Rodriguez-Herrera, A., Herring, S., Stevens, K., Dahl, G.V., Borgos, H., Marino, S., Paddock, D., 2013. Development of a heterogeneous earth model in unconventional reservoirs, for early assessment of reservoir potential. In: ARMA 13–667, American Rock Mechanics Association 47th US Rock Mechanics/Geomechanics Symposium, June 23–26, 2013.

Waters, G.A., Lewis, R.E., Bentley, D.C., 2011. The effect of mechanical properties anisotropy in the generation of hydraulic fractures in organic shales. In: SPE 146776, SPE Annual Technical Conference and Exhibition, Denver, CO USA, October 30–November 2, 2011.

Wei, Y.N., Holditch, S.A., 2009. Multicomponent advisory system can expedite evaluation of unconventional gas reservoirs. In: SPE 124323, SPE Annual Technical Conference and Exhibition, New Orleans, LA USA, October 4–7, 2009.

Weiss, N., Hassett, M., 1982. Introductory Statistics. Addison-Wesley Publishing Co., Philippines.

Woodhouse, R., 2002. Statistical line-fitting methods for the geosciences: pitfalls and solutions. In: Lovell, M., Parkinson, N. (Eds.), Geological Applications of Well Logs: AAPG Methods in Exploration No. 13, pp. 91–114.

Zhu, Y., Xu, S., Payne, M., Martinez, A., Liu, E., Harris, Ch, Bandyopadhyay, K., 2012. Improved Rock-Physics Model for Shale Gas Reservoirs. SEG Annual Meeting, Las Vegas, NV, USA. http://dx.doi.org/10.1190/segam2012-0927.1.

APPENDICES
APPENDIX 1: PITFALLS IN USING LINEAR REGRESSION

Figure 11.13(a) shows an example of bivariate linear regression. The best-fit linear regression is derived from the data, which shows an R^2 of 0.73. The best-fit equation could be used on subsequent data sets containing the explanatory variable for prediction of the response variable.

Given the high R^2 value of 0.73, many interpreters would consider this to be a good model for predicting the values of Y from input X. However, a better solution could be gained through further analysis of the data. Figure 11.13(b) shows the histogram of the input values on the X-axis. What becomes apparent is that the data are neither approximately Gaussian distributed nor unimodal. We can also see from Fig. 11.13(a) that the data are not symmetric about the regression line (violation of the average residual being zero) and that the residuals are not homoscedastic (Fig. 11.13(c)). These observations should call into question the appropriateness of the regression in Fig. 11.13(a).

While the histogram in Fig. 11.13(b) suggests at least two or three modes, which is even more apparent looking at the residuals in Fig. 11.13(c). Consequently, a better mechanism for predicting the values in Y from X could consist of dividing the data into three groups, as shown in Fig. 11.14, and building three separate linear models. Furthermore, Fig. 11.14(b) shows that indeed each of the three classes isolated in the scatter of Fig. 11.14(a) have more approximately normal distributions than the whole (Fig. 11.13(b)).

Figure 11.14(c) shows the residuals calculated from the three separate models of Fig. 11.14(a). It should be noted that the residuals are much more homoskedastic within each class than for the global model shown in Fig. 11.13(c), and each group of residuals is approximately centered about zero. The spread of absolute residual values is also much less for the three-class model; between −2 and 2

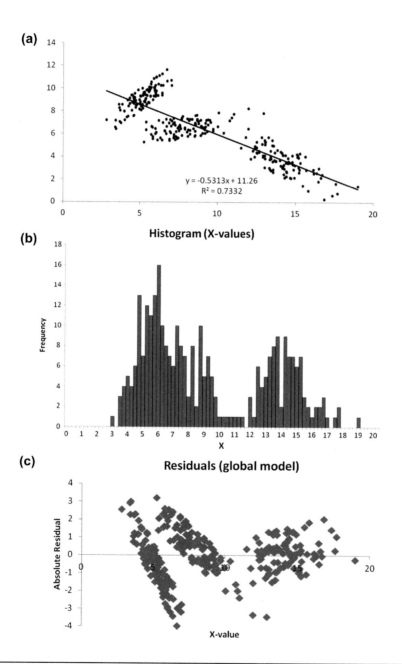

FIGURE 11.13

(a) Linear regression model for some generic inputs. The independent (predictor or explanatory) variable is on the X-axis, and the dependent (response) variable is on the Y-axis. (b) Distribution of the inputs along the X-axis of panel A. (c) Residuals of the global model. Note that the residuals suggest that regression fit to the data in Fig. 11.13 violate each of the assumptions listed in the text for linear regression fits (residuals average zero for any X, the residuals for any X are approximately normally distributed and the residuals are homoscedastic).

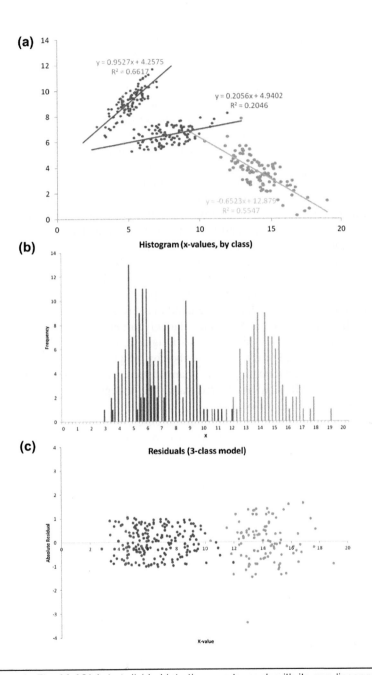

FIGURE 11.14

(a) The same data as in Fig. 11.13(a), but divided into three parts, each with its own linear regression.
(b) Distributions of the inputs along the X-axis for each group isolated from the scatter in Fig. 11.14(a). Note that each of these is much more approximately normally distributed than the whole (see Fig. 11.13(b)).
(c) Absolute residuals of the data set when the three separate models are applied. Note that the residuals here show no trends and are much less than the single global model residuals (see Fig. 11.13(c)). The residuals are also much more homoscedastic within each class than the entire population for the single global model (Fig. 11.13(c)).

(except one bad data point) instead of −4 and 4, and thus the prediction is doing a much better job of predicting the extreme values in the overall data set. Lastly, the grouping bias shown in Fig. 11.13(c) is mitigated.

It should also be noted that the single model (Fig. 11.13(a)) shows inverse correlation between the predictor and response variables, yet when the three groups are isolated (Fig. 11.14(a)), two of them show positive correlation (blue and red) and only one shows an inverse trend (green). This is an example of Simpson's Paradox (Simpson, 1951; Blyth, 1972; Huber, 2011; Ma et al., 2014b). Furthermore, the R^2 of the single model in Fig. 11.13(a) is higher than the R^2 of any of the individual models in Fig. 11.14(a). However, the three separate models are more appropriate predictors than the single model, as the residuals are lower when the three-model approach is used (Fig. 11.14(c)). This poses a quandary to conventional thinking, but also points out the danger in simply relying on R^2 to be the sole arbiter of model quality. The R^2 value describes the percent of the variance in Y explained by the functional dependence on X (Weiss and Hassett, 1982), but this does not necessarily address the question of how absolutely spread that variance in Y is. If the variance about a predicted trend is low, as can be implied by the isolated red group trend (Fig. 11.14(a)), then the difference between the defined trend and a hypothetical trend with higher R^2 (smaller data spread about such a trend) may not be all that significant overall, and residual analysis, instead of purely R^2 could illuminate this. In fact, these two concepts go hand in hand, as the slope of the trend in the red-colored data is quite low; suggesting that there is overall limited variance in Y for the relatively larger spread in X. The absolute residuals for red are between −1 and 1 unit, which suggests a tight fit, especially when we consider that the residuals for the red-class portion of the data described by the global trend spread between −3 and 3, despite the global trend having a much higher R^2 than the red class alone. As a result, another perhaps better question is whether the R^2 supports the red class-specific regression as a legitimate descriptor of the trend in the red-colored data (see below).

Therefore, the single model is technically inappropriate because the data population as a whole violates the assumptions inherent in linear regression modeling. Figure 11.15 shows another, more extreme example of the same concept, with a (weak) Simpson's Reversal. In this obviously bimodal distribution, the (inappropriately applied) best-fit line has a high R^2, but few would argue that this is an appropriate trend. Indeed, the R^2s of the best-fit lines that describe the individual subgroupings suggest both poor predictive value for each trend through low explained variance (R^2) and that each regression fails to describe a valid trend in the data.

In addition to "variance explained," correlation coefficients also show whether or not, for a given number of samples, the trend described by the best-fit line is statistically different from a null result within a given confidence level. In other words, for a reported R^2, one can assess whether the best-fit line that comprises the regression model is a legitimate descriptor of the trend in the data, rather than a product of random chance. Figure 11.16 shows the relationship between the number of sample points used to define a trend, and the R (square root of R^2) value necessary to suggest that the trend is a valid description of the data to the 95th and 99th percentiles, derived from the t-test *(adapted from Fisher and Yates, 1963)*. This is a profound figure, in that it suggests that for few data points, one needs to have a very high R^2 to suggest that the best-fit line is a legitimate model of the data trend, but with many data points, low R^2 can still yield a valid trend. However, this interpretation also requires the assumptions inherent to regression models be met for a particular case.

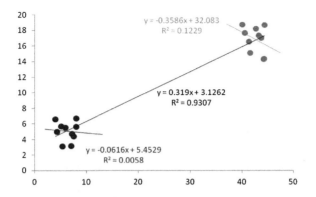

FIGURE 11.15

Regression fits to obviously bimodal data. This illustrates that you can get a very high R^2 to a fit, but the interpretation of that R^2 as being "good" requires an assumption that the data are approximately unimodal and Gaussian, which is clearly not the case here. The R^2 values of the individual modes suggest no correlations intrinsic to each mode, as the Fisher and Yates threshold R^2 (Fig. 11.16) for correlation significant to $>P_{95}$ for 10 data points (the number per class) is 0.4.

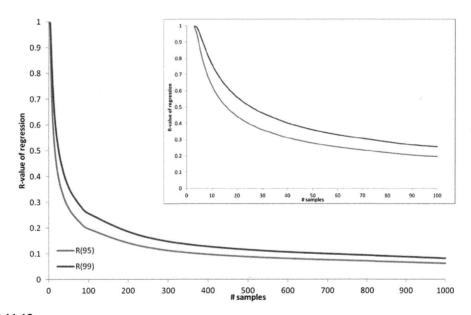

FIGURE 11.16

Relationship between the number of points that define a regression and the R-value that needs to be exceeded in order to claim that the trend is a legitimate descriptor of the data variability to either 95% (blue; gray in print versions) or 99% (red; black in print versions). The inset is an expansion of the main figure, from 0 to 100 samples instead of 0–1000.

Based on Fisher and Yates (1963).

APPENDIX 2: CENTRAL LIMIT THEOREM (CLT)

Let $X_1, X_2, X_3,...X_n$ be a set of n independent and identically distributed random variables having finite values of mean, μ, and variance, σ^2. The central limit theorem states that as the sample size n increases, the distribution of the sample average approaches the normal distribution with the mean μ and variance σ^2/n irrespective of the shape of the original distribution. The classical CLT, however, is only defined on an asymptotic limit with the assumptions of the independent and identically distributed random variables that have a finite variance. Some of these assumptions are sometimes not realistic for natural phenomena. The extension of the CLT to dependent variables and finite sample size (Louhichi, 2002; Bertin and Clusel, 2006; and Kaminski, 2007) provides theoretical supports.

The CLT applies further in that it also states that a random sampling of a normally distributed population will also have a normal distribution with the same $\bar{x}$ and variance as the overall population (Weiss and Hassett, 1982). This implies that if one can establish that the expected property population is normally distributed, then a smaller random sampling of it should represent its mean and variance. The establishment of normality in the property population distribution is assumed through the normality in the distributions of the logs affected by it.

CHAPTER 12

INTEGRATED HYDRAULIC FRACTURE DESIGN AND WELL PERFORMANCE ANALYSIS

Mei Yang[1], Aura Araque-Martinez[1], Chenji Wei[2], Guan Qin[3]

Weatherford International, Houston, TX, USA[1]; PetroChina Coalbed Methane Company Limited, Beijing, China[2]; University of Houston, Houston, TX, USA[3]

A hydraulic fracture design methodology is presented in this chapter, integrating petrophysical, geomechanical, and well performance analysis, and completions optimization. The key components of the presented workflow are basic hydraulic fracture design, dynamic calibration of this design based on performance analysis, and completion optimization. We first present an overview of classical fracture design, including rock mechanics, fracture propagation theory, pump schedule design, and fracture diagnosis. Then, we will discuss the well performance analysis to understand the effective fracture geometry, and the various factors affecting well productivity. Finally we will discuss optimal completion strategies and refrac opportunities. A refrac is currently considered the best practice to mitigate production decline in unconventional reservoirs. Therefore, it is extremely important to select the right candidates for increasing the probability of success for the refrac jobs. A systematic approach to select appropriate refrac candidates is also presented. The integrated hydraulic fracture design aggregates reservoir data from multiple sources to optimize well completion design.

12.1 OVERVIEW OF HYDRAULIC FRACTURE PROPAGATION AND MODELING

This section covers the basic concepts of rock mechanics, fracture design, and optimization. Hydraulic fracture is created as a result of formation pressurization. As fracture fluid is injected into a formation, a differential pressure between the wellbore pressure and the original reservoir pressure is generated. The differential pressure increases as rate increases, which causes additional stress around the wellbore. If this differential pressure increase is significant enough, a fracture will form when the induced stress exceeds the stress needed to break the rock. Proppants are injected into the fracture afterward to keep the fractures open. A path of increased conductivity is created from the reservoir to the wellbore. Fracture treatment monitoring and postfracture analysis are of great importance in understanding the effectiveness of fracture placement.

12.1.1 LINEAR ELASTIC FRACTURE MECHANICS

The stress–strain behavior of rock is quite complex. The most commonly used form to idealize and simplify the stress–strain relationship for rocks is that of linear elasticity (Jaeger et al., 2007). Linear elastic fracture mechanics (Rice, 1968; Ahmed, 1985; Settari and Cleary, 1986; Fung et al., 1987; Rahim and Holditch, 1995) predicts how much stress is required to propagate a fracture. It assumes that linear elastic deformation was followed by brittle fracture, which means there is no energy lost due to plastic deformation or other effects and that all energy in the material is transferred to fracture propagation.

Fracture will stop if the stress (energy) reaches equilibrium; in other words, fracture toughness at the tip equals the stress intensity factor. A number of factors, such as in situ stress contrast, elastic properties, fracture toughness or stress intensity factor, ductility, permeability, and the bonding at the interface, impact whether an adjacent formation will act as a fracture barrier. The stress intensity factor can be calculated as Eq. (12.1) where a crack extends from $-a$ to $+a$ on the y-axis, as shown in Fig. 12.1 (Yang, 2011). The equilibrium height satisfies the condition that the computed stress intensity factors at the vertical tips (top and bottom) are equal to the fracture toughness of the layer. Equation (12.2) should be satisfied at the two fracture tips (Anderson, 1981; van Eekelen, 1982; Warpinski and Teufel, 1987; Clifton and Wang, 1991; Barree and Winterfeld, 1998; Smith and Shlyapobersky, 2000; Jeffrey and Bunger, 2009; Cipolla et al., 2011).

$$K_I = \frac{1}{\sqrt{\pi a}} \int_{-a}^{a} p(y_m) \sqrt{\frac{a+y_m}{a-y_m}} dy_m \qquad (12.1)$$

$$K_I = K_{IC} \qquad (12.2)$$

12.1.2 CLASSICAL FRACTURE PROPAGATION MODELS

For a two-dimensional fracture model, the Perkins-Kern-Nordgren (PKN) (Perkins and Kern, 1961) and Kristonovich-Geertsma-de Klerk (KGD) (Kristianivch-Zheltov, 1955; Geertsma and deKlerk, 1969) are two commonly accepted models. The PKN model (Appendix A.1) uses the vertical plane strain assumption and is more appropriate for long fractures; the KGD model (Appendix A.2) uses the horizontal plane strain and is applicable for short fractures (Fig. 12.2).

12.1.3 PUMP SCHEDULE DESIGN

The pump schedule is designed to deliver the desired fracture geometry according to the following method.

Solve the material balance for injection time, t_i:

$$\frac{q_{i,1wing}}{h_f x_f} t_i - 2C_L \sqrt{t} - 2S_p - w_{avg} = 0 \qquad (12.3)$$

Then, the injected slurry volume can be calculated:

$$V_{i,1wing} = q_i t_i \qquad (12.4)$$

12.1 OVERVIEW OF HYDRAULIC FRACTURE PROPAGATION AND MODELING

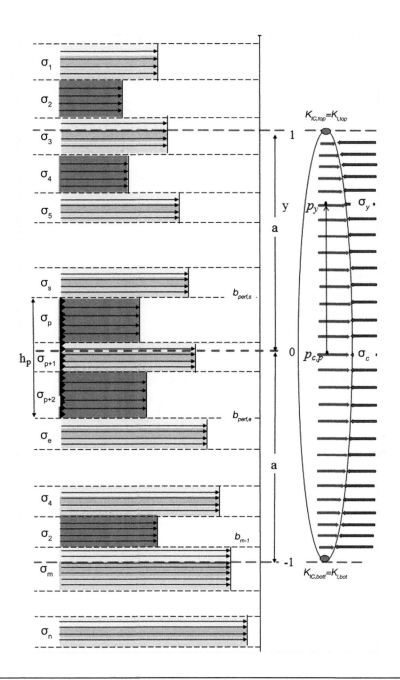

FIGURE 12.1

Fracture height growth in an *n*-layer reservoir and it stops at the stress equilibrium.

Yang et al. (2012).

CHAPTER 12 INTEGRATED HYDRAULIC FRACTURE DESIGN

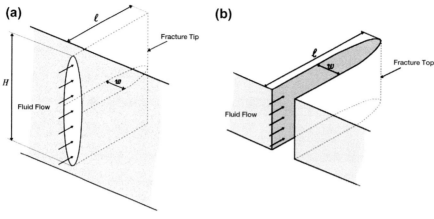

FIGURE 12.2

Fracture models (a) PKN model, (b) KGD model.

Yang (2011).

Therefore, the fluid efficiency is:

$$\eta = \frac{h_f x_f w_{avg}}{V_{i,1wing}} \tag{12.5}$$

The pad injection time will be determined according to Nolte (1986).

$$c = c_e \left(\frac{t - t_{pad}}{t_i - t_{pad}} \right)^{\varepsilon} \tag{12.6}$$

where $t_{pad} = \varepsilon t_i$, $\varepsilon = \frac{1-\eta}{1+\eta}$ and $c_e = \frac{m_{i,1wing}}{\eta V_{i,1wing}}$

The concentration for dirty fluid which, after clean fluid, went through the blender is:

$$c_{added} = \frac{c}{1 - \left(\frac{c}{\rho_p}\right)} \tag{12.7}$$

We can use material balance to relate the input proppant concentration during the injection time to determine the concentration of proppant in the fracture that exists at the end of the pump schedule. The end of job proppant concentration can be calculated under the assumption of uniform concentration at end of pumping. Or a more "accurate" end of job proppant concentration value can be calculated by including advection effects to model various possible proppant transport mechanisms and proppant settling.

With either a fixed aspect ratio or net pressure to stress contrast and fracture toughness, the fracture height, proppant placement efficiency should be recalculated until convergence.

12.1.4 FRACTURE DIAGNOSIS AND FRACTURE GEOMETRY CONSTRAINTS

Fracture diagnosis tests are common practices to have a fairly good understanding of the reservoir stress field, leaking off (Economides and Martin, 2007).

12.1 OVERVIEW OF HYDRAULIC FRACTURE PROPAGATION AND MODELING

12.1.4.1 Nolte–Smith Analysis
The Nolte–Smith test uses a log–log plot of net pressure versus pump time to predict fracture geometry. A negative slope usually indicates unconfined height growth, resulting in "penny" shape fracture geometry. A shallow, positive slope indicates fair to good fracture height confinement, and unrestricted fracture extension. A steep, positive slope indicates that something has restricted fracture half-length growth, essentially a tip screen-out, which does not necessarily occur at the fracture tip, but can be near the wellbore. Operationally, screen-out can lead to a condition where continued injection of fluid inside the fracture requires pressures in excess of the safe limitations of the wellbore or wellhead equipment and results in wellbore cleaning before resumption of operations. Finally, a flat slope indicates that fracture extension has been drastically slowed, which could be the start of extensive height growth, or an acceleration of fluid loss due to the opening of natural fractures.

12.1.4.2 Step Rate Tests
A fracture step rate test is used to determine the maximum injection rate possible without fracturing the reservoir. The test is conducted by injecting into the formation at a series of increasing rates, allowing each to stabilize, and noting the stabilized injection pressure for each rate. When an increase in rate does not result in a proportionate increase in injection pressure, fracture pressure has been reached. A plot of injection pressure versus injection rate is then made to identify the fracture pressure.

12.1.4.3 Minifracs
Minifrac analysis is designed to determine initial stresses, minimum in situ stress, maximum in situ stress, and the leak-off coefficient. The fracture fluid is injected into the well and then pressurized to create a fracture in the reservoir. To initiate the crack in the reservoir, the downhole pressure must overcome the breakdown pressure (the peak of the first cycle). After the crack is created, the downhole pressure decreases while fracturing continues to propagate into the reservoir. The fracture closure pressure can be evaluated after injection is stopped. The observation of the closure pressure is shown in Fig. 12.3. The second cycle almost seems identical to the first. However, it requires lower downhole pressure to reopen the fracture (reopening pressure, p_r) in the reservoir than it does for fracture creation ($p_b > p_r$).

12.1.4.4 Microseismic
Microseismic observation is integrated to optimize fracture design and constrain the fracture modeling results. Microseismic can help understand the fracture propagation, fracture geometries

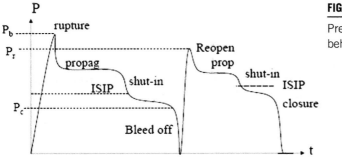

FIGURE 12.3

Pressure profile of fracture propagation behavior.

Economides et al. (2002).

(including length, height, and width), complexity, interaction between wells, stages, and formations. It is also a tool to identify the influence of preexisting fracture, fault, in situ stress, etc. that can be used to calibrate the fracture model, and optimize fracture design. The fracture design can be adjusted after each stage, well, or cluster wells. Real -time microseismic monitoring data can be utilized to change fracture pump schedule during a stimulation treatment.

12.2 WELL PERFORMANCE ANALYSIS

Well performance analysis (WPA) is a key component for reservoir analysis and characterization. It is used to calibrate the reservoir properties and also to understand the impact of changes in these properties on the production forecast, as well as defining hypotheses to explain the well's behavior. Performing WPA on several wells in a particular area of interest provides a better understanding of the area and allows identifying not only similarities but also the contrast in well performances.

In general, WPA will answer the following questions:

- How does the well perform and what is the individual estimated ultimate recovery?
- What are the lessons learned from the particular studied well?
- What are the uncertainties governing the ultimate recovery and what is their impact?
- Is there any abnormal behavior associated with the well performance?
- Is there any room for improvement in the future wells?
- Is there any best practice for a given study area/play?
- What do the special tests, such as pressure transient analysis, minifrac, microseismic, and Production Logging Tool (PLT), indicate about the well performance?

The key inputs are mainly reservoir data (i.e., geology, petrophysics, geophysics, and geomechanics), and wells data (i.e., rates and pressure history, fluid properties, completion, stimulation, and drilling reports).

The WPA workflow (Fig. 12.4) starts with data Quality Assessment/ Quality Control (QA/QC). The first step is performed to identify anomalous well behavior, data inconsistency, and data pedigree. The data has been broadly classified into four main groups: geology and petrophysics, geomechanics, fluid properties, and well data.

- Geological and petrophysical modeling is used to describe the rock quality (brittleness) and reservoir quality (original hydrocarbons in-place (OHIP)).
- Fluid properties are required to explain the phase behavior and convert surface fluid rates into reservoir conditions.
- Well data includes drilling, completion, and performance data, which is key to providing hypotheses for a well's behavior: actual historical data will be calibrated against modeled data during the well performance analysis.
- Geomechanical/fracture modeling is performed to define the completion characteristics (fracture geometry) as input to performance modeling.

Depending on the nature of the resource play and fluid type, the available tools to analyze the well performance will vary. The most commonly used tools are: rate transient analysis, decline curves, and analytical and numerical modeling.

12.2 WELL PERFORMANCE ANALYSIS

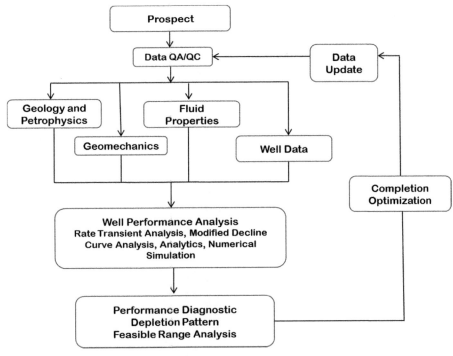

FIGURE 12.4

Well performance analysis workflow.

Weatherford Internal Report.

Each approach has certain limitations on the assumptions, time required and output. Hence, it is left up to the end users to use their judgment in selecting a particular tool to analyze the well data.

Once the well history is calibrated, a performance diagnostic is provided with the most likely explanation for the well's behavior, along with the stimulated reservoir volume, OHIP, and feasible range analysis for permeability and fracture geometry.

Upscaling the individual well performance analysis to a larger set of wells provides an overall trend, if any. Mapping of those parameters from each well is also a useful tool for a broader understanding of the study area.

12.2.1 SOME MECHANISMS AFFECTING WELL PERFORMANCE

This chapter focuses on three mechanisms affecting well productivity in unconventional plays:

- Pressure-dependent permeability
- Fracture conductivity degradation with pressure
- Fracture geometry complexity

12.2.1.1 Pressure-Dependent Permeability

All sensitivity results from synthetic models show that the higher the compaction exponent (alpha), the greater the deviation from the original line in the diagnostic plots as shown in Fig. 12.5. This deviation shows up in the plot as a departure from the linear flow even though the well is not showing internal depletion or boundary dominated flow. As a result, a misinterpretation of the transition time is likely to occur when permeability is dependent upon pressure (Araque-Martinez, 2010).

12.2.1.2 Pressure-Dependent Fracture Conductivity

Fracture conductivity degradation shows a dramatic, instantaneous reduction in productivity (blue curve in Fig. 12.6). Results suggest that fracture conductivity degradation is a local phenomenon (Araque-Martinez, 2010).

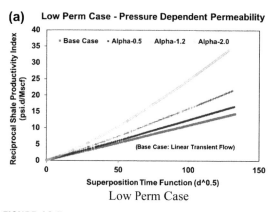
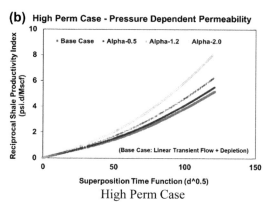

FIGURE 12.5

Pressure-dependent permeability.

Araque-Martinez (2010).

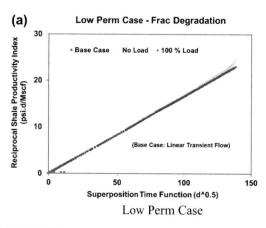
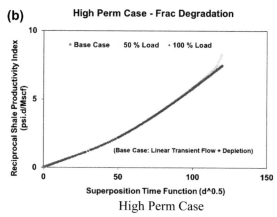

FIGURE 12.6

Fracture conductivity degradation.

Araque-Martinez (2010).

12.2.1.3 Fracture Complexity

The complexity was added by just reducing the frac spacing and preserving the frac area. These models include four fractures with 25% frac length and 25% frac spacing compared to the base case. Results show two consecutive linear flow periods that are more evident in the high permeability case (Fig. 12.7). The first linear flow period is associated to microfractures or complex network and the second one is associated to the created hydraulic fractures (Araque-Martinez, 2010).

12.2.1.4 Comparison of Affecting Mechanisms

Fracture conductivity degradation does not greatly affect the estimated ultimate recovery (EUR). However, a large reduction in 30-year recovery resulted when compaction or complexity is present in both the low and high permeability cases (Fig. 12.8) (Araque-Martinez, 2010).

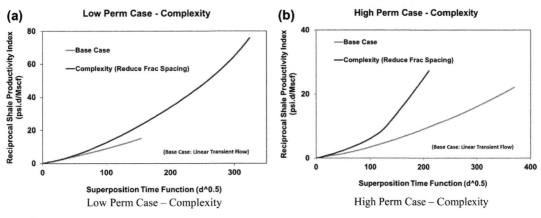

FIGURE 12.7

Fracture complexity.

Araque-Martinez (2010).

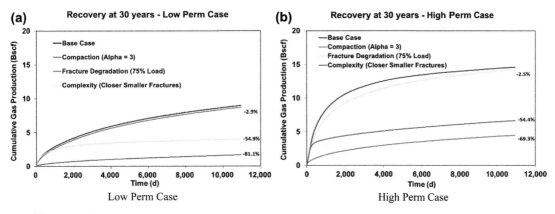

FIGURE 12.8

30-year recovery comparison.

Araque-Martinez (2010).

12.2.2 TRANSIENT LINEAR FLOW IN STRATIFIED RESERVOIRS

Unconventional plays are characterized by extremely low permeability, where transient effects are more predominant. In fact, in most of these unconventional wells, transient linear flow regime could last for long periods of time (>1 year) and capturing those transient effects is very important.

For a hydraulically fractured vertical well, Cinco-Ley and Samaniego (1981) identified four different flow regimes as shown in Fig. 12.9.

A commonly used diagnostic plot to identify the well's flow regime is the so-called log–log plot, (Fig. 12.10). The productivity index is presented as a function of time, showing three regimes occurring during the early flow: bilinear, linear, and boundary dominated flow. Each of them shows a distinctive signature that can be differentiated from the slope of the lines.

Unconventional plays are very heterogeneous in nature; therefore, the completed wells are expected to produce from different flow units within the same target zone. For those cases where one or more of the flow units shows a strong difference in permeability, the drainage pattern will be significantly different in the vertical direction. This is mainly because each flow unit will have a different transient flow behavior. Under these conditions, upscaling the permeability to reproduce the flow behavior for all flow units becomes difficult and common averaging methods do not comply. In fact, the usual "kh" permeability average does not reproduce the multilayer case results when high perm streaks are present. This problem has been previously studied for pressure transient analysis. Actually, several authors (Almehaideb, 1996; Fetckovich, 1990; Frantz, 1992; Gao, 1987; Lefkovits, 1961 and Lolon et al., 2008) have studied and proposed solutions for the performance of bounded stratified reservoirs to apply it to well test analysis in multilayer reservoirs. In general, they have shown the importance of partial depletion, and its effect on ultimate recovery. They have also suggested that,

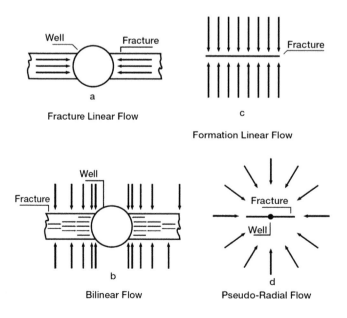

FIGURE 12.9

Flow regimes for a vertical fractured well.

Weatherford Internal Report.

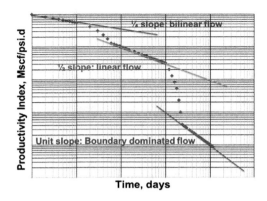

FIGURE 12.10

Log–log diagnostic plot.

in the presence of partial depletion, the use of a single average layer from the well test analysis can lead to optimistic forecasts.

Here, we present an average permeability during linear transient flow and quantify the error by comparing the results from multilayer and single average layer models for different permeability ratios (Araque-Martinez, 2011).

Derivation for the transient permeability average is described below:

12.2.2.1 Transient Permeability Average

Assuming an infinite-acting reservoir with constant pressure at the wellbore, we have the following early time approximation for the diffusivity equation in dimensionless form:

$$q_D = \frac{1}{\sqrt{\pi t_D}} \qquad (12.8)$$

For radial flow (low pressure approximation) in a vertical well:

$$q_D = \frac{p_{sc} T \mu_g z}{\pi k h T_{sc} \Delta \psi} q_{sc} \qquad (12.9)$$

and

$$t_D = \frac{kt}{\phi \mu C_t r_w^2} \qquad (12.10)$$

The total rate for the two-layer case with commingled production without cross-flow:

$$q_{sc}^T = q_{sc}^1 + q_{sc}^2 \qquad (12.11)$$

Combining Eq. (12.9) and Eq. (12.11) and assuming constant gas properties, Eq. (12.11) becomes

$$q_D^T = \frac{k_1 h_1}{\overline{kh}} q_D^1 + \frac{k_2 h_2}{\overline{kh}} q_D^2 \qquad (12.12)$$

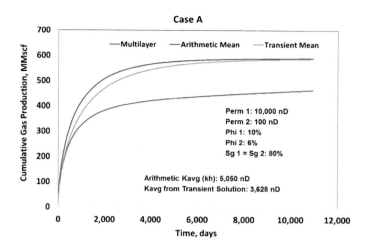

FIGURE 12.11

Cumulative gas production comparison—vertical well—constant pressure-case A.

Araque-Martinez (2011).

where the dimensionless rate (q_D) superscripts refer to total (T), layer 1 (1), and layer 2 (2).

Combining Eqs (12.8) and (12.10) into Eq. (12.12) and rearranging terms:

$$\bar{k} = \frac{\phi_1 \phi_2 \left(k_1 h_1 \sqrt{\frac{k_2}{\phi_2}} + k_2 h_2 \sqrt{\frac{k_1}{\phi_1}}\right)^2}{\bar{\phi}\, \bar{h}^2 k_1 k_2} \tag{12.13}$$

which is the transient permeability average. Eq. (12.13) shows that during transient flow, productivity depends not only on the flow capacity, but also on the conductivity–storability ratio of each layer (Araque-Martinez, 2011). In case of n layers, Eq. (12.13) becomes:

$$\bar{k} = \frac{1}{\bar{\phi}\, \bar{h}^2}\left(\sum_{i=1}^{n} \frac{k_i h_i}{\sqrt{\frac{k_i}{\phi_i}}}\right)^2 \tag{12.14}$$

In the following, we will describe how to evaluate the effect of different layers' permeability with constant fracture geometries. Different scenarios were evaluated varying the permeability ratio in a synthetic two-layer model. Here we only present the most relevant cases, using the so-called transient permeability average (Araque-Martinez, 2010, 2011) as opposed to the commonly used "kh", to generate average permeability in the single-layer model.

Cases A and B have permeability ratios of 100, with a lower permeability in case B. Therefore, the transient flow period is longer in case B than in case A. As a consequence, the single-layer model for case B with the transient permeability average matches the multilayer model results for longer time (Fig. 12.12). On the contrary, the single-layer case A with transient average permeability greatly overpredicts the recovery after 260 days when the linear flow period ends (Fig. 12.11). Finally, case C has a

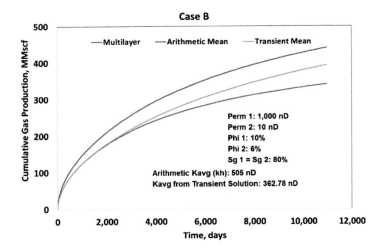

FIGURE 12.12

Cumulative gas production comparison—vertical well—constant pressure-case B.

Araque-Martinez (2011).

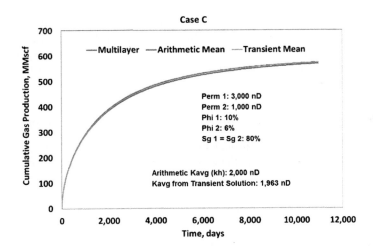

FIGURE 12.13

Cumulative gas production comparison—vertical well—constant pressure-case C.

Araque-Martinez (2011).

permeability ratio of 3 with similar permeability by layers. In this case, the transient and arithmetic average permeability are very similar to each other and to the individual layer's permeability; therefore, both single-layer models match well with the multilayer model results as no major partial depletion is observed (Fig. 12.13).

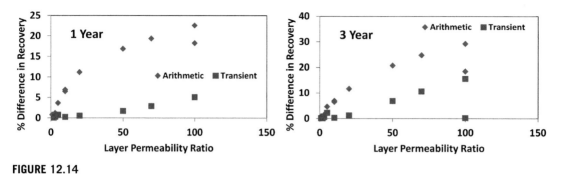

FIGURE 12.14

Difference in 1 and 3-year recovery from multilayer case—vertical well.

Araque-Martinez (2011).

Figure 12.14 presents the percentage difference between the multilayer and single average layer models results versus perm ratio at 1 and 3 years of production, when the economics can be more impacted by the wells' productivity. They show that the differences increase as the permeability ratio increases and also as time increases. However, using the transient average permeability decreases the error by about 15% during the first year and by about 13% at 3 years of production. For permeability ratios up to a value of 5, there is not much difference between multilayer and single-layer results using the arithmetic or transient average. For permeability ratios greater than 5 showing substantial partial depletion, the errors greatly increase but in all cases, using the transient average permeability reduces the errors. This is because the partial depletion in the multi-layer case cannot be reproduced in the average layer model.

12.2.3 UNCERTAINTY ANALYSIS DURING TRANSIENT LINEAR FLOW

As mentioned previously, the linear transient flow might last for a long time (>1 year) in unconventional wells. During this flow regime there is great uncertainty in the flow capacity of the wells since the only thing we can determine from performance data is the product of fracture area and the square root of permeability, $A_f \sqrt{k}$. In other words, the solution is nonunique and there are different possible combinations of fracture area and reservoir permeability that will calibrate the historical performance data. Despite the nonuniqueness the maximum value of permeability can be determined by assuming that the transition time from linear to depletion flow would occur at the end of history or beyond that point in time. Different values of permeability and its corresponding fracture area are tested in the flow model to check for the solution that diverts from historical behavior. Figure 12.15 shows an example from a field case, in which, by testing different fracture areas and permeability calculations, the upper bound permeability range and lower bound fracture half-length was determined by means of solution diversion from historical calibration. Those nonpossible solutions showed that depletion flow had occurred and the well was in linear transient flow regime. Therefore, during transient linear flow, it is crucial to include ranges of possible values for specific parameters, such as: reservoir permeability and fracture area that will lead to possible EUR ranges for economic analysis. As the asset is developed and more performance data is added, the ranges of such properties will be

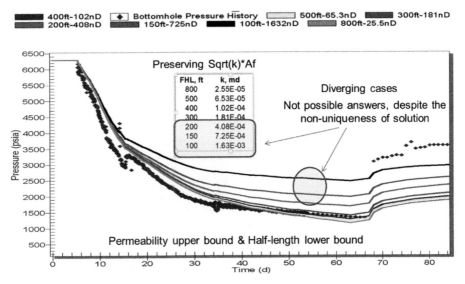

FIGURE 12.15

Example for permeability upper bound value during linear transient flow.

Weatherford Internal Report.

reduced, adding more confidence to the estimated forecasts. In the example case, EUR was calculated using permeabilities between 1.8×10^{-4} and 2.55×10^{-5} md.

12.3 INTEGRATED HYDRAULIC FRACTURE DESIGN WORKFLOW

This section describes the fracture design workflow to either optimize future hydraulic fracture designs or refracture treatments. Although both workflows are very similar, the refrac design includes a key step to select the best candidates to increase the probability of the refrac success. These workflows integrate dynamic and static data into an optimal completion design.

12.3.1 HYDRAULIC FRACTURE DESIGN OPTIMIZATION

The following integrated hydraulic fracture design workflow (described in this section shown in Fig. 12.16) provides intelligence for completion optimization of future wells after investigation into fracture design, completion operation, and fracture characteristics (Yang et al., 2014).

As shown in Fig. 12.16, the framework begins with the reservoir characterization step based on logs and core/cuttings analysis and interpretation. This is followed by building the preliminary geomechanical model. Next, the fracture model is compared with microseismic data if available. Finally, the constrained fracture model is calibrated against performance data to determine the effective fracture area. The effective fracture geometry is determined based on well production calibration. Wellbore and completion problems could be diagnosed in this analysis, including damage of fractures, fluid behavior, and well interference.

376 CHAPTER 12 INTEGRATED HYDRAULIC FRACTURE DESIGN

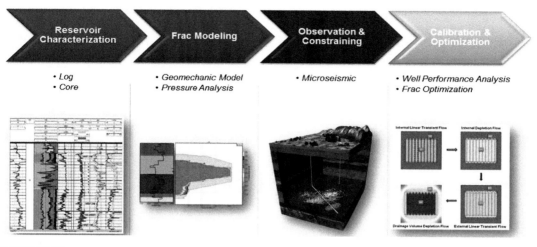

FIGURE 12.16

Integrated hydraulic fracture design workflow.

Yang et al. (2014).

It is a systematic approach to optimize single well design as well as field development (multiple wells, pad drilling, etc.).

12.3.2 REFRACTURE OPTIMIZATION

Refrac is a remedial production operation often done either because the original fracturing failed to contribute any significant amount of flow, or the initial completion's performance has degraded over time below operationally or economically acceptable limits or significant unfractured pay exists in the well. Any fracture that is not optimum with respect to the formation requirements potentially is a refracture candidate. Those factors include inadequate or inappropriate fracture design, low on-site fracture job efficiency and inappropriate fracture fluid selection. Refrac well candidate selection plays a major part in a successful refracturing treatment. To evaluate the potential of refracturing, we should examine factors including reservoir potential, the original fracturing completion efficiency, production history, and well issues (Araque-Martinez and Boulis, 2014).

Figure 12.17 presents a description of a refrac workflow, which spans across the complete lifecycle of a refracture treatment.

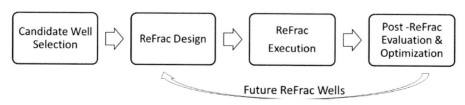

FIGURE 12.17

Refrac workflow.

12.3.2.1 Candidate Selection

Refrac candidate selection is a key part for the refrac workflow. It starts with data collection and data QA/QC. The data required to perform the analysis consists of geological, petrophysical and geomechanical data, drilling and completion reports, postfrac report, production history, and well historical events report. All data needs to be reviewed and validated. This process is laborious but very crucial. Once the data is validated, the screening process starts. This is performed in two levels of diagnostics following a funnel-type structure.

Preliminary Diagnostics

A large number of fractured wells are initially checked for operational conditions (cementing and casing integrity), as well as a quick performance evaluation through diagnostic plots (linear flow and log–log plots). This initial stage is the "basic diagnostic" and it is mainly to:

- Assure that the wells selected for refracturing are in good condition so the job operations will not compromise well integrity and/or stages isolation.
- Provide a qualitative assessment for potential productivity increase

Advanced Diagnostics

A much more comprehensive selection is performed on a smaller number of wells to evaluate undepleted reserves and initial completion effectiveness as follows:

Formation Evaluation. This first step is to determine the rock and reservoir quality by means of formation brittleness and hydrocarbon content respectively. Petrophysical and geomechanical analysis are performed to evaluate parameters such as: Total Organic Carbon (TOC), porosity, water saturation, brittleness index, Young's modulus, Poisson's ratio, and stress profile.

Hydraulic Fracture Modeling. A postfrac report and observations/constraints are applied to determine fracture geometry. This modeling step iterates until the fracture geometry falls into constraint data ranges. Constraining data includes: microseismic, anisotropy from dipole sonic logs, PLT, and/or tracer data. The purpose of this step is to understand the current fracture placement and fracture geometry.

Performance Analysis. With petrophysical interpretation, well historical events analysis and initial fracture geometry from previous steps, performance data is evaluated to calibrate pressure and production profiles. This step integrates dynamic data to determine the possible ranges for effective fracture geometry and reservoir permeability as discussed in Section 12.2.3.

Refrac Candidates

The refrac candidates are selected based on the results from the advanced diagnostic. Parameters such as: initial completion effectiveness, reservoir and rock quality, current completion characteristics, as well as well historical events are combined and normalized to generate a well index.

Well Index has two dimensionless components, reservoir index and completion improvement index. Only the well with good reservoir potential (reservoir index) and good completion improvement margin will be picked when all wells are compared, as shown in Fig. 12.18.

12.3.2.2 Refrac Design

Once the refrac wells are identified, those wells are subject to a sensitivity analysis to evaluate different fracture treatments and their respective production profile. The final refrac design will be determined

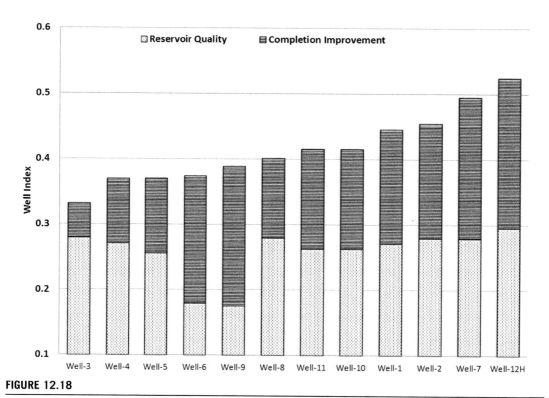

FIGURE 12.18

Example for refrac candidates selection.

based on net present value (NPV) analysis, including the operational and logistics practices. This design will include: pumping schedule, fluid type, initiation point, proppant type and proppant concentration for propped fractures, and/or acid/gels type and concentration for any other frac type.

12.3.2.3 Refrac Execution

Isolation of existing perforation clusters is critical and working inside existing tubulars provides some challenges. Mechanical or chemical isolation tools are available to pick. Completion options include coil tubing frac straddle, inner string recompletion, ball drop sleeve and packer, coil shift sleeve, and reelfrac. Chemical diverting balls have become interesting due to their lower cost and lower complicity as mechanical isolation tools. The biodegradable balls are programmed to cement the undesired perforations and dissolve after few days. All options are evaluated from cost, operational complexity, and effectiveness to ensure the success of refrac treatment. Then optimal refrac pump schedule will be executed. Well will be ready to produce after flow-back and cleanup.

12.3.2.4 Evaluation and Calibration

Production rates are monitored after the wells are refractured. Performance analysis will determine the flow regime, stimulated fracture area, and the effectiveness of the refrac treatment.

If new data (logs, microseismic, PLT, tracers) is acquired during the intervention work, all models are updated and recalibrated. This process will rely on all learnings from each refrac job executed and well performance analyzed to improve the reservoir and rock understanding to reduce uncertainty. The uncertainty reduction is the key to place better wells with better completions and reduce costs.

The concept for refracture optimization is that of "dynamic fracture modeling into dynamic completion designs" and it will be based on advanced surveillance, modeling, and optimization technologies.

12.4 CONCLUSION

The integrated hydraulic fracture design workflow provides thorough investigation into hydraulic fracture modeling, well performance and dynamic constraints for unconventional reservoirs. The hydraulic fracture model provides preliminary information, such as fracture geometry and proppant concentration distribution. Then this model will be constrained by results from Nolte–Smith analysis, minifrac and step rate test, and microseimic. Afterward, the outcomes from well performance analysis are essential to determine the effective fracture area and to optimize completion design. This chapter discusses key factors that affect the interpretation of well performance as well. A transient permeability average technique is recommended to avoid optimistic forecasts. Also the chapter introduces refracturing as a technique for production decline remediation, where refracture candidates selection is the key to success. It evaluates reservoir potential, well completion improvement margin, and well production history. The final refracture design applies operational constraints to achieve optimal completion design with highest NPV. This integrated fracture design workflow is a systematic approach to optimal well(s) completion design.

NOMENCLATURE

asp Fracture aspect ratio
A Reservoir drainage area, L^2, acre
A_f Fracture surface area, L^2, ft^2
b Layer's dimensionless location
$b_{perf,s}$ Perforation start layer's dimensionless location
$b_{perf,e}$ Perforation end layer's dimensionless location
c Proppant concentration, m/L^3, ppg
c_e Proppant concentration at the end of the job, m/L^3, ppg
c_{added} Added proppant concentration, m/L^3, ppga
C_{fD} Dimensionless fracture conductivity
C_L Leak-off coefficient, L/t$^{0.5}$, ft/min$^{0.5}$
d true vertical depth
e end
E Young's modulus, m/Lt^2, psi
E' Plane strain modulus, m/Lt^2, psi
h_f Fracture height, L, ft
h_n Thickness of net pay, L, ft
h_p Thickness of perforation interval, L, ft

Δh_d Fracture growth into lower bounding formation, L, ft
Δh_u Fracture growth into upper bounding formation, L, ft
I_x Penetration ratio
J Well productivity index, $L^4 t^2/m$, bbl/psi
J_D Well dimensionless productivity index
k Reservoir permeability, L^2, md
k_{00} Pressure at center of crack, m/Lt^2, psi
k_1 Hydrostatic gradient, $m/L^2 t^2$, psi/ft
k_f Propped fracture permeability, L^2, md
K Rheology consistency index, m/Lt^2, lbf s^{npr}/ft^2
K_I Stress intensity for opening crack, $m/L^{0.5}t^2$, psi-in$^{0.5}$
$K_{I,bottom}$ Stress intensity at bottom tip of crack, $m/L^{0.5}t^2$, psi-in$^{0.5}$
$K_{I,top}$ Stress intensity at top tip of crack, $m/L^{0.5}t^2$, psi-in$^{0.5}$
K_{IC} Fracture toughness, $m/L^{0.5}t^2$, psi-in$^{0.5}$
K_{IC2} Fracture toughness of upper layer, $m/L^{0.5}t^2$, psi-in$^{0.5}$
K_{IC3} Fracture toughness of lower layer, $m/L^{0.5}t^2$, psi-in$^{0.5}$
K' Modulus of cohesion, $m/L^{0.5}t^2$, psi-in$^{0.5}$
M_{prop} Proppant mass, m, lbm
$M_{prop,stage}$ Proppant mass required for each stage, m, lbm
n Rheology flow behavior index
N_{prop} Proppant number
Δp Pressure difference, m/Lt^2, psi
p_b Breakdown pressure or rupture pressure, m/Lt^2, psi
p_c Fracture closure pressure, m/Lt^2, psi
$p_{c,p}$ Pressure at center of perforation, m/Lt^2, psi
$p_{c,y}$ Pressure at any location y, m/Lt^2, psi
p_r Fracture reopening pressure, m/Lt^2, psi
p_{net} Net pressure at center of perforation, m/Lt^2, psi
p_{nw} Net pressure at center of crack, m/Lt^2, psi
$p_n(x)$ Net pressure at any location in x-direction, m/Lt^2, psi
$p_n(y)$ Net pressure at any location in y-direction, m/Lt^2, psi
q_i Slurry injection rate for one-wing, L^3/t, bbl/min
q_p Production rate, L^3/t, bbl/min
r_e Reservoir drainage radius, L, ft
S_f Fracture stiffness, $m/L^2 t^2$, psi/in
S_p Spurt loss coefficient, L, ft
t_e Pumping time, t, min
t_{pad} Padding time, t, min
T_0 Tensile strength, m/Lt^2, psi
u_{avg} Average velocity of slurry in fracture, L/t, ft/s
V_f Fracture volume, L^3, ft^3
V_i Total slurry injection volume, L^3, ft^3
V_{pad} Padding volume, L^3, gal
V_{prop} Proppant volume, L^3, ft^3
V_{res} Reservoir volume, L^3, ft^3
V_{stage} Liquid volume required for each stage, L^3, gal

w Propped fracture width, L, in
$\bar{w}$ Average hydraulic fracture width, L, in
$w_0(x)$ Max. hydraulic fracture width at any location, L, in
$w_{w,0}$ Max. hydraulic fracture width at wellbore, L, in
x_e Reservoir length, L, ft
x_f Fracture half-length, L, ft
y Dimensionless vertical position
y_d Dimensionless vertical position of bottom perforation
y_u Dimensionless vertical position of top perforation

Greek
γ Shape factor
γ_w Surface energy of fracture, mL/t², psi-ft²
ε Exponent of the proppant concentration curve
ϵ Strain
κ Nolte's function at $\Delta t = 0$
η Slurry efficiency
η_0 Ratio of fracture volume in net pay to total fracture volume
ϕ_p Fracture packed porosity
ρ_p Proppant density, m/L³, lbm/ft³
σ Normal stress, m/Lt², psi
$\sigma(y)$ Normal stress at any location in y-direction, m/Lt², psi
σ_h Minimum horizontal in-situ stress, m/Lt², psi
σ_H Maximum horizontal in-situ stress, m/Lt², psi
$\Delta\sigma_{avg}$ Average stress difference, m/Lt², psi
$\Delta\sigma_d$ Stress differential of reservoir and lower formation, m/Lt², psi
$\Delta\sigma_u$ Stress differential of reservoir and upper formation, m/Lt², psi
τ Shear stress, m/Lt², psi
μ Viscosity, m/Lt, cp
μ_e Equivalent Newtonian viscosity, m/Lt, cp
μ_f Friction coefficient, L, in
ν Poisson's ratio

REFERENCES

Almehaideb, R.A., October 1996. Application of an Integrated Single Well Model to Drawdown and Buildup Analysis of Production from Commingled Zones. SPE 36987.

Ahmed, U., 1985. Hydraulic Fracture Treatment Design of Wells with Multiple Zones. SPE paper 13857.

Anderson, G.D., 1981. Effects of Friction on Hydraulic Fracture Growth Near Unbounded Interfaces in Rocks. http://dx.doi.org/10.2118/8437-PA. SPE 8347-PA.

Araque-Martinez, A., December 2010. Mechanistic Study on Haynesville Shale. White Paper, Object Reservoir.

Araque-Martinez, A., March 2011. Single Average Layer Investigations and its Effect on Ultimate Recovery When High Perm Streaks Are Present. White Paper, Object Reservoir.

Araque-Martinez, A., Boulis, A., November 2014. An innovative approach for refracturing optimization in shale and tight reservoirs. In: Presented at CONEXPLO 2014, Mendoza, Argentina.

Barree, R.D., Winterfeld, P.H., 1998. Effects of Shear Planes and Interfacial Slippage on Fracture Growth and Treating Pressure. SPE 48926.

Cinco-Ley, H., Samaniego, V.F., 1981. Transient pressure analysis on fractured wells. Journal of Petroleum Technology 33 (9), 1749–1766.

Cipolla, C., Maxwell, S., Mack, M., Downie, R., 2011. A Practical Guide to Interpreting Microseismic Measurements, SPE 144067.

Clifton, R.J., Wang, J.J., 1991. Modeling of Poroelastic Effects in Hydraulic Fracturing. SPE 21871.

Economides, M.J., Martin, T., 2007. Modern Fracturing, pp. 20–425.

Economides, M.J., Oligney, R., Valko, P.P., 2002. Unified Fracture Design: Bridging the Gap between Theory and Practice. Orsa Press, Alvin Texas, pp. 23–129.

van Eekelen, H.A.M., 1982. Hydraulic Fracture Geometry: Fracture Containment in Layered Formations. SPE.

Fetkovich, M.J., et al., September 1990. Depletion performance of layered reservoirs without crossflow. SPE Formation Evaluation 310–318.

Frantz, J.H., et al., 1992. Using a Multi-layer Reservoir Model to Describe a Hydraulically Fractured, Low-permeability Shale Reservoir. SPE 24885.

Fung, R.L., Vilajakumar, S., Cormack, D.E., 1987. Calculation of vertical fracture containment in layered formations. SPE Formation Evaluation 2 (4), 518–523.

Gao, C., March 1987. Determination of parameters for individual layers in multi-layer reservoirs by transient well tests. SPE Formation Evaluation 43–65.

Geertsma, J., de Klerk, F., 1969. A rapid method of predicting width and extent of hydraulically induced fractures. Journal Petrol Technology 21, 1571–1581.

Jaeger, J.C., Cook, N.G.W., Zimmerman, R., 2007. Fundamentals of Rock Mechanics, fourth ed. 106–144.

Jeffrey, R.G., Bunger, A.P., 2009. A detailed comparison of experimental and numerical data on hydraulic fracture height growth through stress contrasts. SPE Journal 14 (3), 413–422.

Khristianovich, S.A., Zheltov, V.P., 1955. Formation of vertical fractures by means of highly viscous liquid. In: Proc. 4-th World Petroleum Congress, Rome, pp. 579–586.

Lefkovits, H.C., et al., March 1961. A study of the behavior of bounded reservoirs composed of stratified layers. Society of Petroleum Engineers Journal 43–58.

Lolon, E.P., Blasingame, T.A., June 2008. New Semi-analytical Solutions for Multi-Layer Reservoirs. SPE 114946.

Nolte, K.G., 1986. Determination of Proppant and Fluid Schedules from Fracturing Pressure Decline. http://dx.doi.org/10.2118/13278-PA. SPE 13278-PA.

Perkins, T.K., Kern, L.R., 1961. Width of hydraulic fractures. JPT 937(49). Trans. AIME 222, 937–949.

Rahim, Z., Holditch, S.A., 1995. Using a three-dimensional concept in a two-dimensional model to predict accurate hydraulic fracture dimensions. Journal of Petroleum Science and Engineering 13, 15–27.

Rice, J.R., 1968. Fracture: an advanced of treatise. In: Liewbowitz, H. (Ed.), Mathematical Analysis in the Mechanics of Fracture. Academic Press, New York City New York, pp. 191–311.

Settari, A., Cleary, M.P., November 1986. Development and testing of a pseudo-three-dimensional model of hydraulic fracture geometry. SPE Production Engineering 449–466.

Smith, M.B., Shlyapobersky, J.W., 2000. Basic of Hydraulic Fracturing. In: Reservoir Stimulation, 5. Wiley, Malden Massachusetts, 13–26.

Warpinski, N.R., Teufel, L.W., 1987. Influence of geologic discontinuities on hydraulic fracture propagation. Journal of Petroleum Technology 39 (2), 209–220.

Yang, M., 2011. Hydraulic Fracture Optimization with a Pseudo-3D Model in Multi-layered Lithology. Texas A&M Univ. Lib, pp. 29–35, 44–45.

Yang, M., Valkó, P.P., Economides, M.J., 2012. Hydraulic Fracture Production Optimization with a Pseudo-3D Model in Multi-layered Lithology. SPE 150002.

Yang, M., Araque-Martinez, A., Abolo, N., November 2014. Constrained hydraulic fracture optimization framework. In: Presented at CONEXPLO 2014, Mendoza, Argentina. IAPG, p. 165.

APPENDIX
A.1 PKN-TYPE FRACTURE GEOMETRY

The PKN model (Perkins and Kern, 1961) assumes that the condition of plane strain holds in every vertical plane normal to the direction of propagation. Also, there is no slippage between the formation boundaries; the width is proportional to the fracture height, which is constant in a well-confined zone. The fracture cross-section is elliptical, with the maximum width at the center proportional to the net pressure at the point. Net pressure is zero at tip.

The maximum width can be calculated using Eq. (12.15)

$$w_0 = \frac{2h_f p_n(x)}{E'} \quad (12.15)$$

where h_f is fracture width, $p_n(x)$ pressure normal to the fracture, and E' the plane strain modulus which is given by Eq. (12.16)

$$E' = \frac{E}{1-v^2} \quad (12.16)$$

Because the net pressure at the tip of the fracture is zero, and the fluid pressure gradient in the propagating direction is determined by the flow resistance in a narrow elliptical flow channel:

$$\frac{\partial p_n(x)}{\partial x} = -\frac{4\mu q_i}{\pi w_0^3 h_f} \quad (12.17)$$

Combine Eqs (12.15) and (12.17) integrating with the zero net pressure condition at the tip, the maximum fracture width profile at any location in the direction of propagation can be derived as shown in Eq. (12.18)

$$w_0(x) = w_{w,0}\left(1 - \frac{x}{p_n}\right)^{1/4} \quad (12.18)$$

where $w_{w,0}$ is the maximum hydraulic fracture width at the wellbore which is given in consistent units system (see nomenclature table at the end) by:

$$w_{w,0} = 3.27\left(\frac{\mu q_i x_f}{E'}\right)^{1/4} \quad (12.19)$$

The above equation gives the maximum fracture width at the wellbore. The average fracture width $\overline{w}$ is equal to the maximum width $w_{w,0}$ multiplied by a shape factor, $\gamma = \pi/5$, which is the product of $\pi/4$, a factor for averaging the ellipse width in the vertical plane, and a laterally averaging factor equal to 4/5. It comes:

$$\overline{w} = \gamma w_{w,0} = \frac{\pi}{5} w_{w,0} \quad (12.20)$$

Assuming that q_i, x_f and E' are known, the only unknown in Eq. (12.5) for maximum fracture width calculation is the dynamic fluid viscosity μ. It can be estimated using the formula for equivalent Newtonian Viscosity for Power law fluids flowing in a limited ellipsoid cross-section:

$$\mu_e = K\left[\frac{1+(\pi-1)n}{\pi n}\right]^n \left(\frac{2\pi\mu_{avg}}{w_0}\right)^{n-1} \quad (12.21)$$

μ_{avg} is the linear average velocity:

$$\mu_{avg} = \frac{q_i}{h_f w} \quad (12.22)$$

combining Eqs (12.18)–(12.21), the maximum fracture width at the wellbore is given by:

$$w_{w,0} = 3.98^{\frac{n}{2+2n}} 9.15^{\frac{1}{2+2n}} K^{\frac{1}{2+2n}} \left(\frac{1-(\pi-1)n}{n}\right)^{\frac{n}{2+2n}} \left(\frac{h_f^{1-n} q_i^n x_f}{E'}\right)^{\frac{1}{2+2n}} \quad (12.23)$$

A.2 KGD-TYPE FRACTURE GEOMETRY

Kristianovich and Zheltov (1955) derived an analytical solution for the propagation of a hydraulic fracture in which the horizontal plane strain is held. As a result, the fracture width does not depend on the fracture height, except through the boundary condition at the wellbore. The fracture has rectangular cross-section, as (b), and its width is constant in the vertical plane because the theory is based on the plane strain condition that was applied to derive a mechanically satisfying model in individual horizontal planes. The fluid pressure gradient in the propagating direction is determined by the flow resistance in a narrow rectangular slit of variable width along the horizontal direction.

The maximum fracture width profile is the same as the PKN model and the KGD width equation is:

$$w_{w,0} = 3.22 \left(\frac{\mu q_i x_f^2}{E' h_f}\right)^{1/4} \quad (12.24)$$

The average fracture width of this model is (with no vertical component):

$$\overline{w} = \gamma w_{w,0} = \frac{\pi}{4} w_{w,0} \quad (12.25)$$

The final equation to determine the maximum fracture width of the KGD model is:

$$w_{w,0} = 3.24^{\frac{n}{2+2n}} 11.1^{\frac{1}{2+2n}} K^{\frac{1}{2+2n}} \left(\frac{1+2n}{n}\right)^{\frac{n}{2+2n}} \left(\frac{q_i^n x_f^2}{E' h_f^n}\right)^{\frac{1}{2+2n}} \quad (12.26)$$

A.3 UNIFIED FRACTURE DESIGN

Unified Fracture Design (Economides et al., 2002) offers a method to determine the fracture dimensions providing the maximum reservoir performance after fracturing with the given amount of proppant. The dimensionless productivity, J_D, represents the production rate very well:

$$J_D = \frac{141.2 \, q_p B \mu}{kh \Delta p} \quad (12.27)$$

The Proppant Number, N_{prop}, is a dimensionless parameter:

$$N_{\text{prop}} = I_x^2 C_{fD} \qquad (12.28)$$

where I_x is the penetration ratio and C_{fD} is the dimensionless fracture conductivity:

$$I_x = \frac{2x_f}{x_e} \qquad (12.29)$$

where x_f is the fracture half-length, and x_e is the equivalent reservoir length.

$$C_{fD} = \frac{k_f w}{k x_f} \qquad (12.30)$$

$$A = r_e^2 \pi = x_e^2 \qquad (12.31)$$

Substituting Eqs (12.30) and (12.31) with Eq. (12.29), the correlation to determine the Proppant Number can be written as:

$$N_{\text{prop}} = \frac{4 k_f x_f w}{k x_e^2} = \frac{2 k_f}{k} \frac{2 x_f w h_n}{x_e^2 h_n} = \frac{2 k_f V_{\text{prop}}}{k V_{\text{res}}} \qquad (12.32)$$

where V_{prop} is the volume of the propped fracture in the net pay. In order to use the mass of proppants to estimate V_{prop}, it requires multiplying the ratio of the net height by the fracture height:

$$V_{\text{prop}} = \frac{M_{\text{prop}} \eta_0}{(1 - \phi_p) \rho_p} \approx \frac{M_{\text{prop}} \left(\frac{h_n}{h_f}\right)}{(1 - \phi_p) \rho_p} \qquad (12.33)$$

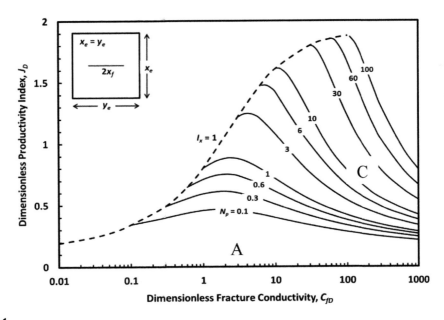

FIGURE A.1

Dimensionless productivity index as a function of dimensionless fracture conductivity.

From the calculated proppant number, the maximum dimensionless productivity index can be computed using the correlation shown in Fig. A.1 (Yang et al., 2012). Then, the penetration ratio, the fracture half-length and the propped fracture width can be calculated using Eqs (12.29)–(12.31). It is important to note that the proppant number includes only that part of the injected proppant volume that reaches the pay layers.

Once the optimum dimensionless fracture conductivity is known, the optimum fracture dimensions, i.e., propped fracture half-length ($x_{f,opt}$) and propped fracture width ($w_{p,opt}$), are set:

$$x_{f,opt} = \sqrt{\frac{k_f V_f}{C_{fD,opt} k h_n}} \tag{12.34}$$

$$w_{p,opt} = \sqrt{\frac{C_{fD,opt} k V_f}{k_f h_n}} \tag{12.35}$$

For hard rock, $k \ll 1$ md, Eqs (12.34) and (12.35) are used to estimate the fracture geometry while for soft formation, $k \gg 1$ md, Eqs (12.36) and (12.37) are appropriate to estimate the values, where they replace the optimum fracture conductivity in Eqs (12.34) and (12.35) with 1.6. As permeability rises, it becomes increasingly difficult to produce sufficient width without also generating excessive length. Tip screenout may be employed for conventional reservoirs to artificially generate extra width.

$$x_{f,opt} = \sqrt{\frac{k_f V_f}{1.6 \, k h_n}} \tag{12.36}$$

$$w_{p,opt} = \sqrt{\frac{1.6 \, k V_f}{k_f h_n}} \tag{12.37}$$

CHAPTER 13

IMPACT OF GEOMECHANICAL PROPERTIES ON COMPLETION IN DEVELOPING TIGHT RESERVOIRS

S. Ganpule[1], K. Srinivasan[2], Y. Zee Ma[2], B. Luneau[2], T. Izykowski[2], E. Gomez[2], J. Sitchler[1]

SPE Member, Denver, CO, USA[1]; Schlumberger, Denver, CO, USA[2]

13.1 INTRODUCTION

An increasingly important tool in the exploration and production of unconventional resources is the mechanical earth model (MEM). An MEM is an estimation of elastic properties, rock strength, pore pressure, and stresses; it is crucial for assessing near-wellbore conditions. The iterative process involved in construction and calibration of MEM is discussed in Chapter 7 (Higgins-Borchardt et al., 2015). Here, we discuss how some of these geomechanical properties impact the completion. In general, the completion design in developing a shale reservoir should be based on a combination of geomechanical and reservoir properties. Some of the most significant factors for fracture design include: (1) Young's modulus, which controls hydraulic fracture width; (2) stress contrast, which affects fracture containment and length; (3) presence of natural fractures, which influences hydraulic fracture complexity; (4) stress orientation, which determines the orientation of the fracture propagation; and (5) permeability, which helps determine the optimal fracturing fluid. Stress magnitudes and reservoir quality also influence completion designs, as lower stress intervals with increased natural fractures and higher reservoir quality are preferable.

A primary driver for completion optimization is the accurate prediction of pore pressure (Couzens-Schultz et al., 2013; Xia and Michael, 2015). Pore pressure acts to alleviate some of the forces transmitted through the structural framework of the rock. The difference between stress and pore pressure is referred to as the effective stress and is defined by Terzaghi's principle:

$$\sigma_e = \sigma_T - Pp$$

where σ_e is effective stress, σ_T is total stress, and Pp is pore pressure.

Although the low-permeability of unconventional reservoirs makes measuring pore pressure difficult, a number of data sources and methods for estimating pore pressure exist. Data from drilling events (e.g., kicks, influxes, and connection gas), fracture injection tests, and build-up curves from production tests can all serve to estimate the bounds of pore pressure.

The Eaton method estimates pore pressure using a porosity-vertical effective stress relationship along a compaction trend (Eaton, 1975). The Bowers method is an effective stress computation from sonic velocity and incorporates main parameters related to overpressure, including unloading (Bowers, 1995). High-tier sonic logs are used to calculate stresses and calibrate to closure estimates from mini fall off tests.

Pore pressure has a direct impact on stress variability and becomes increasingly important in unconventional plays, where overpressure is common due to hydrocarbon generation and expulsion in ultra-low permeability rock (Meissner, 1978; Xia and Michael, 2015). The increased pressure is unable to dissipate due to the low-permeability and subsequently leads to overpressure and increased stress state within the formation. Therefore, characterizing the pore pressure variability in a basin becomes an important prerequisite for analyzing the impact of stress distribution on well completion effectiveness.

13.1.1 OVERPRESSURE SYSTEM

In unconventional plays, common causes of overpressure include undercompaction (or "compaction disequilibrium") and fluid expansion, such as heating during burial, hydrocarbon maturation, and processes associated with clay digenesis. Meissner (1978) introduced the observations and concepts relating petrophysical properties, thermal maturation, and overpressure in the Bakken Formation. Meissner (1978) and other investigators observed the coincidence of thermally mature source rocks, overpressure, and the necessity of fractures to achieve economic production in the early vertical Bakken wells. The extent of development of the Bakken and other unconventional overpressured plays has validated Meissner's premise that hydrocarbon generation within the zone of maturity has caused the creation of abnormally high fluid pressure within the Bakken and closely adjacent zones. The process of oil generation and expulsion incurs anomalous fluid pressures as a result of increasing effective stress due to increasing depth of burial, and the conversion of solid overburden-supporting kerogen to liquid, nonoverburden-supporting mobile oil and associated volume changes. The ultra-low permeabilities of the overlying and underlying formations have enabled this disequilibrium to be maintained through time.

13.1.2 BAKKEN PORE PRESSURE EXAMPLE

While pore pressures in the overlying Lodgepole formation and the underlying formations are normally pressured, the Upper, Middle, and Lower Bakken and the upper stratigraphic packages of the Three Forks are all overpressured. The observed variations in pore pressure across the basin are controlled by geological, geochemical, and geomechanical processes. A multidomain integration of geology, geochemistry, petrophysics, geomechanics, completions and reservoir engineering domains is therefore needed to characterize pore pressure and stress regime for completion optimization in unconventional plays.

The earliest observations from sonic velocities (Meissner, 1978) indicated that the vertical pore pressure profile transitions from normal pressure in the overlying Lodgepole to overpressure in the three Bakken stratigraphic packages. This has been confirmed by numerous well measurements. Based on multiple sources of pressure calibration data (i.e., diagnostic fracture injection tests or DFIT and instantaneous shut in pressure or ISIP), the upper benches of the Three Forks also exhibit overpressure signatures that return back to normal pressure regime in the lower benches and subsequent older formations.

13.1.3 SCOPE OF STUDY

This chapter mainly addresses a practical question: can we apply one stimulation design throughout a field or basin without impacting production, reserves or recovery factors? Or should we make changes to stimulation design to reflect variability in in situ stresses? This question is rudimentary in assessing the risk of understimulating or overstimulating the target reservoir. The understimulation would result from underestimating stimulation treatment size when a treatment design is applied in a "copy-paste" fashion without modifications to account for variation in pore pressure and in situ stress across a basin. Although the examples illustrated here are mostly tight oil reservoirs, the concept can be applied to other tight reservoirs that have a variable in situ stress environment.

Specifically, we discuss how geomechanics influences well completion. In our investigation, we consider a number of pore pressure profiles and their corresponding stress profiles to assess the impact of stress on modeled hydraulic fracture geometries. For this end, we first give an overview on hydraulic fracturing and production modeling; we then discuss the impact of geomechanical properties on well completion and asset development for shale oil reservoirs. Furthermore, we discuss relationships among geomechanical properties, reservoir quality and completion strategies to optimize production and field development. Finally, we draw conclusions for completing shale reservoirs based on geomechanical properties and reservoir quality.

13.1.4 DATA

Figure 13.1(a) shows the four different pore pressure profiles used in calculating horizontal stress in the Bakken petroleum system. Profile A maintains normal pore pressure (0.435 psi/ft) from the surface to TD. Profiles B, C, and D assume normal pressure from the surface before ramping up

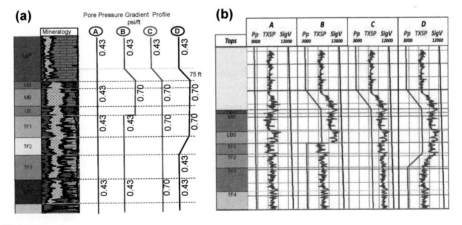

FIGURE 13.1

(a) Schematic of pore pressure profiles used to assess the impact of differences in horizontal stress calculations in hydraulic fracture geometry modeling in the Bakken petroleum system. (b) Impact on minimum horizontal stress calculation due to variations in pore pressure estimates in the Bakken petroleum system. TXSP, minimum horizontal stress (green; gray in print versions); SigV, overburden stress (red; dark gray in print versions); *Pp*, pore pressure (pink; light gray in print versions).

to an overpressured state (0.7 psi/ft) beginning 75 ft above the top of the Upper Bakken. Profile B ramps back to normal pressure immediately at the base of the Lower Bakken. Profile C maintains the overpressure through the Bakken units as well as the underlying Three Forks (TF) benches. Profile D gradually ramps back down from overpressure to normal pressure through the Three Forks-2 bench (TF2). In this chapter the following abbreviations are used to denote various zones: Lodgepole (LgP), Upper Bakken Shale (UBS), Middle Bakken (MB), Lower Bakken Shale (LBS), Three Forks-1 (TF1), Three Forks-2 (TF2), Three Forks-3 (TF3) and Three Forks-4 (TF4).

Static elastic properties combined with pore pressure information in a poro–elastic horizontal strain model provide estimates of the minimum and maximum horizontal stresses. Fig. 13.1(b) shows the minimum horizontal stress (green) as calculated using the four pore pressure profiles (pink) as previously outlined, as well as the overburden stress (red).

13.2 OVERVIEW OF HYDRAULIC FRACTURE AND PRODUCTION MODELING

An adequate understanding of geomechanical and petrophysical properties is key in optimizing completion designs in unconventional reservoirs. Although there is no one perfect model that can predict completion and production characteristics from these horizontal wells, the workflow shown in Fig. 13.2 will work best in reducing uncertainties in properties that otherwise cannot be measured directly. For example, the hydraulic fracture modeling loop combines multiple field measurements such as DFIT, 3D geomechanical log measurements, fracturing pressure data from the field and numerical fracture modeling to determine fracture geometries and conductivities with a high level of confidence.

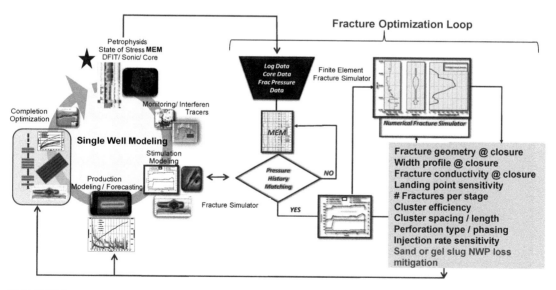

FIGURE 13.2

Workflow for completion optimization.

Modified after Ganpule et al. (2015).

13.2 OVERVIEW OF HYDRAULIC FRACTURE AND PRODUCTION MODELING

Petrophysical measurements in the form of magnetic resonance, mineralogy identification, and free fluid volume estimations can provide reasonable estimates of effective permeabilities, porosities, and water saturations in these formations. Fracturing fluid leak-off mechanisms and the type of fluids to be used for completions will be significantly impacted by these measurements. These petrophysical properties can be calibrated with production history matching using fracture geometries from pre-evaluated hydraulic fracture models and then sensitizing on reservoir properties to increase confidence levels in these parameters.

Stimulation job designs in the Middle Bakken and Three Forks have evolved with time since the North Dakota oil boom began in 2008. Most operators have come to an understanding of the optimum limits on the number of fracturing stages, number of perforation clusters per stage and fluid/proppant volumes per stage. The pumping schedule shown in Table 13.1 is a typical design used by most operators today. While the number of dominant fractures propagating per stage is hugely dictated by stress regimes, thicknesses of stress barriers, and depletion effects, it is generally assumed to be in the range of two to four fractures per stage depending on the type of fracturing fluid and the pumping rates. For the purpose of this example, we will assume two fractures per stage to evaluate the impact of geomechanics on the fracture geometries.

Figure 13.3 shows a typical fracture geometry profile from a finite element numerical fracture model using an MEM (in situ stresses and pore pressure gradients) as input (shown on the left side). The outputs from the model include fracture width at the wellbore (w), propped length (X_f prop) and hydraulic length (X_f hydraulic) of the fracture.

Although the fracture model used for the study is fully numerical and results of the model are accurate enough to provide estimates of fracture geometries, low confidence levels in inputs require the models to be calibrated with real field measurements to make the models repeatable. For example, net pressures calculated from the model can be compared with net pressures calculated from fracturing pressures, hydrostatic pressures, and fluid frictions. Fine resolution grids can be used to accurately estimate upper bounds of fracture height growth. Near-wellbore pressure calculations from step-down tests can be compared with net pressures to determine the number of dominant fractures propagating for each stage. Microseismic measurements can be overlaid on fracture geometries from the numerical models to tune fluid efficiencies. Additional real-time measurements such as distributed acoustic sensing (DAS), distributed temperature sensing (DTS), and tracers can be evaluated after the frac job to further reduce uncertainties in fracture geometries that will be fed to a production model eventually.

Once the fracture model is built and calibrated, a combination of proppant distribution and conductivity plots is used to identify upper bounds of productive half lengths and effective fracture conductivities. By feeding these geometries to a production model and obtaining a production history match, petrophysical measurements can be calibrated. Many of the rock properties that go into a production model require PVT measurements, special core analysis and basin-wide knowledge to get a good match on long-term production history. Uncertainties surrounding effective permeabilities, effective porosities, relative permeability end points, and capillary pressures make history matching in unconventional reservoirs even more challenging. It is important to pick a candidate well that has been producing long enough to traverse through multiple flow regimes starting from bilinear flow all the way to pseudoradial flow. Rate transient analysis may be performed to evaluate effective fracture conductivity and production half lengths from bilinear and formation linear flow regimes respectively, to compare with those from the production model. Once a repeatable model is available, a variety of sensitivity runs and forecasts can be done to quantify impact of fracturing parameters on

Table 13.1 Hydraulic Fracture Stimulation Design Used in the Study

Stage Name	Pump Rate (bbl/min)	Fluid Name	Gel Concentration (Lb/mgal)	Fluid Volume (gal)	Prop Name	Prop. Concentration (ppa)	Prop. Mass (lbm)	Slurry Volume (bbl)	Time (min)
WB volume	35	WF130	30	12,000	<None>	0	0	286	8
PAD	35	#130HTD	30	16,000	<None>	0	0	381	11
0.5 PPA	35	#130HTD	30	10,000	EconoProp 30/50	0	5000	243	7
1 PPA	35	#130HTD	30	10,000	EconoProp 30/50	1	10,000	249	7
1.5 PPA	35	#130HTD	30	10,000	EconoProp 30/50	1	15,000	254	7
2 PPA	35	#130HTD	30	10,000	EconoProp 30/50	2	20,000	259	7
2.5 PPA	35	#130HTD	30	8000	EconoProp 30/50	2	20,000	212	6
3 PPA	35	#130HTD	30	8000	EconoProp 30/50	3	24,000	216	6
4 PPA	35	#130HTD	30	6000	EconoProp 30/50	4	24,000	168	5
5 PPA	35	#130HTD	30	4400	EconoProp 30/50	5	22,000	117	3
Shut In	0	<None>	0	0	<None>	0	0	0	30
				94,400			140,000	2385	98

Note: WB volume, Well bore volume; PAD, injected fluid without proppant; PPA, pounds of proppant added per gallon.

13.3 IMPACT OF GEOMECHANICAL PROPERTIES ON WELL COMPLETIONS

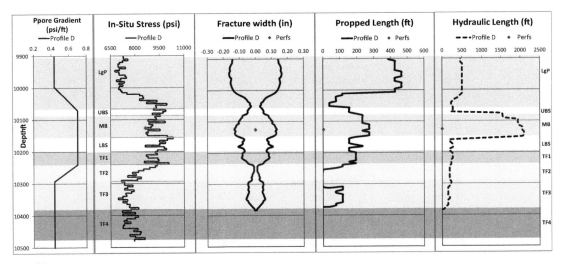

FIGURE 13.3

Fracture geometry estimated from fracture simulation for a given stress profile.

recovery factors, which can be combined with detailed economics and risk analyses to optimize completions.

Production history matching provides further validation and is performed by incorporating rock and reservoir properties, fracture geometry, and production conditions (i.e., bottom hole flowing pressure obtained via fluid measurement level). Several iterations and revisions of the input data may be needed until results converge into a common value that satisfies all the methods involved in single well modeling. Results from the production history match process are shown in Fig. 13.4.

Grid pressure from the simulation model at the end of one year of production is illustrated in Fig. 13.4(b) and gives a visual cue on the extent of depletion around a Middle Bakken well. The impact of this depletion on the estimates of drainage area for different stress profiles is discussed in the following section.

13.3 IMPACT OF GEOMECHANICAL PROPERTIES ON WELL COMPLETIONS

A primary component of completion optimization lies in the reduction of completion costs while increasing production. As oil and gas companies operating in unconventional plays move toward efficient and faster drilling, trial and error experiments with completion designs can be expensive in mature basins. The Bakken and Three Forks formations have been quite forgiving in terms of rock heterogeneity going from one field to another and hence a "cookie-cutter" approach—replicating a successful completion design from a neighboring operator—has been mostly successful throughout the basin. However, these designs should not be considered optimized; in these cases, the operator is happy with returns above a certain oil price and the need to move to work toward a more optimized approach will have to happen at oil prices lower than break-even points.

Fracture placement is controlled by multiple variables, whose combined impact on the final geometry can be very nonlinear. These variables typically include, but are not limited to: thickness of

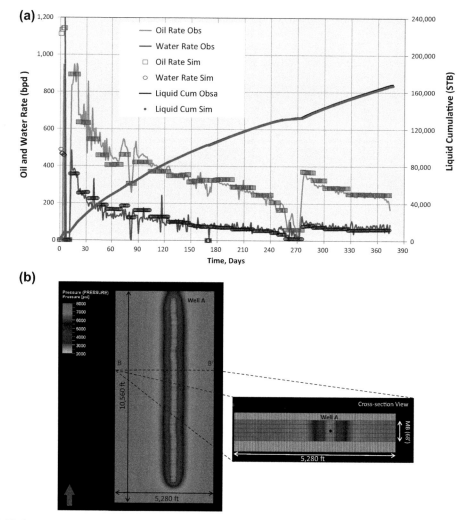

FIGURE 13.4

(a) Production history match for Well A with productive fracture length of 150 ft calibrated to the production data from Well A *(after* Ganpule et al., 2015*)*. (b) Production history matched model showing pressure depletion around Well A in the Middle Bakken (MB) zone *(after* Ganpule et al., 2015*)*.

the formation of interest, stress regimes, magnitude and thickness of stress barriers, pore pressure, and the presence of thin-beds/laminations. It is very critical that we have a good understanding of the vertical stress variation along the depth to be able to capture the effects of stress changes on the fracture height and width, which subsequently affects fracture length by way of mass balance. Often, source rocks act as stress barriers and, even when fractured through, do not contribute significantly to production due to their low-permeability, high-clay content nature. In such cases, it is important to

13.3 IMPACT OF GEOMECHANICAL PROPERTIES ON WELL COMPLETIONS 395

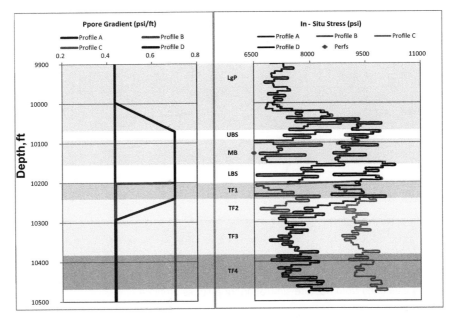

FIGURE 13.5

Pore pressure and in situ stress profiles used for modeling fracture geometry variability in the Middle Bakken and Three Forks.

achieve longer productive half lengths in the zone of lateral landing and avoid uncontained height growth or loss of proppant to nonproductive formations. These factors also influence well-spacing considerations and CAPEX estimations, which will be discussed in the following sections.

As part of this study, several pore pressure ramps and their effects on in situ stresses and fracture geometries using a standard pumping design are evaluated (Fig. 13.5). For each case, the fracture is assumed to initiate from the Middle Bakken (MB) and the resulting half length, fracture height and width are examined at closure conditions.

Profile A assumes a normal pore pressure gradient of 0.435 psi/ft from the surface to total depth. Such an assumption is not very realistic as it does not take into account any of the overpressure that is widely observed due to hydrocarbon generation. Nevertheless, the stress profile is included as part of the study to show the drawback of not incorporating higher stresses in over pressured zones on completions.

Fracture geometry using profile A is shown in Fig. 13.6. The fracture is well contained and propped half lengths greater than 400 ft are observed. Most of the fracture height growth happens in the downward direction into the Three Forks benches, which may work to our advantage if the objective is to connect the two productive zones and keep them open when the fracture closes. However, microseismic experiments in the Williston basin show that most of the height growth from fractures initiated from the Middle Bakken happens in the upward direction breaking into the Lodgepole.

Profile B assumes normal pore pressure from surface until the Lodgepole and a smooth ramping into an over pressured gradient state (0.7 psi/ft) from the bottom of Lodgepole (roughly 75 ft) to the

396 CHAPTER 13 IMPACT OF GEOMECHANICAL PROPERTIES

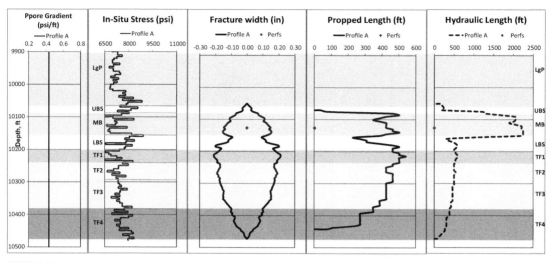

FIGURE 13.6

Fracture geometry estimated from fracture simulation for a stress profile A.

top of the Upper Bakken Shale. The pore pressure stays at 0.7 psi/ft until the bottom of the Lower Bakken Shale and goes back to normal pore pressure immediately from the top of the Three Forks benches. This is also not a very realistic stress assumption as the Three Forks benches contain hydrocarbons sourced by the Lower Bakken Shale. However, due to lack of pore pressure data in the Three Forks during the early stages of exploration, they were assumed to be normally pressured initially.

Fracture geometries using Profile B are shown in Fig. 13.7. The fracture height growth is very similar to that from profile A, however most of the length extension happens in the Three Forks

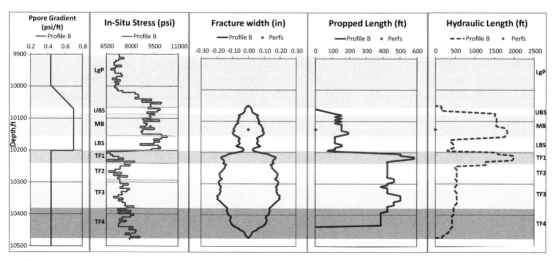

FIGURE 13.7

Fracture geometry estimated from fracture simulation for a stress profile B.

13.3 IMPACT OF GEOMECHANICAL PROPERTIES ON WELL COMPLETIONS

benches. Fracture width at the interface of Lower Bakken Shale and Three Forks-1 is very low at closure indicating that the fracture will eventually close at that point with time. This also means that, with very short propped half lengths in the Middle Bakken, production from the well that is completed in Middle Bakken will decline rapidly as production contribution from the Three Forks benches will be minimal to nothing.

Profile C has a stress profile very similar to that of B but the over pressure state extends into the Three Forks benches all the way to the bottom. Because all the Three Forks benches are predominantly sourced by the Lower Bakken Shale, the water saturation increases as it gets deeper. Historically, there are not too many horizontals that have been completed in the lower Three Forks benches due to high water cuts and hence not too many data sets are available to understand the pore pressure regime in these lower benches.

Fig. 13.8 shows fracture geometry estimated using profile C. The X_f prop in MB is approximately 280 ft for a fracture that is not contained within the MB and TF benches. Fracture simulation shows excessive fracture growth in Lodgepole and no growth in the TF4.

Unlike profiles A and B, the fracture grows into the Three Forks initially and breaks through the Upper Bakken Shale growing uncontrolled in height into the Lodgepole. Due to lower pore pressure assumption in the Lodgepole, this height growth is unavoidable as the net pressure in the fracture exceeds the differential stress between the zone of fracture initiation (Middle Bakken) and the stress barriers (Lower and Upper Bakken Shales). The fracture grows reasonably in length in the Middle Bakken, but nearly half the proppant pumped with the fracture is lost to the Lodgepole formation (which is not productive) and pinches out at the Upper Bakken Shale interface.

Profile D assumes normal pressure gradient from surface and a smooth ramping into an over pressure state roughly 75 ft above the Lower Bakken Shale that runs into the upper Three Forks bench and ramps down to normal pressure in the lower Three Forks benches. Fracture geometries

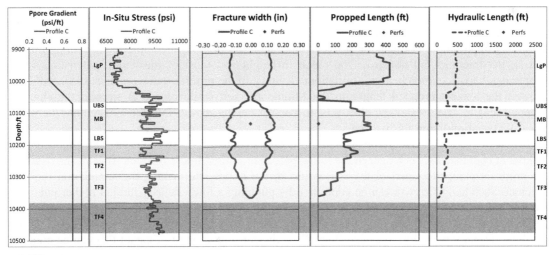

FIGURE 13.8

Fracture geometry estimated from fracture simulation for a stress profile C.

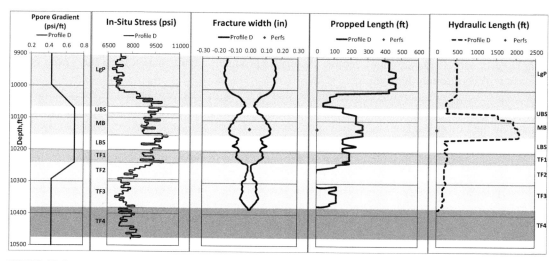

FIGURE 13.9

Fracture geometry estimated from fracture simulation for a stress profile D.

resulting from profile D are shown in Fig. 13.9. Very similar to profile C, after breaking into the upper Three Forks bench initially, the fracture grows in height into the Lodgepole. The Lower Bakken Shale between Middle Bakken and Three Forks stays open at closure which will eventually pinch out as the well continues to produce. Connectivity between Middle Bakken and Three Forks benches in such cases is always a challenge because the Lower Bakken Shale separating the two zones is at a higher stress regime and exhibits very high-clay content which results in significant proppant embedment and rapid loss of fracture conductivity if there is any.

To summarize all the fracture profiles seen above, understanding the pore pressure regimes is extremely important to evaluate the stress profiles and their impact on completions. With profiles A and B, the fractures did not grow too much in height and most of the length extensions were happening in hydrocarbon bearing formations. Among all the profiles that were discussed, profile A showed the longest propped length in the Middle Bakken while profile B showed the shortest propped length. The greater the ability is, to control height growth into the Lodgepole, the more effective the stimulation job will be.

Profiles C and D show uncontrolled height growth into the Lodgepole and the propped half lengths in both the cases is longer in the Lodgepole than in the Middle Bakken. This means that half the proppant pumped as part of the frac is lost to the Lodgepole making the stimulation job less efficient. Such height growth can be controlled by pumping a low viscosity fluid at higher pumping rates to keep the fracture contained, in the zone of initiation and transport the proppant as much in length as possible into the fracture to provide a conductive path for the hydrocarbons to flow into the wellbore. Operationally, low viscosity fluids limit the concentration of proppant that can be pumped with them due to faster settlement of the proppant. Fracture geometries and zone connectivity from these profiles are summarized in Table 13.2.

Table 13.2 Summary of Fracture Modeling Results

Profile (Ppore Grad. & stress)	X_f Hydraulic in MB (ft)	X_f Prop in MB (ft)	MB & TF Connectivity*	Growth in Lodgepole
A	2200	400	High	Nil
B	1700	125	Low	Nil
C	2200	280	Moderate	Highest
D	2100	225	Pinch-off	High

*Connectivity is a qualitative flag and cannot be easily measured under current technical environment.

13.4 IMPACT OF GEOMECHANICAL PROPERTIES ON ASSET DEVELOPMENT

Well spacing is undoubtedly a key driver in asset development of tight oil resources such as the Bakken. Well spacing dictates the number of wells needed per section to develop the asset, hence misjudging well spacing could easily make a project seem uneconomic. In this study we estimated well spacing from drainage information gleaned from the production modeling phase of our workflow that integrates long term production data. In an ideal case, long-term production should be long enough to ensure that the flow regime has reached pseudo radial, though this may not be practical in tight reservoirs. Figure 13.10 shows a cross-section from a production model for Well A after the well had produced 160,000 bbl of fluid in one year. The cross-section shown here is across a fracture stage in

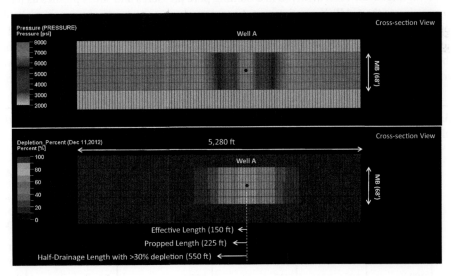

FIGURE 13.10

Cross-sections from production model of Well A showing drainage around the well. Pressure and depletion percentages are indicative of drainage around the well after 160,000 bbl of fluid production in one year.

horizontal Well A (well path is orthogonal to plane of paper). Two properties, pressure and depletion percent, are indicative of drainage, as illustrated below.

$$Depletion\ \% = [(P_i - P_t)/(P_i)] * 100$$

where, P_i is cell pressure at time zero (pressure at initial condition) and P_t is cell pressure at a certain production time t (i.e., pressure after the well had produced for 6 months or 1 year).

The fracture model used for Well A assumes profile D for the pore pressure gradient and stress profile that results in 2100 ft of hydraulic length of which only 225 ft is propped. The production model shows that out of 225 ft propped length only 150 ft is effectively participating in fluid flow toward the well. However, after 160,000 bbl of fluid is withdrawn from the well in one year, the drainage region extends 550 ft (1100 ft if accounting for drainage on either side of the well) from the well considering 30% depletion cut off as shown in Fig. 13.10. Well spacing estimated in this case is 1200 ft and includes 1100 ft of drainage length plus 100 ft offset distance between Well A and the immediate offset well completion. For this example, if an operator pumps a standard fracture treatment design (Table 13.1) in Well A that follows profile D for the pore pressure gradient and stress profile then the well spacing is roughly 1200 ft after one year of production. Consequently a development section will need approximately four to five wells to effectively drain and boost hydrocarbon recovery. Assuming a conservative estimate of five million USD as CAPEX requirement per well, the development cost can range from 20 to 25 million USD.

The results for the other stress profile scenarios can be seen in Table 13.3. The results emphasize a wide range in the number of wells needed to develop a section depending on the pore pressure gradient and stress profile scenario. Profile A requires only two wells while profile B needs as many as seven wells. Accordingly, the CAPEX required to develop a section would range from 10 to 35 million USD.

Table 13.3 highlights the risks involved in assuming pore pressures and stress profiles as it may either under or overestimate the CAPEX requirement. For example an operator might consider profile A as a possible option in his assets due to lack of data, when in fact profile D is valid, and would estimate development cost per section of 10–15 million USD. This will result in production shortfall and lower recovery because he will drill two to three wells when four to five wells are needed to effectively drain the reservoir. On the other hand, if profile B is considered, when profile D is valid, then the price tag of

Table 13.3 Summary of Production Modeling Results for Four Stress Profiles along with Expected Wells Per Section and CAPEX Range

Profile (Ppore Grad. & Stress)	X_f prop in MB (ft)	X_f effective in MB (ft)	Half-Drainage Length with >30% Depletion (ft) (A)	Well Spacing (ft) (B)=(A)*2 + 100 ft	Number of MB Wells Needed Per Section (C)= (5280− 100 ft)/(B)	CAPEX $ (USD million) (D) = (C)* 5 MM USD
A	400	265	950	2000	2 to 3	10 to 15
B	125	95	350	800	6 to 7	30 to 35
C	280	185	725	1550	3 to 4	15 to 20
D	225	150	550	1200	4 to 5	20 to 25

30–35 million USD per section could make the project seem uneconomical. Thus, improper characterization of stress variations could prematurely truncate further development activity in a resource basin.

An integral part of an effective asset development strategy in unconventional resources is the ability to forecast the capital requirement for multiple years. Usually such forecasts are required very early into the development phase. Accurate and timely assessment of capital expenditures is imperative to efficiently manage the costs of a well and subsequently those of a project as a whole. Controlling these costs may be the deciding factor between a prospect being deemed economical or uneconomical.

13.5 DISCUSSION: GEOMECHANICAL PROPERTIES, RESERVOIR QUALITY, AND COMPLETION STRATEGY

In evaluation and development of unconventional resources, it is generally useful to distinguish completion quality from reservoir quality. Geomechanical properties are more related to completion quality, but they can be also related to reservoir quality. Often, production is positively correlated to reservoir quality, completion quality, and completion effort (hydraulic fracture treatment), but the correlations may not always be high. An example of correlation between production and the composite reservoir quality is shown in Fig. 13.11(a), in which for very low reservoir quality, the production is low. When the reservoir quality is moderate or high, the production is positively correlated with the completion quality and effort. In this example, not much completion effort was applied to the very low reservoir quality formations because even if the completion quality and effort is high, production will be likely low. In other words, reservoir quality is the basis for both completion strategy and production; it has to be moderate to high in order for completion to be effective.

Production is not always highly correlated to individual reservoir and completion variables (Jochen et al., 2011; Miller et al., 2011; Gao and Du, 2012; Ma, 2015). The weak or no correlation is often due

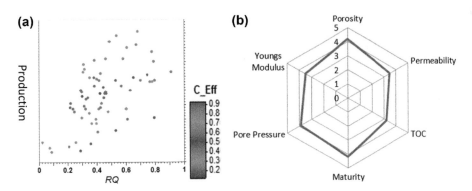

FIGURE 13.11

(a) Example of relationships among reservoir quality, completion efficiency (C_Eff) and production. Both reservoir quality and completion efficiency are positively correlation to the production, but the reservoir quality has to reach a certain level in order for completion to be effective. Reservoir quality and completion efficiency range from 0 to 1. (b) Example of radar plot for analyzing the qualities of individual reservoir and completion variables. Ranking quality index is scaled between 0 and 5 in this example, with 5 being the highest quality.

Reservoir Quality

Property	Desired Condition	Behavior Effect
Structural Dip	Low	Lateral in zone
Source Quality	High	HC generation
Thermal Maturity	Within range	HC generation
HRA Rock Type	High quality	Favorable reservoir properties
PHI/S$_{wd}$/k	High	HC volume

Completion Quality

Property	Desired Condition	Behavior Effect
Pressure	Overpressure	charge & containment
Min Hz Stress	Low	initiation
Min Hz Stress	Variable	containment
Stess Aniso	Low	complexity
Fracability	High	initiation & complexity

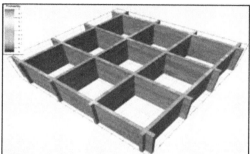

Relative Quality

FIGURE 13.12

Example workflow for combining reservoir and completion quality in a three-dimensional model. The resulting relative quality indicator can then be used for drilling and completion optimization.

to the interaction among the variables in a multiple-variable system. For everything else being equal, better completion quality leads to easier completion and higher production, but everything else is generally not equal, and thus the correlation between production and individual reservoir or completion variables can be weak or even reversed (Ma, 2015).

Production from shale reservoirs is highly variable as a function of the location in a field because of the heterogeneities in both reservoir quality and completion quality. Reservoir quality, completion quality, and completion effort all impact production. To understand the heterogeneous productions (Baihly et al., 2010), not only the impact of reservoir quality and completion quality on production need to be analyzed, but often it is important to analyze individual physical variables, such as kerogen, TOC, porosity, fluid saturation, and permeability for reservoir quality; stress, mineralogy, and mechanical properties for completion quality; and lateral length, proppant tonnage, and stage count for completion effort. In practice, only a small number of these parameters may be evaluated because of limited data. Notice also that the reservoir and completion quality parameters may not be linearly combined; it is not straightforward to derive a composite score of quality. Radar plots provide a useful way to visually analyze qualities as an integrated approach (Fig. 13.11(b)).

Geoscientists may attribute geomechanical properties to completion quality, but in fact, they are sometimes also indicative of reservoir quality because of their correlation. For example, Young's modulus is often used as a criterion for completion quality, but Britt and Schoeffler (2009) showed an example of separating prospective shales from nonprospective shales using static and dynamic Young's modulus correlation. It also happens that reservoir and completion qualities are not significantly or even inversely correlated. Defining the key production drivers and their interrelationships is highly important for completion optimization. Figure 13.12 shows an example workflow where reservoir and completion quality drivers are coupled in three dimensions to provide a relative quality indicator.

13.6 CONCLUSIONS

The multidisciplinary integration in this study underscores the need to characterize the pore pressure regime and its impact on stress profile for efficient completion strategy in tight oil plays. In summary of this study, pore pressure and the in situ stresses often vary both vertically and laterally in the shale formations. Accurately characterizing pore pressure profile and stress profile is vital for designing and optimizing fracture treatments. We should not generally apply one stimulation design throughout a basin irrespective of the stress and mechanical property heterogeneity. Applying a single standard treatment design in a "copy-paste" mode across a basin could lead to understimulation or overstimulation of a shale reservoir. Adapting treatment designs for a specific pore pressure and stress profile scenario addresses shortcomings of applying the "copy-paste" treatment design, and often provides better drainage coverage.

Pore pressure and resulting stress variability control fracture geometry, which in turn influences drainage lengths and well spacing estimates. In the Middle Bakken example, the number of wells per section required could increase by 100% if an improper pore pressure profile is assumed; such miscalculation could have significant economic consequences.

ACKNOWLEDGMENTS

The authors acknowledge Gary Forrest, Bilu Cherian, Veronica Gonzales, Hongxue Han, Shannon Higgins–Borchardt, Rob Marsden and John Zachariah for useful comments and suggestions.

REFERENCES

Baihly, et al., 2010. Unlocking the Shale Mystery: How Lateral Measurements and Well Placement Impact Completions and Resultant Production. Paper SPE 138427 presented at the SPE Annual Technical Conference and Exhibition.

Bowers, G.L., June 1995. Pore pressure estimation from velocity data: accounting for overpressure mechanisms besides undercompaction. SPE Drilling & Completion 89–95.

Britt, L.K., Schoeffler, J., September 23–25, 2009. The Geomechanics of a Shale Play. Paper SPE 125525 presented at the SPE Annual Technical Conference and Exhibition, New Orleans, LA, USA.

Couzens-Schultz, B.A., Axon, A., Azbel, K., Hansen, K.S., Haugland, M., Sarker, R., Tichelaar, B., Wieseneck, J.B., Wilhelm, R., Zhang, J., Zhang, Z., 2013. In: Pore Pressure Prediction in Unconventional Resources. International Petroleum Technology Conference, Beijing, China, March 26–28, 2013. IPTC paper #16849, 11 p.

Eaton, B.A., 1975. In: The Equation for Geopressure Prediction from Well Logs. SPE 50th Annual Fall Meeting, Dallas TX, September 28–October 1, 1975. SPE paper #5544, 11 p.

Ganpule, S.V., Srinivasan, K., Izykowski, T., Luneau, B., Gomez, E., 2015. In: Impact of Geomechanics on Well Completion and Asset Development in the Bakken Formation, SPE Hydraulic Fracture Technology Conference, the Woodlands, Texas, USA, Febraury 3–5, 2015. SPE paper #173329.

Gao, C., Du, C., October 8–10, 2012. Evaluating the Impact of Fracture Proppant Tonnage on Well Performances in Eagle Ford Play Using the Data of Last 3–4 Years. Paper SPE 160655 presented at the SPE Annual Technical Conference and Exhibition, San Antonio, TX, USA.

Higgins-Borchardt, S., Sitchler, J., Bratton, T., 2015. Geomechanics for unconventional reservoirs. In: Ma, Y.Z., Holditch, S., Royer, J.J. (Eds.), Unconventional Resource Handbook: Evaluation and Development (2015). Elsevier.

Jochen, V., et al., 2011. Production Data Analysis: Unraveling Rock Properties and Completion Parameters. Paper SPE 147535 presented at the…, Calgary, AB, Canada.

Ma, Y.Z., 2015. Unconventional resources from exploration to production. In: Ma, Y.Z., Holditch, S., Royer, J.J. (Eds.), Unconventional Resource Handbook: Evaluation and Development (2015). Elsevier.

Meissner, F.F., 1978. Petroleum geology of the bakken formation, Williston Basin, North Dakota and Montana. In: Rehrig, D. (Ed.), The Economic Geology of the Williston Basin. Montana Geological Society, Montana, North Dakota, South Dakota, Saskatchewan, Manitoba, pp. 207–227.

Miller, C., Waters, G., Rylander, E., June 14–16, 2011. Evaluation of Production Log Data from Horizontal Wells Drilled in Organic Shales. Paper SPE 144326 presented at the SPE Americas Unconventional Conference, The Woodlands, Texas, USA.

Xia, X., Michael, G.E., 2015. Pore pressure in unconventional petroleum systems. In: Ma, Y.Z., Holditch, S., Royer, J.J. (Eds.), Unconventional Resource Handbook: Evaluation and Development (2015). Elsevier.

CHAPTER 14

TIGHT GAS SANDSTONE RESERVOIRS, PART 1: OVERVIEW AND LITHOFACIES

Y. Zee Ma[1], W.R. Moore[1], E. Gomez[1], W.J. Clark[1], Y. Zhang[2]

Schlumberger, Denver, CO, USA[1]; *University of Wyoming, Laramie, WY, USA*[2]

14.1 INTRODUCTION AND OVERVIEW

Although shale gas plays have been in the spotlight recently, natural gas in tight sandstones is actually also an important hydrocarbon resource. In many cases, tight gas sandstone resources can be developed more easily than shale gas reservoirs as the rocks generally have higher quartz content, and are more brittle and easier to complete for production.

This chapter first gives an overview of evaluating tight gas sandstone reservoirs, including their depositional environments, and other reservoir characteristics. Then, it presents an integrated methodology for lithofacies analysis and modeling. Because identification of lithofacies and rock typing is a key to characterizing these types of reservoirs (Rushing et al., 2008), we discuss lithofacies classification using wireline logs. We present methods on how to populate the lithofacies data from wells to a three-dimensional (3D) model while discussing the advantages and disadvantages of each method for modeling tight gas formations.

The next chapter is the second part of tight gas sandstone reservoirs, in which petrophysical analysis, formation evaluation, and 3D modeling of petrophysical properties are presented.

14.1.1 BACKGROUND

Tight gas sandstone reservoirs are natural extensions of conventional sandstone reservoirs, but with lower permeability and generally lower effective porosity. Hydrocarbons have traditionally been produced from sandstone and carbonate reservoirs with high porosity and permeability. Sandstone reservoirs with permeability lower than 0.1 milliDarcy (mD) were historically not economically producible, but advances in stimulation technology have enabled production from these tight formations. There is some confusion regarding the definition of tight gas sandstone reservoirs; they are sometimes referred to as deep-basin, basin-centered gas accumulation, or pervasive sandstone reservoirs (Meckel and Thomasson, 2008). The United States Gas Policy Act of 1978 classified tight gas formations as those that have in situ permeability less than 0.1 mD (Kazemi, 1982). Thus, sandstone gas reservoir in which the formation has average permeability lower than 0.1 mD is a tight gas play regardless of its depositional environment. These reservoirs can occur in numerous settings, including channelized fluvial systems (e.g., the Greater Green River basin, Law, 2002; Shanley, 2004; Ma et al., 2011), alluvial fans, delta fan, slope and submarine fan channels deposits (e.g., Granite Wash, Wei and Xu, 2015), or shelf margin (e.g., Bossier sand, Rushing et al., 2008), among others. Some tight gas sandstones contain different depositional facies; the Cotton Valley formation, for example, includes stacked shoreface/barrier bar deposits, tidal channel, tidal delta, inner shelf, and back-barrier deposits. Because of the variety of depositional environments for sandstones and other variations, there is no typical tight gas sandstone reservoir (Holditch, 2006a). While drilling, well design, and completion techniques for producing a tight sandstone reservoir are often similar to producing a shale gas reservoir, exploration and resource evaluation for them generally are quite different (Kennedy et al., 2012).

Tight gas production was first developed in the San Juan Basin of the western United States; large-scale development of tight gas sandstone reservoirs has a longer history than large-scale development of shale gas reservoirs. Approximately 1 trillion cubic feet (Tcf) was produced per year by 1970 in the United States (Naik, 2003). Meckel and Thomasson (2008) distinguished three periods of evolution for the evaluation and production of tight gas sandstone reservoirs: preparadigm period (1920–1978), paradigm period (1979–1987), and mop-up period (1988–present). The paradigm period was marked by Master's article (1979), arguing the extensive existence of hydrocarbon resources in tight formations. Many of the basins that contain gas in tight sandstones in North America have already been in exploration or production phase, playing an important role in the North American energy equation. In fact, as clastic formations are found in many parts of the world, tight gas sands make up an important resource worldwide. Rogner (1997) has estimated the worldwide tight gas resource to be approximately 7500 Tcf, but other estimates carry a much larger number for tight gas plays (Naik, 2003). In comparison, the worldwide shale gas resource has been estimated to be 16,100 Tcf (Rogner, 1997).

14.1.2 BASIN-CENTERED EXTENSIVE DEPOSITS OR CONVENTIONAL TRAPS

Two schools of thought exist regarding the geologic control of tight gas sandstone reservoirs: continuous basin-centered gas accumulations or BCGAs (Law, 2002; Schmoker, 2005), and gas accumulation in low-permeability tight sandstones of a conventional trap (Shanley et al., 2004). The difference between these two theories can have a huge impact on the strategy for gas exploration in tight sandstones and the estimate of worldwide gas resources in these formations (Aguilera and Harding, 2008).

As a matter of fact, conventional traps represent more limited, favorable structural and/or stratigraphic setups that enable natural gas accumulation after its generation and migration. They are thus rather geologically circumstanced. On the other hand, the BCGA theory implies a continuous basin-wide gas accumulation, or at least widely spread within a basin. Law (2002) argued that BCGAs contained extremely large gas resources, and were one of the more economically viable unconventional gas resources, but there was generally a poor understanding of BCGAs despite a significant body of work on defining characteristics of gas-saturated reservoirs. These include studies in the deep basin of Alberta, the San Juan basin of New Mexico and Colorado by Masters (1979), and the Greater Green River basin and Great Divide basin of the western United States (Law and Spencer, 1989; Law, 1984; Spencer, 1989). According to Law (2002), the BCGA must meet the following five criteria (Camp, 2008): (1) large regional extent, measuring tens of miles in diameter; (2) low permeability, less than 0.1 mD; (3) abnormal pressure (either overpressured or underpressured); (4) gas saturated; and (5) absence of downdip water contacts. Figure 14.1 illustrates the

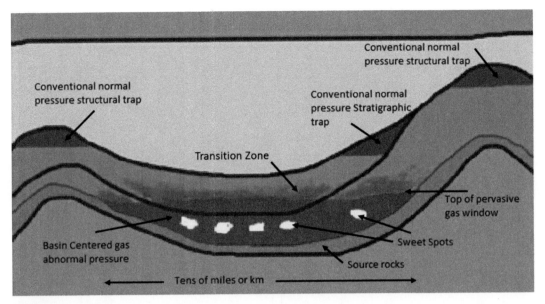

FIGURE 14.1

A schematic cross-section view illustrating BCGA model.

Modified from Schenk and Pollastro (2002).

main characteristics of BCGS theory. One important characteristic in this theory is that the trapping mechanism is a diffused capillary-pressure seal instead of conventional structural or stratigraphic controls.

Shanley et al. (2004) argued that low-permeability reservoir systems such as those found in the Greater Green River basin were not examples of BCGAs, and they suggested that only truly continuous-type gas accumulations were to be found in hydrocarbon systems in which gas entrapment was dominated by adsorption. Studies on several tight gas sandstone reservoirs in the Greater Green River Basin have suggested that subtle structural and stratigraphic traps better explain the primary controls for the hydrocarbon accumulations (Camp, 2008).

14.1.3 GENERAL PROPERTIES OF TIGHT GAS SANDSTONE RESERVOIRS

Although the two theories, the BCGA and conventional trapping mechanism, have significant differences regarding the exploration strategies of tight gas sandstone reservoirs, none of them denies the importance of formation evaluation in any given tight gas play.

14.1.3.1 Source rock

Source rock is important for all hydrocarbon resource accumulations. For tight gas sands, the source rock should be in proximity to the relatively porous deposit so that the expulsion can drive the gas into the porous formation and form the reservoir (Meckel and Thomasson, 2008). The total organic carbon of the source rock should be large enough for generation of a significant amount of hydrocarbon.

CHAPTER 14 TIGHT GAS SANDSTONE RESERVOIRS, PART 1

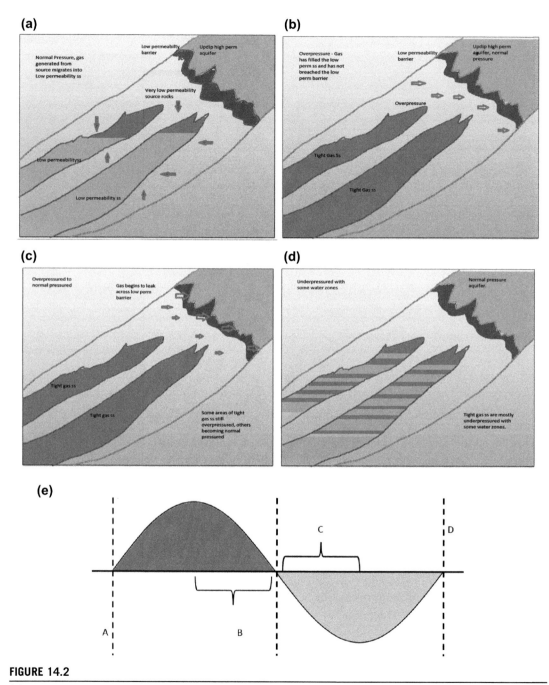

FIGURE 14.2

(a) Low permeability sands are charged by hydrocarbons produced by source shales. A very low permeability barrier prevents the hydrocarbons from escaping into the updip aquifer. Pressure varies from normal to

Moreover, the source rock should be subjected to heat transformation within the gas-generation window under burial history. Therefore, it is good practice to select exploration targets proximal to organically rich intervals (Coleman, 2008).

14.1.3.2 Abnormal Pressures

The productive intervals of a tight gas sandstone reservoir, in general, are abnormally pressured, either overpressured or underpressured (Meckel and Thomasson, 2008). The formation pressure depends on the structure, stratigraphy, and basin history. In general, when the source rock generates a large amount of gas in tight formations, gas cannot escape easily and causes overpressure in the system (Meckel and Thomasson, 2008). When some amount of the generated gas escapes from the updip margins of the overpressured area, an underpressured rim around the overpressured area may be generated. Measurements need to be acquired to characterize the pressure before development.

Burnie et al. (2008) discussed the mechanism for over and underpressure of tight gas sandstone reservoirs. Low permeability sandstones encased in shaly source rocks can be charged with gas. The pressure will build and the tight sandstones will be overpressured, but a low permeability barrier above them will contain the gas. Once a certain pressure threshold is reached, the gas will begin leaking through the low permeability barrier and escape updip. This will reduce the pressure to a normal and ultimately underpressure state. Figure 14.2 shows this pressure evolution, gas generation, and maturity model. Before the organic matter becomes mature, the pressure of the system is essentially normal. As the organic matters matures, i.e., during the active gas generation, the pressure builds up and the system becomes overpressured. In the postgeneration, the pressure of the system decreases and eventually becomes underpressured.

14.1.3.3 Stacking Patterns

Two basic types of tight gas sandstone reservoirs can be distinguished: stacked sandstones and blanket sandstones; they are related to the depositional environments. The stacked sandstones are often turbidites, deltas, or braided streams while blanket sandstones are usually extensive, laminated shallow-marine deposits. The stacked sandstones can have thousands of feet of 5–15 ft thick sand bodies of limited extent, while the blanket sandstones can be thin or thick, but typically thicker (20–30 ft) with greater areal extent. The stacked sandstones generally need to be drilled vertically or near vertically to

---

◀

overpressured. In favorable areas, hydrocarbons can be produced. (b) Tight gas sands are fully charged and overpressured. Hydrocarbons with minor water are producible. (c) The pressure reaches a point where it breaches the overlying low permeability zone and hydrocarbons begin to leak through the barrier into the overlying aquifer. Pressure drops from over to underpressure. Hydrocarbons can be mostly produced, but water will begin to encroach. (d) Leakage has stopped and the sands are underpressured, and some hydrocarbons may be left in favorable areas along with producible water. (e) Pressure evolution model for tight gas reservoirs (pressure as a function of time; red (black in print versions) is overpressure, and pink (gray in print versions) is underpressure. A, normal pressure, hydrocarbon generation begins; B, overpressure, hydrocarbons contained; C, seal breached by overpressure, hydrocarbons migrate until pressure stabilizes; D, normal to underpressured, hydrocarbons contained.

Modified from Meckel and Thomasson (2008) and Burnie et al., 2008.

contact as many sandstone bodies as possible for better stimulation, while the laminated blanket sandstones need to be drilled horizontally to contact as much of the sandstone bodies as possible for better stimulation. Combinations of these types can also exist, along with conventional reservoir intervals. Examples of stacked sandstone reservoirs with high pressure include Pinedale (Webb et al., 2004; Ma et al., 2011), Jonah (Cluff and Cluff, 2004; Jennings and Ault, 2004), and Wamsutter (Barrett, 1994) fields, and examples of the stacked sandstones with moderate pressure include Williams Fork formation in the Piceance Basin (Hood, and Yurewicz, 2008) and Travis Peak formation in the East Texas Basin. The Frontier formation in the Green River basin and the Mancos B sandstones in the Piceance Basin are rather blanket sandstones (Finley, and O'Shea, 1985).

14.1.3.4 Reservoir Quality
Tight gas sandstones can range from mostly quartz to a variable mixture of quartz, feldspars, clays, carbonates, and pyrite. Matrix values can range from textbook sandstone values to very high or low values, so it is important to use actual rock data to determine the complexity of the reservoir. In some reservoirs, a single porosity calculation (density or sonic) may be adequate to get a useable interpretation, but in others a multimineral model with a variable number of clay and nonclay constituents may be needed. Average porosity for many known tight gas sandstones ranges between 7 and 10%, but lower or higher average porosities are also possible (Meckel and Thomasson, 2008). Average permeability often is in the order of 0.01 mD. Natural fractures may be important to producibility in tight gas sandstone reservoirs, but they may increase water production as well.

Based on the lithofacies characterization and petrophysical evaluation, completion technique and quality will drive the economic viability of each play. Reservoir quality is discussed in detail in the next chapter.

14.1.4 DRILLING, COMPLETION, AND DEVELOPMENT SCENARIOS
Because of the low permeability in tight gas sandstones, more wells are drilled to develop them compared with developing a conventional gas reservoir. A well life is typically longer as the initial high production rate may drop off fast to a plateau. Over the life time of a tight gas development, modest investments are made over a longer period than for developing a conventional gas reservoir (David and Stauble, 2013).

Typically, the stacked sandstone reservoirs are best drilled with vertical wells while the blanket sandstones are best drilled with horizontal wells. However, some stacked sand plays have many successful horizontal wells (Wei and Xu, 2015). In practice, understanding the spatial distribution of sand bodies, including their size and geometry, and using an integrated approach is critical in developing a tight gas sandstone reservoir. Pranata et al. (2014) presented an example of achieving high production rate using an integrated reservoir characterization approach and the dual-lateral horizontal technology without using the hydraulic fracturing.

Drilling challenges in the tight sandstones include lost circulation due to natural fractures or low pressure, sloughing of shale layers, formation damage, mud invasion, and drilling bit abrasion (Pilisi et al., 2010). A number of drilling methods are available for developing tight gas sandstone reservoirs, including conventional drilling, casing drilling, coiled tubing drilling, underbalanced drilling, overbalanced drilling, and managed pressure drilling, each of which have advantages and limitations (Table 14.1, also see Pilisi et al., 2010).

Table 14.1 Comparing Different Drilling Methods

Drilling Method	Conventional	Casing	Coiled Tubing	Overbalanced	Underbalanced	Managed Pressure
Drilling problem: lost circulation, stuck pipe, etc.	May increase	Lower	No effect	May be high	Lower	Much lower
Reduce formation damage	No	Little	No	No	Yes	Yes
Kick detection				Yes	Yes	Yes
Equipment complexity	Low	Medium	Medium	Low	High	High
Rate Of Penetration (ROP) improvement	No	Little	Yes (smaller diameter)	No	Yes	Yes

Modified from Pilisi et al. (2010).

For stacked sandstone reservoirs, designing the hydraulic fracture treatment should be based on the formation evaluation, especially the lithofacies layering geometry as the completion is based on the producing zones that are separated by vertical flow barrier layers (Holditch, 2006). The number of stages, for example, can be determined based on the stacking pattern of the geometry of sand bodies and shaly barriers. Depending on the thickness of the sand and shale layers and formation in situ stress profile, a single fracture treatment can be sometimes used to stimulate multiple layers, the well can be completed and stimulated with a single stage, and gas will be produced by commingling the different layers. On the other hand, when a thick shale barrier separates two productive layers, multiple hydraulic fractures should be created, especially when the in situ stress contrast is high.

One important decision to make in completing a tight gas sandstone reservoir is to select a diversion method (Table 14.2, also see Holditch and Bogatchev, 2008; Wei et al., 2009). A number of diversion techniques are available and each of them has certain advantages, disadvantages, and limitations (Wei et al., 2010). The selection of diversion technique should be based on the following parameters:

- number of layers,
- depth of each layer,
- net pay thickness in each layer,
- effective porosity in each layer,
- water saturation in each layer,
- drainage area for each layer,
- pressure and temperature in each layer, and
- gas gravity.

Table 14.2 Issues and Decisions in Various Stages of Developing Tight Gas Sandstone Reservoirs

Evaluation	Drilling	Completion	Production
Basin analysis and regional geology	Method selection	Stage count	Gas flow rate
Stratigraphy/structural analysis	Lost circulation	Diversion technique	Water production
Seismic interpretation and attribute analysis	Stuck pipe	Perforation design	Artificial lift
Formation evaluation with well logs and core data	Formation damage	Fluid selection	Tubing design
3D reservoir model	Sloughing, kicks	Proppant selection	Casing diameter

14.2 LITHOFACIES AND ROCK TYPING

As tight gas sandstone reservoirs reside in siliciclastic formations, their lithofacies typically include sandstone, siltstone, and shale based on grain size; alternatively, they can be divided into sand, sandy shale, shaly sand, and shale. Modeling lithofacies is generally sufficient for characterizing tight gas sandstone reservoirs although special care may be needed to assess the impact of minerals on the logs (Ma et al., 2011, 2014a). For example, based on depositional characteristics, fluvial deposits often include sand-dominant channel facies, shale-dominant floodplain, and a mixture of sand and shale in the crevasse and splay facies. Rushing et al. (2008) summarized a list of rock type definitions from 26 studies on sandstone reservoirs. There is much debate regarding definition of rock types, depending on the classification scheme used: depositional, petrographic, log-based, or hydraulic. Depositional rock types are defined in the context of the large-scale geological framework and represent the rock properties at deposition, and they are generally termed depositional facies, such as channel, overbank, crevasse, and splay. Petrographic rock types are based on small-scale, microscopic rock properties, typically defined with the aid of various imaging tools (Rushing et al., 2008). Log-based rock types are typically called electrofacies, and they represent the signature of wireline logs (Wolff and Pelissier-Combescure, 1982). Hydraulic rock types are mainly based on the rock flow properties, such as flow zone indicator or FZI (Amaefule et al., 1993). Because the subsurface formations are subject to multiple geological processes, including deposition, erosion, redeposition, compaction, and diagenesis, these definitions of rock types are generally very different; it is often impossible to establish a consistent match between two different definitions. The three most commonly used definitions include lithofacies that have a connotation of depositional facies and lithology, electrofacies that are defined using logs, but not necessarily calibrated to geology, and rock types based on FZI or other petrophysical and/or engineering criteria. Lithofacies are closely related to geology because lithology represents compositional mineral content and geologic facies are highly related to depositional environments. Thus the calibration of electrofacies to lithofacies can relate the logs to geology more closely, but it is not always straightforward. The advantage of using lithofacies is that they can be modeled using geologic propensity analysis so that the 3D lithofacies model is consistent with the conceptual depositional model based on sedimentary and sequence-stratigraphic analyses (Ma, 2009). This is especially important when data are limited.

We discuss methods of electrofacies and lithofacies classification with an emphasis on lithofacies. Electrofacies are clusters defined from wireline logs, and as such, they may or may not be lithofacies before calibration to geologic interpretations. Moreover, smaller-scale electrofacies can also be modeled within a lithofacies in a hierarchical order. This can be carried out using a cascaded methodology (Ma, 2011).

14.2.1 LITHOFACIES IN TIGHT GAS SANDSTONES AND WIRELINE LOGS

Wireline logs provide basic data sources for evaluating petrophysical properties as they measure some characteristics of the rock formations. Histograms of wireline logs can be used to assess the possible separations of various lithofacies (Ma et al., 2014a). The most striking characteristic of a mixture of multiple lithofacies is the multimodality in the histograms of logs. This also explains why logs rarely show a normal distribution, but rather they are characterized by either a multimodal histogram or a skewed long-tail in the histogram. Depositional or lithological facies are often a dominant factor that causes the nonnormality, including multimodality and skewness, in the histogram of a wireline log.

More specifically, logs in a tight sandstone formation often show bimodality (sometimes more than bimodal) in their histograms (Fig. 14.3), but notice that bimodality may conceal the existence of more than two mixtures of lithofacies. For instance, the GR (Gamma Ray) histogram (Fig. 14.3(a)) cannot be decomposed into two quasinormal histograms, but rather into three quasinormal histograms (Ma et al., 2014b). This is because the GR log conveys a three-lithofacies mixture. Resistivity and porosity logs in Fig. 14.3 also reflect the mixtures of three lithofacies. Typically, many wireline logs carry information about the lithofacies, but none of them is individually capable of accurately discriminating them. Two or more logs are typically needed to accurately separate the individual lithofacies because of the overlaps of lithofacies mixtures on the logs.

A basic exploratory tool for multivariate analysis is the correlation or correlation matrix that includes correlation coefficients between any two variables of concern. Graphic displays, such as cross plots, and two-dimensional (2D) histograms, often enable an insightful analysis of the relationship between two variables and the characteristics of the lithofacies mixtures. For example, the 2D histogram of the GR and logarithm of the resistivity (Fig. 14.3(d)) shows three modes while the monovariate histograms of GR and resistivity show only two modes (Fig. 14.3(a)–(c)). Similarly, the 2D histogram of porosity and GR reveals three modes, albeit discreetly (Fig. 14.3(e)). In theory, a multidimensional joint probability histogram reveals even more clearly the data structures of a mixture, but there is no effective way to display it graphically; a 2D histogram matrix can be used to gain insights into multivariate relationships of numerous wireline logs (Ma et al., 2014b).

14.2.2 LITHOFACIES CLASSIFICATION USING WIRELINE LOGS IN TIGHT GAS SANDSTONE FORMATIONS

The importance of classification of lithofacies for tight gas reservoirs lies in identifying potential gas zones using wireline logs because recoverable gas resides mostly in sand-dominated lithofacies, and marginally in silty lithofacies.

14.2.2.1 Problem of Cutoff Methods

The two modes in the histogram of individual logs (Fig. 14.3) often suggest more than two lithofacies. In the well-log GR histogram (Fig. 14.3(a)), two modes are pronounced with the smaller mode at

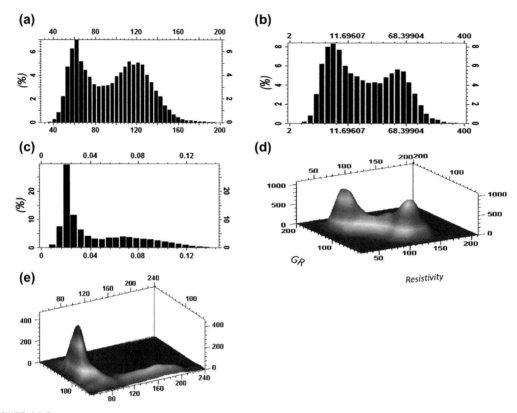

FIGURE 14.3

Histogram examples of wireline logs in tight gas sandstones. (a) GR; (b) logarithmic resistivity; (c) porosity; (d) 2D histogram of GR-logarithmic resistivity; (e) 2D histograms of porosity-GR (porosity ranges between 0 and 15%, and was rescaled to between 0 and 250 for display purpose).

approximately 60 API and the larger mode at approximately 120 API. Traditionally, a cutoff was used to generate the sandy lithofacies from low GR values, silty lithofacies from moderate GR values, and the shaly lithofacies from high GR values, such as shown by the example in Fig. 14.4(a). However, these three lithofacies exhibit significant overlaps in the GR distributions based on core data (Fig. 14.4(b)). Only the smallest and largest GR values (less than 60 API or more than 120 API) do not have significant overlaps of the three lithofacies, but GR values between 60 and 120 API are a mixture of samples from all the three lithofacies (Ma et al., 2011, 2014a).

Therefore, the method of applying cutoffs to the GR log is not optimal as it results in the three lithofacies at different GR bands. Using GR alone misclassifies many samples into the wrong lithofacies. Many sandy and shaly samples with GR values between 85 and 95 API are incorrectly classified as siltstones. Many siltstone samples with GR values below 80 API or above 100 API are incorrectly classified as either sand or shale samples. As sandy lithofacies are hydrocarbon-bearing and shale lithofacies are nonreservoir rocks, misclassification of shale lithofacies into sand lithofacies leads to false positives, and the converse leads to false negatives. When two or more logs are used, the

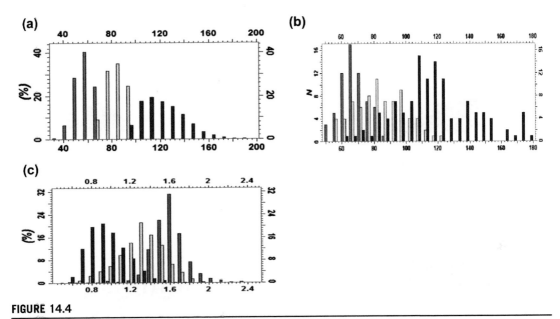

FIGURE 14.4
(a) Decomposition of the histogram in Fig. 14.3(b) into three histograms based on the GR cutoffs. Color code: orange (dark gray in print versions), channel facies; green (gray in print versions), crevasse-splay; and black, overbank. (b) GR histograms by lithofacies from core data from one well in the Greater Green River basin. (c) Resistivity histograms for the three lithofacies clustered using GR cutoffs. Note that the histograms of the different lithofacies on the same plots in (b) and (c) are not normalized together, but each histogram is normalized by itself. Therefore, the display does not show the relative proportion of each histogram over the total population.

cutoff method uses a gated logic for lithofacies classification, but the gated logic is not optimal as it causes many boundary effects. A different approach is needed to more accurately discriminate the overlapped GR or other log samples into different lithofacies.

14.2.2.2 Lithofacies Classification from Mixture Decomposition of Wireline Logs

Although the histograms of many logs show only two modes, they cannot be modeled by two normal or quasinormal distributions, but some of them can be aptly fitted into three quasinormal distributions. This is because the frequency distribution of an individual log signature by the silty lithofacies often overlaps completely with the log signature of the sandy and shaly lithofacies, and their modes are hidden in the histogram. The main idea of lithofacies mixture decomposition lies in use of multiple logs to separate the overlaps in individual logs.

Resistivity is often a good discriminator for classification of the lithofacies when no aquifer is present, which is quite common for tight gas sandstone reservoirs (Law et al., 1986, 2002; Naik, 2003; Aguilera and Harding, 2008). Whereas shaly lithofacies typically have lower resistivity responses because of the absence of gas and more bounded water, sandstones exhibit higher resistivity as a result of bearing hydrocarbon and lower bounded water content. The GR and resistivity logs are generally correlated negatively; their correlation is -0.821 using the logarithm of the resistivity

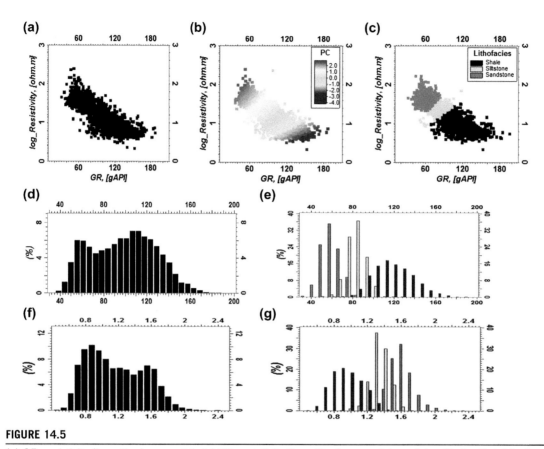

FIGURE 14.5
(a) GR–resistivity (logarithm) cross plot. (b) GR–resistivity (logarithm) cross plot overlain with the first PC of PCA of the two logs. (c) GR–resistivity (logarithm) cross plot overlain with the three clustered lithofacies. (d) GR histogram. (e) GR histograms for three clustered lithofacies. (f) Logarithmic resistivity histogram. (g) Logarithmic resistivity histogram for the three clustered lithofacies. Orange (gray in print versions) represents the sandstones; black, shale; and green (light gray in print versions), siltstones. For (f) and (g), see the note in Fig. 14.2.

in the example (Fig. 14.5). The resistivity log also shows two modes similar to the GR histogram, but the larger mode corresponds to smaller resistivity values, and the smaller mode corresponds to large resistivity values. This is because the two logs are negatively correlated and there is significantly more shale than sand in the formation.

Principal component analysis, or PCA, (a tutorial is found in the appendix of Chapter 7) can be used to combine the two or more logs for classification of these lithofacies. Here, we show examples of classifying the three lithofacies using PCA over two logs: GR and resistivity. The three lithofacies clustered using PC1 are shown in Fig. 14.5. Notice the three component histograms are similar to the experimental histograms based on the core data except that the core data histograms are less normal because of the limited data. Similarly, the three lithofacies overlap between 9 and 40 Ω on the resistivity; overlapping resistivity values between the three lithofacies implies that these lithofacies cannot be correctly classified using the resistivity log alone, but the resistivity histogram was decomposed into three component histograms in the classification of the three lithofacies (Fig. 14.5(e)). Statistically, this can be considered as a mixture decomposition that separates the component histograms from the original histogram (Ma et al., 2014b). Although this example uses only two logs, three or more logs can be used in the same workflow as PCA is very effective to handle multiple input variables for classification of the lithofacies. Readers can find a lithofacies classification example using three logs in Ma et al. (2014a).

Moreover, principal components can be rotated to emphasize importance of one log versus the others (Ma, 2011). Three or more logs can be used for classification as well using PCA (Ma, 2011; Ma et al., 2014b).

14.2.2.3 Determining Proportions of Lithofacies in Classification

The determination of the proportion of each lithofacies in classification is important because it impacts the assessment of the overall reservoir quality of the field. In tight gas sandstone reservoirs, a higher proportion of sandy lithofacies from the clustering than the true proportion will cause an optimistic view of the reservoir, and a higher proportion of shaly lithofacies will lead to a pessimistic view. Ideally, lithofacies data from core are abundant and representative without sampling bias, and then they can be used as a reference for the proportions. Unfortunately, core lithofacies are generally limited and statistically are rarely representative. Purely automatic methods, including unsupervised artificial neural network (ANN) and statistical techniques, are not robust enough to generate accurate proportions of lithofacies. In sandstone formations, ANN typically yields more siltstone in classification because automatic methods have a tendency to make similar proportions for all the clusters, unless the clusters are highly distinct. Moreover, siltstone typically has intermediate values for many logs, such as GR, resistivity and effective porosity, but the intermediate values do not always correspond to siltstone. The relatively abundant overlapping intermediate values easily cause an over classification of siltstone than its true proportion for most automatic methods.

PCA has an advantage over ANN as it enables determining the proportion of each clustered lithofacies using the cumulative histogram of a selected PC or rotated PC. As ANN makes the lithofacies classification purely based on the data structures, it does not incorporate the significance of each input log. For example, ANN clustering based on GR and resistivity in the previous example gives unrealistic clusters, as shown in Fig. 14.6(a), in which some samples with lower GR and higher resistivity are classified to be siltstone while higher GR and lower resistivity samples are classified as sandstone. It is possible to overcome the problem by combining PCA and ANN, i.e., ANN using only the first PC, such as shown in Fig. 14.6(b).

The method presented here includes: (1) generating lithofacies by applying cutoffs on some selected but representative principal component(s), and (2) using the generated data as training data for applying the supervised ANN. In classifications using PCA, the number of quasinormal distributions decomposed from the initial histogram can be used as an initial estimate of the number of lithofacies that have similar petrophysical characteristics. The proportion of each lithofacies is inferred from the decomposition of the histogram. The ratio of the cumulative frequency of each of the three component histograms to the cumulative frequency of the composite GR histogram is the proportion of the

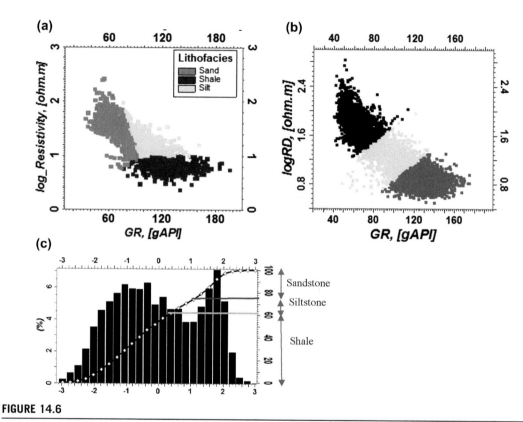

FIGURE 14.6

(a) GR-resistivity (logarithm) cross plot, overlaid with the lithofacies clustered using ANN. (b) GR–RD (Gamma ray-resistivity) cross plot overlain with the lithofacies clustered by ANN using PC1. (c) Illustrating the cutoff method and the proportions of the three clustered lithofacies using cumulative histogram of a PC (histogram is displayed in the background).

corresponding lithofacies. When one component, such as PC1, is used for clustering, it is straightforward to find the cutoff values that give a predetermined proportion for each lithofacies, as illustrated in Fig. 14.6(c). In this example, the clustered shale represents 60.3%, siltstone 13.9%, and sandstone 25.8%. In comparison, the clustering by ANN (Fig. 14.4(a)) gave 37.4% shale, 35.6% siltstone, and 26.9% sandstone. The latter proportions are highly inconsistent with the outcrop observation and other analogs.

This method of using the cumulative histogram to define the proportions of the lithofacies clusters is very general; it can be used with any variable or component. For example, it can be used even for the cutoff method with a single wireline log, but we have already shown the limitation of using a single log. Notice also that the example in Fig. 14.6(c) is an original PC, but this component can be a rotated component that represents information from a number of the original PCs.

14.2.3 IMPACT OF LITHOFACIES CLASSIFICATION ON STACKING PATTERNS AND DEPOSITIONAL INTERPRETATION

Different methods for lithofacies classification have a significant impact on the interpretation of the deposition, net-to-gross (NTG), and stacking patterns. Typically, the cutoff methods generate too much intermediate-quality rocks, such as siltstone or crevasse–splay facies. Figure 14.7(a) compares the lithofacies vertical sequences by the cutoff method, PCA, and ANN for the example discussed above (Figs 14.4 and 14.5) for one well. The GR cutoff method generated much more siltstone, and less sandstone than the PCA. On the other hand, ANN generated much more sandstone, and less shale than the PCA. This is also true for all the wells in the model area. The vertical proportion profiles (VPPs) of the lithofacies generated by the three methods are also compared (Fig. 14.7).

The three lithofacies proportions are 55.3% shale, 22.3% siltstone, and 22.4% sandstone for the GR cutoff method; 60.3% for shale, 13.9% for siltstone, and 25.8% for sandstone for the PCA; 35.6% shale, 26.9% siltstone, and 37.4% sandstone for ANN. For each VPP, each stratigraphic unit had different fractions of those lithofacies. Relative fractions of the three facies can differ from layer to layer. Lithofacies VPP represents an "average" stacking pattern and can be used to analyze and model vertical sequence succession of depositional facies (Ma et al., 2009).

Variogram of lithofacies often shows a certain so-called "hole" effect, which reflects cyclicity of the lithofacies (Jones and Ma, 2001). For blanket sandstones, cyclicity is mainly in the vertical direction as the lateral sandstone layers are generally extensive. For stacked, lenticular sandstones, cyclicity can appear both vertically and laterally, but generally stronger in the vertical direction because of the lithofacies depositional sequence. The vertical cyclicity in the lithofacies is highly related to the stacking pattern and vertical proportion profiles.

14.3 THREE-DIMENSIONAL MODELING OF LITHOFACIES IN TIGHT SANDSTONE FORMATIONS

Populating a 3D lithofacies model using the lithofacies data at the wells commonly requires upscaling the lithofacies samples into the 3D grid because well logs typically have a half-foot sampling rate and 3D grid cells have a larger size. All known methods for upscaling a categorical variable are statistically biased as they use the "winner-take-all" approach (Ma, 2009). The method of most-abundant-value, which is the most reasonable and most common one, tends to favor the major lithofacies while reducing minor lithofacies. This problem can be bypassed if the lithofacies data at wells are predicted from wireline logs. Instead of creating lithofacies at the well-log scale, the continuous variables used for the lithofacies prediction can be upscaled into the 3D grid with an unbiased method, and then they can be used for lithofacies classification directly in the grid-cell scale (Ma et al., 2011). Detail of this workflow is beyond the scope here, we assume availability of upscaled lithofacies data at wells before the 3D modeling.

For blanket sandstone reservoirs, stratigraphic correlation is relatively simpler when enough wells are available, and the lithofacies model may be constructed in a relatively straightforward way, for example using sequential indicator simulation (SIS) with a large horizontal variogram range because of the large lateral continuity. Here, we compare four common methods of modeling lithofacies for a stacked, fluvial sandstone reservoir. These four methods for modeling categorical variables are presented in the Appendix. Among these techniques, object-based modeling (OBM) generally is more

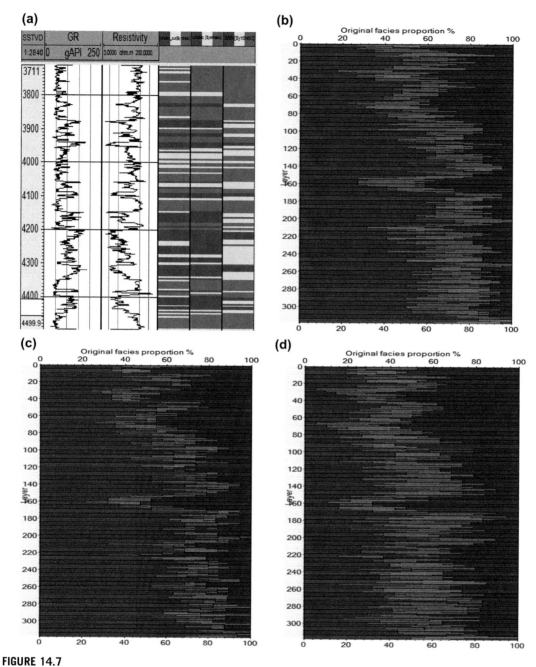

FIGURE 14.7

(a) Well section showing GR (Track 1), resistivity (Track 2), zonation of the clustered lithofacies by the GR cutoff (Track 3), zonation of lithofacies by the PCA using both GR and resistivity (Track 4), and zonation of lithofacies generated by ANN (Track 5). (b) Vertical profile of the clustered lithofacies by GR cutoffs using all the 20 wells in the model area. (c) Vertical profile of the clustered lithofacies by PCA using all the 20 wells in the model area. (d) Vertical profile of the clustered lithofacies by ANN using all the 20 wells in the model area. Orange (gray in print versions) represents sandy lithofacies, yellow (light gray in print versions) the silty lithofacies, and black the shaly lithofacies (Greater Green River basin).

14.3 THREE-DIMENSIONAL MODELING OF LITHOFACIES

suitable for clear shape definitions of geologic bodies, such as channels and bars. The truncated Gaussian method is more suitable when the order of the facies is clearly definable. SIS is one of the most commonly used geostatistical methods to model lithofacies because of its capabilities in integrating various data. In particular, lithofacies probability maps or volumes can be integrated more easily in the SIS model than in the other modeling methods to constrain the positioning of the lithofacies bodies. The model by the method with user-defined objects generally has spatial characteristics between the models generated by OBM and SIS.

Figure 14.8(a) shows a relatively small area with 20 wells, wherein the lithofacies data were obtained by PCA classification using GR and resistivity (as discussed earlier). Without other information,

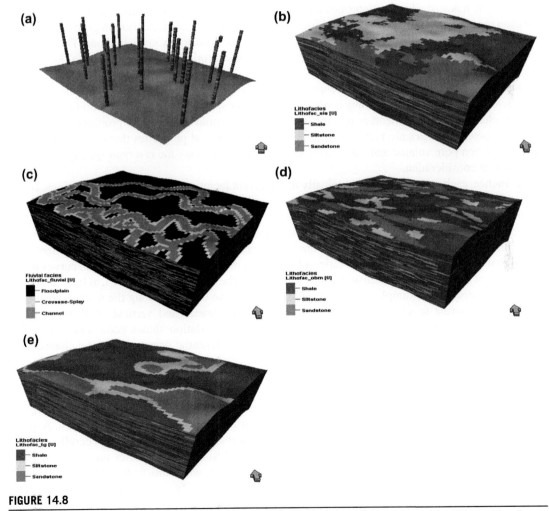

FIGURE 14.8

(a) Lithofacies data at 20 wells, classified by PCA. (b) 3D lithofacies model constructed by SIS honoring the data in (a). (c) 3D lithofacies model constructed using fluvial OBM. (d) 3D lithofacies model constructed using user-defined objects (ellipses with rounded base). (e) 3D lithofacies model constructed using truncated Gaussian simulation. Notice the halos of the siltstone (intermediate lithofacies) around the sand lithofacies.

the knowledge of fluvial system as the depositional environment by itself cannot always determine the best method for populating the lithofacies in the 3D model. Given the fluvial example with 20 wells, four 3D lithofacies models using different methods are shown in Fig. 14.8(b)–(e). As no probability is used for conditioning, the SIS model (Fig. 14.8(b)) is driven by the indictor variogram and data from the wells. As a result of the scarcity of the data, the facies objects are often distributed randomly, especially in the areas of no control points. The model using fluvial OBM method (Fig. 14.8(c)) mimics the fluvial depositional characteristics, including meandering of the channels, and amalgamation and cutting between the channels. Figure 14.8(d) shows a 3D lithofacies model constructed using ellipses with rounded base. The lithofacies model using this type of defined objects often has characteristics between fluvial OBM and SIS models.

Notice the halos of siltstone (intermediate lithofacies) around the sand lithofacies in the truncated Gaussian model (Fig. 14.8(e)). This is because the truncated Gaussian simulation imposes a sequence of transitions between different lithofacies, which can be a drawback in some situations. This problem can be mitigated by an enhanced method, termed truncated pluri-Gaussian simulation (Galli et al., 1994; Hu et al., 2001).

One limitation of the fluvial OBM technique is that the geometry of crevasse–splay facies may not be modeled realistically because many modeled crevasses or splays do not form as small fans breaking off from channels. For sedimentary process modeling, this would be a serious problem. In reservoir modeling, however, facies are highly linked with the petrophysical properties that directly determine the hydrocarbon pore volume and fluid flows. Using facies to indicate the reservoir quality of rocks is an important consideration.

In practice, each stratigraphic unit typically has different facies proportions and associations. In the case of fluvial channel deposition, the channel characteristics differ from one stratigraphic package to another, including orientation, width, and thickness. Uncertainties in channel characteristics, including orientations, sinuosity, width, and thickness can be expressed as probabilistic distributions of a triangle, uniform, Gaussian function, or truncated Gaussian function. Lithofacies stacking patterns are related to the facies fraction for each stratigraphic zone in the model. Vertical proportion curves in the facies vertical profile, object dimensions, and facies data at the wells can be honored, at least to certain extent. In the discussed example, more continuity occurs in the channels along the main depositional direction (north-west to south-east). Channel sinuosity and lateral and vertical amalgamations of channel and crevasses facies are evident (Fig. 14.8). Well correlation shows good sand continuity, especially from the north-west to south-east, which is the preferential orientation of the channels.

14.4 CONCLUSION

A variety of depositional environments exist for tight gas sandstone reservoirs. A realistic lithofacies model is important as it impacts the spatial distribution of the main reservoir properties. A good lithofacies model should realistically capture the important reservoir heterogeneity. Generally, there is a lack of core lithofacies data for field-wide reservoir analysis, and lithofacies are obtained from wireline logs. We combine the classical petrophysical analysis with statistical methods and neural networks for lithofacies clustering. These methods enable classifying lithofacies by integrating geological and petrophysical interpretation.

All the categorical-modeling techniques can be hierarchized in two or more levels for multilevel facies modeling. For example, facies may be first modeled using fluvial OBM or truncated Gaussian method, and then modeled again using SIS based on the model already constructed by OBM or truncated Gaussian method. This is because OBM and truncated Gaussian methods generally produce larger facies objects in the model, and SIS can be further used to model small-scale heterogeneities. In fact, the workflow of modeling lithofacies by SIS following a truncated Gaussian modeling or OBM has been applied to a number of hydrocarbon field-development case studies, including carbonate ramps and shallow-marine depositional environments. An example of constructing a facies model using OBM based on the model constructed by SIS can be found in Cao et al. (2014), in which a facies model of sand-shale was first built using SIS with facies probability maps. Then a fluvial OBM method was used to generate channels with splays.

Because of the complexity of geological deposits, modeling lithofacies is simultaneously indeterministic and causal. In a large-scale model where depositional trends are geologically definable, lithofacies probability maps or cubes can be constructed while incorporating the depositional trends (Ma, 2009); these probability maps or cubes can be used to constrain the 3D lithofacies model, along with the data at wells.

ACKNOWLEDGMENT

The authors thank Schlumberger Ltd for permission to publish this work. They also thank Dr Shujie Liu and Mi Zhou for reviewing the manuscript.

REFERENCES

Amaefule, J.O., Altunbay, M., Tiab, D., Kersey, D.G., Keelan, D.K., 1993. Enhanced reservoir description: using core and log data to identify hydraulic (flow) units and predict permeability in uncored intervals/wells: SPE 26436. In: 68th Annual Technology Conference and Exhibition Houston, Texas.

Aguilera, R., Harding, T.G., 2008. State-of-the-art tight gas sands characterization and production technology. Journal of Canadian Petroleum Technology 47 (12).

Barrett, F.J., 1994. Exploration and development of almond tight gas sands along the Wamsutter/Creston Arch, Wasahakie-red Desert basins, Southwest Wyoming. AAPG Search and Discovery. Article #90986, presented at AAPG Annual Convention, Denver, Colorado, June 12–15, 1994.

Burnie, S.W., Maini, B., Palmer, B.R., Rakhit, K., 2008. Experimental and empirical observations supporting a capillary model involving gas generation, migration, and seal leakage for the origin and occurrence of regional gasifers. In: Cumella, S.P., Shanley, K.W., Camp, W.K. (Eds.), Understanding, Exploring, and Developing Tight Gas Sands, 2005 Vail Hedberg Conference, AAPG Hedberg Series, vol. 3, pp. 29–48.

Camp, W.K., 2008. Basin-centered gas or subtle conventional traps? In: Cumella, S.P., Shanley, K.W., Camp, W.K. (Eds.), Understanding, Exploring, and Developing Tight Gas Sands, 2005 Vail Hedberg Conference, AAPG Hedberg Series, vol. 3, pp. 49–61.

Cao, R., Ma, Y.Z., Gomez, E., 2014. Geostatistical applications in petroleum reservoir modeling. Southern African Institute of Mining and Metallurgy 114, 625–629.

Clement, R., et al., 1990. A computer program for evaluation of fluvial reservoirs. North Sea Oil and Gas Reservoirs-II.

Cluff, S.G., Cluff, R.M., 2004. Petrophysics of the Lance Sandstone Reservoirs in Jonah Field, Sublette County, Wyoming, in AAPG Studies in Geology #52, Jonah Field: Case Study of a Tight-Gas Fluvial Reservoir pp. 215–241.

Coleman, J.L., 2008. Tight gas sandstone reservoirs: 25 years of searching for "The Answer". In: Cumella, S.P., Shanley, K.W., Camp, W.K. (Eds.), Understanding, Exploring, and Developing Tight gas Sands, AAPG Hedberg Series, vol. 3, pp. 221–250. Tulsa, OK.

David, F., Stauble, M., 2013. Developing a sustainable unconventional business in China. In: SPE Paper 167051, Presented at the SPE Unconventional Resources Conference and Exhibition-Asia Pacific, Brisbane, Australia, November 11–13, 2013.

Deutsch, C.V., Journel, A.G., 1991. Geostatistical Software Library and User's Guide. Oxford Univ. Press, 340 p.

Finley, R.J., O'Shea, P.A., 1985. Geologic and engineering analysis of blanket-geometry tight gas sandstones. In: SPE-Doe Joint Symposium on Low Permeability Gas Reservoirs, Denver CO, March 13–16, 1985.

Galli, A., Beucher, H., Le Loc'h, G., Doligez, B., 1994. The pros and cons of the truncated Gaussian method. In Geostatistical Simulation. Quantitative Geology and Geostatistics 7, 217–233.

Holden, L., et al., 1997. Modeling of fluvial reservoirs with object models. In: AAPG Computer Applications in Geology, vol. 3.

Holditch, S.A., 2006a. Tight gas sands. Journal of Petroleum Technology (June) 86–93.

Holditch, S.A., 2006b. Optimal Simulation Treatments in Tight Gas Sands. SPE 96104.

Holditch, S.A., Bogatchev, K.Y., 2008. Developing Tight Gas Sand Adviser for Completion and Stimulation in Tight Gas Sand Reservoirs Worldwide. Paper SPE 114195, presented at the CIPC/SPE Gas Technology Symposium 2008 Joint Conference, Calgary Alberta, Canada, 16–19, June.

Hood, K.C., Yurewicz, D.A., 2008. In: Assessing the Mesaverde Basin-Centered Gas Play, Piceance Basin, Colorado. AAPG Hedberg Series, vol. 3, pp. 87–104. Tulsa, OK.

Hu, L.Y., Le Ravalec, M., Blanc, G., 2001. Gradual deformation and iterative calibration of truncated Gaussian simulations. Petroleum Geosciences 7, S25–S30.

Jennings, J.L., Ault, B.P., 2004. Jonah Field Completions: An Integrated Approach to Stimulation Optimization with an Enhanced Economic Value, in AAPG Studies in Geology #52, Jonah Field: Case Study of a Tight-gas Fluvial Reservoir, pp. 269–279.

Jones, T.A., Ma, Y.Z., 2001. Geologic characteristics of hole-effect variograms calculated from lithology-indicator variables. Mathematical Geology 33 (5), 615–629.

Journel, 1983. Nonparametric estimation of spatial distribution. Mathematical Geology 15 (3), 445–468.

Kazemi, H., October 1982. Low-permeability gas sands. Journal of Petroleum Technology.

Kennedy, R.L., Knecht, W.N., Georgi, D.T., 2012. Comparison and contrasts of shale gas and tight gas developments, North American experience and trends. In: SPE Paper 160855, Presented at the SPE Saudi Arabia Section Technical and Exhibition, Al-khobar, Saudi Arabia, 8–11 April.

Law, B.E., 1984. In: Law, B.E. (Ed.), Structure and Stratigraphy of the Pinedale Anticline, Wyoming, USGS Open-File 84–753.

Law, et al., 1986. Geologic Characterization of Low Permeability Gas Reservoirs in Selected Wells, Greater Green River Basin, Wyoming, Colorado, and Utah, AAPG SG 24, Geology of Tight Gas Reservoirs, pp. 253–269.

Law, B.E., Spencer, C.W., 1989. Geology of Tight Gas Reservoirs in Pinedale Anticline Area, Wyoming, and Multiwell Experiment Site, Colorado. In: US Geologic Survey Bulletin 1886.

Law, B.E., 2002. Basin-centered gas systems. AAPG Bulletin 86 (11), 1891–1919.

Ma, Y.Z., 2009. Propensity and probability in depositional facies analysis and modeling. Mathematical Geosciences 41, 737–760. http://dx.doi.org/10.1007/s11004-009-9239-z.

Ma, Y.Z., 2011. Lithofacies clustering using principal component analysis and neural network. Mathematical Geosciences 43, 401–419.

Ma, Y.Z., et al., 2014a. Identifying hydrocarbon zones in unconventional formations by discerning Simpson's paradox. Paper SPE 169496 presented at the SPE Western and Rocky Regional Conference, April, 2014.

Ma, Y.Z., Wang, H., Sitchler, J., et al., 2014b. Mixture decomposition and lithofacies clustering using wireline logs. Journal of Applied Geophysics 102, 10–20. http://dx.doi.org/10.1016/j.jappgeo.2013.12.011.

Ma, Y.Z., Gomez, E., Young, T.L., Cox, D.L., Luneau, B., Iwere, F., 2011. Integrated reservoir modeling of a Pinedale tight gas reservoir in the Greater Green River Basin, Wyoming. In: Ma, Y.Z., LaPointe, P. (Eds.), Uncertainty Analysis and Reservoir Modeling. AAPG Memoir 96, Tulsa.

Ma, Y.Z., Seto, A., Gomez, E., 2009. Depositional facies analysis and modeling of Judy Creek reef complex of the Late Devonian Swan Hills, Alberta, Canada. AAPG Bulletin 93 (9), 1235–1256.

Masters, J.A., 1979. Deep basin gas trap, Western Canada. AAPG Bulletin 63 (2), 152.

Meckel, L.D., Thomasson, M.R., 2008. Pervasive tight gas sandstone reservoirs: an overview. In: Cumella, S.P., Shanley, K.W., Camp, W.K. (Eds.), Understanding, Exploring, and Developing Tight Gas Sands, AAPG Hedberg Series, vol. 3, pp. 13–27. Tulsa, OK.

Naik, G.C., 2003. Tight gas reservoirs–an unconventional natural energy source for the future. Retrieved from: www.sublette-se.org/files/tight_gas.pdf.

Pilisi, N., Wei, Y., Holditch, S.A., 2010. Selecting Drilling Technologies and Methods for Tight Gas Sand Reservoirs. Society of Petroleum Engineers. http://dx.doi.org/10.2118/128191-MS. Paper SPE 128191, presented at the 2010 IADC/SPE Drilling Conference in New Orleans, Louisiana, USA, 2–4 Febraury.

Pranata, H.M., Su, W., Huang, B., Li, J., et al., 2014. A Horizontal Drilling Breakthrough in Developing 1.5 m Thick Tight Gas Reservoir. Paper IPTC 17486, presented at the IPTC, Doha Qatar, January 20–22, 2014.

Rogner, H.H., 1997. An assessment of world hydrocarbon resources. Annual Review of Energy and the Environment 22, 217–262.

Rushing, J.A., Newsham, K.E., Blasingame, T.A., 2008. Rock Typing – Keys to Understanding Productivity in Tight Gas Sands. SPE 114164.

Schenk, C.J., Pollastro, 2002. Natural Gas Production in the United States. US Geological Survey Fact Sheet FS-113-01, 2 p.

Schmoker, J.W., 2005. U.S. Geological survey assessment concepts for continuous petroleum accumulations. In: Chapter 13 of Petroleum Systems and Geologic Assessment of Oil and Gas in the Southwestern Wyoming Province, Wyoming, Colorado and Utah. U.S. Geological Survey Digital Data Series DDS-69-D.

Shanley, K.W., 2004. Fluvial Reservoir Description for a Giant Low-permeability Gas Field, Jonah Field, Green River Basin, Wyoming, in AAPG Studies in Geology #52, pp. 159–182.

Shanley, K.W., Cluff, R.M., Robinson, J.W., 2004. Factors controlling prolific gas production from low-permeability sandstone reservoirs. AAPG Bulletin 88 (8), 1083–1121.

Spencer, C.W., 1989. Review of characteristics of low permeability gas reservoirs in western United States. AAPG Bulletin 73 (5), 613–629.

Webb, J.C., Cluff, S.G., Murphy, C.M., Pyrnes, A.P., 2004. Petrology, Petrophysics of the Lance Formation (Upper Cretaceous), American Hunter, Old Road Unit 1, Sublette County, Wyoming, in AAPG Studies in Geology #52 pp.183–213.

Wei, Y.N., Cheng, K., Holditch, S.A., 2009. Multicomponent advisory system can expedite evaluation of unconventional gas reservoirs. In: SPE 124323, SPE Annual Technical Conference and Exhibition, New Orleans, LA, USA, October 4–7, 2009.

Wei, Y., Cheng, K., Jin, X., Wu, B., Holditch, S.A., 2010. Determining Production Casing and Tubing Size by Satisfying Completion Stimulation and Production Requirements for Tight Gas Sand Reservoirs. Paper SPE 132541, presented at the SPE Tight Gas Completions Conference, San Antonio, Texas, USA, November 2–3, 2010.

Wei, Y., Xu, J., 2015. Development of liquid-rich tight gas sand plays - granite wash example. In: Ma, Y.Z., Holditch, S., Royer, J.J. (Eds.), Unconventional Resource Handbook: Evaluation and Development. Elsevier.

Wolff, M., Pelissier-Combescure, J., 1982. FACIOLOG - automatic electrofacies determination. In: Proceeding of Society of Professional Well Log Analysts Annual Logging Symposium, Paper FF.

APPENDIX: METHODS FOR 3D LITHOFACIES MODELING
A1 INDICATOR KRIGING AND SEQUENTIAL INDICATOR SIMULATION (SIS)

Because an indicator variable represents a binary state with two possible outcomes: presence or absence, the indicator variable for three or more lithofacies can be defined in terms of one lithofacies and all others combined to indicate the absence of that selected lithofacies. Each of the lithofacies is analyzed in its turn so that all the lithofacies can be modeled accordingly.

Indicator kriging was initially developed as a nonparametric estimation of spatial distribution (Journel, 1983); it provides a least-square estimate of the conditional cumulative distribution function (ccdf) at a cutoff value for a continuous variable. However, it is now more often used for modeling categorical variables, such as lithofacies. Consider k mutually exclusive lithofacies or rock types; as any location can only have one lithofacies, k, its probability can be estimated using simple indicator kriging (Deutsch and Journel, 1991, p. 73, p. 148):

$$I^*(x) = Prob\{I_k(x) = 1\} = p_k + \sum w_i [I_k(x_i) - p_k] \tag{A.1}$$

where p_k is the overall or global proportion of the lithofacies k, w_i is the weight of the datum i, and $I_k(x_i)$ is the state of the indicator variable at x_i, i.e., the presence or absence of lithofacies, k. In other words, the probability of a lithofacies is estimated by a linear estimator, including its global proportion, and the neighboring data. This equation can be solved using simple kriging system:

$$\sum w_i c_{ij} = c_{0j} \quad \text{for } i = 1, \ldots, n \tag{A.2}$$

where C_{ij} and C_{0j} are the indicator covariances of the indicator variables for the lag distances between x_i and x_j, and x_0 and x_j, respectively. The relationship between indicator covariance and indicator variogram for a stationary indicator random function is the same as the general relationship between covariance function and variogram, i.e.,

$$Covariance\ (h) = Variance - Variogram\ (h) \tag{A.3}$$

While indicator kriging is rarely used in reservoir modeling, its stochastic simulation counterpart is commonly used for modeling lithofacies and rock type. This section discusses indicator variogram and sequential indicator simulation for modeling lithofacies.

A1.1 Indicator Variogram

Indicator variograms of lithofacies often show a second-order stationarity with a definable plateau (Jones and Ma, 2001). A lithofacies variogram observed across stratigraphic formations is typically cyclical as a function of lag distance. This has been termed a hole-effect variogram in geostatistics (Jones and Ma, 2001). Cyclicity and amplitudes in hole-effect variograms are strongly affected by relative abundance of each lithofacies and by the variation in the sizes of lithofacies bodies.

Sample density is very important for accurately describing an experimental variogram (Ma et al., 2009). In a reservoir with low sandstone fraction, if individual sandstone or other lithofacies bodies are sampled densely enough, the experimental variogram will likely show spatial correlations, possibly with cyclicity. However, if individual lithofacies bodies are sampled with only one observation each, the experimental indicator variogram will likely fluctuate around a plateau even for short lag distances, and thus appear as nugget effect. Geology is not random, but a sparse sampling can make it appear random. An experimental variogram calculated from data typically is fitted into a theoretical model, such as a spherical or exponential function, possibly with a partial nugget effect. The larger the relative proportion of the nugget effect, the more random will be the lithofacies model. The cyclicity of lithofacies in spatial distributions, vertical or horizontally, can be better modeled using a hole-effect variogram.

A1.2 Sequential Indicator Simulation (SIS)

SIS simulates the indicator variable sequentially, and it is basically the categorical-variable simulation counterpart of sequential Gaussian simulation, which is for continuous-variable simulation (Deutsch and Journel, 1991). For simulating lithofacies, at each grid cell, indicator kriging is first carried out according to Eqs (A.1) and (A.2); an ordering of the lithofacies is then defined, which also defines a cdf-type scaling of the probability interval between 0 and 1; this is followed by drawing a random number and determining the simulated lithofacies at the location. This process is repeated following a random path until all the grid cells are simulated. The previously simulated data and original data are both used in indicator kriging (Eq. (A.2)), when they are within the defined neighborhood, for conditioning the simulation.

A2 OBJECT-BASED MODELING

One of the most commonly used object-based modeling (OBM) methods is the fluvial object-based modeling, which generates channels with defined ranges in width, thickness, and sinuosity (Clement et al., 1990; Holden et al., 1997). Channels can also be modeled in association with attached crevasses/splays that can have defined ranges of width and thickness. The fluvial channel parameters defined for OBM include uncertainty ranges, based on the regional geology, seismic attribute analysis, fluvial depositional analogs, and sedimentology of the basin. An example of defining parameters in modeling meandering fluvial channels in Pinedale of the Greater Green River Basin can be found in Ma et al. (2011).

More general OBM uses the user-defined objects. The shape of the objects typically include ellipse, half ellipse, quart ellipse, pipe, lower half pipe, upper half pipe, box, fan lobe, Aeolian sand dune, and oxbow lake. Some of these objects can be modeled with a specified profile, such as rounded, rounded base, rounded top, or sharp edges. The object shapes should be determined based on the depositional characteristics.

A3 TRUNCATED GAUSSIAN SIMULATION

Some depositional environments show a clear orderly lithofacies transition; SIS can attempt to model the transition through the use of probability maps, but the SIS models often do not replicate the transition satisfactorily. The truncated Gaussian simulation (TGS) can be used for modeling the lithofacies transition more aptly. This method can ensure consistency of the indicators variograms and cross variograms in the model (Galli et al., 1994). Often a facies transition represents a nonstationarity, but truncated Gaussian method can deal with the nonstationarity of lithofacies both laterally and vertically.

Before the TGS simulation, the facies codes are transformed into continuous normal score space using cutoffs on a cumulative distribution function. The transform is based on the facies code and the probability of the facies (the target fraction for the facies if no trend is specified). TGS then performs a Gaussian random function simulation or sequential Gaussian simulation (SGS) on normal-scored continuous values. In theory, the spherical variogram, which is the most commonly used variogram, cannot be used for this method.

CHAPTER 15

TIGHT GAS SANDSTONE RESERVOIRS, PART 2: PETROPHYSICAL ANALYSIS AND RESERVOIR MODELING

W.R. Moore[1], Y. Zee Ma[1], I. Pirie[1], Y. Zhang[2]

Schlumberger, Denver, CO, USA[1]; *University of Wyoming, Laramie, WY, USA*[2]

15.1 INTRODUCTION

We discuss petrophysical analysis of wireline logs, methods for formation evaluation and issues in evaluating tight sandstone formations. The main concerns of log interpretation in tight gas sandstones include porosity interpretation, understanding the effect of clay on the log responses, accurate computation of water saturation, and permeability determination (Kukal et al., 1985). Moreover, validation of wireline-based petrophysics with routine and special core analysis, well tests, and petrography is often needed to develop a reliable interpretation model.

Petrophysical analysis can help evaluate the hydrocarbon potential, estimate gas in place and, to some extent, producibility of tight gas sands; completion technique and effort drive the economic viability for each play. Many tight gas sandstone formations are known as "tease" intervals—there is a gas show or gas kick but no or little production occurs on a conventional test. Improved technology in drilling and completion has made it possible to produce gas from many tight sandstone reservoirs. An appropriate stimulation program is critical to make wells economic in these types of reservoirs.

For optimal development of a field, a reservoir model that integrates all the data can be extremely valuable for planning the drilling, completion, and production design. The integrated data should include geological (including depositional and lithofacies such as discussed in the previous chapter), petrophysical, and engineering data. In this chapter, we first review some common issues in petrophysical analysis of wireline logs in tight gas sandstones. We then discuss the three main reservoir properties based on well logs and core data, including porosity, permeability, and water saturation. The three-dimensional (3D) modeling of these properties is presented to populate the reservoir model based on data from the wells. Issues in dynamic modeling of tight-gas sandstone reservoirs are also discussed.

15.2 COMMON ISSUES IN PETROPHYSICAL ANALYSIS OF TIGHT GAS SANDSTONES

Wireline logs in tight gas sandstones may include basic and high-tier logs. Basic logs are the triple or quad combination of neutron porosity, density porosity, resistivity, sonic velocity, and γ-ray logs. As in conventional reservoirs, the basic logs can be quite effective in determining some reservoir parameters. However, light hydrocarbon and clay effects may be exacerbated because of the abnormal pressure and low porosity. These problems need to be corrected for in characterizing tight gas sandstone reservoirs.

The spontaneous potential log (SP) is a basic measurement that may be available and useful, but often the deflection from baseline values is minimal or difficult to interpret. The SP deflection from a baseline can indicate permeability and be used to estimate connate water resistivity. In oil-based muds, typical of drilling, the spontaneous potential log is rendered useless by the mud system.

High tier logs may include spectral γ-ray, elemental spectroscopy, two-dimensional (2D) and 3D acoustic dipole velocity, nuclear magnetic resonance, and dielectric measurements. Spectral γ-ray measurements can identify the presence of uranium and help in clay typing and detection of reservoir zones. Elemental spectroscopy tools can be used to determine the percentage of minerals in the rock composition so that porosity and clay volume can be calculated more accurately. 2D and 3D acoustic dipole logs can show anisotropy and the shear-compressional data can be used as a direct hydrocarbon indicator. Nuclear magnetic resonance (NMR) data can be used as an independent porosity measurement, but can be affected by gas and light hydrocarbons. Density magnetic resonance processing is commonly used in tight gas to correct the NMR porosity and subsequent permeability estimate in the presence of gas (low hydrogen index). Dielectric logs can be used to determine water saturations using an external porosity. Work has shown that nuclear magnetic resonance and dielectric dispersion logging can be used to determine fluid types, permeability, and residual saturations (Al-Yaarubi et al., 2014).

In tight-gas sandstones, borehole washouts and rugosity can be a problem that affects all logs. In exploration areas, the log data may be old, the tight gas interval was not a zone of interest, and key data may be missing or difficult to interpret. Figure 15.1 is an example that shows washouts and a tension pull producing invalid readings in a well. Approximately 120 ft out of the 200 ft interval has been affected by the log pull or washouts. These need to be identified and corrected by using repeat runs that have useable data over the same interval.

The γ-ray measurement can be affected by kerogen and radioactivity not related to clay content (Ma et al., 2014a). Due to shallow invasion in many tight gas sands, neutron porosity deficit can be a very good indicator of formation gas. Excessive neutron porosity or neutron-density cross plots can be very useful as a shale/clay indicator. An overlay of the neutron porosity and γ-ray measurement can be used as a clay/pore fluid indicator. In the gas bearing part of the reservoir, the neutron porosity is usually lower than the γ-ray measurement. At the top of the reservoir, the presence of water can flip this relationship and can be used to pick the top of the pervasive or sustained gas. This technique can be very useful for thick intervals of stacked sandstones, but it is not really applicable for individual or limited sandstone deposits. While this technique may not always be precise, it can be a quick way to evaluate a thick stacked sandstone interval. Figure 15.2 illustrates the procedure.

15.2 COMMON ISSUES IN PETROPHYSICAL ANALYSIS OF TIGHT GAS 431

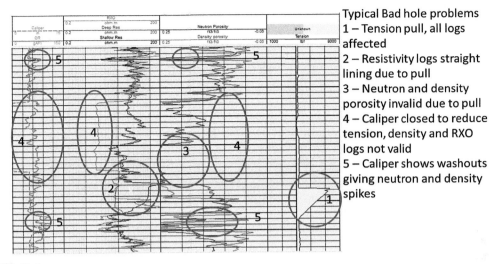

FIGURE 15.1

Typical bad hole problems.

Because there can be limited invasion in tight gas reservoirs due to low permeability and possibly higher pressure, the neutron and density porosities will need to be corrected for light hydrocarbon and gas effects. If there are differing data types from well to well, the data may need to be standardized to get a basis of comparison. Environment of deposition, rock type, and borehole size should all be considered in performing the standardization.

Figures 15.3–15.6 are plots of γ-ray, resistivity, and neutron and density porosity over stacked sandstone pay and blanket sandstone for some example intervals. They illustrate a variety of log signatures in tight gas sandstone intervals.

Figure 15.3 shows a 5000 ft interval of overpressured stacked sandstones in the Lance formation from Pinedale in the Greater Green River basin, Wyoming. The γ-ray, neutron porosity, resistivity, and density porosity all show fairly large deflections due to the moderate to high clay content.

Figure 15.4 is a 3000 ft interval of a normal to underpressured stacked sandstone in the Mesa Verde formation from the Piceance basins in Colorado. The γ-ray, resistivity, and neutron porosity deflections reflect less clay content.

Figure 15.5 illustrates a 1300 ft interval of a slightly over to under-pressured stacked sandstones in the Travis Peak formation from the East Texas Basin in Texas. The formation is fairly clean, and there is much less clay effect on the logs.

Figure 15.6 shows a 300 ft interval of blanket sandstones in the Frontier formation of the Powder River Basin in Wyoming. The formation is slightly overpressured and contains a moderate amount of clay.

432 CHAPTER 15 TIGHT GAS SANDSTONE RESERVOIRS, PART 2

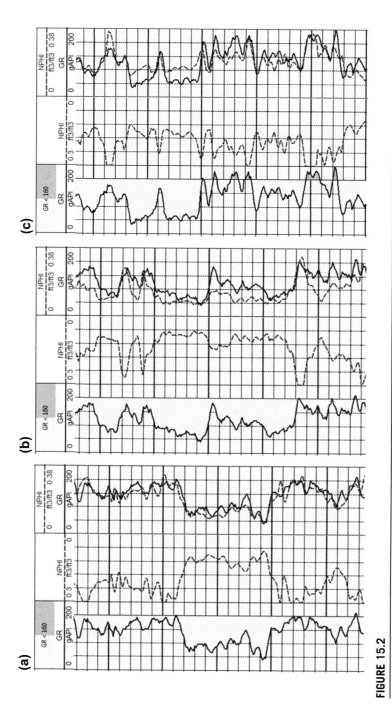

FIGURE 15.2

Example of using GR–NPHI to pick possible top gas. (a) Base of well, GR and NPHI set to overlie. (b) Middle of well—NPHI showing less shale effect than GR. (c) Top of interval—GR showing less shale effect than NPHI, probably above pervasive gas. Approximately 3000 ft between (a) and (c).

15.2 COMMON ISSUES IN PETROPHYSICAL ANALYSIS OF TIGHT GAS

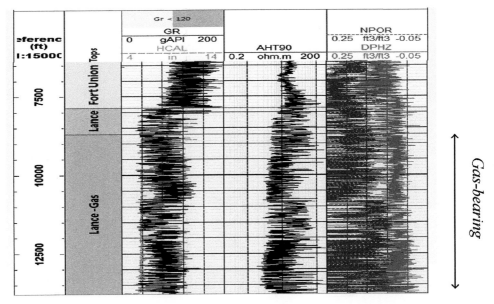

FIGURE 15.3

Log signatures of stacked sands in a productive interval at Pinedale Anticline, Wyoming (moderate clay content, overpressured zone).

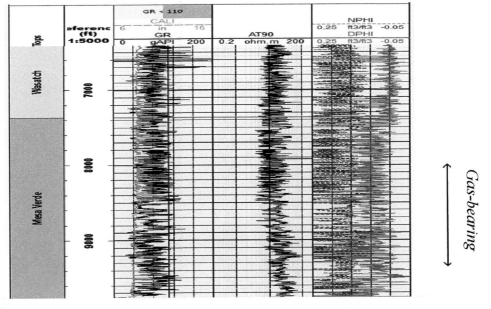

FIGURE 15.4

Log signatures of stacked sandstones in a productive interval, Piceance basin, Colorado (moderate clay content, normal to underpressured zone).

434 CHAPTER 15 TIGHT GAS SANDSTONE RESERVOIRS, PART 2

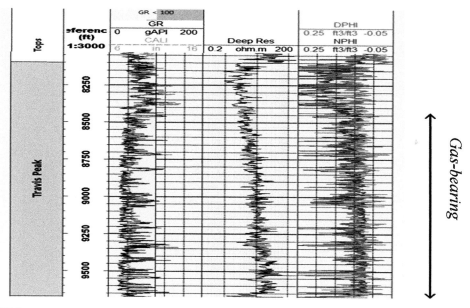

FIGURE 15.5

Log signatures of stacked sands in a productive interval of the Travis Peak formation in the East Texas Basin (minimal clay content, somewhat overpressured to underpressured zones).

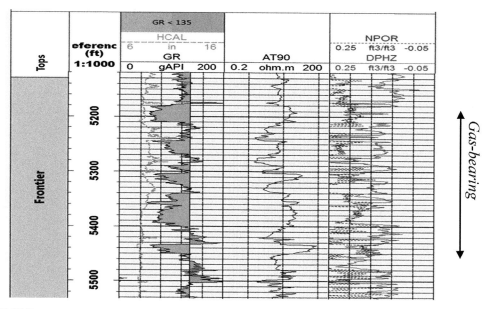

FIGURE 15.6

Log signatures of blanket sands in a productive interval of the Frontier formation of the Powder River basin, Wyoming (moderate clay, slightly overpressured zone).

15.3 PETROPHYSICAL ANALYSIS FOR RESERVOIR PROPERTIES

The main reservoir properties calculated from wireline logs typically include porosity, fluid saturation, and permeability. Interpretations of these properties using wireline logs and core are discussed here.

15.3.1 POROSITY

Total porosity is considered to be the combination of intergranular connected porosity, isolated (nonconnected) porosity, and apparent clay porosity. Effective porosity is considered to be the intergranular connected porosity. Porosity in tight gas sandstones usually ranges from 2% to 12%. Through diagenesis, the primary porosity may be reduced in the primary pore system by quartz overgrowths, and secondary porosity may be produced in feldspars or clays. The best reservoir may not equate to the cleanest because the cleanest formations often represent low porosity and producibility. (Byrnes, 1997).

The use of logs in tight gas sand analysis is in general similar to those of conventional reservoirs. Two major methods are used to derive porosity from wireline logs. The traditional method is to calculate effective porosity and water saturation through the environmentally corrected logs using a clay volume log, usually calculated from γ-ray, neutron porosity, or a combination of logs. The combinations of basic logs (density porosity–neutron porosity–sonic velocity–γ-ray measurements) can be quite effective in evaluating the porosity, but the interpretation can be enhanced by the use of core data and high tier logs. Many analysts use density porosity alone calculated from the bulk density with a variable grain matrix to correct for mineral composition (pyrites and clay minerals) and a fluid density lower than 1 to compensate for incomplete flushing in the measurement zone. This procedure can be used to generate acceptable values for effective formation porosity (Byrnes and Castle, 2000). However, only a limited number of minerals (pyrite, kerogen, quartz) can be confidently modeled using a basic log suite. In the multimineral method, the environmentally corrected logs are used in the model that defines the mineral and fluid types, and their log end points. Mineral types and volumes, porosity, and saturation are calculated to fit the input data in the model. The traditional method is simpler to implement but it does not explicitly account for mineral variations. The multimineral method is more complex as it requires more information to define mineral types and end points.

Total porosity from log interpretations in tight gas sandstones can appear to be high due to the effect of clay on the neutron porosity and sonic velocity measurements. In many areas, washouts and rugose boreholes resulting from over or under pressure can affect the log readings and make interpretation difficult, especially for density porosity. The typical response to borehole problems or increased mud gas is to raise mud weight, which can lead to even more washouts and borehole problems. Therefore, environmental corrections (borehole size, pressure, fluid type, salinity, and temperature) should be performed on all logs to get the most appropriate data set for analysis (Holditch, 2006; Moore et al., 2011). Figure 15.7 summarizes the general relationships between log measurements, core measurements, pore types, clay, rock framework, and fluid types based on an earlier study by Eslinger and Pevear (1988).

Core data can be used to validate the porosity interpretation from logs. In clean zones, the log and core should give similar readings, but these are often low porosity intervals. Sometimes, the better reservoirs could be the "shalier" zones. The γ-ray could be higher due to uranium from kerogen, or if porosity could be developed from the degradation of higher γ-ray rocks like feldspar. The core can guide building the interpretation model and the parameter selection used in these zones. Core porosity can be between the total and effective porosities depending on how the core was treated during the acquisition and analysis. Core analysis results are not absolute and can vary by laboratory (Luffel and Howard, 1987) or by the technique used (Morrow et al., 1991). Incidentally, how cores are treated prior to analysis can significantly affect the measured absolute and relative permeability (Morrow et al., 1991). If

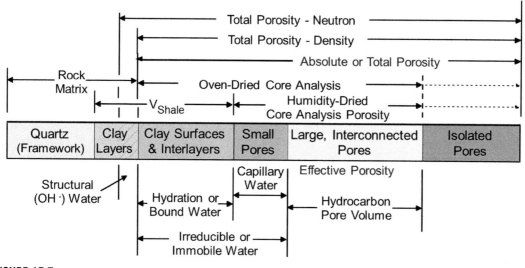

FIGURE 15.7

Porosity relationships among log measurements, core, pores, and fluids (modified from Eslinger and Pevear, 1988).

the available data is from older, publically available reports with unknown or undocumented methods, the error in core porosity may be unacceptable.

In tight gas sandstones, the total porosity calculated from logs is typically a little higher than the measured core porosity and the effective porosity calculated from logs is close to or a little lower than the measured core porosity. However, a number of variables may affect the core-log porosity relationships. When clay content is low, the total porosity calculated from logs is usually higher than measured core porosity and the effective porosity calculated from logs can be a little higher to somewhat lower than measured core porosity. When the clay content (feldspar decomposition—high illite, chlorite, and swelling clays) is high, the total porosity calculated from logs tends to be much higher than the measured core porosity and the effective porosity calculated from logs may be higher than the measured core porosity. The significant presence of heavy minerals (pyrite, siderite, dolomite and calcite) in the matrix or cement can affect the above stated core-log porosity relationships as well. A multimineral model can be used to correct the variable compositions, but often there is not enough special core analysis data (such as XRD–FTIR–XRF) or enough log data (such as spectroscopy) to differentiate the minerals.

In summary, common issues related to porosity interpretation in tight gas sandstones include:

1. Invasion can be shallow due to low permeability and generally high formation pressure, so the correction of logs for gas effect can be important.
2. Presence of heavy minerals, even at low volumes, can reduce the calculated porosity if not accounted for.
3. The cleanest intervals may be the tightest due to diagenetic factors, and may not be the best reservoir.

4. In stacked sandstone intervals, it is important to identify the top of overpressured gas when it is present, because the important parameters in the analysis (water salinity, Archie m and n, shale/clay points) may change at that point.
5. Washouts, rugose hole, and log-pull blind spots are common in tight gas intervals, and bad hole models need to be used over them where repeat passes fail to provide usable data.

15.3.2 FLUID SATURATIONS

Common issues in analyzing fluid saturations for tight gas sandstones include:

1. Irreducible water saturations can be high and may vary greatly depending on rock type.
2. Often, there is not much water production from tight gas sandstones even when estimated water saturations are high, and the water produced may be low salinity vapor from the gas phase. Using the parameters derived from traditional methods in a tight shaly sandstone, the Archie saturation equation may give abnormally high water saturation values. In this case, core capillary pressure versus water saturation curves correlated to actual production can be used to back out a salinity value that may be more representative.
3. While most tight gas sandstones do not produce a lot of water even for high water saturation zones, there can be higher porosity and permeability intervals that can.
4. Salinity may be variable over thick sections of tight gas sandstones.

Irreducible water saturation in tight gas sandstones can be quite high, and connate water resistivity can be variable. Because the Archie equation for computing water saturation was developed in the lab from empirical data in high porosity clean sandstones, using it to calculate water saturation in tight gas sandstone intervals can give inaccurate results. The exponents in the Archie equation are usually unknown, and the clay content used to correct the logs to use in the equation can be overestimated or underestimated. Clay corrected Archie equations (like Simandoux, Indonesian, and Dual Water) are often used but suffer from the uncertainties. Using log data alone for irreducible water saturation is less reliable; it is better to use core analysis (mercury injection capillary pressure measurements) to guide what the particular capillary characteristics of a producible zone are. Adjusting saturation parameters to align with capillary pressure measurements and production results can be very useful in interpreting the formation interval. The saturation and cementation exponents, m and n, from core data can be very helpful when available but should be used with caution due to core handling/analysis inconsistencies as discussed above. Figure 15.8 is an example of core measured Archie exponent m versus core porosity in a stacked sandstone interval. A linear or nonlinear fit can be used to calculate a variable, m, but confidence in the correlation, as indicated by the data scatter, will be low. More accurate relationships could be developed for different rock types, flow regimes, and lithofacies.

In stacked sandstone intervals, it is common to have a top sustained gas point at which abnormal pressure exists below it and normal or under pressure exists above it. Using different analysis parameters for the zone above the sustained gas point (salinity, Archie's m and n, clay points) can give more plausible results. Often, the formation water salinity will decrease, and Archie's m and n parameters will increase above the gas zone; thus keeping the parameters constant may result in false indications of hydrocarbons. Different methods can be used to pick this point—an increase in mud gas, an increase in connection gas, and divergence of normalized GR/NPHI (gamma ray and neutron porosity) overlay are commonly used. There are many parameters for the analysis of tight gas intervals

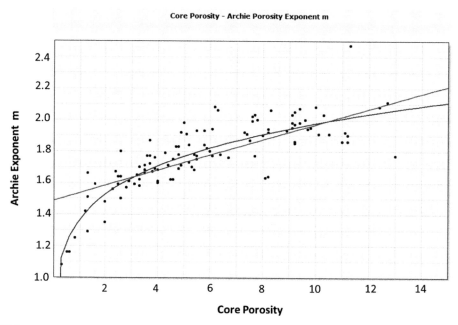

FIGURE 15.8

Cross plot of core porosity (horizontal axis) versus measured core Archie m (vertical axis) in a tight gas stacked sandstone interval. Red (gray in print versions) line shows a linear fit, and blue (dark gray in print versions) shows a nonlinear fit.

that depend on the fluid, pressure and rock type. Among them are neutron porosity corrections, density and sonic fluid corrections, and Archie m and n values.

Saturation history and not just current calculated saturation, can also affect production. The reservoir model may need to incorporate static (rock types) and dynamic (flow units) conditions to get a true picture of its viability (Kaye et al., 2013; Spain et al., 2013). If the area involved has a history of multiple burials or diagenetic events, saturation history should be investigated to understand production. Multiple instances of imbibition and drainage can affect the saturation and relative permeability of the reservoir, and may be important in reservoir development.

Capillary pressure curves can be important in defining reservoir types. Figure 15.9 illustrates capillary pressure—water saturation relationships that are typical in tight gas sandstones. There are three capillary pressure-water saturation relationships. The sweet spot sandstones are conventional reservoir sandstones within the thicker interval of tight gas sandstones, and show the lowest irreducible water saturation (15–20%) along with the lowest capillary pressure profile. The tight gas sandstones have higher irreducible water saturation (50%) and a higher capillary pressure. The shale and siltstones show the highest irreducible water saturation (more than 60%) and highest capillary pressure.

The tight gas sandstones could have relatively high water saturation but still be capable of water free gas production.

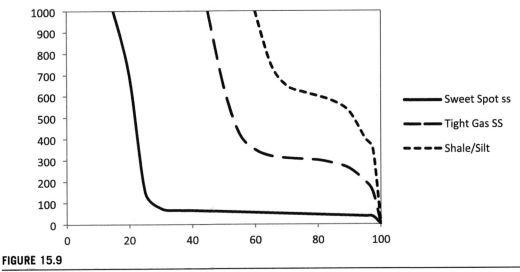

FIGURE 15.9

Conceptual models of capillary pressure versus water saturation (horizontal axis) for three different reservoir types.

Adapted from Burnie et al. (2008).

15.3.3 PERMEABILITY

Low permeability is a characteristic of tight gas sandstones, and it is generally less than 0.1 mD, often between 0.0001 and 0.01 mD. The permeability may correlate to porosity, rock type, mineralogy, stratigraphy, and other variables. Core permeability data can be used to derive a correlation between permeability and other petrophysical variables, but there is often a complex interplay among these variables in the low permeability of tight gas sandstones. Pressure, stress, rock type, diagenesis and natural fractures can all impact permeability. Sometimes, multiple porosity and permeability relationships may be necessary to correctly characterize the permeability (Kukal and Simons, 1985; Wells and Amaefule et al., 1985; Luffel et al., 1989; Davies et al., 1991; Deng et al., 2013).

Small scale tests (wireline or drill stem) can be useful in validating permeability and porosity–permeability relationships, but it may take a thorough fracture treatment, extended flow and buildup analysis to properly evaluate the permeability and porosity–permeability relationships of a tight gas sandstone interval. Higher permeability and porosity zones within a larger interval may also be water productive.

The measured gas permeability from core needs to be corrected for Klinkenberg effect for gas slippage. In a two phase gas–water reservoir fluid system, relative permeability as a function of fluid saturation drives the fluid production, which is especially pronounced when the absolute matrix permeability is in the microDarcy range, such as in tight sandstones. The accurate measurement or prediction of gas effective permeability as a function of water saturation can maximize gas production and control water cut.

Cluff and Cluff (2004) illustrated how to use core permeability measurements at reservoir net confining stress versus core measurements at some minimum confining stress for their permeability correlation. More typically, however, data sets contain only air permeability and Klinkenberg corrected

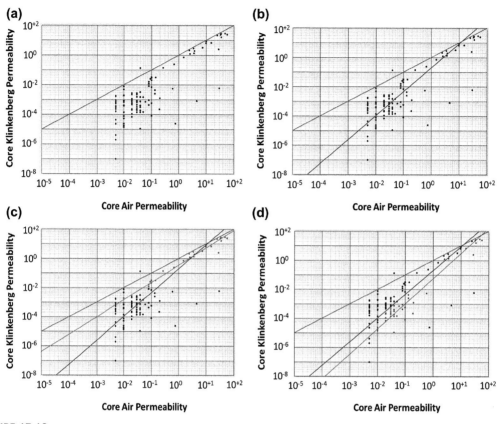

FIGURE 15.10

(a) A typical core air permeability to Klinkenberg permeability dataset. (b) Using all the data for a correlation. (c) Nonmagenta points are tight gas rocks, as a subset of the data for a correlation. (d) Green (light gray in print versions) points are tight-gas rocks, as a subset of the data for correlation. Note: red lines (dark gray in print versions) are one-to-one fits as a reference.

permeability values. Figure 15.10(a) is a multiwell core data set showing the spread that is present in many real datasets. The air permeability has not been reported with the same precision as the Klinkenberg permeability, and there is uncertainty in the correlation. Figure 15.10(b) uses a best fit line through all the data which tends to give an optimistic permeability. Figure 15.10(c) eliminates the lower permeability data, and is probably only representative of sweet spots within the tight gas interval. The selected low permeability data give a more realistic interpretation (Fig. 15.10(d)). The most appropriate solution is a combination of the models in Figs 15.10(c) and (d) using rock typing or another modeling technique for a more accurate correlation. Without enough knowledge, the correlations can be subjective, and care needs to be taken to select the appropriate data. If a permeability cutoff is used for pay determination, it should be recognized that the correlation may carry a large uncertainty in the model, leading to an uncertain estimate of the recovery, and possibly impacting the completion design.

15.3.4 DISCUSSION ON PETROPHYSICAL INTERPRETATIONS

There are a number of strategies for planning the evaluation of exploration or delineation wells. If washouts and rugose borehole are a problem, a geomechanical evaluation can be made to generate a safe mud weight window, which can help keep the borehole pressures balanced and the washouts to a minimum so that the traditional triple combo logs are not so adversely affected by the washouts. Repeat logs or down logs over these intervals may be necessary to get useable data. High tier logs can be acquired for better reservoir analysis. If core data is lacking or the existing core data measurements are of uncertain quality, more core data will be needed.

The interpretation of porosity from wireline logs can be reconciled with core data, which can then be used for model development. Along with routine core analysis, special core analysis should be performed to build a model: XRD–FTIR–XRF for mineralogy, mercury injection for rock typing, geomechanics testing to help calibrate the mechanical earth models and improve hydraulic stimulation design. For stacked sandstones over a thick interval, it may be important to analyze as many data points as possible, as the pressure regime may change over the interval.

Often there are just a few wells with core data, common and specialty logs, wherein either the traditional or multimineral method can be used to get the best analysis possible. The petrophysical model can then be simplified and applied to wells with less complete log suites and no core data. It is important that the core data distribution in both areal and vertical directions be as complete as possible to ensure adequate rock description. Rock typing can be very important to petrophysical parameter selection (Chapin et al., 2009; Liu et al., 2012). There can be sweet spots dispersed throughout the tight gas interval, and the tight gas zone parameters may not all be similar. Using the core data to constrain the calculated log porosity is generally a good practice, especially in cases where log data are limited. If there is not adequate core data for rock types, errors in porosity can be quite substantial. Where there is adequate delineation and validation, rock typing can enhance the reservoir characterization.

In summary, the analyst needs to tailor the log interpretation to fit the available data. Gross reservoir properties are fairly easy to determine, but it is the determination of what is net that requires the integration of the reservoir and petrophysical data to derive a relationship to production. High tier log data, if present, can be used to further refine the log interpretation model. While there is no absolute combination of basic and high tier logs that works all the time, each can be important and should be investigated for a given reservoir to determine its effectiveness. Using a comprehensive suite of log measurements can improve the interpretation and evaluation of the reservoir.

15.4 THREE-DIMENSIONAL MODELING OF RESERVOIR PROPERTIES

Some of the most important reservoir properties include porosity, fluid saturation, and permeability, of which porosity is the most basic variable that describes the pore space for fluid storage in the subsurface formation. Analysis and interpretation of core and well-log data describe reservoir characteristics at or near the wellbores, but hydrocarbon resource and production also depend on the distribution of reservoir properties in the field away from the wells. 3D modeling of main reservoir properties enables the calculations of field-wide pore volume and hydrocarbon pore volume, and evaluation of the heterogeneities of reservoir properties.

Because porosity is one of the most basic reservoir variables and its data are generally more available and reliable than fluid saturation and permeability data, porosity is typically modeled before modeling water saturation and permeability. When lithofacies or depositional facies models are available (discussed in the previous chapter), the porosity model should be constrained to the lithofacies or depositional facies model. This is because in the hierarchy of multiple scales of reservoir heterogeneities, characteristics of petrophysical properties are controlled by geologic facies or lithofacies (Ma et al., 2009). Geostatistical methods for modeling porosity include kriging and stochastic simulation, and they can be used with the lithofacies model as a constraint (Cao et al., 2014).

In tight gas sandstones, porosity, fluid saturation, and permeability are generally correlated with the lithofacies, and they are also correlated between themselves (Ma et al., 2011). Because of the correlation with porosity, fluid saturation, and permeability should be modeled in relation to porosity. Researchers often focus on using correlation for prediction; in fact, an accurate modeling of the correlation between fluid saturation and porosity is not just for prediction, but it has an impact on the estimation of the in-place volumetrics. Similarly, an accurate modeling of the porosity–permeability relationship is not just for better prediction of permeability, but has an impact on the hydrocarbon recovery rate.

15.4.1 CONSTRUCTING STATIC MODELS

15.4.1.1 Modeling Porosity

Geostatistical methods for modeling porosity include kriging and stochastic simulation. Kriging produces smoother results as the variance of the kriging model is smaller than the variance of the data. Commonly used stochastic simulation methods include sequential Gaussian simulation or SGS (Deutsch and Journel, 1992) and Gaussian random function simulation or GRFS (Gutjahr et al., 1997).

In early development of a field, few data are available and kriging may be a method of choice to generate the porosity model, and the moving average method can be a valid alternative technique. When more wells are drilled with a thorough formation evaluation program using well logs and geological analysis, stochastic methods may be a better choice to model porosity, especially for stacked sandstone reservoirs. Some academies have argued that stochastic simulation is preferable for modeling reservoir properties in early field developments because of limited data and high uncertainty. This can be true for the sake of a general analysis of uncertainty. In practice, however, when data is very limited, stochastic models generally have no operational value. On the other hand, when more densely sampled seismic data are available and can be calibrated with porosity, cosimulation of porosity with seismic data can be useful (Cao et al., 2014).

In addition, the lithofacies model can be used to constrain the spatial distribution of porosity using SGS or GRFS because depositional facies or lithofacies govern spatial and frequency characteristics of porosity to a large extent. Even though porosity can still be variable within each lithofacies, the porosity statistics by lithofacies generally exhibit less variation (Ma et al., 2008). Figure 15.11 compares two porosity models constructed with four different lithofacies models presented in the previous chapter.

Typically, a histogram of the effective porosity from the well logs exhibits a bimodal distribution (Fig. 15.11(a)), but a bimodal appearance often conceals some components from three or more lithofacies, as shown by the example (Fig. 15.11(b)). The hidden and nonhidden modalities can be modeled by mixture decomposition (Ma et al., 2014b). In this example, the lithofacies include

15.4 THREE-DIMENSIONAL MODELING OF RESERVOIR PROPERTIES

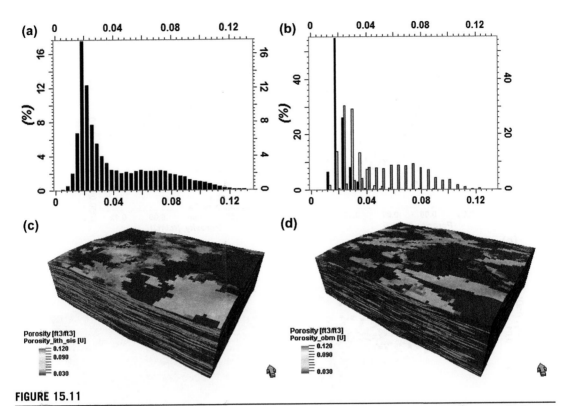

FIGURE 15.11

(a) Histogram of porosity from well logs. (b) Component histograms by lithofacies. Black is the porosity for shale, green is for siltstone, and red (dark gray in print versions) is for sandstone. (c) Porosity model constrained to the SIS lithofacies model (Fig. 14.8(b)) in the previous chapter). (d) Porosity model constrained to the defined object lithofacies model (Fig. 14.8(d)) in the previous chapter).

sandstone, siltstone and shale, and the lithofacies models constructed using SIS and defined object-based modeling techniques were used to constrain the porosity model. GRFS was used in modeling porosity by lithofacies, honoring its well log data, histogram and variogram. The models are shown in Figs 15.11(c) and (d), in which the shale was assigned zero effective porosity even though it has some total porosity.

15.4.1.2 Modeling Water Saturation and Permeability

Porosity, water saturation (Sw) and permeability in tight gas sandstone reservoirs are correlated, as shown in Fig. 15.12. As a result, Sw and permeability should be modeled in relation to porosity as porosity has more reliable data and its model is constructed first. Sometimes the correlation between porosity and Sw may only appear to be moderate; researchers may decide to model them independently because the statistical literature generally predicates the use of correlated variables for prediction. How to model the correlation between fluid saturation and porosity impacts the estimation of the in-place volumetrics. Unlike for predictions, when two physical variables are correlated, even

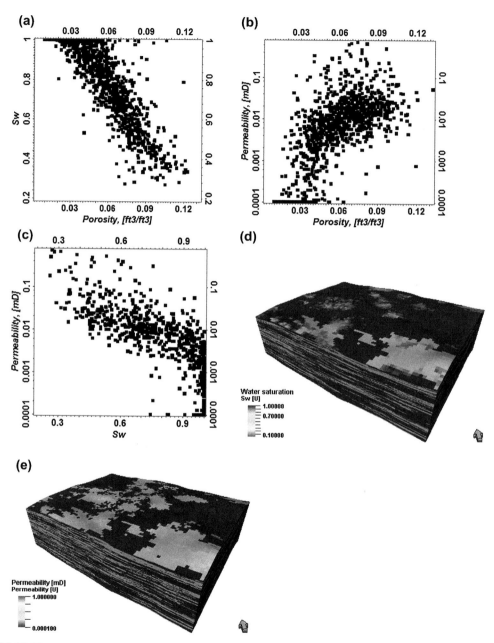

FIGURE 15.12

Reservoir property relationships based on the well-log data from a tight gas sandstone. (a) Porosity–Sw cross plot (correlation = −0.894), (b) porosity–permeability (logarithm) cross plot (correlation = 0.843). (c) Sw–Permeability (logarithm) cross plot (correlation = −0.855). (d) Sw model constructed using CocoSim that is constrained to the lithofacies model and honors the Sw data at the wells and correlation between porosity and Sw. (e) Permeability model constructed using CocoSim that is constrained to the lithofacies model and honors the permeability data at the wells and correlation between porosity and permeability.

moderately, their correlation may need to be modeled because the correlation impacts other physical properties. In the case of fluid saturation and porosity, how to model their correlation impacts the estimation of the in-place volumetrics, and thus should be modeled.

Sw and permeability can be modeled using collocated cokriging or collocated cosimulation (CocoSim) to honor the relationship between porosity and Sw or permeability (Ma et al., 2008; Cao et al., 2014). CocoSim can model Sw, honoring the well-log Sw data, its histogram, variogram, and its correlation with the porosity based on the well logs data. An example of a Sw model constructed using CocoSim is shown (Fig. 15.12(d)); the model is constrained to the SIS lithofacies model (Fig. 15.8(b) in the previous chapter) while honoring the Sw data at the wells, correlation between porosity and Sw, and the variogram synchronized between porosity and Sw.

Similarly, a 3D permeability model can be constructed using CocoSim that is constrained to a lithofacies model while honoring the well-log's permeability data, correlation between porosity and permeability, and the variogram synchronized between porosity and permeability (Fig. 15.12(e)). Other advantages of modeling the permeability using CocoSim have been discussed elsewhere (Ma et al., 2008; Cao et al., 2014). Notice that the porosity-permeability (logarithm) relationship in real data is a nonlinear correlation (Fig. 15.12). A linear transform, such as regression of the logarithm of permeability from porosity, reduces the permeability because the exponential of the mean is smaller than the mean of the exponential (Vargas-Guzman, 2009; Cao et al., 2014).

Because of the high correlations between porosity, Sw, and permeability, the Sw and permeability models generally are highly correlated as well. This correlation can be either implicitly modeled as a result of modeling the correlation between porosity and Sw and the correlation between porosity and permeability, or explicitly modeled using CocoSim.

15.4.2 DYNAMIC MODELING

Typically, static models are constructed at a high resolution to convey the geological heterogeneity, especially important in stacked sandstone reservoirs. These high-resolution models are upscaled into a coarser grid for dynamic simulation. In order to preserve the heterogeneities in the fine-grid model, the upscaling needs to select an appropriate method. Relatively robust upscaling techniques to preserve heterogeneity include the residual optimization method (Li and Beckner, 1999), and constrained optimization approach (King et al., 2006). For a relatively small sector model, upscaling may not be necessary (Apaydin et al., 2005).

One of the main tasks in dynamic simulation is the history match of the model to the production data, including pressure data from monitoring wells, historical flowing bottom hole pressures and historical productions of water and gas from producing wells (Iwere et al., 2009; Diomampo et al., 2010). Alternatively, completion, historical production and pressure data can be consolidated and directly input into a flow simulator. Boundary conditions derived from field operation can be used as production controls for the wells, and natural and hydraulic fracture properties can be assigned to the fracture cells for each well in the model (Apaydin et al., 2005). Streamline simulation can be also performed to analyze the reservoir connectivity, sweep efficiency, and other reservoir characteristics.

Forecasting performance of planned infill wells for tight gas sandstone reservoirs can be carried out by maintaining the existing well locations in the model while drilling down to the chosen well pattern and density or removing the existing well locations in the model while placing new wells with a chosen pattern and density (Diomampo et al., 2010).

Studies have shown a significant impact of the lithofacies modeling method on forecasting well performance (Apaydin et al., 2005). Typically, the SIS model has an overall higher connectivity than object-based models, and "produces" more gas and water. On the other hand, the lithofacies model using fluvial object-based modeling tends to have higher anisotropy: high connection in the fluvial direction and low connection in the perpendicular direction. The lithofacies model using object-based modeling with defined objects with ellipse-based geometry tends to have an overall spatial connectivity between the sequential indicator simulation and fluvial object-based models as this approach offers the flexibility to generate channel bodies spanning a spectrum of geometries from individual point bars to stacked and amalgamated sheets.

15.5 CONCLUSION

Petrophysical analysis based on well logs is the cornerstone for formation evaluation of tight gas sandstone reservoirs. A number of issues related to analyzing well logs in this type of formation are discussed in this chapter. Porosity is one of the most important reservoir variables in hydrocarbon resource evaluation as it describes the subsurface pore space for fluid storage. Deriving accurate porosity data based on the available well logs is thus highly important for estimation of the effective pore volume.

Volumetrics, including field-wide pore volume and hydrocarbon pore volume, depend on not only the data at wells, but more importantly the distributions of the porosity and fluid saturation of the formation in the field. The correlation between porosity and fluid saturation should be modeled not necessarily for the sake of the prediction, but for the sake of the physical nature and impact on the accuracy of in-place resource estimate. Similarly, the correlation between porosity and permeability and the correlation between fluid saturation and permeability impact the recovery of fluids in production. In a reservoir model, pathways characterized by high reservoir quality and connectivity are drained early, and more isolated and heterogeneous sandstone bodies may be drained later or remain undrained.

Because of the uncertainty and risk associated with development of tight gas sandstone reservoirs, all the development stages, including drilling, completion, stimulation, and production, should be optimized.

ACKNOWLEDGMENT

The authors thank Schlumberger Ltd for permission to publish this work and Dr Shujie Liu for reviewing and commenting on the manuscript.

REFERENCES

Al-Yaarubi, A., Onyeije, R., Lukmanov, R., Faivre, O., 2014. Advances in tight gas evaluation using improved NMR and dielectric dispersion logging. In: SPWLA-2014-PPPP, 68th Annual Technology Conference and Exhibition, Abu Dhabi.

Apaydin, O., Iwere, F.O., Luneau, B.A., Ma, Y.Z., 2005. Critical Parameters in Static and Dynamic Modeling of Tight Fluvial Sandstones. Paper SPE 95910, presented at the 2005 SPE Annual Technical Conference, Dallas.

REFERENCES

Burnie, S.W., Maini, B., Palmer, B.R., Rakhit, K., 2008. Experimental and empirical observations supporting a capillary model involving gas generation, migration, and seal leakage for the origin and occurrence of regional gasifers. In: Cumella, S.P., Shanley, K.W., Camp, W.K. (Eds.), Understanding, Exploring, and Developing Tight-Gas Sands, 2005 Vail Hedberg Conference: AAPG Hedberg Series, vol. 3, pp. 29–48.

Byrnes, A.P., 1997. Reservoir characteristics of low-permeability sandstones in the rocky Mountains. Mountain Geologist 34, 39–51.

Byrnes, A.P., Castle, J.W., 2000. Comparison of core petrophysical properties between low-permeability sandstone reservoirs: Eastern U.S. Medina Group and Western U.S. Mesaverde Group and frontier formation. In: 60304-MS SPE Conference Paper - 2000.

Cao, R., Ma, Y.Z., Gomez, E., 2014. Geostatistical applications in petroleum reservoir modeling. Southern African Institute of Mining and Metallurgy 114, 625–629.

Chapin, M.A., Govert, A., Ugueto, G., 2009. Examining Detailed Facies and Rock Property Variation in Upper Cretaceous. Tight Gas Reservoirs, Pinedale Field, Wyoming. AAPG Search and Discovery Article #20077.

Cluff, S.G., Cluff, R.M., 2004. Petrophysics of the lance sandstone reservoirs in Jonah field, Sublette County, Wyoming. In: AAPG Studies in Geology #52, Jonah Field: Case Study of a Tight-gas Fluvial Reservoir, pp. 215–241.

Davies, D.K., Williams, B.P.J., Vessell, R.K., April 15–17, 1991. Reservoir geometry and internal permeability distribution in fluvial, tight, gas sandstones, travis peak formation, Texas. In: SPE Rocky Mountain Regional Low Permeability Reservoirs Symposium and Exhibition, Denver, CO.

Deng, J., Hu, X., Liu, X., Wu, X., 2013. Estimation of porosity and permeability from conventional logs in tight sandstone reservoirs of North Ordos Basin. In: 163953-MS SPE Conference Paper - 2013.

Deutsch, C.V., Journel, A.G., 1992. Geostatistical Software Library and User's Guide. Oxford Univ. Press, 340 p.

Diomampo, G.P., Roach, H., Chapin, M., Ugueto, G.A., Brandon, N., Fleming, C.H., 2010. Integrated Dynamic Reservoir Modeling for Multilayered Tight Gas Sand Development. Society of Petroleum Engineers. http://dx.doi.org/10.2118/137354-MS.

Eslinger, E., Pevear, D., 1988. Clay Minerals for Petroleum Geologists and Engineers. SEPM Short Course 22.

Gutjahr, A., Bullard, B., Hatch, S., 1997. General joint conditional simulation using a fast Fourier transform method. Mathematical Geology 29 (3), 361–389.

Holditch, S.A., 2006. Tight gas sands. Journal of Petroleum Technology (June) 86–93.

Iwere, F.O., Gao, H., Luneau, B., 2009. Well Production Forecast in a Tight Gas Reservoir—Closing the Loop with Model-Based Predictions in Jonah Field, Wyoming. Society of Petroleum Engineers. http://dx.doi.org/10.2118/123296-MS.

Kaye, L., Webster, M., Spain, D.R., Merletti, G., 2013. The importance of saturation history for tight gas deliverability. In: 163958-MS SPE Conference Paper – 2013.

King, M.J., Burn, K.S., Wang, P., Muralidharan, V., Alvarado, F.E., Ma, X., Datta-Gupta, A., 2006. Optimal Coarsening of 3D Reservoir Models for Flow Simulation. Society of Petroleum Engineers. http://dx.doi.org/10.2118/95759-PA.

Kukal, G.C., Biddison, C.L., Hill, R.E., Monson, E.R., Simons, K.E., March 13–16, 1985. Critical problems hindering accurate log interpretation of tight gas sand reservoirs. In: SPE-DOE Joint Symposium on Low Permeability Gas Reservoirs, Denver, CO.

Kukal, G.C., Simons, K.E., May 19–22, 1985. Log analysis techniques for quantifying the permeability of sub-millidarcy sandstone reservoirs. In: SPE-DOE Joint Symposium on Low Permeability Gas Reservoirs, Denver, CO.

Li, D., Beckner, B., 1999. A Practical and Efficient Uplayering Method for Scale-Up of Multimillion-Cell Geologic Models. Society of Petroleum Engineers. http://dx.doi.org/10.2118/57273-MS.

Liu, S., Spain, D.R., Dacy, J.M., June 16–20, 2012. Beyond volumetrics: petrophysical characterization using rock types to predict dynamic flow behavior in tight Gas sands. In: SPWLA 53rd Annual Logging Symposium.

Luffel, D.L., Howard, W.E., May 18-19, 1987. Reliability of laboratory measurement of porosity in tight gas sands. In: SPE-DOE Joint Symposium on Low Permeability Gas Reservoirs, Denver, CO.

Luffel, D.L., Howard, W.E., Hunt, E.R., March 6–8, 1989. Travis peak core permeability and porosity relationships at reservoir stress. In: SPE Joint Rocky Mountain Regional Low Permeability Reservoirs Symposium and Exhibition, Denver, CO.

Ma, Y.Z., et al., 2014a. Identifying Hydrocarbon Zones in Unconventional Formations by Discerning Simpson's Paradox. Paper SPE 169496 presented at the SPE Western and Rocky Regional Conference, April, 2014.

Ma, Y.Z., Wang, H., Sitchler, J., et al., 2014b. Mixture decomposition and lithofacies clustering using wireline logs. Journal of Applied Geophysics 102, 10–20. http://dx.doi.org/10.1016/j.jappgeo.2013.12.011.

Ma, Y.Z., Gomez, E., Young, T.L., Cox, D.L., Luneau, B., Iwere, F., 2011. Integrated reservoir modeling of a Pinedale tight-gas reservoir in the Greater Green River Basin, Wyoming. In: Ma, Y.Z., LaPointe, P. (Eds.), Uncertainty Analysis and Reservoir Modeling. AAPG Memoir 96, Tulsa.

Ma, Y.Z., Seto, A., Gomez, E., 2009. Depositional facies analysis and modeling of Judy Creek reef complex of the Late Devonian Swan Hills, Alberta, Canada. AAPG Bulletin 93 (9), 1235–1256. http://dx.doi.org/10.1306/05220908103.

Ma, Y.Z., Seto, A., Gomez, E., 2008. Frequentist meets spatialist: A marriage made in reservoir characterization and modeling. SPE 115836, SPE ATCE, Denver, CO.

Moore, W.R., Ma, Y.Z., Urdea, J., Bratton, T., 2011. Uncertainty analysis in well log and petrophysical interpretations. In: Ma, Y.Z., LaPointe, P. (Eds.), Uncertainty analysis and reservoir modeling, memoir, 96. AAPG, Tulsa, Oklahoma, pp. 17–28.

Morrow, N.R., Cather, M.E., Buckley, J.S., Dandge, V., April 15–17, 1991. Effects of drying on absolute and relative permeabilities of low-permeability gas sands. In: SPE Rocky Mountain Regional Low Permeability Reservoirs Symposium and Exhibition, Denver CO.

Spain, D.R., Merletti, G., Webster, M., Kaye, L., 2013. The Importance of Saturation History for Tight Gas Deliverability. SPE 163958.

Vargas-Guzman, J.A., 2009. Unbiased estimation of intrinsic permeability with cumulants beyond the lognormal assumption. SPE Journal 805–809.

Wells, J.D., Amaefule, J.O., May 19–22, 1985. Capillary pressure and permeability relationships in tight gas sands. In: SPE-DOE Joint Symposium on Low Permeability Gas Reservoirs, Denver, CO.

GRANITE WASH TIGHT GAS RESERVOIR

CHAPTER 16

Yunan Wei, John Xu
C&C Reservoirs Inc., Houston, TX, USA

16.1 INTRODUCTION

Substantial volumes of hydrocarbons are accumulated in low-permeability reservoirs. These low quality hydrocarbon resources are called unconventional reservoirs, which include tight gas sands. In the 1970s, the United States (US) government defined a tight gas reservoir as a reservoir with an expected value of permeability to gas flow of 0.1 mD or less. However, this definition is rather a "political" definition that was used by both state and federal government agencies to establish incentives for operators who choose to produce gas from unconventional reservoirs. Holditch, in his distinguished author series article for The Society of Petroleum Engineers (SPE) (Holditch, 2006), defined a tight gas reservoir as "a reservoir that cannot produce at economical rates nor recover economic volumes of natural gas unless the well is stimulated by large-scale hydraulic fracture treatment or produced by use of a horizontal wellbore or multilateral wellbores. The Granite Wash (GW) tight gas sand play is a good example for the development of tight gas sand reservoirs.

The Granite Wash Play is a liquid-rich tight sands play about 160 miles long by 30 miles wide, covering Hemphill, Roberts, and Wheeler counties in the Texas Panhandle and Beckham, Custer, Roger Mills, and Washita counties in south-west Oklahoma (Figs 16.1 and 16.2). The former is often referred to as the Texas Panhandle Granite Wash, while the latter is referred to as the Colony Granite Wash. The total recoverable resources potential is 500 trillion cubic feet equivalent (TCFE), including gas, condensate, and oil. The play is located in the Anadarko Basin, which is one of the deepest basins in the US and contains a significant section of Cambrian to Permian strata. The Granite Wash reservoir rocks were deposited during the Pennsylvanian period from the weathering of ancestral Wichita/Amarillo uplift and subsequently compacted into stacked layers of discontinuous lobes of sandstones and siltstones separated by shales or amalgamated sheet sandstones. There are typically 3 to 20 stacked layers, with the upper three layers typically oil-rich. The 9000 to 15,000 ft true vertical depth (TVD) (12,000 ft TVD average) deep Granite Wash Formation has a total gross thickness of 1500–3500 ft (2000 ft average). Single reservoir layer thickness ranges from 20 to 200 ft (100 ft average) with a net/gross ratio of 0.05–0.70 (0.56 average). Permeability varies from 0.0005 to 100 mD (0.001 mD average) and the porosity range is 1–16% (8% average).

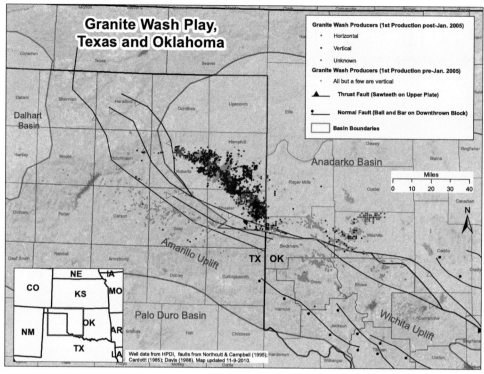

FIGURE 16.1

Map showing structural elements of the Granite Wash Play. Also shown is the well location and well type in the play as of November 2010. Most of the wells are concentrated in Roberts, Hemphill and Wheeler counties in Texas and Washita/Roger Mills counties in Oklahoma (US. Energy Information Administration, 2010).

In 1956 the first two exploration wells were drilled in the Granite Wash Formation, the Price F No. 1 in Hutchinson County of Texas by Phillips Petroleum Company and the Lard No. 1 to 3 in Roberts County of Texas by Union Oil Company, drilled vertically to total depths of 8070 ft and 9362 ft, respectively. Upon testing, the Price F No. 1 well flowed at 81 barrel oil per day (BOPD) and the Lard No. 1 well produced 39 BOPD. The play has been producing oil and gas since then, primarily as a modest oil/gas play with vertical wells targeting fairly tight reservoir rocks, which were stimulated with traditional single-stage hydraulic fracturing where necessary. Before 2006, almost all reported Granite Wash wells were vertical (Rajan and Moody, 2011). Since then, new technologies such as horizontal wellbores with long-reach laterals, and multistage hydraulic fracturing have sparked renewed interest and have brought new life on the mature Granite Wash Play. The Granite Wash Formation has become a perfect opportunity of applying new technologies to better drain individual zones, and to bring new life to this mature liquid-rich gas play.

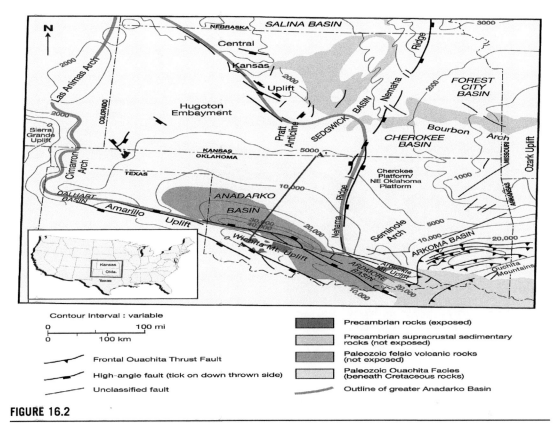

FIGURE 16.2

Regional tectonic elements of the Anadarko Basin and surrounding areas showing depth structure of the top of basement *(modified from Forster et al., 1998)*. Cross-section A–A' is shown in Fig. 16.3.

16.2 BASIN EVOLUTION

The Anadarko Basin is a foreland basin formed during the early Pennsylvanian Wichitan Orogeny as a result of collision of the South American and African plates with the southern margin of the North American craton (Thomas, 1991) (Fig. 16.2). This caused folding and uplift of the Amarillo–Wichita fold and thrust belt, cutting through what had been a thermal sag basin above a rift during earlier Paleozoic times. The basin has a broad, gently-dipping shelf area to the north and a southward-thickening sedimentary section that comprises up to 40,000 ft of Upper Cambrian–Permian strata resting on >7000 ft of Middle Cambrian rift volcanics in the foredeep, which abut the fold and thrust belt in the south (Fig. 16.3). The Anadarko Basin is bounded by the Sierra Grande Uplift and Las Animas Arch in the west, the Central Kansas Uplift to the north and the Nemaha Ridge in the east (Fig. 16.2) (Clement, 1991; Keighin and Flores, 1993).

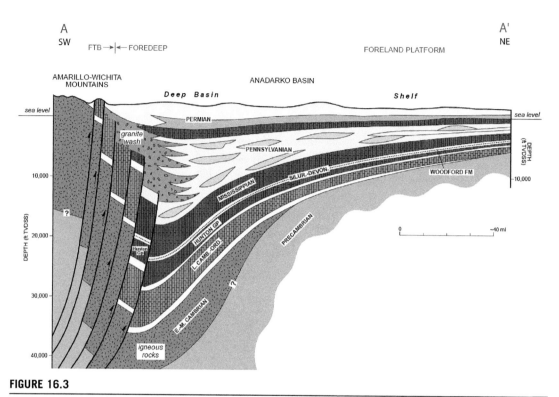

FIGURE 16.3

SW–NE schematic geological section across the Anadarko foredeep and platform *(modified from Johnson, 1989)*. Section location is shown in Fig. 16.2.

Following the cessation of Middle Cambrian rifting, the Anadarko Basin was filled with an almost complete Paleozoic sedimentary section and has produced hydrocarbons in some parts of the basin from every system present. The succession begins with up to 8000 ft of Upper Cambrian-Lower Ordovician deposits of mainly platform carbonates (Timbered Hills and Arbuckle groups) (Fig. 16.4). Alternating units of marine-shelf sandstones and carbonates characterize the Middle-Upper Ordovician. These are followed by carbonate deposits of the uppermost Ordovician–Devonian Hunton Group. Regional unconformities of end-Early Devonian and particularly end-Middle Devonian age, associated with uplift and basin–ward tilting during the Acadian Orogeny, truncate progressively older units toward the basin margins (Adler, 1971). Marine transgression in the latest Devonian-earliest Carboniferous resulted in accumulation of the organic-rich Woodford Shale, the primary hydrocarbon source rock in the basin. Mississippian sediments are largely carbonates and shales deposited on a marine shelf that gradually differentiated into a stable northern part and a rapidly subsiding southern depocenter. prePennsylvanian reservoirs are largely carbonates containing hydrocarbons in structural traps. At the beginning of the Pennsylvanian, clastic sediment flooded the basin from the north to form the sandstone reservoirs and shales of the Springer and Morrow groups (Clement, 1991; Davis and Northcutt, 1991).

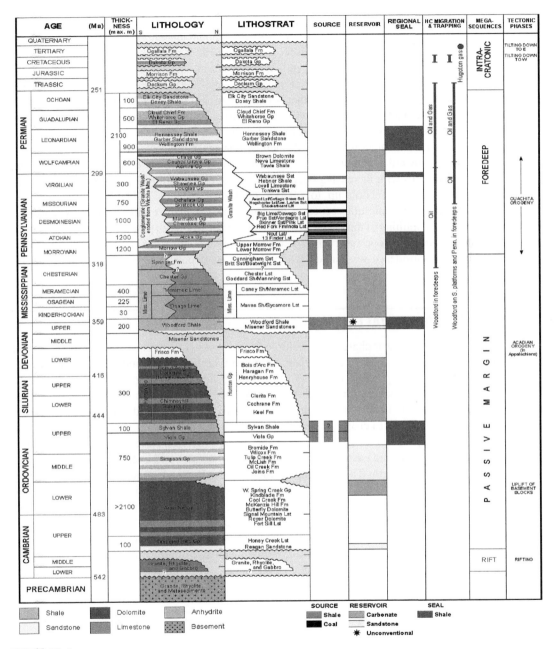

FIGURE 16.4

Generalized stratigraphy of the Anadarko Basin showing megasequences, petroleum system elements and tectonic history.

Compiled from Johnson (1989), Clement (1991), and other sources.

The Wichitan Orogeny began to drastically reshape the basin from latest Morrowan time with emergence of the east to southeast (ESE)-trending Amarillo–Wichita mountain belt, which provided a southern source of coarse clastic sediment (Fig. 16.3) (Rascoe and Adler, 1983). A thick wedge of conglomeratic sandstone, detrital dolomite and "granite wash" spilled northward into a deep trough adjoining the uplift and interfingered with basinal shales and cherty carbonates (Lyday, 1990). On the northern shelf, tectonic pulses and frequent eustatic sea-level changes that occurred throughout the Pennsylvanian led to cyclic sedimentation of predominantly carbonates with lesser fluvio-deltaic to shallow-marine shale and sandstone units that form stratigraphic traps. Some structural traps occur over rejuvenated deeper structures. The Permian sedimentation began with shelf carbonates at the center and marginal-marine to continental sandstones toward the margins of the basin. An arid climate established itself during Middle–Late Permian times with increasing occurrence of evaporites, dolomites, continental red beds, and aeolian sandstones in a section up to 6000 ft thick that eventually blanketed the denuded Amarillo–Wichita Uplift. Tectonic quiescence in the late Pennsylvanian resulted in more widespread sandstone deposition and infilling of the southern trough. Permian carbonates produce hydrocarbons from stratigraphic traps, while sandstone reservoirs more commonly contain structurally trapped hydrocarbons. Triassic-Jurassic subsidence was minor, while Cretaceous deposits were possibly up to 4000 ft thick.

16.3 SOURCE ROCK EVALUATION

The Granite Wash gas system was charged by multiple source rocks (LoCricchio, 2012). The two most likely sources are the latest Devonian to earliest Mississippian Woodford Shale and the flooding surfaces of the Pennsylvanian shales separating the various Granite Wash intervals (Fig. 16.4) (Gilman, 2012).

The Woodford Shale, which underlies the Granite Wash Formation, is a phenomenal regional source rock and has generated the vast majority of hydrocarbons in the Anadarko Basin (Wang and Philp, 1997). It contains Type II kerogen with average total organic content (TOC) of 4–6% (with a maximum of 26%). The present thermal maturity of the Woodford Shale varies from early oil mature around the northern, western, and eastern basin margins to gas mature and overmature in the deepest parts toward the southern margin. The flooding surfaces of the Pennsylvanian shales could also be the source rocks that charged the Granite Wash Formation (Fig. 16.4). However, the source rock qualities of these flooding surfaces are different. Source rock analyses performed on several flooding surfaces in the Desmoinesian and Missourian units show that the Missourian Shale Unit contains up to 7% TOC with Type II and III kerogens and were mature for oil generation while the Desmoinesian Shale Unit is not a good source rock. The Pennsylvanian Morrow Shale contains Type III kerogen with TOCs of 1.4–2.8% with maximum organic richness developed close to the Amarillo–Wichita mountain front.

16.4 TRAP AND SEAL

The Granite Wash Play covers approximately 4800 square miles in the southern part of the Anadarko Basin (Figs. 16.1 and 16.5). The basin dips gently to the north but is separated from the Amarillo/Wichita Uplift on the south by major faults. The Granite Wash Formation in the basin is 1500–3500 ft (2000 ft average) thick and comprises layered tight sandstones and interbedded shales.

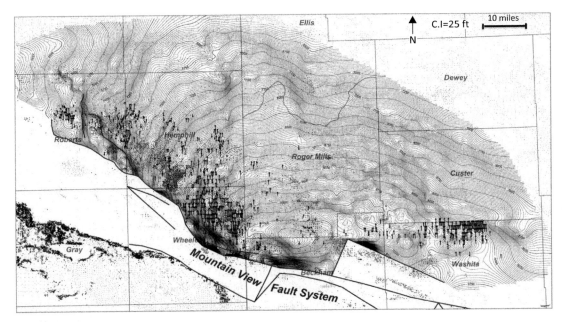

FIGURE 16.5

Regional structure map of the Marmaton Wash, which is a unit of the Desmoinesian age Granite Wash Formation.

LoCricchio (2012); AAPG@2012 reprinted by permission of the AAPG whose permission is required for further use.

It occurs at 9000 to 15,000 ft TVD (12,000 ft TVD average), and contains numerous structural and stratigraphic traps for oil and gas in multiple reservoir zones. In certain parts of the play the gas is found as deep as 17,000 ft (Dar, 2010). Several trapping mechanisms, such as lateral depositional pinchout, large-scale unconformities, overlaps, folds, and faults provide appropriate geological conditions for significant, commercially attractive hydrocarbon accumulations (Srinivasan et al., 2011). One example is the trapping of nine Pennsylvanian reservoir intervals (Granite Wash) at Cheyenne West Field. The main reservoir, the Puryear, produces from a stratigraphic trap on a gently-dipping homocline. The trap is formed by lateral pinchout to the north, east and west of the distal end of a sandstone-conglomerate fan-delta lobe (Puckette et al., 1996). Although a fault associated with the regional Amarillo–Wichita fault system cuts the southern part of Cheyenne West Field, it does not affect production.

The trap in the tight sand reservoir can have either partially or fully-enclosing stratigraphic seals. The lateral seal to the trap is provided by lateral depositional pinchout into shales, while the top seal is provided by marine shales on top of each reservoir unit.

16.5 STRATIGRAPHY AND DEPOSITIONAL FACIES

The Early Permian to Pennsylvanian Granite Wash is a stacked heterogeneous series of sandstones, shales and siltstones derived from the ancestral Wichita/Amarillo Uplift. Geologically, this play is a combination of the Virgilian, Missourian, Desmoinesian, and Atokan formations. The entire play is characterized as an extremely complex, heterogeneous rock with a diverse mineralogy. The name Granite Wash is actually a catch-all term that refers to a number of oil and gas producing formations, mostly Pennsylvanian in age. As a result, stratigraphic nomenclature for productive formations in Granite Wash fields varies considerably. Categorizations vary across a number of company schemes, with designations including the Marmaton, Kansas, Cherokee, Red Fork, Cleveland, and Atoka washes (Table 16.1). For tight-gas classification, the Granite Wash-A, or Cherokee Marker horizons are commonly used to designate the top of the interval on gamma ray logs. The bottom of the Granite Wash is typically marked by the 13 Finger Lime or the top of the Morrow Shale (LoCricchio, 2012). Some operators also designate the formation as Virgilian (Tonkawa), Missourian (including Lansing, Kansa and Cleveland), Desmoinesian Granite Wash (including Carr, Britt, A, B, C, D, E, and F), and Atoka Wash (including A, B, C, D, E) (Table 16.1) (Shipley, 2013).

The Granite Wash was deposited in stacked alluvial fan, fan delta, slope, and submarine fan channel environments (Fig. 16.6), with stacked successions of coarse clastics interfingering with marine shales and carbonates (Rothkopf et al., 2011). Before the Granite Wash was deposited, units of Lower Paleozoic were deposited on a stable, shallow shelf that was periodically covered by epicontinental seas. The Wichita Orogeny led to the uplift of a large crustal block known as the Amarillo–Wichita Uplift along the southern boundary of the NW–SE trending Anadarko foreland basin. The Amarillo–Wichita mountain front became the source of deposition as erosion of the mountains occurred. The resultant deposits represent a stratigraphic inversion of the originally uplifted strata. The Granite Wash detrital grains were derived from granite, rhyolite, gabbro, sandstone, chert, limestone, and dolomite. This is evidenced by the amount of quartz, K-feldspar, plagioclase feldspar, calcite, chert, and rock fragments within these rocks (Crawford et al., 2013). These Granite Wash rocks were deposited in a primarily SW–NE direction into the Anadarko Basin as fan delta and submarine fan accumulations coming off the mountains (Natural Gas Intelligence, 2014). Erosion of earlier Paleozoic sediments and Precambrian granitic basement atop the uplift resulted in the deposition of a thick wedge of fine to very course-grained, poorly sorted, arkosic sandstone in the basin adjacent to and north of the bounding Amarillo–Wichita fault. The majority of Granite Wash sediments were deposited during the Atokan through Desmoinesian time periods. Upon cessation of tectonic uplift, this area underwent subsidence and burial followed by the deposition of Permian-age evaporites and redbeds that serve as a cap for the Granite Wash sequence (Rothkopf et al., 2011).

16.6 RESERVOIR ARCHITECTURE AND PROPERTIES

The Granite Wash Formation consists of a wedge of discontinuous sandstones and siltstones that thins northward. The depositional sandbodies are lenticular in geometry with reservoir properties varying significantly. Type logs demonstrate that the formation can be divided into a number of reservoir zones separated by high γ-ray shale intervals (Fig. 16.7). These shales are widespread in the basin but pinch out toward the uplift (Rothkopf et al., 2011). There are 3–20 stacked productive intervals. An individual sand package is 20–200 ft (100 ft average) thick while the entire Granite Wash Formation can be 1500–3500 ft thick (2000 ft average). The net/gross ratio is 0.05–0.7 (0.56 average) (Casero et al., 2013). Detailed analysis based on well logging and effective permeability shows the dynamically calibrated net pay thickness is 244.5 ft from 1130 ft net thickness of 1580 ft gross interval (Rushing et al., 2009).

16.6 RESERVOIR ARCHITECTURE AND PROPERTIES

Table 16.1 Stratigraphic Nomenclature for Productive Formations in Granite Wash Fields by Different Companies, Showing Varying Categorizations across a Number of Company Schemes

Epoch	Age	Category 1 (LoCricchio, 2012) (From Cordillera Energy)	Category 2 (Shipley, 2013) (From Linn Energy)	Category 3 (Farris, 2014) (From Apache)
Lower Permian	Wolfcampian	Hugoton/Pontotoc (Brown Dolomite) Chase/Council Grove Admire		Hugoton Brown Dolomite Chase/Council Grove Admire
Pennsylvanian	Virgilian	Wabaunsee Shawnee Douglas Tonkawa	Tonkawa	Upper Virgil Douglas Tonkawa
	Missourian	Cottage Grove Hoxbar/Hogshooter Checkboard Cleveland	Lansing Kansas (Hogshooter) Cleveland Carr Britt	Cottage Grove Hogshooter Checkboard Cleveland
	Desmoinesian (Des Moinesian)	Marmaton group (Glover/Big Line/Oswego) Cherokee (Skinner/ Pink Lime/Red Fork)	A/GW A B/GW B C/GW C D/GW D E/GW E F/GW F	Marmaton Oswego Cherokee Skinner Red Fork
	Atokan	Atoka Lime 13 Finger Lime	"A" to "C" LWR "C" through "E"	Atoka 13 Finger Lime
	Morrowan	Morrow Shale/Dornick Hill shale		Morrow SH Morrow SD
Mississippian	Chesterian	Springer		
	Meramecian	Meramec Lime/St Louis		
	Osagean	Osage Lime/Osage chert		
	Kinderhookian	Kinderhook/Sycamore Lime		
Upper Devonian		Woodford Hunton		

Lower (LWR); Granite Wash (GW); Shale (SH); Sand (SD)
Compiled from LoCricchio (2012), Shipley (2013) and Farris (2014).

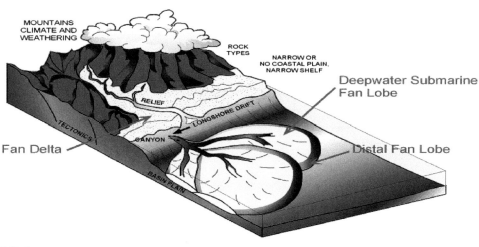

FIGURE 16.6

Depositional model of Granite Wash.

Casero et al. (2013), reprinted with written permission from Core Lab.

The dynamic nature of Granite Wash deposition resulted in highly variable grain sizes and permeabilities. Generally, reservoir rock composition varies from quartz and feldspar-rich at the top to more carbonaceous and finer grained down section (Srinivasan et al., 2013). The lower section of Granite Wash (Morrowan) consists typically of chert. The Atokan Granite Wash varies from chert to carbonate, while the lower and middle Cherokee Granite Wash comprises typically carbonates. The Red Fork Granite Wash consists of carbonates with granitic material, the amount of which was controlled by local drainage areas during deposition (Grieser and Shelley, 2009). The grain size, sorting and lithology can change dramatically over a short vertical distance within a well bore. For example, the grain size can vary from bolder down to silt-sized particles in a few feet (Fig. 16.8).

The Granite Wash tight gas sand has porosity ranging from 1 to 16% ($\sim$8% average) and gas permeability from 0.0005 to 100 mD (0.001 mD average). The low permeability is attributed to significant diagenesis, including compaction, cementation and the presence of chlorite clay, which can fill up to 65% of the available pore space (Rothkopf et al., 2011). Detrital quartz and feldspar as well as lesser quantities of lithic fragments make up the majority of framework grains. Authigenic pore-filling components including quartz overgrowths, calcite and chlorite clay account for about 15% of bulk rock volume (Rothkopf et al., 2011).

16.7 RESOURCES AND FLUID PROPERTIES

There are no reported data on original in-place resources, but a total recoverable resources potential of 500 TCFE was estimated in 2012 (LoCricchio, 2012). Tight gas sand reservoirs may present challenges to reservoir engineers in reserve estimation. Owing to the long period to reach the pseudo-steady stage flow pattern, it is problematic to apply classic reservoir engineering techniques to

16.7 RESOURCES AND FLUID PROPERTIES

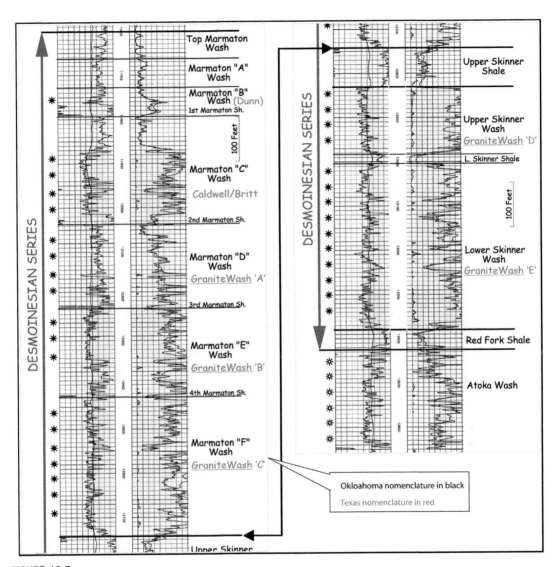

FIGURE 16.7

Type log of the up to 1500 ft thick Desmoinesian Granite Wash series from the Devon #16-4 Truman–Zybach Well in Sec. 16, Block R&E Survey, Wheeler, Texas.

Mitchell (2012); reprinted with written permission.

these reservoirs. As a result, it is difficult to accurately estimate the ultimate recovery from the wells. Both decline curve and material balance methods were found to have drawbacks when applied to tight gas reservoirs that had not established a constant drainage area (Cox et al., 2002). Holditch (2006) concluded that the best reserves evaluation techniques for tight gas sands were careful application of hyperbolic decline curves and reservoir modeling with adequate layers. Having recognized the steep

FIGURE 16.8

Image log over a 75 ft interval in the Wheeler County of Texas, demonstrating the wide range of grain size in the Granite Wash Formation (Ingram et al., 2006).

initial decline rates and long periods of transient flow in tight gas reservoirs, Kupchenko et al. (2008) concluded that using transient production data would result in inaccurate forecasts. As a solution, they suggested using Arps' original equation with certain exponent restrictions to achieve better forecasts. Generally, estimated ultimate recovery (EUR) gas equivalent per well in the Granite Wash Play is 3.14–17 billion cubic feet equivalent (BCFE) (6.49 BCFE average) for horizontal wells, compared with 0.67–1.5 BCFE (1.06 BCFE average) for vertical wells.

The Granite Wash Play is liquid-rich and produces both gas and condensate, the latter accounting for 30–40% of the total production (Natural Gas Intelligence, 2014). The condensate comprises

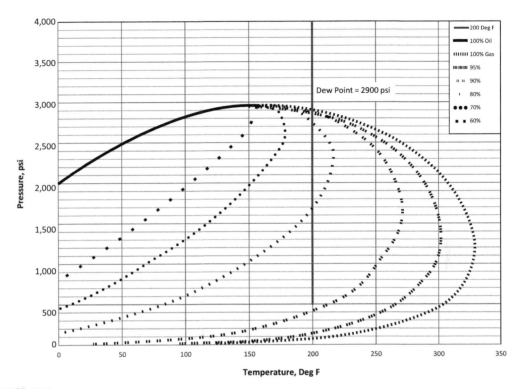

FIGURE 16.9

PVT chart of the Granite Wash gas (Rothkopf et al., 2011).

propane, butane, pentane, hexane, and heptane (Rothkopf et al., 2011). With a pressure gradient of 0.47–0.7 psi/ft, the reservoir is normally pressured in most parts of the basin but with some overpressured areas. The produced condensate has an American Petroleum Institute (API), gravity of 43–61.3°API (56° API average). The condensate yield varies from 10 to 100 barrel condensate per million cubic feet gas (BC/MMCFG) (30 BC/MMCFG average) depending on reservoir depth and location. For example, condensate yield in the Texas Panhandle is 60–70 BC/MMCFG in the upper Granite Wash, but decreases to 10 BC/MMCFG below. PVT analysis shows a 2900 psia dew point (Fig. 16.9). As shown in the figure, condensate yield of the reservoir decreases with decreasing reservoir pressure from the dew point to about 2300 psia, and increases again later in the life of the well (Rothkopf et al., 2011). Gas-water relative permeability curves show the Granite Wash rock to be water-wet, with ~70% gas-water cross point water saturation (Fig. 16.10). Typical reservoir temperature is ~200 °F.

16.8 PRODUCTION HISTORY

The Granite Wash has been producing oil and gas since 1956, with almost exclusively vertical wells before 2006 (Rajan and Moody, 2011). These vertical wells targeted fairly tight reservoir rocks and

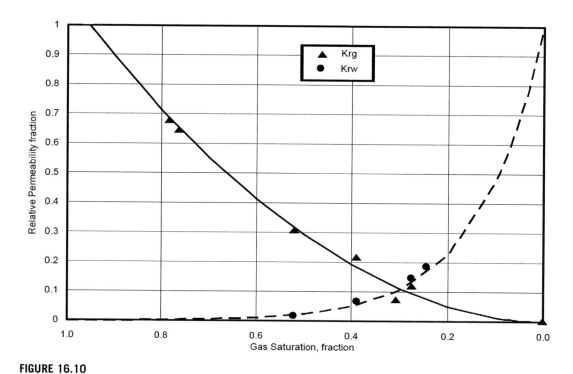

FIGURE 16.10

Water-gas relative permeability curves of the Granite Wash reservoir in the Texas Panhandle area (Rushing et al., 2009).

were stimulated with traditional single-stage hydraulic fracturing where necessary. The development accelerated rapidly after November 2004 when 20-acre downspacing was approved in the Buffalo Wallow Field. In 2003–2008, vertical infill drilling was actively pursued with operators using an assembly line methodology that concentrated on cost reduction to achieve economic success. However, vertical well counts dropped rapidly in 2009 as a result of the plummeting gas price. Meanwhile, advances in horizontal drilling and completions made horizontal wells the preferred development option. As of 2012, more than 70% of the wells in the Texas Panhandle area were horizontal which were completed by multistage hydraulic fracturing in order to connect as much pay as possible (Kennedy et al., 2012).

Owing to the huge area, large number of operators, two different states (Texas and Oklahoma), and different definitions of condensate and natural gas liquids (NGL), only gas production rates from different sources can approximately match with each other. Therefore, production history is analyzed based on gas rate. The gas production history can be divided into five stages, including development (1956–1985), plateau (1986–1988), decline (1989–1993), mature (1994–2003), and rejuvenating (2004–present) (Fig. 16.11). In 1964–1974, there were only a relatively flat number of Granite Wash completions. The number of producing wells increased to almost 2000 through the last boom period of 1977–1985. As a result, production entered the plateau stage in 1986–1988. The next surge in well completion occurred

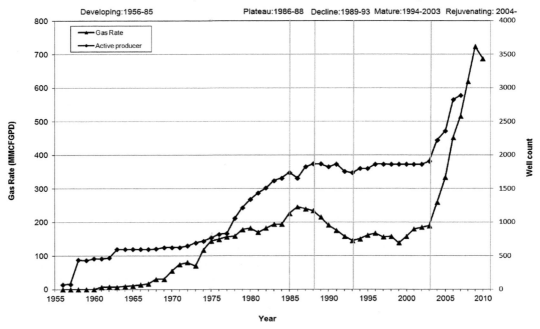

FIGURE 16.11

Gas production history of Granite Wash Play from 1956 to 2010.

Compiled from various sources.

during the period of 2003–2008 with more than 1000 primarily vertical wells added, owing to the high gas price. As a result, gas production from the Granite Wash development increased sharply, rising from ~190 million cubic feet of gas per day (MMCFGPD) in 2003 to ~619 MMCFGPD in 2008. By 2007, the play had produced a total of 1.88 trillion cubic feet gas (TCFG) plus 61 million barrel condensate (MMBC). In the first eight months of 2012 alone, the Granite Wash Play produced >1.8 MMBC and 250 billion cubic feet gas (BCFG). IHS Cambridge Energy Research Association Inc. (IHS CERA) projects that by 2025 the production from the Granite Wash will exceed three billion cubic feet gas per day (BCFGPD) gas with 140,000 barrel condensate per day (BCPD) (Rajan and Moody, 2011).

16.9 HORIZONTAL WELLS

As the Granite Wash Formation has multiple stacked pay zones, the reservoir was originally drilled with vertical wells to contact as many pay zones as possible, with an emphasis on minimizing well cost. However, the laterally continuous nature of the south-west to north-east trending submarine-fan channel deposits make the Granite Wash ideal for horizontal well development. Due to the heterogeneity in mineralogy and pore pressure and the varying hydrocarbon type, production potential varies greatly throughout the formation. As a result, determining the best completion method can be difficult and must be evaluated carefully. In the past few years, operators have been improving well designs and operations, such as using new drilling bit technology and cementing procedures to reach target depths more quickly, replacing diesel with cleaner and less expensive natural gas for rig fuel, and pad drilling to reduce trucking and rig mobilization costs.

Since 2006, operators have moved away from vertical wells to multistage hydraulically fractured horizontal wells to better drain each individual zone (Fig. 16.12) (Rajan and Moody, 2011). In the Texas Panhandle area, for example, more than 70% of the wells are horizontal (Kennedy et al., 2012). Owing to the low permeability, small well spacing is required (such as 80 acre, representing a typical 7.0 BCFG EUR/well). As a result, a large number of wells are required to develop the Granite Wash Play. It was realized that the ~3300 vertical wells can be worked over to take advantage of new technology, such as horizontal sidetracking, to increase production while reduce cost.

The complex mineralogy of the sandstones makes it difficult to analyze open-hole log data for quantifying hydrocarbon pore volume. By contrast, vertical well production logs can be used to identify target zones which have the highest gas flow rates with minimal water production (Rothkopf et al., 2011).

The key to the successful rejuvenation of the mature Granite Wash Play is the application of multistage hydraulic fracturing technique in long-reach horizontal wells (4500 ft average lateral length). The initial production rates of many Granite Wash horizontal wells were as high as 20 times offset

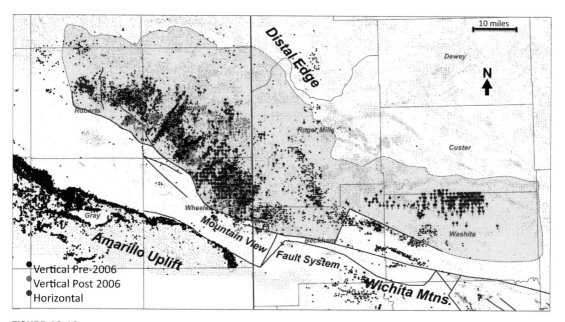

FIGURE 16.12

Granite Wash Play vertical and horizontal completions, showing horizontal wells being the major well type since 2006.

LoCricchio (2012); AAPG@2012, reprinted by permission of the AAPG whose permission is required for further use.

vertical wells (Fig. 16.13) (Rothkoft et al., 2011). Average first month production rates in typical horizontal wells varies from 5 to 8 MMCFGPD, while EUR gas per well in the range of 3–17 BCFG (6.49 BCFG average). Depending on the depth of targeted Granite Wash members, those wells were located approximately in 9000 to 15,000 ft TVD (12,000 ft TVD average). The maximum in situ stress is normally oriented in a west–east direction. As a result, horizontal wells are typically drilled in a north–south orientation, so that the hydraulic fractures can grow transversely to the wellbore (Fig. 16.12). Most wells have a lateral length of 700–9500 ft (4500 ft average). All reported horizontal wells are successes, with 2–27 million cubic feet equivalent per day (MMCFEPD) IP flow rates in some top performance wells (averaged 22 MMCFEPD from seven New Field top performance wells). Apache reported that the first month oil equivalent rates amount to 2500 to 4700 BOEPD (3200 BOEPD average) with 48% liquid. The single most productive well is the Black 50-1H well in the Stiles Ranch area of the Texas Panhandle, which had a 24-h initial production (IP) flow rate of 27.0 MMCFGPD with 3190 BCPD and 3530 barrel per day (bbl/d) of NGL, representing 60.2 MMCFEPD (or 10,035 barrel oil equivalent per day (BOEPD)) (Linn Energy, 2010). However, like all wells in unconventional reservoirs, the wells exhibit high decline rates in the first few production years (Fig. 16.14).

An example of a horizontal well operated by Linn Energy was designed to drill to 12,000 ft TVD with a 5500 ft lateral, which would be hydraulically fractured in 20 stages. Wellbore construction

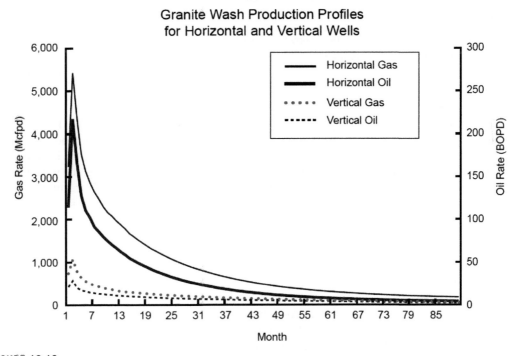

FIGURE 16.13

Typical production profiles of horizontal well and vertical well in Granite Wash Play, showing horizontal well greatly outperforms vertical well.

Rajan and Moody (2011); reprinted with written permission.

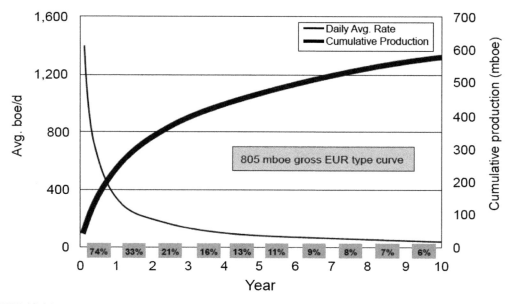

FIGURE 16.14

A typical well decline curve analysis of the Colony Granite Wash from Chesapeake Energy, showing steep initial decline rate and cumulative production.

Chesapeake Energy (2014); reprinted with written permission.

would typically consist of a 3-string or 4-string casing design, depending on mud loss severity encountered in the Brown Dolomite overlying the Granite Wash reservoir rocks. The production liner was to be 4 1/2" 13.5 lb/ft, with the Open Hole multistage system as an integral part of the liner. For the ease of tubing installation and future well intervention operations, the production liner top was to be set as close to the vertical section as possible (Casero et al., 2013). The initial hydraulic fracturing design was a slick-water fracturing approach assisted with nitrogen: with 4000–5000 barrels slick-water and 100,000 lbs of 20/40 white sand for each stage. To maximize the near wellbore conductivity, this original design was subsequently revised to a hybrid of slick-water and linear gel, and later further optimized to a hybrid of linear gel and cross-linked gel with proppant concentration as high as six pound proppant per gallon (ppg) (Casero et al., 2013). As a result, water volumes have steadily decreased to 3000 barrels (bbls)/stage, while proppant amount has increased to 150,000–175,000 lbs of 20/40 mesh white sand per stage based on reservoir thickness.

The performances of the Granite Wash horizontal wells are continually improving as a result of technology advancement. For example, the best well in the Western Oklahoma Granite Wash in 2007 produced 1.1 BCFG with 73,092 stock tank barrel (STB), compared with 2.8 BCFG and 234,672 STB in 2008 and 4.5 BCFG and 270,996 STB in 2009 (Srinivasan et al., 2011). The same trend also occurred in Texas Panhandle area. For example, initial production rates from the best well in 2007 were 4.67 MMCFGPD and 127.8 STB, compared with 20.5 MMCFGPD and 851 STBD in 2008 and 22.7 MMCFGPD and 1696 STBD in 2009 (Srinivasan et al., 2011).

16.10 HYDRAULIC FRACTURING

Due to the variation in mineralogy, pore pressure, and hydrocarbon type in the Granite Wash, production potential varies greatly throughout the formation. Determining the best completion method can be difficult and must be evaluated carefully.

Completion optimization requires determining the optimum number of fractures that can maximize recovery and economics. Reservoir simulation shows the optimum number of fractures is 20–40 for a 4000 ft long horizontal well with average Granite Wash Formation reservoir properties (Fig. 16.15) (Rothkopf et al., 2011).

Two main completion methods involving multistage hydraulic stimulation have been applied in the Granite Wash: (1) Cemented Liner with Plug-and-Perf Diversion (PnP); and (2) Open Hole multistage fracturing system (OHMS). Some wells are also fractured using coiled tubing technology. The goals of these methods are to increase access to the reservoir through the induction of fractures along the entire length of the horizontal wellbore (Edwards et al., 2010). The PnP completion involves cementing and casing in the horizontal wellbore, and plug perforation and stimulation. Mechanical isolation in the liner is accomplished by setting bridge plugs using pump down wireline or coiled tubing (CT) and followed by perforation and fracturing. The cement provides the

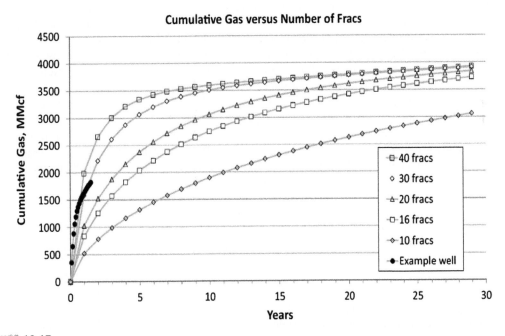

FIGURE 16.15

Plot of cumulative gas versus production years correlated with number of fractures based on reservoir simulation, showing the optimum number of fractures in a Granite Wash horizontal well being in 20–40 range (Rothkopf et al., 2011).

mechanical diversion in the annulus while the bridge plug provides the mechanical diversion in the liner. The process is then repeated for each stage. After the completion of all stages, CT is used to drill out the composite plugs to produce. The OHMS use external packers to isolate the wellbore. These packers typically have elastomer elements that expand to seal against the wellbore and do not need to be removed (Edwards et al., 2010). The system has sliding sleeve tools to create ports in between the packers. These tools can be opened hydraulically at a specific pressure or by dropping size-specific actuation balls into the system to shift the sleeve and expose the port. OHMS permits the fracturing process to be performed in a single continuous pumping operation without the need for a drilling rig. Once the stimulation is completed, the well can be immediately flowed back and produce. Studies have concluded that there is no appreciable difference in well IPs by using either method (Kennedy et al., 2012). Both methods have advantages and disadvantages. However, the PnP method is more widely used in the Granite Wash (some 70% of the horizontal wells), owing to the benefits of the PnP method including: (1) simple to implement from an operational standpoint; (2) having numerous vendors on location; and (3) allowing pumping at higher rates because the fracturing job is pumped down the casing (Castro et al., 2013).

Along with improved well performance, the main parameters of hydraulic fracturing increased significantly between 2007 and 2009 (Fig. 16.16). For example, in the Texas Panhandle area, the average perforated lengths increased from 1500 ft in 2007 to ∼2250 ft in 2008, and 3100 ft in 2009. The proppant mass increased from ∼700,000 lb in 2007 to ∼1,200,000 lb and >2,000,000 lb in 2008 and 2009, respectively. The average proppant concentration increased from ∼0.3 to ∼0.5 pound proppant added per gallon (ppa) and ∼0.9 ppa from 2007 to 2009. Analysis of these data shows that with increasing average lateral length, total proppant, proppant concentration, and the number of stages also increased. The exception is total fluid volume which dropped somewhat in 2008 and then increased a little in 2009. A trend of increased proppant and proppant concentration indicates improved fracture conductivity. Well performance appears to be responding more to the changes in proppant concentration and total proppant mass than to fluid volume (Srinivasan et al., 2011). The same trend also occurred in the Western Oklahoma area. The effectiveness of increasing main stimulation parameters can also be proved by comparing the well performance of Western Oklahoma and Texas Panhandle areas. The stimulation parameters of the Western Oklahoma Area were larger than that of the Texas Panhandle Area (except for total fluid). Consequently, the Western Oklahoma wells in general have better performances than Texas Panhandle wells (Fig. 16.16).

In practice, the Granite Wash horizontal wells are fractured in 7–20 (16 average) stages, with 270–550 ft (400 ft average) stage length that contains about three perforation clusters with ∼3 spf perforation density (Castro et al., 2013). Proppant mass amount ranges from 1,500,000 to 3,680,000 lbs of 20/40, 30/50. 40/70 and 100 + mesh sand or resin-coated sand (Castro et al., 2013; Rajan and Moody, 2011). Fracture fluid is injected at 40–100 barrel per minute (bpm). The generated artificial fractures have a fracture half length of ∼300 ft. Fractures in the Granite Wash Formation tend to have a singular, linear geometry (Rothkopf et al., 2011). A typical fracture treatment uses a nitrogen assisted slick-water fracturing approach with 4000–5000 bbls slick-water and 100,000 lbs of 20/40 white sand pumped per fracturing stage. This design has evolved into a hybrid of slickwater/linear gel and more recently linear gel/cross-linked hybrid design with proppant concentration as high as six ppg. Water volumes have steadily decreased with the application of the hybrid designs to 3000 bbls/stage and proppant amount has increased to 150,000–175,000 lbs of 20/40 white sand per stage depending upon the reservoir thickness (Casero et al., 2013).

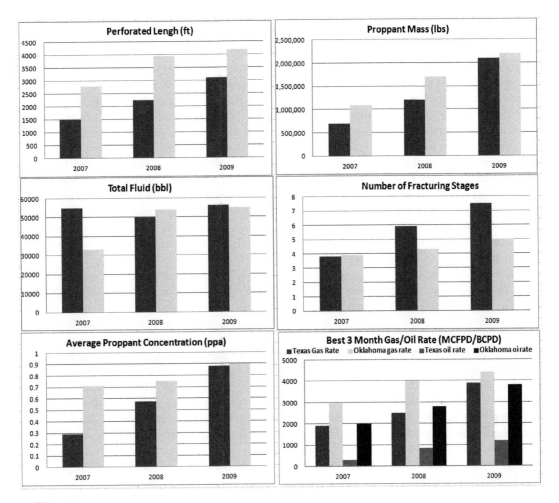

FIGURE 16.16

Trend of stimulation parameters in the Texas Panhandle Area (in black) and Western Oklahoma Area (in gray), showing increasing tread and improving well performance.

Modified from Srinivasan et al. (2011).

16.11 MULTILATERAL WELL

Multilateral wells have been used to reach multiple pay zones in both conventional and unconventional reservoirs with marked success. They may also be used to reach multiple pay zones in the Granite Wash Formation from a single surface location. A typical multilateral well in Granite Wash can double the reservoir exposure compared with a single horizontal well, while still allow for selective fracturing of both the main bore and lateral legs. The well can be designed to have two

5000 ft long laterals with each lateral being fractured in 10–15 stages (Durst and Vento, 2013). A study was conducted by Linn Energy to evaluate the stimulation of two stacked laterals in the Granite Wash to determine the extent of created fracture overlap and treatment connectivity between two productive intervals: the upper lateral targeting the Carr Granite Wash and the lower lateral targeting the Britt Granite Wash. The two zones were separated by the clay-rich Caldwell interval, which may act as a barrier to fracture height growth and prevent effective drainage of the reservoirs not contained in the lateral (Crawford et al., 2013). The purpose was to determine if the Caldwell barrier could be overcome by fracture stimulation or if adequate drainage of both the Britt and Carr zones would require individual laterals. Various completion techniques were employed, including Zipper Fracs, volume adjustments, perforation density and phasing changes, and pump rate variations to determine if the fracture geometry would be affected by these changes. Treatment evaluations were analyzed using microseismic monitoring in conjunction with pumping treatment and pressure data from multiple offset locations. The microseismic data indicated that overall treatment was limited to the target laterals. Chemical tracing records confirmed the lack of communication and interference. The analyses suggested that multiple laterals are required to adequately drain all reserves of the two Granite Wash layers (Crawford et al., 2013).

Two multilateral wells in the Hemphill County of the Texas Panhandle area were drilled by Cimarex Energy, including the Flowers 84-11H well and the Washita Ranch 22-1H well. The Flowers 84-11H well was designed to have two 3000 ft long laterals with each lateral being fractured in 10–15 stages (Goodlow et al., 2009). The lower lateral of this well was completed using an open-hole packer system. Each stage was pumped to the planned water and sand volumes. Then a ball was pumped to the next sleeve up hole which diverts the fracture treatment to the following interval. The lower main wellbore was fractured in 11 stages with a total fluid volume of 6,510,000 gallons and total proppant mass of 2,200,000 lb. Fracture fluid was injected at 72 bpm in an average 7200 psi pressure. Proppant concentrations started at 0.25 lb/gal and were increased to a maximum of 1.5 lb/gal, with multiple sweeps run throughout for an average pump time of 3.75 h per stage. The upper lateral was fractured in 10 stages, of which nine were completed successfully. Total fluid of the treatment was 5,544,000 gallons with 2,150,000 pounds proppant, representing an average of 546,000 gallon fluid and 215,000 pounds proppant per stage, respectively. Initial rate of the well amounted to 6 MMCFGPD per lateral, which is approximately equivalent to two separate horizontal wells in the same area. Cost savings were realized over drilling and completing two separate wells. The drilling cost of the Flowers well was 17% less than the average cost of drilling two separate wells, while the completion cost was 15% less than that of stimulating two separate wells. With an average 13% cost savings over drilling and completion, the two wells proved to be a success (Goodlow et al., 2009).

16.12 CONCLUSIONS

The following conclusions can be drawn from the analysis of geological and production data of the Granite Wash Play:

1. The Lower Permian–Pennsylvanian Granite Wash Formation in the Anadarko Basin is a lithologically complex unconventional liquid-rich gas play, consisting of stacked heterogeneous series of sandstones, siltstones, carbonates and shales deposited in a series of stacked alluvial

fans, fan delta, slope and submarine fan channels. Understanding reservoir architecture and heterogeneity paves the ground for subsequent application of drilling and completion technology.
2. The ~2000 ft thick formation consists of 3–20 stacked productive intervals occurring at ~12,000 TVD. With porosity of 1–16% (~8% average) and gas permeability of 0.0005–100 mD (0.001 mD average), the reservoir is representative of a tight gas sand. It provides a real-case analog for benchmarking similar tight gas plays in other basins worldwide.
3. The key to the successful rejuvenation of the mature Granite Wash Play since 2006 is the application of horizontal wells with multistage hydraulic fracturing. EUR gas equivalent per horizontal well is 3.14–17 BCFE (6.49 BCFE average). The successful application of new technology in Granite Wash offers calibers for future application of multistage hydraulic fracturing in horizontal wells in similar plays.

ACKNOWLEDGMENT

The authors would like to thank C&C Reservoirs, Inc., for allowing the publication of this paper.

REFERENCES

Adler, F.J., 1971. Future Petroleum Provinces of the Mid-continent, Region 7. AAPG Memoir, No. 15–2, pp. 985–1042.

Casero, A., Adefashe, H., Phelan, K., 2013. Open hole multi-stage completion system in unconventional plays: efficiency, effectiveness and economics. In: SPE Middle East Unconventional Gas Conference and Exhibition, Muscat, Oman, SPE 164009.

Castro, L., Bass, C., Pirogov, A., Maxwell, S., 2013. A comparison of proppant placement, well performance, and estimated ultimate recovery between horizontal wells completed with multi-cluster plug & perf and hydraulically activated frac ports in a tight gas reservoir. In: SPE Hydraulic Fracturing Technology Conference, The Woodlands, Texas, SPE 163820.

Chesapeake Energy, 2014. Focus on Value Delivering Growth- Analyst Day. Oklahoma City, Oklahoma, 135p.

Clement, W.A., 1991. East Clinton. In: Foster, N.H., Beaumont, E.A. (Eds.), Stratigraphic Traps II, AAPG Treatise of Petroleum Geology, Atlas of Oil and Gas Fields. AAPG, Tulsa, OK, pp. 207–267.

Cox, S.A., Gilbert, J.V., Sutton, R.P., Stoltz, R.P., 2002. Reserve analysis for tight gas: SPE Eastern Regional Meeting, Lexington, Kentucky. SPE 78695.

Crawford, E.M., Tehan, B., Launhardt, B., 2013. Examination of treatment connectivity between Granite Wash layers using microseismic, tracer, and treatment data. In: SPE Production and Operations Symposium, Oklahoma City, Oklahoma, SPE 164496.

Dar, V., 2010. The Granite Wash: An Emerging Tight Sands Natural Gas Play in the U.S. seekingalpha.com/article/181187-the-granite-wash-an-emerging-tight-sands-natural-gas-play-in-the-u-s.

Davis, H.G., Northcutt, R.A., 1991. Anadarko Basin. In: Gluskoter, H.J., Rice, D.D., Taylor, R.B. (Eds.), Economic Geology, U.S.: The Geology of North America, Decade of North American Geology, vol. P-2. Geological Society of America, pp. 325–338.

Durst, D.G., Vento, M., 2013. Unconventional shale play selective fracturing using multilateral technology. In: SPE Middle East Unconventional Gas Conference and Exhibition, Muscat, Oman, SPE 163959.

Edwards, J.W., Braxton, D.K., Smith, V., 2010. Tight gas multi-stage horizontal completion technology in the Granite Wash. In: SPE Tight Gas Completions Conference, San Antonio, Texas, USA, Paper SPE 138445, 9 p.

Farris, S., 2014. Apache Investor Day. Houston, TX, 175 p.

Forster, A., Merriam, D.F., Hoth, P., 1998. Geohistory and thermal maturation in the Cherokee Basin (Mid-Continent, USA): results from modeling. AAPG Bulletin 82, 1673–1693.

Gilman, J., 2012. Depositional Patterns, Source Rock Analysis Identify Granite Wash Fairways. The American Oil & Gas Reporter.

Goodlow, K., Huizenga, R., McCasland, M., Neisen, C., 2009. Multilateral completions in the Granite Wash: two case studies. In: SPE Production and Operations Symposium, Oklahoma City, Oklahoma, SPE 120478.

Grieser, B., Shelley, B., 2009. What can injection falloff tell about job placement and production in tight-gas sand. In: SPE Eastern Regional Meeting, Charleston, West Virginia, SPE 125732.

Holditch, S.A., 2006. Tight gas sands. JPT 58 (6), 86–93. http://dx.doi.org/10.2118/103356-MS. SPE 103356-MS.

Ingram, S., Paterniti, I., Rothkoft, B., Stevenson, C., 2006. Granite wash field Study—Buffalo wallow field, Texas Panhandle. In: SPE Eastern Regional Meeting, Canton, Ohio, Paper SPE 10456, 13 P.

Johnson, K.S., 1989. Geological evolution of the Anadarko Basin. In: Johnson, K.S. (Ed.), Anadarko Basin Symposium, 1988: Oklahoma Geological Survey Circular, No. 90, pp. 3–12.

Keighin, C.W., Flores, R.M., 1993. Petrology and sedimentology of Morrow/Springer rocks and their relationship to reservoir quality. In: Anadarko Basin, Oklahoma: Oklahoma Geological Survey Circular, No. 95, 25 pp.

Kennedy, R.L., Knecht, W.N., Georgi, D.T., 2012. Comparisons and contrasts of shale gas and tight gas developments. In: SPE Saudi Arabia Section Technical Symposium and Exhibition, AlKhobar, Saudi Arabia, SPE-sas-245.

Kupchenko, C.L., Gault, B.W., Mattar, L., 2008. Tight gas production performance using decline curves. In: CIPC/SPE Gas Technology Symposium 2008 Joint Conference, Calgary, Alberta, Canada, Paper SPE 114991.

Linn Energy, 2010. Linn Energy announces 60.2 MMcfe per day horizontal Granite Wash well.

LoCricchio, E., 2012. Granite Wash Play Overview, Anadarko Basin-stratigraphic Framework and Controls on Pennsylvanian Granite Wash Production, Anadarko Basin, Texas and Oklahoma. AAPG, Search and Discovery Article #110163, 43 p.

Lyday, J.R., 1990. Berlin Field-USA, Anadarko Basin, Oklahoma. In: Beaumont, E.A., Foster, N.H. (Eds.), Stratigraphic traps I: AAPG treatise of petroleum geology. Atlas of oil and gas fields, pp. 39–68.

Mitchell, 2012. The Anadarko Basin Oil and Gas Exploration-past, Present, and Future. SM Energy Co., Tulsa, Oklahoma, 61 p.

Natural Gas Intelligence, 2014. Information on the Granite Wash. www.naturalgasintel.com/granitewashinfo.

Puckette, J., Abdalla, A., Rice, A., Al-Shaieb, Z., 1996. The Upper Morrow reservoirs: Complex fluviodeltaic depositional systems: Oklahoma Geological Survey Circular. no 98, 47–84.

Rajan, S., Moody, W., 2011. Rejuvenation of mature plays. In: Granite Wash Reborn, IHS CERA Decision Brief, September.

Rascoe, B., Adler, F.J., 1983. Permo-Carboniferous hydrocarbon accumulation. Mid-Continent, USA: AAPG Bulletin 67, 979–1001.

Rothkopf, B., Christiansen, D.J., Godwin, H., Yoelin, A.R., 2011. Texas Panhandle Granite Wash formation: horizontal development solutions. In: SPE Annual Technical Conference and Exhibition, Denver, Colorado, SPE 146651.

Rushing, J.A., Newsham, K.E., Pergo, A.D., Comisky, J.T., Blasingame, T.A., 2009. Beyond decline curves lifecycle reserves appraisal using an integrated workflow process. In: Unconventional Reservoirs Workshop, Oklahoma Geology Survey, 29 p.

Shipley, K.A., 2013. Dynamic Mud Cap Drilling in the Granite Wash Formation downloaded from: http://www.aade.org/app/download/7129857204/Dynamic+MudCap+Drilling+-+LINN+Energy.pdf.

Srinivasan, K., Dean, B., Olukoya, I., Azmi, Z., 2011. An overview of completion and stimulation techniques and production trends in Granite Wash horizontal wells. In: SPE North American Unconventional Gas Conference and Exhibition, The Woodlands, Texas, SPE 144333.

Srinivasan, K., Dean, B., Belobraydic, M., Azmi, Z., 2013. Evolution of horizontal well hydraulic fracturing in the Granite Wash—understanding well performance drivers of a liquids-rich Anadarko Basin formation. In: SPE Hydraulic Fracturing Technology Conference, The Woodlands, Texas, SPE 163857.

Thomas, W.A., 1991. The Appalachian-Ouachita rifted margin of southeastern North America. Geological Society of America Bulletin 103, 415–431.

U.S. Energy Information Administration, 2010. Granite Wash Play, Texas and Oklahoma downloaded from: http://www.eia.gov/oil_gas/rpd/shaleusa10.pdf.

Wang, H.D., Philp, R.P., 1997. A geochemical study of Viola source rocks and associated crude oils in the Anadarko basin, Oklahoma. In: Johnson, K.S. (Ed.), Simpson and Viola Groups in the Southern Midcontinent, 1994 Symposium: OGS Circular 99, pp. 87–101.

CHAPTER

COALBED METHANE EVALUATION AND DEVELOPMENT: AN EXAMPLE FROM QINSHUI BASIN IN CHINA

17

Yong-shang Kang[1,2], Jian-ping Ye[3], Chun-lin Yuan[1], Y. Zee Ma[4], Yu-peng Li[5], Jun Han[6], Shou-ren Zhang[3], Qun Zhao[7], Jing Chen[1], Bing Zhang[3], De-lei Mao[6]

College of Geosciences, China University of Petroleum, Beijing, China[1]; State Key Laboratory of Petroleum Resources and Prospecting, China University of Petroleum, Beijing, China[2]; CNOOC China Limited, Unconventional Oil & Gas Branch, Beijing, China[3]; Schlumberger, Denver, CO, USA[4]; EXPEC ARC, Saudi Aramco, Dhahran, Saudi Arabia[5]; PetroChina Coalbed Methane Company Limited, Beijing, China[6]; Langfang Branch of Research Institute of Petroleum Exploration and Development, PetroChina, Hebei, China[7]

17.1 INTRODUCTION AND OVERVIEW
17.1.1 OVERVIEW

Coalbed methane (CBM), also referred to as natural gas from coal or coal seam gas, is generated either from methanogenic bacteria or thermal cracking of the coal. The worldwide resources from coalbed methanes are immense, with estimates exceeding 9000 Tcf (Rogner, 1997; Jenkins and Boyer, 2008). Because much of the gas generated in coal remains in the coal, primarily because of the sorption of gas in the coal matrix, coal acts as both the source rock and the reservoir for the gas. Exploration for and exploitation of coalbed methane requires knowledge of the unique coal-fluid storage, transport processes and well completions to extract commercial quantities of gas (Clarkson and Barker, 2011).

The interest in production of coalbed methane was initially motivated by the need to predrain methane prior to mining of coals for safety reasons (Clarkson and Bustin, 2010). The 1970s "energy crisis" led to studies of the feasibility of producing the gas from CBM for commercial use in the USA (Flores, 1998). The first CBM wells in the Black Warrior Basin of Alabama were drilled in 1971. The production also started in Pennsylvania with the drilling of degasification holes prior to mining. A larger-scale experimental drilling project began in 1976 and commercial production from bituminous coals began in 1980 (Schraufnagel and Schafer, 1996; Bodden and Ehrlich, 1998; Markowski, 1998). Commercial production from subbituminous and lignite coals of Paleocene Fort Union and Eocene Wasatch formations in the Powder River Basin in the mid-1990s demonstrated that CBM in low-rank coals were commercially exploitable (Jenkins and Boyer, 2008). Commercial CBM development was extended to Canada and Australia, but generally limited to lignite, subbituminous, and bituminous coals.

Coal is a source, reservoir, and trap for significant quantities of methane and minor amounts of other gases. CBM, unlike conventional gas resources, is unique in that gas is retained in a number of ways, including: (1) adsorbed molecules within micropores—2 nm in diameter; (2) trapped gas in matrix porosity of coals; (3) free gas in cleat and fractures; and (4) as a solute in ground water within coal fractures (Bustin and Clarkson, 1998). However, the majority of the methane is adsorbed to the internal surfaces of the microporous coal matrix, and a small fraction of the total methane is stored as free gas in the secondary porosity (Jamshidi and Jessen, 2012).

CBM is generated along with coalification, and most methane and other gases are retained in coal by the process of adsorption. Free gas also occurs in fissures and pore systems, and it is believed to be in equilibrium with the gas in the adsorbed phase. If the coalbed is capped by an impermeable layer, free gas in the fissure–pore network is also in equilibrium with the hydrostatic pressure of the surrounding strata. However, if the coalbed is overlain by permeable strata, free gas will percolate upward, escaping from the coalbed. This loss of free gas destroys the equilibrium between free and adsorbed gas, allowing desorption to begin. Desorption is the opposite of adsorption and occurs when the free gas pressure in the fracture–pore system is reduced, which leads to a disequilibrium between the free and adsorbed gas phases. Pressure changes in the free gas occur either by gradual percolation upward through permeable strata, or lateral migration through the coal caused by a pressure difference. Both of these processes require an interconnected fracture system, including fissures (Ulery, 1988).

The amount of coalbed methane is highly related to coal petrographic character, preservation, and accumulation conditions. CBM preservation and accumulation are controlled by geological and hydrologic conditions (Kaiser et al., 1994; Bachu and Michael, 2003; Liu et al., 2006; Hamawand et al., 2013), as well as by tectonic history (Fan et al., 2005; Song et al., 2005, 2007; Wang et al., 2006), and paleo–hydrogeological cycles (Mao et al., 2012). CBM generally increases with increasing pressure and coal rank (Ulery, 1988; Li et al., 2010). In coal bearing strata, local structures, such as faults, anticlines, and synclines, control gas accumulations. Gas accumulation is generally higher in the axes of synclines than in the axes of anticlines (Li et al., 2005). Faults, especially normal faults, generally have a negative effect to gas preservation (Qin et al., 2013).

Coalbeds are commonly classified as dual porosity-type reservoirs, with gas primarily in the matrix as an adsorbed phase while a secondary porosity system (cleats) provides the permeability for the flow to a production well (Jamshidi and Jessen, 2012). Production of CBM generally requires dewatering and depressurizing the coal formations, which enables the coals to release their adsorbed methane (Colmenares and Zoback, 2007; Jenkins and Boyer, 2008; Kang et al., 2008). For dry coals, such as the Cretaceous Horseshoe Canyon in western Canadian, gas can be produced directly without predewatering (Clarkson and Bustin, 2007; Jenkins and Boyer, 2008; Palmer, 2010).

CBM potential and producibility of coal bearing strata are strongly affected by the hydrogeological regime of formation waters and coal permeability, which in turn depends on the effective stress regime of the coals. The flow of formation waters plays an important role for the producibility of CBM (Bachu and Michael, 2003). Gas-saturation is also a primary factor influencing gas producibility from coal reservoirs (Jenkins and Boyer, 2008; Li et al., 2008). In coal bearing strata, local structures, such as faults, anticlines or synclines, may strongly control the performance of a coalbed methane well. Wells situated at the high position of anticlines often have higher gas producibility and low water production rate (Zhao et al., 2012a). Faults, especially normal faults, generally have a negative effect to gas producibility due to their role as high pathway for water supply from surrounding aquifers (Kang et al., 2013a).

Another factor, sorption time of coalbeds reservoirs, should also be taken into account. Sorption time influences the timing and the volume of peak gas rate, and therefore the recoverable resources (King and Ertekin, 1986; Li et al., 2008).

17.1.2 BACKGROUND FOR CBM IN THE QINSHUI BASIN

CBM development in China began in 1989. The original in-place CBM resources in China are estimated to be 36.81×10^{12} m^3 in the coalbeds above a depth of 2000 m (Liu et al., 2009a). To date, the Qinshui Basin is the most mature basin for CBM exploration and development in China, with an original in-place resource of approximately 3.85×10^{12} m^3 (Ye, 2013), attracting state and private companies' interest for exploration and development. Production from anthracite coals in the southeastern Qinshui Basin demonstrated its economic value (Liu et al., 2004; Rao et al., 2004) essentially at the same time low-rank coals were demonstrated commercially exploitable in the Powder River Basin in the mid-1990s. Since then, CBM wells in China have increased significantly. By the end of 2006, more than 1183 wells had been drilled (including 19 multilateral and horizontal wells) and the number increased to 12,547 wells by the end of 2012, with most of these wells being drilled in the Qinshui Basin and the Ordos Basin. These two basins both belong to the Northern China Craton Basin.

The high rank coal reservoirs in the Qinshui Basin have relatively low permeability. The coalbeds are interbedded with mudstones, sandstones and carbonates; three coal layers are generally exploitable for CBM. The sandstones contain water and influence the drainage of water from coalbeds and production performance of coalbed methane wells. The permeability of coal reservoirs is stress dependent, and a careful dewatering schedule control is an important aspect for cost-effective development of these low permeability reservoirs.

This chapter discusses the geological conditions, gas generation and coalbed methane of the Qinshui Basin. The challenges of coalbed methane development in the basin are also discussed.

17.2 BASIN EVOLUTION AND GAS GENERATION

The Qinshui Basin, with an area of 32,000 km^2, is located in Shanxi Province, Central China (Fig. 17.1). The Qinshui basin is bounded by the Taihang Mountain uplift in the east and southeast, Huoshan uplift in the west, Wutaishan uplift in the north, and Zhongtiaoshan uplift in the southwest. The Qinshui basin is a syncline, striking north-northeast to south-southwest (Su et al., 2005). Folds are commonly present in the basin and they are mainly high-angle normal faults with large extensions; one set of faults strike from south–southwest to north–northeast, and the other set strike from west–southwest to east–northeast.

The formations in the basin include the Cambrian, Ordovician, Carboniferous, Permian, Triassic, Jurassic, Tertiary, and Quaternary sequences. The dip of the strata is relatively flat in the north and south margins and steep in the center of the basin. The Pennsylvanian Taiyuan Formation and the Permian Shanxi Formation are the main coal-bearing sequences (Fig. 17.2, also see Su et al., 2005).

During Pennsylvanian (or Taiyuan) stage and Permian (or Shanxi) stage, the Qinshui basin was a part of the Northern China Cratonic Basin. The Taiyuan coals were deposited in an extensive area of fluvial–deltaic environments with shallow marine–limnic influence; the Permian Shanxi coals are typically limnic. After deposition, the coalification process began as a gradual, progressive alteration (metamorphism) of plant material to peat, lignite, bituminous, and higher-rank coals.

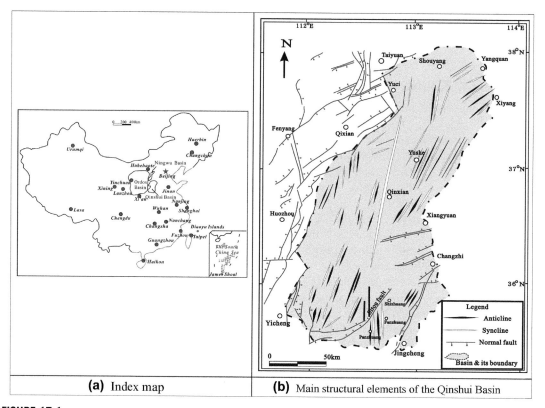

FIGURE 17.1

Map showing the location and the main structural elements of the Qinshui Basin, central China.

During the Triassic Indosinian orogeny and Jurassic to Cretaceous Yanshanian orogeny, many tectonic basins, including the Qinshui basin, and uplifts developed on the late Paleozoic Northern China platform. The Pennsylvanian-Permian coals were buried to the maximum depth of 3000–4000 m until late Triassic (Fig. 17.3, also see Song et al., 2005), reaching their first thermo-maturity peak with the highest vitrinite reflectance, R_o, of about 1.2% (medium volatile bituminous) under a normal paleo-geothermal gradient (2–3 °C/100 m; 1.1–1.7 °F/100 ft). Gas generation is estimated to be 81.45 m^3/t in Jincheng area, southeastern Qingshui Basin (Liu et al., 2004).

A tectonic thermal event took place during the late Jurassic to early Cretaceous Yanshanian orogeny, especially active in the southern and northern parts of the Qinshui Basin, wherein many Mesozoic igneous intrusions are found (Sun et al., 2005). Radioactive isotope K–Ar dating of the black granite in the southwestern Qinshui Basin showed that the black granite was formed 91–138 million years ago, and the paleo-geothermal gradient was estimated to be greater than 5.5 °C/100 m in the Qinshui Basin (Wang et al., 2006). Igneous intrusions caused the coal maturity to be higher in the southern, northern, and eastern parts of the Qinshui basin than in the central and western parts. The highest R_o value was equal or greater than 4% in the Jincheng area in the southeastern Qinshui

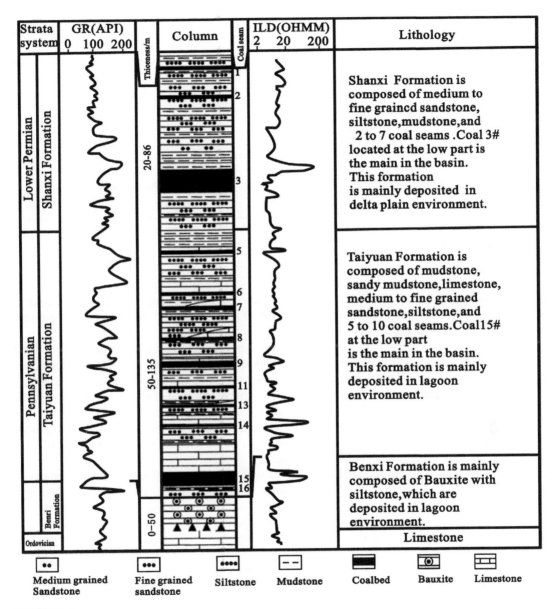

FIGURE 17.2

Stratigraphic column of the Permo–Carboniferous coal-bearing sequence and geophysical log curves from well Js-1 in the southeastern Qinshui Basin.

According to Su et al. (2005).

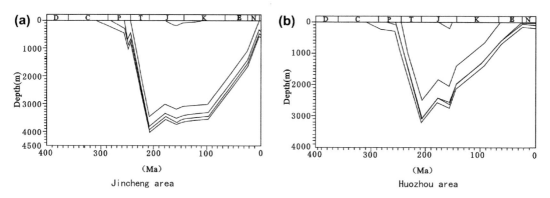

FIGURE 17.3

Coal measures burial history in Jincheng area. (a) South-eastern Qinshui Basin and in Huozhou area. (b) Western Qinshui Basin.

According to Song et al. (2005).

Basin (Rao et al., 2004). This thermal event enhanced the coal maturity and caused a second peak of gas generation with an estimated gas generation of 359.10 m³/t in Jincheng area, located in the south-eastern Qinshui Basin (Liu et al., 2004).

From late Cretaceous period to present-day, the Permian–Pennsylvanian coals in the Qinshui Basin were further uplifted and their burial depth decreased due to the Himalayan orogenic movement and pacific plate subduction under the Euro–Asian plate; in some places, the burial depth increased slightly due to Neogene to Quaternary depositions (Fig. 17.3) and the geothermal gradient dropped down to be normal (2–3 °C/100 m; 1.1–1.7 °F/100 ft, see Liu et al., 2004; Wang et al., 2006).

The present-day maturity of the Permian–Pennsylvanian coals shows its variability in the Qinshui Basin, with R_o generally higher than 2.13% in the northern part and higher than 2.97% in the southern part, with coals rank being semianthracite to anthracite.

Igneous intrusions or high geothermal anomaly have a very positive effect on CBM development (Eddy, 1984; Gurba and Weber, 2001). In the Qinshui Basin, the Permian–Pennsylvanian high rank coals experienced hypozonal metamorphism until the late Triassic period, and then heat-affected metamorphism during late Jurassic to early Cretaceous periods. These activities facilitated the gas generation and CBM development (Rao et al., 2004; Wang et al., 2006).

17.3 CBM RESERVOIR CHARACTERIZATION

The main target coals for CBM exploration and development are coal #3 of the Permian Shanxi formation and coal #15 of the Pennsylvanian Taiyuan formation (Fig. 17.2). Other coals, including coal #8 and coal #9 of the Pennsylvanian Taiyuan formation (Fig. 17.2), are secondary targets in the northern part of the basin. Many studies have been undertaken in the Qinshui Basin to understand the gas accumulation mechanism (Chi, 1998; Ye et al., 2001; Hu et al., 2004; Li et al., 2005; Qin et al., 2005; Liu et al., 2006; Chen et al., 2007), gas content (Ye et al., 2002; Li et al., 2005; Liu et al., 2009b), permeability (Fu et al., 2004; Ye et al., 2014), and characteristics of adsorption and desorption

(Li et al., 2008; Liu et al., 2010; Ma et al., 2011). Main characteristics of the CBM in the Qinshui Basin, including the formation thickness, gas content, permeability, gas saturation, and sorption time of coal reservoirs, are discussed below.

17.3.1 THICKNESS AND GAS CONTENT

The total thickness of Permian–Pennsylvanian coals in the Qinshui Basin, on average, is above 10 m in most parts of the basin, but locally some areas have the thickness in the range of 5.8–9 m (Table 17.1). The regional distributions of coal #3 and coal #15 show some differences. Coal #3 is thicker in the south and thinner in the north of the basin, and the distribution of Coal #15 thickness is the opposite (Table 17.1).

Gas content is highly related to vitrinite reflectance, and it is influenced by other geologic factors, such as lithology of the surrounding rocks, local fold structures (Ulery, 1988) and faults (Dou et al., 2013). Within different ranks of coal, the variation in gas accumulation is due to the differences in matrix pore size distribution, coal maceral composition, and coal chemistry (Decker et al., 1986). Gas generally increases with increasing depth and rank, and the determining factors are paleo–depth, paleo–geothermal gradient (temperature), and time near maximum temperature (Markowski, 1998). Enrichment of gas is controlled by tectonic location, geothermal events, effective thickness of overlaying rocks and hydrogeological conditions (Wang et al., 2006).

The Permian–Pennsylvanian coals in the Qinshui Basin are mainly semianthracite to anthracite coals (Table 17.1) with significant gas contents, being greater than 10 m^3/t in most areas (Table 17.2 and Fig. 17.4). The gas content has a tendency to increase toward the basin center in the Qinshui Basin.

17.3.2 PERMEABILITY

The two most important variables in coalbed methane reservoir characterization are permeability and gas saturation. Permeability will largely determine the flow rates of gas and water. As noted by Markowski (1998), the gas production rate from test holes in eastern Pennsylvania is low mainly due to the lack of a good cleat system and low permeability. Gas saturation determined from desorption and adsorption measurements also impact the gas rate and production.

Without sufficiently high gas saturation, gas flow will not be economic. Coalbed methane reservoirs generally should have a gas saturation greater than ~70% to be economic (Moore, 2012), but some proposed a threshold of 60% with consideration of other factors (Li et al., 2010).

The magnitude and orientation of the present-day stress field, in combination with the orientation of cleats and fractures in the coals, are two of the most important factors controlling permeability

Table 17.1 Thickness, in meter, of Permian−Pennsylvanian Coals in the Qinshui Basin									
Thickness	Shouyang	Yangquan	Qinyuan	Hezuo–Xiangyuan	Tunliu–Zhangzi	Gaoping–Fanzhuang	Qinbei–Huodong	Jincheng
Total	10.64	13.71	5.79	11.21	11.05	11.51	8.98	13.03
Coal#3	1.5	1.8	1.00	1.60	2.00	5.00	5.50	6.00
Coal#15	3.5	5.00	3.00	3.50	2.50	2.00	1.80	2.60

Source: Ye (2013).

Table 17.2 Main Coalbed Methane Geological Characteristics of Permian–Pennsylvanian Coals in Shouyang Block, the Northern Qinshui Basin, and in Shizhuang Block, the Southern Qinshui Basin

Block	Parameter	Coalbed #3 Range	#3 Average	#9 Range	#9 Average	#15 Range	#15 Average
Shouyang, northern Qinshui basin	Thickness (m)	0.7–3.6	2.1	0.4–3.8	1.67	0.3–5.4	2.8
	$R_{o,max}$ (%)	1.798–2.393	2.128	2.099–2.545	2.38	2.183–2.649	2.405
	Gas content (m^3/t)[a]	2.55–22.23	13.31	3.99–20.32	13.38	1.83–20.28	11.61
	Sorption time (d)	14–155	54.5	17–123	79.67	15–155	73.27
	Sg (%)	12.73–71.52	50.06	46.76–73.39	55.6	20.25–71.02	46.54
	ϕ (%)[b]	5.01–7.32	5.74	5.145–6.548	6.038	5.146–6.68	5.84
	K (mD)[c]	0.02–1.07	0.25	0.02–0.74	0.3	0.03–1.43	0.57
Shizhuang, southern Qinshui basin	Thickness (m)	3.3–9.2	6	/	/	1–6.8	3.8
	R_o (%)	2.37–3.57	3.01	/	/	2.14–3.68	2.97
	Gas content (m^3/t)[a]	9.0–23.0	13.7	/	/	8.1–20.3	15.8
	Sorption time (d)	4.3–59.4	29.1	/	/	1.2–50.6	22.6
	Sg (%)	49.44–85.67	63.38	/	/	50.79–80.69	64.4
	ϕ (%)[b]	4.72–5.96	5.41	/	/	4.97–5.95	5.34
	K (mD)[c]	0.04–1.1	0.33	/	/	0.02–0.81	0.31

[a] On dry-ash free basis.
[b] Lab test results.
[c] In-site injection/fall off test results.

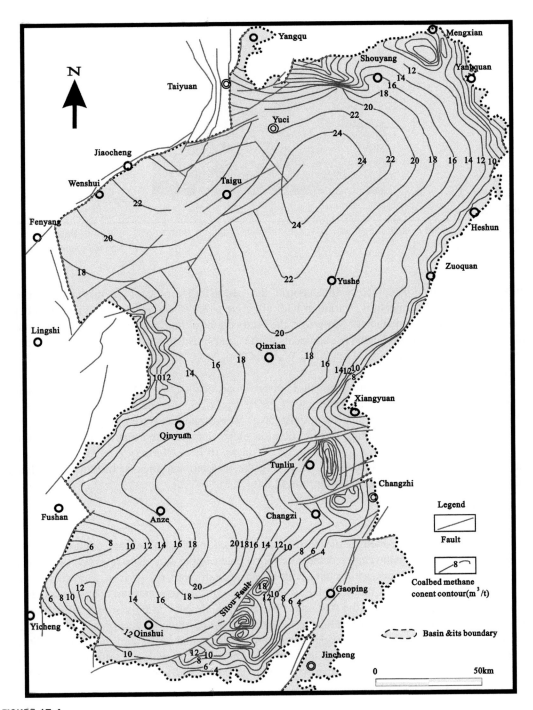

FIGURE 17.4

Gas content (m³/t, on dry-ash free basis) distribution of Permian Shanxi Formation coal #3 in the Qinshui Basin.

According to Ye (2013).

FIGURE 17.5

Cleats filled with calcite in coal of Permian Shanxi formation, Jincheng, south-eastern Qinshui Basin.

According to Su et al. (2001).

(Gentzis et al., 2007). The factors that control the permeability of coalbed methane reservoirs (Fu et al., 2001), and permeability variation with effective stress (Deng et al., 2009) and with depth (Ye et al., 2014) have been studied. Moreover, the factors controlling the recoverability and producibility of gas in the Permian–Pennsylvanian coals of the Qinshui Basin have been analyzed (Ma et al., 1998; Zhou et al., 1999; Ma, 2003; Lou, 2004; Wan and Cao, 2005; Zhang et al., 2006; Qin et al., 2013; Dou et al., 2013).

In specimens, the cleats in the Permian–Pennsylvanian coals of the Qinshui Basin are mainly closed or filled with calcite, clay and rarely with pyrite (Fig. 17.5, also see Su et al., 2001; Liu et al., 2008), with some exceptions in shallow mining areas where open and nonfilling cleats have been observed (Chen, 1998). K–Ar dating of illite filled in cleats demonstrated that the illite was formed during different geological stages from the middle Triassic to Early Tertiary (Liu et al., 2008). The cleats were formed and then closed or filled during the tectonic activities with influence of fluids, resulting in low permeability of coal reservoirs (Table 17.2).

17.3.3 GAS SATURATION

Gas saturation in coal is defined as the percentage or fraction of adsorbed gas content relative to adsorption capacity. This value can be determined for a coal sample at a given reservoir pressure and temperature by comparing desorption data with an adsorption isotherm derived from that sample. Variation of gas saturation is a fundamental aspect of reservoir heterogeneity in coal (Pashin, 2010). Thermogenic gas generation near the maximum burial depth almost certainly supported gas content equivalent to Langmuir volume. Thermogenesis ceases as the basin cools during the uplift. Cooling can also give rise to an increased gas capacity, thereby favoring undersaturated reservoir conditions. The reservoirs are undersaturated in most areas in the Qinshui Basin, with gas saturation varies from 12.75 to 85.67% (Table 17.1). As a comparison, gas saturation from Sydney Basin ranges from 2% to 85% (Bustin and Clarkson, 1998).

17.3.4 SORPTION TIME

Diffusion effects can be quantified by sorption time, τ(in days), as shown in Eq. (17.1) (Zuber, 1996):

$$\tau = \frac{s_f^2}{8\pi D} \qquad (17.1)$$

where, s_f is cleat spacing, ft; D is diffusion coefficient, ft^2/day.

The sorption time is generally gained from canister desorption tests. Coal matrix pore structure is the main factor influencing the sorption time in the condition that the sample volumes of desorption tests are standardized. The sorption time is highly related to the coal ranks, with higher-rank coals having longer sorption time (Li et al., 2008).

The sorption time of the Permian–Pennsylvanian coals in the Qinshui Basin ranges from less than 30 days to more than 70 days (Table 17.1). As a comparison, sorption time in San Juan Basin and middle Appalachian Basin ranges from 1 to 5 days, in Black Warrior Basin 1–30 days, in north Appalachian basin 80–100 days (Masszi,1991; Bodden and Ehrlich, 1998; Li et al., 2008).

17.4 CBM DEVELOPMENT CHALLENGES
17.4.1 WELL COMPLETION

Most commercial coalbed methane plays in the United States have permeability of between 3 and 30 mD. Very few are as low as 1 mD. However, many shallow coals, and deep coals, have permeability that is less than 1 mD. Horseshoe Canyon in the Western Canadian Basin has permeability that ranges from 0.1 to 100 mD and the coal is dry. Gas rates in dry coals are up to 10 times higher than in wet coals, explaining the commercial potential of low permeable (less than 1 mD) coalbed reservoirs in dry coals of Horseshoe Canyon. But in general coals are wet. Palmer (2010) demonstrated that certain well completion types are preferred for certain ranges of coal permeability defined as $k < 3$ mD, 3–20 mD, 20–100 mD, and >100 mD. This approach leads to a decision tree which includes various completion options as a function of the coal permeability. For the range of permeability $k < 3$ mD, multilateral wells or microholes are preferred and for the range from 3 to 20 mD, standard fracs and single-laterals are preferred (Palmer, 2010).

The production of coalbed methane from the Pennsylvanian-Permian semianthracite to anthracite coals in the Qinshui Basin was proved to be economic with standard fracs in Jincheng area, southeastern part of the basin, where the permeability of coal reservoirs is around 1 mD (Table 17.3) and the

Table 17.3 Injection/Fall Off Test Permeability in Coalbed Methane of Jincheng Area, Southeastern Qinshui Basin					
Areas		Fanzhuang North	Fanzhuang Central	Fanzhuang South	Panzhuang
K (mD)	Range	0.004–0.946	0.065–1.095	0.605–2.000	1.600–3.610
	Average	0.47	0.58	1.325	2.61
Modified after Ye et al. (2002).					

gas saturation is 85–95% (Li et al., 2010). The gas rates of standard frac wells range from 2000 to 6000 m^3/d, with a maximum of 16,000 m^3/d (Liu et al., 2004). Recently, some multilaterals in southeastern Qinshui Basin reached daily production over 20,000 m^3/d (with a maximum of 90,000 m^3/d) (Zhang et al., 2010), indicating horizontal well completions might be a good choice.

In the Jincheng area of the southern Qinshui Basin, the permeability of coalbed reservoirs is relatively high due to open tectonic fractures instead of cleats as is the cases in other countries. In most parts of the Qinshui Basin, however, the cleats are highly deteriorated by intense tectonic activity with cleats or joints being filled and thus the permeability of coal reservoirs is generally less than 1 mD (Table 17.2). In these extra low permeable areas, such as in Shizhuang Block, in Shouyang Block, standard frac completions are tested with poor results. Recent horizontal wells gave fair results with maximum gas rate of about 5000 m^3/d in Shizhuang Block. That implies that new completion methodologies should be tested. One major direction may be cased single-lateral with staged fracturing for enhancing the permeability of coals.

17.4.2 SITE/INTERVAL SELECTION

Coalbeds are normally saturated with water and may be in communication with an aquifer or be linked to an aquifer through vertical hydraulic fractures that may extend into adjacent formations. This can result in excess water production and inefficient depressurization of coals caused by water production from the surrounding formations (Colmenares and Zoback, 2007). When a large amount of water is produced during the natural gas extraction, disposal of the produced water is an environmental challenge as harmful impurities must be removed by appropriate purification techniques. Consequently, a reduction of water production in coalbed methane operations is desirable (Jamshidi and Jessen, 2012).

A successful hydraulic fracturing requires that strata separating the coals from water-bearing sandstones have sufficient strength and thickness to constrain fracture-height growth. A simple example illustrates the importance of avoiding these aquifers. At a production rate of 200 Barrel Water Per Day (BWPD), the time required to produce 5% of initial water in place from a 30-ft coal seam of 1% porosity and 160-acre areal extent is 93.1 days. The time required for that same coal in communication with like-sized water-bearing sand of 15% porosity is 4.1 years (Roadifer et al., 2009). The ideal hydrofracturing design is to fracture only the coalbeds and to avoid fracturing the roofs and the floors of the perforated coalbed intervals to obtain an effective local pressure drop zone surrounding the wells (Wang et al., 2013).

To avoid communication with neighboring aquifers due to hydraulic fractures while increasing the efficiency of dewatering, identifying the possible aquifers and understanding the structural relationships between coalbeds and aquifers are fundamentally important.

The Shouyang coalbed methane reservoir lies in the northern part of the Qinshui basin (Fig. 17.1). The overall structure shape shows a monocline extending east-west and dipping south with some secondary folds developed. The main coalbed methane production is from the Pennsylvanian/Taiyuan formation coals #15 and #9, and from Permian Shanxi formation coal #3.

Vertical wells were put on dewatering and production after single or multiple stimulations (perforation and fracturing). Typical daily water production is defined as the average daily water production during the period with stable water level in borehole (Kang et al., 2013b), which is a representative index. The typical daily water production was obtained from well rates records for 30 wells (Table 17.4).

Table 17.4 Typical Daily Water Production of Coalbed Methane Wells in Shouyang Block, Northern Qinshui Basin

Wells	X-07d	X-07d-1	X-14d	X-17d-1	X-04d-2	X-19d-1
Hydraulic fractured coalbed(s)	#3	#3	#3	#9	#15	#15
Typical daily water production (m^3/d)	14.5	14.7	12.6	13.3	3.2	3.9
Wells	X-11d-2	X-05d	X-20d	X-08d-1	X-01d	X-01d-2
Hydraulic fractured coalbed(s)	#15	#15	#15	#15	#3、#15	#3、#15
Typical daily water production (m^3/d)	2	11.4	6	18.5	120.3	68.7
Well no.	X-06d	X-08d	X-12d-2	X-14d-1	X-14d-2	X-18d-1
Hydraulic fractured coalbed(s)	3#、15	#3、#15	#3、#15	#3、#15	#3、#15	#3、#15
Typical daily water production (m^3/d)	59.5	17.1	43.6	3.7	10.8	40.6
Wells.	X04d-1	X-04d	X-05d-1	X-06d-2	X-10d-2	X-15d
Hydraulic fractured coalbed(s)	9#、#15	#3、#9、#15	#3、#9、#15	#3、#9、#15	#3、#9、#15	#3、#9、#15
Typical daily water production (m^3/d)	146.5	19.1	52.7	11.7	54.8	13.6
Wells	X-15d-2	X-11d	X-11d-1	X-18d	X-10d	X-19d-2
Hydraulic fractured coalbed(s)	#3、#9、#15	#3、#9、#15	#3、#9、#15	#3、#9、#15	#3、#9、#15	8#、#9、#15
Typical daily water production (m^3/d)	41.5	11.9	14.4	14.5	92.5	14

Data before September 18th, 2013.

A detailed comparison study between water productions and the lithological assemblages concluded that the limestone strata in Pennsylvanian/Taiyuan formation can be regarded as aquifuge, and the sandstones as aquifers (Chen et al., 2014). A lithological correlation section that includes the coal #15 is shown in Fig. 17.6. A lithological correlation section that includes the coal #9 and the coal #3 is shown in Fig. 17.7. This kind of lithological correlation section can help the site/layer selection for coalbed methane development. In well locations where a coalbed is contacted directly with overlying or underlying aquifers, or the aquifuges between the coalbed and the overlying or underlying aquifers are less than 3 m (as a rule of thumb), the coalbed should not be perforated and fractured. In well locations where the aquifuges between the coalbed and overlying or underlying aquifers are 3–6 m (as a rule of thumb), the frac operation should proceed.

17.4.3 PERMEABILITY ENHANCEMENT OPERATIONS

Until now, the most applied permeability enhancement operation is fracturing in China or in other countries. In the Southern Qinshui Basin, increased fracturing volumes showed their limited increases

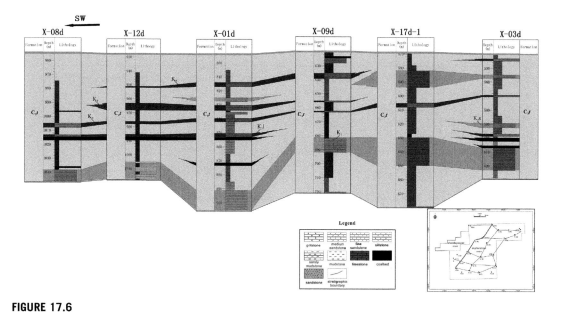

FIGURE 17.6

SW–NE direction lithological correlation section surrounding the coal #15.

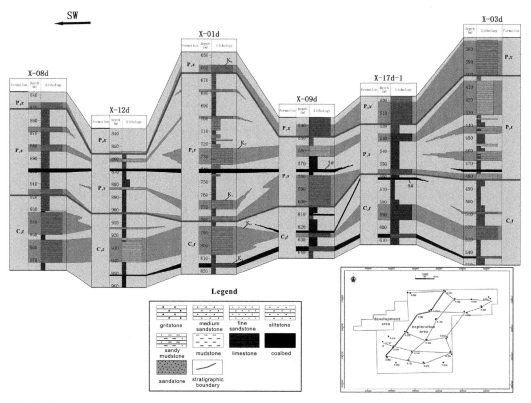

FIGURE 17.7

SW–NE direction lithological correlation section surrounding the coals #9 and #3.

Table 17.5 Average Gas Rate Variations with Frac Fluid Volumes in the Southern Qinshui Basin

Years	2006–2007	2008–2009	2010–2011
Frac fluids, m^3	370–420	550–600	700–750
Average Gas rate, m^3/d	960	980	1100

Source: Zhu et al. (2015).

of the gas rate (Table 17.5). Furthermore, increasing fracturing volume risks building a hydraulic link between coalbeds and neighboring aquifers when the aquifuges between the perforated coalbed and the overlying or underlying aquifers are less than 6 m.

As mentioned earlier, the cleats in coals in the Qinshui Basin are mostly deteriorated or closed due to multiple tectonic activities and present-day stress, or they are filled mainly with calcite. Acidizing–fracturing operations should be more favorable than fracturing operations alone by dissolving the calcite in the cleats. Laboratory studies demonstrated that the permeability of coal cores increased 10–20 times after being immersed at 12 h in chlorhydric acid (Zhao et al., 2012b). Currently, acidizing–fracturing in-site evaluation is ongoing. It is expected to increase the permeability of these reservoirs.

17.4.4 DEWATERING SCHEDULE

Coalbed methane recovery is a cost-effective approach to achieving hydrocarbon production, environmental, and safety goals (Markowski, 1998). When free gas in coals is produced, water is pumped from the secondary porosity, causing the pressure to decline in the formation. This results in the release of adsorbed methanes to the free gas phase via a reduction of the gas pressure. The desorbed methanes subsequently flow through the natural fracture system (secondary porosity) toward the production well (Jamshidi and Jessen, 2012).

Coal is relatively compressible compared to the rock; the permeability of coal is more stress-dependent than rocks (Seidle et al., 1992; Harpalani and Chen, 1997; Shi and Durucan, 2005; Deng et al., 2009; Liu and Harpalani, 2013). Characterization of the hydromechanical properties, including rock strength, deformation, and permeability change, is fundamentally important to the development of a coalbed methane reservoir (Deisman et al., 2013). The absolute permeability of coal reservoirs changes significantly during gas production, often initially decreasing but later increasing as the reservoir pressure and gas are drawn down. The permeability of coal reservoirs behaves in a similar fashion to other fractured reservoirs with respect to effective stress, decreasing exponentially as the effective stress increases (Pan and Connell, 2012). However, a permeability increase of up to 100 times in San Juan Basin (Palmer, 2009) was observed during depletion of coalbed methane wells due to matrix shrinkage (Levine, 1996; Liu and Harpalani, 2013). The stress effect to permeability reduction has a predominant role in the early stage of dewatering, and it is especially harmful to low-permeability coal reservoirs (Zhao et al., 2008). A fast dewatering strategy for commercial consideration may be problematic for low-permeability coal reservoirs. The stress of coal surrounding a coalbed methane production well is affected by the bottom hole flowing pressure (BHFP).

The permeability of coal shows a marked change under compression in the Qinshui basin (Ni et al., 2007; Zhao et al., 2013). The BHFP must be restricted to a specific range to favor higher permeability in the surrounding coal and thus higher productivity of the well. In most parts of the Qinshui Basin, the permeability of coal reservoirs is low, which is the Achilles heel for coalbed methane ventures (Palmer, 2010). The stress-dependent effect should be taken into account, especially in early stage of dewatering of coalbed methane wells for avoiding fast permeability reduction at the beginning, and dewatering schedule control is a key to the development of low-permeability coalbed methane reservoirs.

17.5 CONCLUSION

The Qinshui Basin is the most mature basin for coalbed methane exploration and development in China, with original in-place resources estimated to be 3.85×10^{12} m^3. The Qinshui Basin has experienced a number of tectonic activities after the deposition of Pennsylvanian–Permian coalbeds. Gas generation reached its first peak before the late Triassic period, and its second peak during early to middle Cretaceous period due to a tectonic thermal event, which raised the coal rank to semianthracite or anthracite.

The Pennsylvanian–Permian coalbeds in the Qinshui basin are in a high rank, with high methane content and low permeability (generally less than 1 mD). Low permeability and effective stress-dependent coal reservoirs lead to four challenges for coalbed methane development: (1) selection of well completion types; (2) selection of site/interval for fracturing operations; (3) new technology for permeability enhancement operations and fracturing volume design; and (4) dewatering and production management.

It will take time to overcome these challenges by trial and error and by developing new technologies appropriate to drive the commercial development for the low permeable reservoirs in the Qinshui Basin.

REFERENCES

Bachu, S., Michael, K., 2003. Possible controls of hydrogeological and stress regimes on the producibility of coalbed methane in Upper Cretaceous Tertiary strata of the Alberta basin, Canada. AAPG Bulletin 87 (11), 1729–1754.

Bodden III, W.R., Ehrlich, R., 1998. Permeability of coals and characteristics of desorption tests: implications for coalbed methane production. International Journal of Coal Geology 35 (1), 333–347.

Bustin, R.M., Clarkson, C.R., 1998. Geological controls on coalbed methane reservoir capacity and gas content. International Journal of Coal Geology 38 (1), 3–26.

Chen, Jing, Kang, Yong-shang, Guo, Ming-qiang, Zhang, Bing, Yuan, Chun-lin., 2014. Aquosity analysis of pennsylvanian taiyuan formation limestone in shouyang CBM exploration block, Qinshui Basin, Central China. In: Ye, J-ping, Fu, X-kang, Li, Wu-zhong. (Eds.), Proceedings of National CBM Conference, Geology Press House.

Chen, Lian-wu, 1998. The significance of studying cleat in Hancheng coalbed methane assessment. Journal of Xi'an Engineering University 20 (1), 33–35.

Chen, Zhen-hong, Song, Yan, Wang, Bo, Wang, Hong-yan, 2007. Damage of active groundwater to coalbed methane reservoirs and its physical simulation. Natural Gas Industry 27 (7), 16–19.

Chi, W-guo, 1998. Hydrogeological control on the coalbed methane in Qinshui basin. Petroleum Exploration and Development 25 (3), 15–18.

Clarkson, C.R., Barker, G.J., 2011. Coalbed methane. In: Guidelines for Application of the Petroleum Resources Management System. Sponsored by SPE, AAPG, WPC, SPEE, SEG, pp. 141–153.

Clarkson, C.R., Bustin, R.M., 2007. Production-data analysis of single phase(gas) coalbed-methane wells. SPE Reservoir Evaluation and Engineering 10 (03), 312–331.

Clarkson, C.R., Bustin, R.M., 2010. Coalbed methane: current evaluation methods, future technical challenges. SPE 131791, 1–42.

Colmenares, L.B., Zoback, M.D., 2007. Hydraulic fracturing and wellbore completion of coalbed methane wells in the Powder River Basin, Wyoming: implications for water and gas production. AAPG Bulletin 91 (1), 51–67.

Decker, A.D., Schraufnagel, R., Graves, S., Beavers, W.M., Cooper, J., Logan, T., Horner, D.M., Sexton, T., 1986. Coalbed methane: an old hazard becomes a new resource. In: Montgomery, S.L., Cobban, W.H. (Eds.), Petroleum Frontiers. Petroleum Information Corporation, Denver, CO, pp. 1–65.

Deng, Ze, Kang, Yong-shang, Liu, Hong-lin, Li, Gui-zhong, Wang, Bo, 2009. Dynamic variation character of coalbed methane reservoir permeability during depletion. Journal of China Coal Society 34 (7), 947–951.

Deisman, N., Khajeh, M., Chalaturnyk, R.J., 2013. Using geological strength index (GSI) to model uncertainty in rock mass properties of coal for CBM/ECBM reservoir geomechanics. International Journal of Coal Geology 112, 76–86.

Dou, Feng-ke, Kang, Yong-shang, Qin, Shao-feng, Mao, De-lei, Han, Jun, 2013. The coalbed methane production potential method for optimization of wells location selection. Journal of Coal Science and Engineering (China) 19 (2), 210–218.

Eddy, G.E., 1984. Conclusion on the development of coalbed methane. In: Rightmire, C.T., Eddy, G.E., Kirr, J.N. (Eds.), Coalbed Methane Resources of the United States. AAPG Studies in Geology Series #17, pp. 373–375.

Fan, Ai-min, Hou, Quan-lin, Ju, Yi-wen, Bu, Ying-ying, Lu, Ji-xia, 2005. A study on control action of tectonic activity on CBM pool from various hierarchies. Coal Geology of China 17 (4), 15–20.

Flores, R.M., 1998. Coalbed methane: from hazard to resource. International Journal of Coal Geology 35 (1), 3–26.

Fu, Xue-hai, Qin, Yong, LI, Gui-zhong, 2001. An analysis on the principal control factor of coal reservoir permeability in Central and Southern Qinshui Basin. Coal Geology and Exploration 29 (3), 16–19.

Fu, Xue-hai, Qin, Yong, Jiang, Bo, Wei, Chong-tao, 2004. Study on the "bottle-neck" problem of coalbed methane producibility of higy-rank coal reservoirs. Geological Review 50 (5), 507–513.

Gentzis, T., Deisman, N., Chalaturnyk, R.J., 2007. Geomechanical properties and permeability of coals from the Foothills and Mountain regions of western Canada. International Journal of Coal Geology 69 (3), 153–164.

Gurba, L.W., Weber, C.R., 2001. Effects of igneous intrusions on coalbed methane potential, Gunnedah Basin, Australia. International Journal of Coal Geology 46 (2), 113–131.

Hamawand, I., Yusaf, T., Hamawand, S.G., 2013. Coal seam gas and associated water: a review paper. Renewable and Sustainable Energy Reviews 22, 550–560.

Harpalani, S., Chen, G., 1997. Influence of gas production induced volumetric strain on permeability of coal. Geotechnical and Geological Engineering 15 (4), 303–325.

Hu, Guo-yi, Guan, Hui, Jiang, Deng-wen, Du, Ping, Li, Zhi-sheng, 2004. Analysis of conditions for the formation of a coal methane accumulation in the Qinshui coal methane field. Geology in China 31 (2), 213–217.

Jamshidi, M., Jessen, K., 2012. Water production in enhanced coalbed methane operations. Journal of Petroleum Science and Engineering 92, 56–64.

Jenkins, C.D., Boyer, C.M., 2008. Coalbed-and shale-gas reservoirs. Journal of Petroleum Technology 60 (02), 92–99.

Kaiser, W.R., Hamilton, D.S., Scott, A.R., Tyler, R., Finley, R.J., 1994. Geological and hydrological controls on the producibility of coalbed methane. Journal of the Geological Society 151 (3), 417–420.

Kang, Yong-shang, Deng, Ze, Liu, Hong-lin, 2008. Discussion about the CBM well draining technology. Natural Gas Geoscience 19 (3), 423–426.

Kang, Yong-shang, Dou, Feng-ke, Zhang, Bing, Ye, Jian-ping, 2013a. Performance statistical method for CBM well production prediction. In: Ye, Jian-ping, Fu, Xiao-kang, Li, Wu-zhong (Eds.), Proceedings of National CBM Conference. Geology Press House, pp. 160–168.

Kang, Yong-shang, Qin, Shao-feng, Han, Jun, Diao, Shun, Mao, De-lei, Dou, Feng-ke, 2013b. Typical dynamic index analysis method system for drainage and production dynamic curves of CBM wells. Journal of China Coal Society 38 (10), 1825–1830.

King, G.R., Ertekin, T., Schwerer, F.C., 1986. Numerical simulation of the transient behavior of coal-seam degasification wells. SPE Formation Evaluation 1 (02), 165–183.

Levine, J.R., 1996. Model study of the influence of matrix shrinkage on absolute permeability of coal bed reservoirs. In: Gayer, R., Harris, I. (Eds.), Coalbed Methane and Coal Geology. Geological Society Special Publication No 109, pp. 197–212.

Li, Gui-zhong, Wang, Hong-yan, Wu, Li-xin, Liu, Hong-lin, 2005. Theory of syncline controlled coalbed methane. Natural Gas Industry 25 (1), 26–28.

Li, Jing-ming, Liu, Fei, Wang, Hong-yan, Zhou, Wen, Liu, Hong-lin, Zhao, Qun, Li, Gui-zhong, Wang, Bo, 2008. Desorption characteristics of coalbed methane reservoirs and affecting factors. Petroleum Exploration and Development 35 (1), 52–58.

Li, Wu-zhong, Tian, Wen-guang, Chen, Gang, Sun, Qin-ping, 2010. Research and application of appraisal variables for the prioritizing of coalbed methane areas featured by different coal ranks. Natural Gas Industry 30 (6), 45–47.

Liu, Cheng-lin, Che, Chang-bo, Zhu, Jie, Yang, Hu-lin, Fan, Ming-zhu, 2009a. Coalbed methane resource assessment in China. China Coalbed Methane 6 (3), 3–6.

Liu, Hong-lin, Zhao, Guo-liang, Wang, Hong-yan, 2004. Coalbed methane exploration theory and practice in the high coal rank areas of China. Petroleum Geology and Experiment 26 (5), 411–414.

Liu, Hong-lin, Li, Jing-ming, Wang, Hong-yan, Zhao, Qing-bo, 2006. Different effects of hydrodynamic conditions on coalbed gas accumulation. Natural Gas Industry 26 (3), 35–37.

Liu, Hong-lin, Kang, Yong-shang, Wang, Feng, Deng, Ze, 2008. Coal cleat system characteristics and formation mechanisms in the Qinshui Basin. Acta Geologica Sinica 82 (10), 1371–1381.

Liu, Hong-lin, Li, Gui-zhong, Wang, Guang-jun, 2009b. Coalbed Methane Geological Characteristics and Development Potential. Petroleum Industry Press, Beijing.

Liu, S., Harpalani, S., 2013. Permeability prediction of coalbed methane reservoirs during primary depletion. International Journal of Coal Geology 113, 1–10.

Liu, Yue-wu, Su, Zhong-liang, Zhang, Jun-feng, 2010. Review on CBM desorption/adsorption mechanism. Well Testing 19 (6), 37–44.

Lou, Jiang-qing, 2004. Factors influencing production of coalbed gas wells. Natural Gas Industry 24 (4), 62–64.

Ma, Dong-min, 2003. Analysis of gas production mechanism in CBM wells. Journal of Xi'an University of Science and Technology 23 (2), 156–159.

Ma, Dong-min, Zhang, Sui-an, Wang, Peng-gang, Lin, Ya-bing, Wang, Chen, 2011. Mechanism of coalbed methane desorption at different temperatures. Coal Geology and Exploration 39 (1), 20–23.

Ma, Feng-shan, Li, Shang-ru, Cai, Zu-huang, 1998. Hydrogeological problems in coalbed methane development. Hydrogeology and Engineering Geology 3, 20–22.

Mao, De-lei, Kang, Yong-shang, Han, Jun, Qin, Shao-feng, Wang, Hui-juan, 2012. Influence of hydrogeological cycles on CBM in Hancheng CBM field, Central China. Journal of China Coal Society 37 (S2), 390–394.

Markowski, A.K., 1998. Coalbed methane resource potential and current prospects in Pennsylvania. International Journal of Coal Geology 38 (1), 137–159.

Masszi, D., 1991. Cavity stress relief method for recovering methane from coalbeds. In: Schwochow, S.D., Murray, D.K., Fahy, M.F. (Eds.), Coal Bed Methane of Western North America. Rocky Mountain Association of Geologists, Denver.

Moore, T.A., 2012. Coalbed methane: a review. International Journal of Coal Geology 101, 36–81.

Ni, Xiao-ming, Wang, Yan-bin, Jie, Ming-xun, Wu, Jian-guang, 2007. Reasonable production intensity of coalbed methane wells in initial production. Journal of Southwest Petroleum University 29 (6), 101–104.

Palmer, I., 2009. Permeability changes in coal: analytical modeling. International Journal of Coal Geology 77 (1), 119–126.

Palmer, I., 2010. Coalbed methane completions: a world view. International Journal of Coal Geology 82 (3), 184–195.

Pan, Zhe-jun, Connell, L.D., 2012. Modelling permeability for coal reservoirs: a review of analytical models and testing data. International Journal of Coal Geology 92 (1), 1–44.

Pashin, J.C., 2010. Variable gas saturation in coalbed methane reservoirs of the Black Warrior Basin: implications for exploration and production. International Journal of Coal Geology 82 (3), 135–146.

Qin, Shao-feng, Kang, Yong-shang, Cao, Ai-juan, Mao, De-lei, Wang, Hui-juan, 2013. Weight evaluation of main geological factors affecting drainage of CBM field and geological significance. In: Ye, Jian-ping, Fu, Xiao-kang, Li, Wu-zhong (Eds.), Proceedings of National CBM Conference. Geology Press House.

Qin, Sheng-fei, Song, Yan, Tang, Xiu-yi, Hong, Feng, 2005. The influence on coalbed gas content by hydrodynamics-the stagnant groundwater controlling. Natural Gas Geoscience 16 (2), 149–152.

Rao, Meng-yu, Zhong, Jian-hua, Yang, Lu-wu, Ye, Jian-ping, Wu, Jian-guang, 2004. Coalbed methane reservoir and gas production mechanism in anthracite coalbeds. Acta Petrolei Sinica 25 (4), 23–28.

Roadifer, R.D., Moore, T.R., 2009. Coalbed methane Pilots–Timing design and analysis. SPE Reservoir Evaluation and Engineering 12 (05), 772–782.

Rogner, H.H., 1997. An assessment of world hydrocarbon resources. Annual Review of Energy and the Environment 22, 217–262.

Schraufnagel, R.A., Schafer, P.S., 1996. The success of coalbed methane. In: Saulsberry, J.L., Schafer, P.S., Schraufnagel, R.A. (Eds.), A Guide to Coalbed Methane Reservoir Engineering. Gas Research Institute Report GRI-94/0397. Illinois, Chicago.

Seidle, J.P., Jeansonne, M.W., Erickson, D.J., 1992. Application of matchstick geometry to stress dependent permeability in coals. In: SPE Rocky Mountain Regional Meeting. Society of Petroleum Engineers.

Shi, J.Q., Durucan, S., 2005. A model for changes in coalbed permeability during primary and enhanced methane recovery. SPE Reservoir Evaluation and Engineering 8 (04), 291–299.

Song, Yan, Zhao, Meng-jun, Liu, Shao-bo, Wang, Hong-yan, Chen, Zhen-hong, 2005. Influence of tectonic evolution on coalbed methane enrichment. Chinese Science Bulletin 50 (Suppl. I), 1–5.

Song, Yan, Qin, Sheng-fei, ZHao, Meng-jun, 2007. Two key geological factors controlling the coalbed methane reservoirs in China. Natural Gas Geoscience 18 (4), 545–552.

Su, Xian-bo, Feng, Yan-li, Chen, Jiangfeng, Pan, Jienan, 2001. The characteristics and origins of cleat in coal from Western North China. International Journal of Coal Geology 47 (1), 51–62.

Su, Xian-bo, Lin, Xiao-ying, Liu, Shao-bo, Zhao, Meng-jun, Song, Yan, 2005. Geology of coalbed methane reservoirs in the Southeast Qinshui Basin of China. International Journal of Coal Geology 62 (2), 197–210.

Sun, Zhan-xue, Zhang, Wen, Hu, Bao-qun, Li, Wen-juan, Pan, Tian-you, 2005. Temperature characteristics and their relation to CBM distribution in Qinshui Basin. Chinese Science Bulletin 50 (Suppl. I), 93–98.

Ulery, J.P., 1988. Geologic Factors Infuencing the Gas Content of Coalbeds in Southwestern Pennsylvania. Report of Investigations, 9195.

Wan, Yu-jin, Cao, Wen, 2005. Analysis on production affecting factors of single well for coalbed gas. Natural Gas Industry 25 (1), 124–126.

Wang, Hongyan, Li, Guizhong, Li, Jingming, Liu, Hong-lin, Wang, Bo, 2006. Characteristics of CBM enrichment in China. China Coalbed Methane 3 (2), 7–10.

Wang, Mi-ji, Kang, Yong-shang, Mao, De-lei, Qin, Shao-feng, 2013. Dewatering-induced local hydrodynamic field models and their significance to CBM production: a case study from a coalfield in the southeastern margin of the Ordos Basin. Natural Gas Industry 33 (7), 1–6.

Ye, Jian-ping, Wu, Qiang, Wang, Zhi-he, 2001. Controlled characteristics of hydrogeological conditions on the coalbed methane migration and accumulation. Journal of China Coal Society 26 (5), 459–462.

Ye, Jian-ping, Wu, Qiang, Ye, Gui-jun, Chen, Chun-lin, Yue, Wei, Li, Hong-zhu, Zhai, Zheng-rong, 2002. Study on the coalbed methane reservoir-forming dynamic mechanism in the Southern Qinshui Basin, Shanxi. Geological Review 48 (3), 319–323.

Ye, Jian-ping, 2013. CBM exploration and development in the Qinshui Basin-history and technologies advancement. In: Lecture Presentation in China Petroleum University (First Semester).

Ye, Jian-ping, Zhang, Shou-ren, Ling, Biao-can, Zheng, Gui-qiang, Wu, Jian, Li, Dan-qiong, 2014. Study on variation law of coalbed methane physical property parameters with seam depth. Coal Science and Technology 42 (6), 35–39.

Zhang, Pei-he, Zhang, Qun, Wang, Bao-yu, Li, Guo-fu, Tian, Yong-dong, 2006. Integrated methods of CBM recoverability evaluation: a case study from Panzhuang mine. Coal Geology and Exploration 34 (1), 21–25.

Zhang, Pei-he, Zhang, Ming-shan, 2010. Analysis of application status and adapting conditions for different methods of CBM development. Coal Geology and Exploration 38 (2), 9–13.

Zhao, Bin, Wang, Zhi-yin, Hu, Ai-mei, Zhai, Yu-yang, 2013. Controlling bottom hole flowing pressure within a specific range for efficient coalbed methane drainage. Rock Mechanics and Rock Engineering 46, 1367–1375.

Zhao, Qing-bo, Kong, Xiang-wen, Zhao, Qi, 2012a. Coalbed methane accumulation conditions and production characteristics. Oil and Gas Geology 33 (4), 552–560.

Zhao, Qun, Wang, Hong-yan, Li, Jing-ming, Liu, Hong-lin, 2008. Study on mechanism of harm to CBM well capability in low permeability seam with quick drainage method. Journal of Shandong University of Science and Technology-Natural Science 27 (3), 27–31.

Zhao Wen-xiu, Li Rui, Wu Xiao-ming, Tang Ji-dan, Wang Kun, Wang Yu-jian. 2012b. Preliminary indoor experiments on enhancing permeability rate of coal reservoir by using acidification technology.9(1):10–13.

Zhou, Zhi-cheng, Wang, Nian-xi, Duan, Chun-sheng, 1999. Action of coalbed water in the exploration and development of coalbed gas. Natural Gas Industry 19 (4), 23–25.

Zhu, Qing-zhong, Zuo, Yin-qing, Yang, Yan-hui, 2015. How to solve the technical problems in the CBM development: a case study of a CMB gas reservoir in the southern Qinshui Basin. Natural Gas Industry 35 (2), 106–109.

Zuber, M.D., 1996. Basic reservoir engineering for coal. In: Saulsberry, J.L., Schafer, P.S., Schraufnagel, R.A. (Eds.), A Guide to Coalbed Methane Reservoir Engineering. Gas Research Institute Report GRI-94/0397, Chicago, Illinois.

CHAPTER 18

MONITORING AND PREDICTING STEAM CHAMBER DEVELOPMENT IN A BITUMEN FIELD

Kelsey Schiltz[1], David Gray[2]

Colorado School of Mines, Golden, CO, USA[1]; Nexen Energy ULC, Calgary, AB, Canada[2]

18.1 INTRODUCTION
18.1.1 GENERAL

Although the word "unconventionals" has almost become synonymous with production from shales in the United States, heavy oil and bitumen constitute another category of unconventionals with massive resource potential. According to standards set by the United States Department of Energy, oil is considered heavy if it has an API gravity of less than 22.3 and bitumen if it is less than 10 (Chopra et al., 2010). Approximately one trillion barrels of heavy oil and bitumen are estimated to be recoverable worldwide, with the largest accumulations in Canada and Venezuela (Rigzone, 2014). The overwhelming majority of Canada's heavy oil and bitumen resources are found in three deposits within the province of Alberta. The Albertan government estimates that its oil sands contain 1.8 trillion barrels of original bitumen in place, 80% of which is located in the Athabasca Oil Sands deposit (Fig. 18.1) (Alberta Energy, 2012).

While the resource potential of heavy oil and bitumen is enormous, the cost of producing and refining the crude is high. In particular, bitumen requires special effort to produce because it does not flow under normal subsurface conditions. Although bitumen reservoirs tend to be shallower than conventional reservoirs, surface mining is rarely a viable option. For the majority of bitumen deposits, in situ recovery methods that are more expensive and generally less effective than surface mining must be used. Optimization at every stage in the development process is crucial in order to maintain the economic viability of the project.

Time-lapse seismic surveying has proven to be one of the most useful ways to monitor the effectiveness of in situ recovery methods, particularly those that involve steam flooding. It can be used to improve operating strategies in both producer and injector wells, change well completions, and find infill drilling opportunities. For example, at ConocoPhillips and Total's Surmont Field in Alberta, seismic amplitude changes were used to identify areas along the wellbore where steam injection had been effective and those where it had not (Byerley et al., 2009). This insight led to a modified injection scheme that improved steam efficiency within just four months. The success of time-lapse monitoring

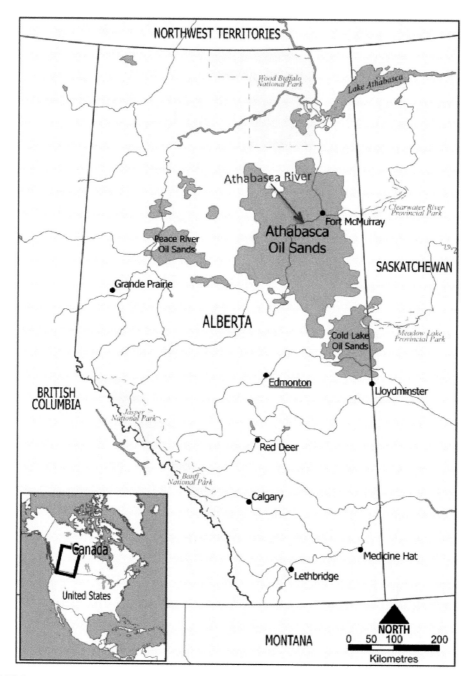

FIGURE 18.1

The field is located within the Athabasca Oil Sands deposit, the largest of three major deposits in Alberta (Wikipeda, 2014).

of bitumen recovery is attributable to several factors, including the shallow depth of the reservoir, the high porosity and low bulk modulus of the sands, and the large change in fluid compressibility during production. These factors all increase the likelihood of seeing a time-lapse seismic signature (Lumley et al., 1997). What has proven more challenging than the time-lapse reservoir monitoring is reservoir characterization aimed at distinguishing good quality sands from shales that can limit production. Some studies have attempted to predict reservoir sands though inversion or neural networks (e.g.,Roy et al., 2008 and Tonn, 2002), but generally the authors do not go as far as to compare their results to observed time-lapse effects for validation. In this study of a bitumen field in Alberta, the distribution of steam in the reservoir determined through time-lapse analysis is leveraged to investigate the accuracy of sand and shale prediction using a neural network.

18.1.2 THE ATHABASCA FIELD

The focus of the study is a bitumen field covering 63,000 acres within the Athabasca Oil Sands deposit in Alberta, Canada (Fig. 18.1). Bitumen has such a high viscosity at in situ conditions that it is immobile and cannot be produced using conventional methods. Production at this field is enabled by a thermal recovery method called steam-assisted gravity drainage, or SAGD. The SAGD process involves two vertically-stacked horizontal wells separated by approximately 5 m. The upper well injects high-pressure, high-temperature steam into the reservoir. The steam heats the bitumen and lowers its viscosity so that it can flow via gravity down to the lower horizontal well where it is produced (Smalley, 2000). As the bitumen drains from the reservoir, it is replaced by steam in the pore space, creating what is known as the steam chamber. The evolution of the steam chamber is highly controlled by the presence of low-permeability intervals within the reservoir (e.g., shale stringers and mud plugs), as depicted in Fig. 18.2.

FIGURE 18.2

Steam chamber growth can be impeded by a low-permeability interval such as shale (McDaniel and Associates Ltd, 2006).

The bitumen (~8 API) is produced from the Lower Cretaceous McMurray Formation. The McMurray Formation was deposited on top of Devonian limestones in a north-west to south-east trending paleo-valley and was subsequently capped by marine shales of the Clearwater Formation (Wightman and Pemberton, 1997). The McMurray is often informally separated into three units known as the lower, middle, and upper McMurray. As a general approximation, these three subdivisions represent a transgressive progression from a fluvial, to an estuarine, to a marine depositional environment. Clean, blocky reservoir sands are generally found toward the base of the McMurray in sand-filled channels or point bars deposited in low-lying estuaries. Toward the top of the formation, increasing marine influence resulted in the deposition of more fine-grained silts, clayey sands, and silty clays (Dusseault, 2001).

Throughout the McMurray, lateral facies variations can be quite abrupt due to the various depositional elements found in the fluvial setting. These depositional elements, including sand-filled channels, mud-filled channels, point bars and counter-point bars, have variable porosities and permeabilities which need to be well-understood in order to optimize production. This study demonstrates how seismic reservoir characterization can be used to create a geologic reservoir model that can predict steam chamber development.

18.1.3 SEISMIC AND WELL DATA

The time-lapse seismic data used in this study cover 2.5 km^2 and encompass four SAGD well pads (Fig. 18.3). The baseline survey was acquired in 2002 before any steam injection had taken place and therefore provides a true snapshot of initial reservoir conditions. The monitor survey was acquired in 2011 after approximately four years of production. One of the most important considerations in the interpretation of time-lapse seismic is the repeatability of the surveys. Repeatability is a measure of how similar two or more seismic surveys are in areas where no subsurface changes have taken place. The more repeatable the surveys, the easier it is to isolate and interpret production-related changes in the subsurface. The first step to ensure good repeatability was to acquire both surveys using similar acquisition parameters. Next, the data were processed in parallel to reduce the chance of introducing spurious differences during the processing stage. Parallel time-lapse processing was performed to create a poststack compressional volume, poststack radial converted-wave volume, and angle stacks for both the baseline and monitor surveys.

Despite taking precautions to ensure repeatability, nonproduction related differences will inevitably exist between the surveys. To further eliminate these differences and increase the repeatability of the data, a cross-equalization workflow was performed on the monitor survey. Cross-equalization is a multistep process that attempts to identify and eliminate unwanted differences between the baseline and monitor, thereby allowing accurate time-lapse interpretation of the production-related changes (Ross et al., 1996).

Throughout the cross-equalization process, the normalized root-mean-square (NRMS) ratio was used as a statistical indicator of repeatability. This ratio, defined in Eq. (18.1), reaches a minimum value of 0 when two traces are identical (i.e., perfectly repeatable) and a maximum of two when two traces are identical but opposite in polarity (Kragh and Christie, 2002). An overall reduction in the NRMS ratio in an area that is expected to be free of production effects indicates that the repeatability has been improved by the cross-equalization process.

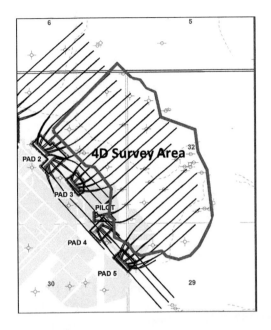

FIGURE 18.3

Map of the SAGD well configuration within the study area. The red outline shows the boundary of the four-dimensional (4D) seismic survey covering well Pads 1 (pilot), 3, 5, and part of Pad 2.

$$\text{NRMS ratio} = \frac{2 \times \text{RMS}(A - B)}{\text{RMS}(A) + \text{RMS}(B)} \qquad (18.1)$$

As a starting point, the mean NRMS was calculated over a window beneath the reservoir in which time-lapse changes were not expected. The initial NRMS in this window was 1.61 (in other words, highly nonrepeatable). Next, a number of cross-equalization steps were applied to the monitor survey in sequence, including phase and time matching, static time shifts, time-variant time shifts, amplitude normalization and a shaping filter. Although the details of these operations are beyond the scope of this paper, the overall effect was a decrease in the mean NRMS in the analysis window from 1.61 to 0.53, which represents a significant improvement in repeatability. A line through each of the cross-equalized time-lapse volumes is shown in Fig. 18.4. The first thing to note is the dramatically lower frequency content of the converted-wave data compared to the compressional data. This is one major limitation of interpreting converted waves. The second observation that can be made, however, is that there are changes in reflectivity within the reservoir (between the blue and green horizons) that are likely related to SAGD production. These cross-equalized time-lapse volumes were used in the subsequent time-lapse analysis and reservoir characterization work.

In addition to the seismic data, there are 43 vertical wells within the study area. Each has a full suite of standard logs, including caliper, gamma ray, density, neutron, and resistivity. Several calculated curves such as neutron porosity, water saturation, and volume of shale were also provided by the operator. All 43 wells within the study area have P-wave (compressional) sonic logs

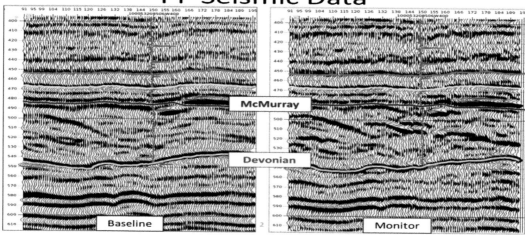

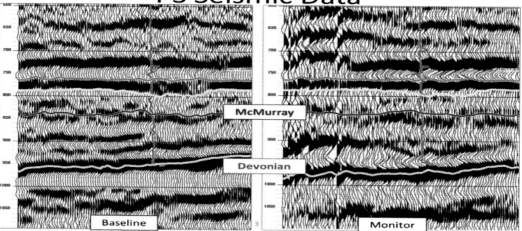

FIGURE 18.4

Line through the cross-equalized time-lapse seismic data. The McMurray reservoir is located between the blue (black in print versions) and green (gray in print versions) horizons.

and four wells also have an S-wave (shear) sonic log. The well control was leveraged extensively for horizon interpretation and for training the neural network.

18.2 MAPPING STEAM

In the theoretical case of a perfectly homogenous reservoir, the steam chamber would develop uniformly along the entire length of the borehole. On the other hand, when heterogeneities such as shale stringers are present, the steam cannot fully penetrate these tight intervals and an irregular steam distribution develops in which certain areas of the reservoir are not being swept. Mapping the distribution of steam is an important first step in determining whether heterogeneities are influencing production from this field.

The SAGD process causes significant fluid and saturation changes that affect the seismic response and enable the visualization of the steam chamber through time-lapse analysis. First, the bitumen transforms from a quasisolid to a liquid as its viscosity is lowered through heating. Then, the bitumen is replaced by steam as it drains from the reservoir rock down to the production well. Overall, these changes result in a decrease in the P-wave velocity within the steamed zones. The lower velocity in turn causes seismic events beneath the steamed reservoir to be shifted downward in the monitor survey relative to the baseline. The map in Fig. 18.5(a) shows the time shifts observed in the Devonian reflector at the base of the reservoir. Significant time shifts of up to 8 ms can be seen where the steam chamber has been effectively developed. Similar time-lapse anomalies were also observed when the amplitude changes within the reservoir were mapped (Fig. 18.5(b)). In both time-lapse maps, the anomalies over well Pad 5 are elongated along the wellbore and it appears that adjacent steam chambers have begun to coalesce. These wells provide an example of good conformance; in other words: a large portion of the borehole has a well-developed steam chamber. The overall lack of time-lapse effects seen along the wells of Pads 1 and 2NE, however, indicates poor steam chamber development that may be related to a higher percentage of low-permeability shales. The results of the reservoir characterization will be compared to these time-lapse anomalies to verify this theory.

Time-lapse analysis of the compressional data indicates that the steam chamber is well-developed in some areas and poorly developed in others. To further solidify the connection between steam chamber development and production, a crossplot was constructed between the length of the steam anomalies in the time shifts map (Fig. 18.5(a)) and cumulative oil production for each well. Wells from Pad 1 were excluded from the plot since these wells have been active over a longer time period. The resulting plot (Fig. 18.6) shows a clear positive correlation between the degree of steam chamber conformance and production (a similar result was found by Byerley et al., 2009 at the Surmont project). This relationship provides confirmation that the 4D seismic anomalies are indicative of steam and that maximizing steam chamber development leads to increased production. Performing reservoir characterization to identify zones of low reservoir quality is one way to optimize well pair placement for improved steam chamber development.

As a final step in the time-lapse analysis, a 4D inversion of the PP (compressional) poststack data was performed to create a three-dimensional (3D) representation of the steam chamber. As previously mentioned, steamed reservoir has a decreased P-wave velocity and bulk density, which in turn lowers the P-impedance. In the 4D inversion process, the baseline and monitor surveys are both inverted for P-impedance and the results are used to calculate the percent change in impedance from 2002 to 2011. A cross-section through the result of this calculation, shown in Fig. 18.7(a), shows that steam-filled reservoir corresponds to an impedance decrease of 10–20%. Observations from wells logged before and after steaming confirm that P-impedance changes on the order of −20% can be expected. Next, regions of the P-impedance difference volume corresponding to a reduction of 15% or greater were extracted to isolate the 3D volume that has been filled with steam. The 3D representations of individual steam chambers are referred to as steam geobodies and can be seen in Fig. 18.7(b). From an aerial perspective, the geobodies have the same distribution as

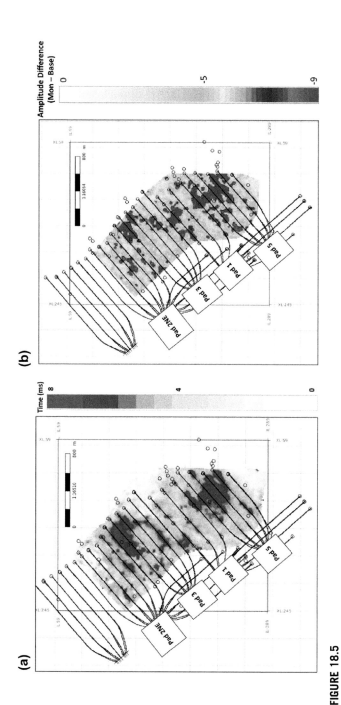

FIGURE 18.5

(a) Time-shifts observed between the baseline and monitor survey in the reflector marking the base of the reservoir (b) Amplitude difference after cross-equalization (monitor-baseline). The map shows the minimum amplitude between the McMurray horizon (McM) and 15 ms below the Devonian horizon (Dev).

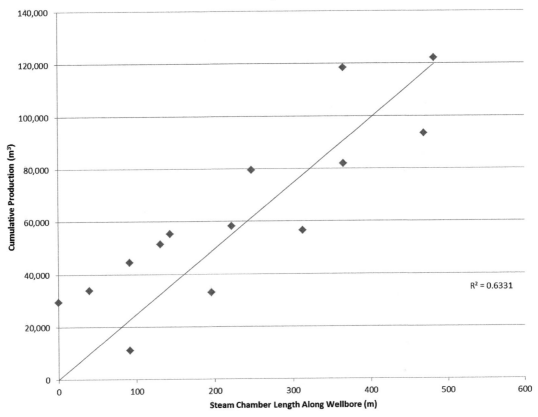

FIGURE 18.6

For each well pair, the length of the borehole associated with significant time shifts in Fig. 18.5(a) is plotted against the cumulative production.

the anomalies in the time-lapse maps in Fig. 18.5, but now the vertical position of the steam chamber has also been defined. Ultimately, the geometry of these steam geobodies will be compared to that of the interpreted shales to determine whether the shales exhibit control over steam chamber development.

18.3 RESERVOIR CHARACTERIZATION USING A PROBABILISTIC NEURAL NETWORK

In an ideal case, one seismic survey acquired before steam injection begins could be used to map shale zones and accurately predict the success of proposed SAGD well pairs. Therefore, the objective of reservoir characterization in this study was to use the seismic data to distinguish between clean reservoir sands and shales that may inhibit the growth of the steam chamber. Seismic inversion is one method commonly used to pursue such goals and was the first method attempted in this study. Prior to initiating the seismic inversion work, a petrophysical analysis was performed using 11 wells

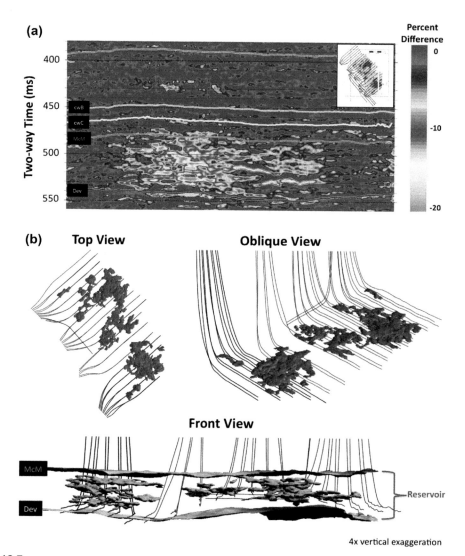

FIGURE 18.7

(a) Percent change in P-impedance between the baseline and monitor surveys. The presence of steam is indicated by P-impedance decreases of 10–20%. (b) 3D steam geobodies created by extracting the areas of the P-impedance percent change volume correspond to impedance decreases of greater than or equal to 15%.

within the 4D study area to determine which of the seismic inversion products, P-impedance, S-impedance, or density, would be the most useful sand-shale indicator. In the analysis, computed logs of each inversion product were plotted against the volume of shale (Vsh) log. While P-impedance was shown to have a weak correlation to Vsh, density had by far the strongest correlation at $R^2 = 0.88$ (Fig. 18.8). Density is notoriously difficult to estimate, however, from the

18.3 RESERVOIR CHARACTERIZATION USING A PROBABILISTIC

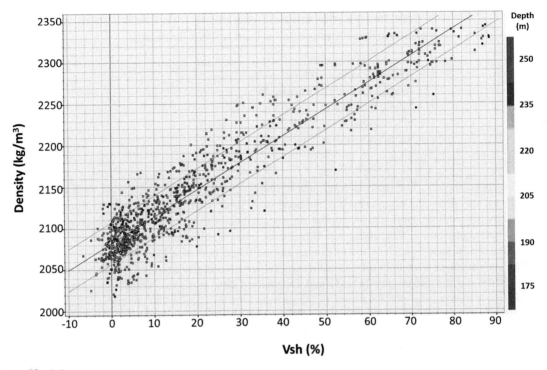

FIGURE 18.8

Crossplot showing a strong correlation ($R^2 = 0.88$) between density and volume of shale well logs.

inversion of seismic data with useable reflection angles of less than 50° (Roy et al., 2008). After attempting both a prestack and joint inversion on the seismic dataset (maximum angle of 45°), it was found that density could not be accurately constrained and another approach would have to be taken. It was decided to use a probabilistic neural network (PNN) to predict density from multiple seismic attributes.

A PNN is an algorithm that simulates the nonlinear way in which the human brain learns (Hampson and Russell, 2012). The PNN tries to determine the relationship between a set of seismic attributes and a target well log. Once the transformation between the attributes and the target log has been determined at the training wells, the same transform is applied to the seismic data to create a volume of the target log property.

The set of seismic attributes with the best correlation to the target density log was determined through multiattribute analysis. Seismic attributes were derived from the full stack compressional and converted-wave volumes, a poststack P-impedance inversion result, and a prestack P-impedance inversion result. Then, the multiattribute list is arrived at using stepwise regression. Stepwise regression is an approximation method that takes much less computation time than testing every possible combination of attributes. The method works by first searching through the list of available attributes to find the single attribute that best correlates with the target log. This attribute is listed as attribute 1. Next, the program finds the best combination of two attributes, assuming that one of these

Table 18.1 The Six Attributes Selected by Multiattribute Analysis to be Used in PNN Training

1. Integrate (PP full stack)
2. Amplitude envelope (PP full stack)
3. Average frequency (PS full stack)
4. Time (PP full stack)
5. Filter 35-40-45-50 (PS full stack)
6. Second derivative (PP full stack)

attributes is attribute 1. The method continues in this way until either all of the attributes have been ranked or a specified validation criterion has been met. Table 18.1 shows the six attributes that were selected by the analysis. Note that two out of the six attributes are calculated from the converted-wave (PS) volume, indicating that converted-wave data adds significant value to density prediction. A detailed description of each of the attributes can be found in Russell (2004).

The six attributes were then used to train the PNN to predict the density logs within the reservoir zone defined by the McMurray (McM) and Devonian (Dev) horizons. Once the relationship was established at the training well locations, the same transform was applied to the entire baseline seismic volume to create a predicted density volume. Figure 18.9 shows the results along an arbitrary line. All the wells shown in this line are blind wells and were not included in the multiattribute analysis or PNN training in any way. For each well, the filtered density log is displayed in color along with the Vsh curve in black. Although the magnitude of the predicted density in this section is not always accurate, the relative density variations are very consistent with the well log data. Ultimately it is the ability to capture the relative highs and lows in the density that is important for distinguishing sands and shales.

The sands and shales interpreted from the PNN can also be related to geologic features. Figure 18.10 shows the predicted density at one inline along with a blind well. The schematic below the inline illustrates some of the common morphologic features in a fluvial–estuarine depositional environment (Leckie et al., 2009). In particular, sandy lower point bars (LPBs) and shaley inclined heterolithic strata (IHS) are often found at the inner bends of meandering channels. A channel-like profile was identified within the density inline, and adjacent features bear strong resemblance to IHS and LPB deposits. The geologic appearance of the PNN density volume provides further support for its accuracy and lends credibility to lithologic interpretations.

The primary goal of reservoir characterization for a bitumen reservoir undergoing SAGD is to accurately identify shales. This is because the distribution of shales highly controls the development of the SAGD steam chamber. Therefore, a test of the utility of the predicted density volume is to see if a relationship exists between the distribution of high density zones interpreted as shales and the distribution of the steam chamber. The map in Fig. 18.11(a) shows the 3D shale geobodies (blue) predicted by the PNN in the lower to middle part of the reservoir. These shales are particularly important because they limit production more severely than shales closer to the top of the reservoir. The map in Figure 18.11(b) shows the same shales along with the steam geobodies (yellow) determined through time-lapse analysis. By comparing the two maps, it can be seen that the steam chamber has developed better in areas that lack shales in the lower half of the reservoir and is poorly developed in areas that have a significant number of shales (Pad 1, for example). In a few areas, most noticeably on Pad 5, the

18.3 RESERVOIR CHARACTERIZATION USING A PROBABILISTIC

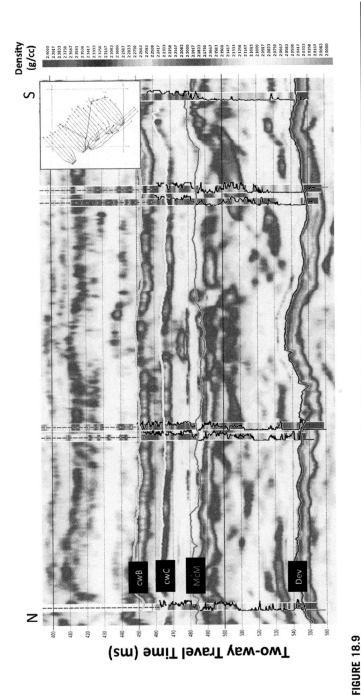

FIGURE 18.9

An arbitrary line through the PNN density volume. Inserted blind wells show the filtered density log in color (0-150-160 Hz) and the Vsh log in black. The reservoir is between the McM and Dev horizons. cwB/C, Clear Water B/C; McM, Top McMurray Formation; Dev, Top Devonian.

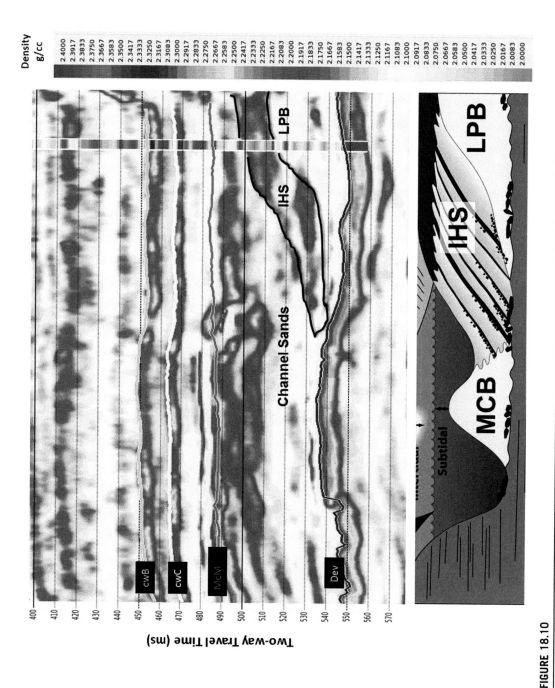

FIGURE 18.10

Inline through the PNN density result along with filtered density at a blind well. The schematic at the bottom shows depositional elements commonly found in this fluvial-estuarine environment (Leckie et al., 2009). Similar elements have been interpreted in the density result.

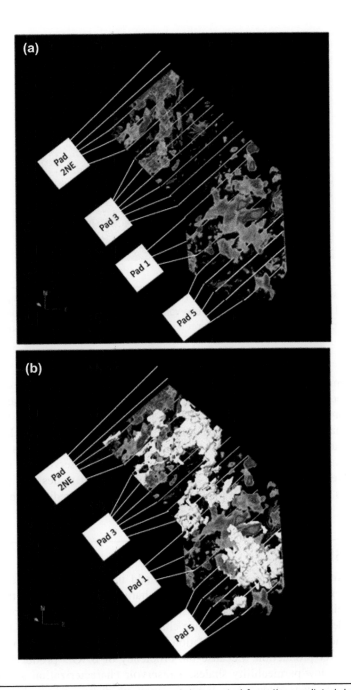

FIGURE 18.11

(a) Distribution of shales in the lower half of the reservoir interpreted from the predicted density volume (b) Interpreted shales (blue, gray in print versions) along with the steam geobodies (yellow, white in print versions). The shales appear to be controlling the development of the steamchamber.

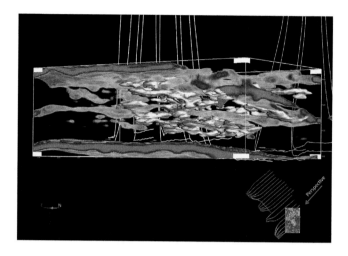

FIGURE 18.12

View looking directly toward the heels of Pad 5 wells. The steam chamber (yellow, white in print versions) has risen and encountered shale in the middle of the reservoir (blue, gray in print versions) but is interpreted to have moved around it over time (red (dark gray in print versions) arrow) to access the good quality reservoir above.

steam chamber seems to have developed regardless of the presence of shales. From a different perspective, however, it appears as if steam that was trapped up against the base of the shale has managed to move around or through it over time to access the reservoir sands above (Fig. 18.12). Whether or not this happens depends on the lateral extent and continuity of the shale body and whether or not good quality sands exist above it. From this view, it can be seen that there is an interval of clean sand that the steam has filled between the shale in the middle of the reservoir and the shales at the top of the reservoir. Overall, the good correlation between interpreted shales and the steam chamber distribution indicates that the PNN density volume is a useful tool for predicting the success of SAGD well pairs.

18.4 CONCLUSION

This study demonstrates the value seismic data can add in the SAGD development of bitumen resources as both a monitoring tool and a reservoir characterization tool. In the first part of the study, time-lapse seismic data were used to visualize steam chamber development by analyzing timeshifts, amplitude changes, and P-impedance changes. The development of the steam chamber was found to be nonuniform and a correlation was observed between poor steam chamber development and lower production. Because the efficiency of SAGD production is known to be greatly influenced by low-permeability shales, a reservoir characterization study of the baseline multicomponent survey was undertaken with the goal of distinguishing between high-quality reservoir sands and low-permeability shales.

A petrophysical analysis of 11 tied wells within the study area identified density as the acoustic parameter most closely associated with the volume of shale. After attempts to constrain density through prestack and joint inversion were unsuccessful, a PNN was used to predict density from multiple seismic attributes. The PNN density prediction was geologic in appearance, accurately captured relative density, and allowed the interpretation of reservoir shale bodies. Perhaps most importantly, a relationship was observed between the interpreted shales and the distribution of the steam chamber in the reservoir. This relationship provided confirmation that the PNN density prediction can be used to predict the success of SAGD in a given area.

Since the study, shale barriers and baffles are being mapped in two ways. The first is by using PNN analysis to predict high density intervals prior to steam injection. The second is by using the boundaries of the 4D steam geobodies to infer thin shales that are below seismic resolution. The net effect is a much better understanding of the behavior of the reservoir.

LIST OF ABBREVIATIONS

IHS Inclined heterolithic strata
LPB Lower point bar
NRMS Normalized root-mean-square
PNN Probabilistic neural network
SAGD Steam-assisted gravity drainage

REFERENCES

Alberta Energy, 2012. Oil Sands- Facts and Statistics. http://www.energy.alberta.ca/oilsands/791.asp.
Byerley, G., Barham, G., Tomberlin, T., Vandal, B., 2009. 4D seismic monitoring applied to SAGD operations at Surmont, Alberta, Canada. SEG Technical Program Expanded Abstracts 2009, 3959–3963.
Chopra, S., Lines, L., Schmitt, D.R., Batzle, M., 2010. Heavy-oil reservoirs: their characterization and production. In: Heavy Oils: Reservoir Characterization and Production Monitoring: Society of Exploration Geophysicists, Geophysical Developments 13 (1), 1–69
Dusseault, M., 2001. In: Comparing Venezuelan and Canadian Heavy Oil and Tar Sands: Proceedings of Petroleum Society's Canadian International Conference. 2001-061, pp. 1–20.
Hampson, D.P., Russell, B.H., 2012. Neural Network Theory. Hampson-Russell Software, 9.1.3.
Kragh, E., Christie, P., 2002. Seismic repeatability, normalized rms, and predictability. The Leading Edge 21 (7), 640–647.
Leckie, D., Fustic, M., Seibel, C., 2009. Geoscience of one of the largest integrated SAGD operations in the World- a case study from the Long Lake, Northeastern Alberta. CSPG Luncheon Talk.
Lumley, D., Behrens, R., Wang, Z., 1997. Assessing the technical risk of 4-D seismic project. The Leading Edge 16 (9), 1287–1292.
McDaniel and Associates Consultants Ltd., 2006. http://www.mcdan.com/oilsands.html.
Rigzone, 2014. How It Works: Heavy Oil. http://www.rigzone.com/training/heavyoil/default.asp.
Ross, C., Cunningham, G., Weber, D., 1996. Inside the cross-equalization black box. The Leading Edge 15 (11), 1233–1240.
Roy, B., Anno, P., Gurch, M., 2008. Imagine oil-sand reservoir heterogeneities using wide-angle prestack seismic inversion. The Leading Edge 27 (9), 1192–1201.

Russell, B., 2004. The Application of Multivariate Statistics and Neural Networks to the Prediction of Reservoir Parameters Using Seismic Attributes (Ph.D. thesis). University of Calgary.

Smalley, C., 2000. Heavy Oil and Viscous Oil in Modern Petroleum Technology. Institute of Petroleum (Chapter 11).

Tonn, R., 2002. Neural network seismic reservoir characterization in a heavy oil reservoir. The Leading Edge 21 (3), 309–312.

Wightman, D., Pemberton, S., 1997. The lower cretaceous (Aptian) McMurray formation: an overview of the Fort McMurray area, Northeastern Alberta. In: Petroleum Geology of the Cretaceous Mannville Group, Western Canada: Canadian Society of Petroleum Geologists, Memoir 18, pp. 312–344.

Wikipedia, 2014. Athabasca Oil Sands. http://en.wikipedia.org/wiki/Alberta-oil-sands (last accessed 27.02.14).

CHAPTER 19

GLOSSARY FOR UNCONVENTIONAL OIL AND GAS RESOURCE EVALUATION AND DEVELOPMENT

Y. Zee Ma, David Sobernheim, Janz R. Garzon
Schlumberger, Denver, CO, USA

19.1 RESERVOIR-RELATED TERMINOLOGY

UNCONVENTIONAL RESOURCES

While different definitions have been proposed to classify unconventional resources, the subsurface hydrocarbon resources that are not conventionally developed are termed unconventional reservoirs. These geological formations include tight gas sandstones, oil or tar sands, heavy oil, gas shales, coalbed methane, oil shales, gas hydrates, and other low-permeability tight formations. Among them, gas and oil shales and coalbed methane are typically source-rock reservoirs.

Conventional hydrocarbon resources typically accumulate in favorable structural or stratigraphic traps in which the formation is porous and permeable, but sealed by an impermeable layer that prevents hydrocarbon from escaping. These favorable subsurface structures possess migration pathways that link the source rocks to the reservoirs, and the formations have good reservoir quality and generally do not require large stimulation to produce hydrocarbon. Unconventional resources reside in tight formations, are of lower reservoir quality, and are more difficult from which to extract hydrocarbons. On the other hand, unconventional resources are more abundant in the Earth.

Some studies consider gas and oil shales, coalbed methane, and gas hydrates as unconventional while putting tight gas sandstones in the conventional category (Sondergeld et al., 2010). From the standpoint of reservoir characterization methodology, tight gas sandstone reservoirs (perhaps other tight formations as well) indeed share many characteristics of conventional reservoirs (Law and Spencer, 1989).

SHALE

For a historical reason, the term "shale" has two connotations: lithological and geological formationwise. In shale reservoirs, shale is generally used as a geological formation or perhaps even more accurately a facies; thus it is important to distinguish shale from clay. As such, shale describes fine-grained rocks that may contain a number of lithological components: clay, quartz, feldspar, heavy minerals, etc., and clay is only one component of the typical shale. In other words, shale should be more properly referred to as a mudstone that has undergone a certain amount of compaction.

KEROGEN

Kerogen is a solid, insoluble organic matter residing in source rocks, and it can yield oil or gas upon heating. Typical organic constituents of kerogen are algae and woody plant material. Kerogens are described as type I, mainly consisting of algal and amorphous; type II, mixed materials of terrestrial and marine source; type III, mainly woody terrestrial source materials; and type IV, containing mostly decomposed organic matter and having no or little potential for petroleum generation. While type III typically generates natural gas, types I and II can generate both oil and gas depending on other conditions, especially the thermal condition.

SOURCE ROCK

Source rock is a sedimentary rock that contains kerogens which can be thermally transformed into petroleum.

TOTAL ORGANIC CARBON

Total organic carbon (TOC) is the amount of carbon bound in an organic compound in a geologic formation, typically source rock (Jarvie, 1991). It is separated from inorganic carbon, which is the carbon present in minerals which form rocks.

ADSORPTION

Adsorption is the adhesion of molecules from gases, solutes or liquids to the surface of solid bodies or liquids with which they are in contact. The adsorption creates a thin film of the adsorbate on the surface of the solid (termed the absorbent), and it is different from absorption in which the adsorbate permeates or dissolved by a liquid or solid. In other words, adsorption is a surface-based physicochemical process and is a consequence of surface energy. The surface concentration of the gas molecules by adsorption depends on pore geometry, surface morphology, pore pressures, and temperatures. Adsorption may include the physical adsorption that originates from the electric dipole interactions, and the chemical adsorption that involves changes in the structure of bonding molecules. In gas shale, the physical adsorption is more important than chemical adsorption.

Two related terms are sorption that encompasses both absorption and adsorption, and desorption that is the reverse of sorption.

VITRINITE REFLECTANCE

Vitrinite reflectance is a measure of the percentage of incident light reflected from the surface of vitrinite particles in a sedimentary rock, and it is commonly used as a measure of thermal maturity of kerogen. In practice, vitrinite reflectance is often an average value based on all vitrinite particles measured in a sample, and thus, when the distribution of the individual measured values on a sample has a widespread, the average is less meaningful.

THERMAL MATURATION AND MATURITY

Thermal maturation is the process of thermal transformation of kerogen into hydrocarbon. Kerogen can be converted into oil or gas upon heating under high temperature (often also under high pressure).

When source rocks are buried deeply over geological time, the kerogen in the rock formation can convert into oil or gas as it is subjected to high temperature.

Thermal maturity is a measure or degree of maturation of kerogen. Vitrinite reflectance is commonly used to describe the thermal maturity of kerogen for hydrocarbon generation. At low thermal maturity, the higher the maturity, the more hydrocarbon is generated. Beyond the certain level of thermal maturity, the higher the maturity, the less liquid hydrocarbons will be generated from the kerogen. At very high thermal maturity, there will be no or little hydrocarbon (gas or liquid) generated.

DIAGENESIS, CATAGENESIS, AND METAGENESIS

Three main stages of evolution of sedimentary rocks include: (1) diagenesis in which no significant thermal transformation of organic matters occurs; (2) catagenesis in which organic matters are under thermal transformation and converted into hydrocarbon; and (3) metagenesis in which organic matters are under thermal transformation and the converted hydrocarbon may be overly cooked or destroyed.

PYROLYSIS

Pyrolysis is a thermochemical decomposition of organic materials at high temperatures in the absence of oxygen, and it involves the simultaneous change of chemical composition and physical phase. The process is irreversible.

HETEROGENEITY

Heterogeneity describes the variation of a physical property or dissimilarity in the composition of a physical property. For example, heterogeneity in mineralogy is the variation of mineral components in the rock, heterogeneity in permeability implies a certain variation of permeability in the formation. Subsurface heterogeneities include differences, dissimilarities and variability in geologic entities, petrophysical properties, and fluids.

RESERVOIR QUALITY

Reservoir quality describes hydrocarbon potential, amount of hydrocarbon in place, and hydrocarbon deliverability of the rock formation. The important variables in reservoir quality include TOC, thermal maturity, organic matter, mineralogical composition, lithology, effective porosity, fluid saturations, permeability, and formation pressure (Passey et al., 2010).

19.2 ROCK MECHANICS-RELATED TERMINOLOGY

COMPLETION QUALITY

Completion quality describes stimulation potential or the ability to create and maintain fracture surface area. In other words, it describes the rock quality for completing an oil or gas well, notably for hydraulic fracturing. Completion quality is highly dependent on geomechanical properties (Waters et al., 2011) and mineral composition of the formation, including rock fracturability, in situ stress regime, and the presence and characteristics of natural fractures.

BRITTLENESS INDEX

Brittleness index is a completion quality index and it has been used as a proxy parameter for completion quality. Similar to completion quality, brittleness index is a relatively loose terminology and a number of definition have been proposed (Wang and Gale, 2009). Notice, however, that there are other more elementary mechanical parameters, such as Young's modulus and Poisson's ratio, that are important. The brittleness index is correlated to these mechanical properties.

STRESS

Stress is defined as the force per unit area; it is an expression of the internal forces that neighboring particles of a continuous material exert on each other. Mathematically, stress is defined as the force per unit area, in which the force is applied to a body with a cross-sectional area.

OVERBURDEN STRESS, MAXIMUM, AND MINIMUM HORIZONTAL STRESSES

The in situ stress condition of a rock particle in the reservoir can be described by three principal stresses acting in compression. These three principal stresses are perpendicular to each other, with the overburden applied in the direction of the gravity forces (vertical direction). The other two principal stresses are in a horizontal plane perpendicular to the direction of the overburden; the lower magnitude stress is the minimum horizontal stress and the higher magnitude stress is the maximum horizontal stress.

IN SITU STRESS

In situ stress is the stress of a subsurface formation. It is commonly characterized by three principal stresses that are perpendicular: overburden stress, maximum horizontal stress, and minimum horizontal stress. In other orientations, stress components devolve into a tensor characterized by directionality and magnitude, and they include normal stress and shear stress components relative to the surface upon which the stress is acting.

STRAIN

Strain is a normalized measure of deformation that represents the displacement between particles in the body relative to a reference length.

POISSON'S RATIO

Poisson's ratio is a measure of the ability of a material to expand in the perpendicular directions to the force applied to the material. When a material is compressed, it tends to expand in the other directions, and this phenomenon is termed the Poisson effect. The Poisson's ratio is defined as the fraction of expansion divided by the fraction of compression. Poisson ratio of rocks ranges typically between 0.2 and 0.35. Because it represents an expansion characteristic, it is an important parameter for determining formation closure stress.

YOUNG'S MODULUS

Young's modulus is a measure of the stiffness of an elastic material, and it is defined as the ratio of stress to strain. Rocks with low Young's modulus tend to be ductile and rocks with high Young's modulus tend to be brittle. Generally, brittle rocks have better completion quality and are better hydraulic fracturing targets.

In practice, the estimated Young's modulus from sonic logs is termed dynamic Young's modulus. A static modulus is directly measured in a deformational experiment. Static moduli are often used in in-situ stress analysis and wellbore stability application to evaluate pore pressure, possible breakouts, and tectonic stress distribution. Static Young's modulus and Poisson ratio are important parameters in hydraulic fracture design.

TENSILE STRENGTH

Tensile strength is a measure of the force required to pull a material to the point where it breaks (Zoback, 2007). In other words, it is the maximum amount of tensile stress that the material can take before failure. This is an important parameter for hydraulic fracturing rocks; hydraulic pressure has to overcome the rocks' tensile strength in order to create fractures.

PORE PRESSURE

Pore pressure is the pressure of the fluid in the pore space of a subsurface formation. Pore pressure has an impact on the state of stress, and thus it is important parameter for developing unconventional resources.

Pore pressure is said to be normally pressured when it is equal to the hydrostatic pressure, underpressured when it is smaller than the hydrostatic pressure, and overpressured when it is greater than the hydrostatic pressure. Overpressure of an unconventional reservoir can be caused by hydrocarbon generation through thermal maturation of kerogen, uplift, and other subsurface processes; it can be an important hydrocarbon production mechanism.

19.3 DRILLING AND COMPLETION-RELATED TERMINOLOGY

COMPLETION

Completion is a process of preparing the drilled well for recovering oil or gas fluids. Completion of a well for unconventional reservoirs may include running the casing, cementing operations, perforating, application of packers, hydraulic fracturing, and flowback/clean-up operations, leading to the running of production tubing string and artificial lift system. The reservoir pressure is an important consideration for all these operations.

STIMULATION

Stimulation is a chemical and/or mechanical process of increasing flow capacity of an oil or gas well. The two most common techniques of stimulation include matrix acidizing and hydraulic fracturing, with hydraulic fracturing being the principal means of stimulation in unconventional reservoirs. Successful stimulation allows a greater flow of hydrocarbon to the wellbore under a similar pressure drawdown.

CASING AND CEMENTING

The installations of casing and cementing are fundamental tasks to provide well integrity. The casing is typically a hollow steel tubular used to line the inside of the drilled hole (wellbore). When cemented in place, it is an essential process to protect groundwater and aquifers in drilling a well because it isolates fresh water from the inside of the drilled well. Setting casing strings at relevant depths is also used to protect the drilling process from abnormal pressure or other mechanical issues which might endanger the wellbore.

Cementing is the process of placing a cement sheath around casing strings. The overlapping cementation of two or more strings of casing is the most important process for protection of groundwater.

ANNULUS

Annulus is the space between the casing strings and the drill-hole (wellbore). It may be filled with cement, drilling mud, or air.

HYDRAULIC FRACTURING

Hydraulic fracturing is not a "drilling process". Hydraulic fracturing is used after the drilled hole has been provided with adequate well integrity. The unconventional oil and gas reservoirs are generally 5000–11,000 ft from surface and due to low flowing reservoir properties it generally requires a stimulation to yield effective production rates.

Hydraulic fracturing is the preferred stimulation for unconventional reservoirs, using fracturing fluid, propping material and pressure to create or restore small fractures in a geological formation and thereby to stimulate hydrocarbon production from oil or gas wells. In this process, hydraulic fracturing breaks the reservoir rocks and creates conductive flow paths between the target reservoir and the wellbore. It is a necessary component of developing nearly all unconventional reservoirs economically. In general, a high volume of fluids is pumped into a formation at a pressure sufficiently high to create or restore fractures in the rocks so that hydrocarbon can move from geologic formations to fractures then to wellbores.

Hydraulic fracturing process includes steps and procedures to protect water supplies. Casing and cementing are some of the procedures to isolate the fracturing fluids and hydrocarbons from water supplies.

FRACTURE HALF-LENGTH

A hydraulic fracture is typically assumed to propagate in two sides of the wellbore in the direction of the least resistance, often termed two fracture wings or biwings; the length of the fracture on each side is termed fracture half length. It can be hundreds or thousands of feet long. The propped fracture half-length is the half-length of the fracture that is propped by the proppant, and it is shorter than the hydraulic fracture half-length. But good use of proppant can make the propped fracture half-length approaching the hydraulic fracture half length. The effective fracture half-length is the half-length of the fracture that is effectively contributing to the flow between the reservoir and wellbore. It is smaller than the propped fracture half-length (Figure 19.1).

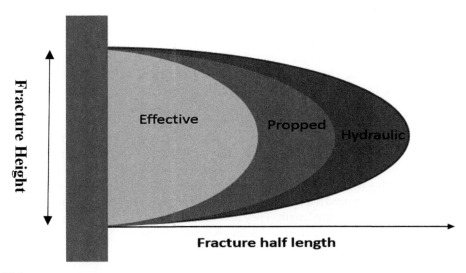

FIGURE 19.1

Illustration of the height and various lengths of a hydraulic fracture.

In many unconventional reservoirs, the hydraulic fracture geometry can be much more complex than simple bi-wing and a more suitable lexicon would describe the geometry in terms of fracture surface area rather than half-length. This complexity is often due to rock fabric characteristics, such as natural fractures, bedding, and other heterogeneous factors within the reservoir.

FRACTURE HEIGHT AND WIDTH

A simple hydraulic fracture is often described by three dimensions: half-length, width, and height. The fracture height is the vertical distance of the fracture (Figure 19.1). The fracture width is the width of the fracture, generally perpendicular to the direction of the fracture length. In practice, neither a hydraulic fracture nor a natural fracture has a perfect prism, but both hydraulic and natural fractures have irregular shapes; fracture half-length, height, and width are used as approximate measures of a fracture dimension.

FRACTURE ORIENTATION

A hydraulic fracture has a tendency to propagate parallel to the maximum principal stress, i.e., perpendicular to the minimum principal stress. In shallow formations, the least principal stress is the overburden stress, and a fracture likely propagates horizontally. In deeper formations, such as development of most unconventional resources, the overburden stress is high, hydraulic fractures tend to propagate vertically, perpendicular to the minimum horizontal stress. When the maximum and minimum horizontal stresses have similar magnitude, fracture orientations tend to change a lot, and rock fabric plays a greater role in creating complex fracture geometry. In highly compressive (overthrust) or strike-slip tectonic settings, greater challenges are often faced in creating effective fracture geometry and much more consideration must be given to geomechanical considerations.

FRACTURING FLUID

Fracturing fluids are typically water-based fluids with a small amount of additives or chemicals (generally less than 1% volume of the fracturing fluid) that are used to treat the subsurface formation to stimulate the flow of oil or gas. The commonly used types of fracturing fluids include slickwater, linear gels, and cross-linked fluids (Montgomery, 2013). Some specialty applications include oil-based fluids, foamed fluids and viscoelastic surfactants (polymer-free fluids).

Fracturing fluid should have stable viscosity during pumping, and following the pumping, it needs to be broken to reduce the viscosity (to essentially water) at the end of fracturing treatment, so as to allow the clean-up of the fluid from the formation prior to production. Fracturing fluids should have a certain degree of viscosity for proppant transport and controlling the fracture net pressure. The viscosity of the fracturing fluid also has an impact on fracture geometry, including width, height, and penetration. Fracturing fluids need to preserve proppant pack conductivity while minimizing formation damage for longer effective fracture length. They also need to control fluid loss from the fracture into the formation so as to maintain sufficient width for proppant placement.

ADDITIVES

An additive is a material that is used to modify fracturing fluid behavior. Some of the common additives for fracturing fluids include buffers (maintaining a desired pH for the fluid), bactericide (mitigating gel degradation), breakers (reducing viscosity, breaking polymer bonds or reducing polymer molecule weight), clay stabilizers (preventing clay swelling and/or destabilization), fluid-loss-preventives, friction reducers, temperature stabilizers, surfactants (reduce interfacial tension), proppant-flowback controllers, nonemulsifying agents (break emulsion between fluid and formation), and proppant-transport enhancers.

Care must be taken when selecting suitable additives for a fracturing fluid formulation to use food-grade or other relatively benign chemical compositions so as to eliminate or effectively reduce potential environmental exposure.

BREAKERS

Breakers are a type of additive in fracturing fluid. The purpose of a breaker at a proper concentration is to increases the retained conductivity of the proppant pack. The typical composition of breakers includes oxidizers or enzymes of which mechanism of action is to break the polymer bonds and reduce the polymer molecular weight under reservoir temperature conditions.

PROPPANT

Proppants are materials that are used to prop the fractures so that the fractures are kept open and thereby oil and gas can flow from the formation into the wellbore at higher flow rates (Holditch, 1979). Proppants are typically riverbank mined and sieved quartz sand grains or factory-made ceramic particles. The main requirements from proppant are uniform grain size with round shape to provide higher proppant pack pore or void space. They should also have good crush resistance to overcome the in situ stresses.

SLICKWATER, WATER FRAC

Slickwater contains mainly water, clay-controlling agents and friction reducer. Water frac has low viscosity, leading to narrow fracture width, and weak proppant transport capability. The advantages of water frac include low cost, and easiness of mixing, recovering and reusing. In certain reservoirs, low-viscosity slickwater or water frac treatments may lead to greater complexity of the fracture geometry, and thus often greater fractured rock flow area, a potentially favorable outcome.

LINEAR GELS OR LINEAR POLYMER GELS

Linear gels are a type of fracturing fluid, composed of water, clay-controlling agents, gelling agents, bactericide and chemical breakers. Similar to water frac, linear gel is a low cost fluid, and its low viscosity generally leads to narrow fracture width.

CROSS-LINKED GELS OR CROSS-LINKED POLYMER GELS

Cross-linked gels contain a chemical agent which binds (or "cross-links") long chain polymer polysaccharide molecules together, resulting in high fluid viscosities. Fracturing polymers are typically composed of guar bean crops which have been ground to a fine powder consistency to facilitate easy hydration of the polysaccharide molecules in water. Cross-linker additives to the hydrated guar polymerinclude borate type, zirconium, or titanium complexes. Borate is the most commonly used cross-linker. It is shear insensitive to the significant turbulence experienced by the fluids while pumping down the casing or tubing string, and can be a time-delayed application so as to limit friction pressure drops in the tubulars. The organometallic cross-linkers are thermally delayed, and normally are the preferred choices for very high temperature applications where the stronger bond provides greater viscosity for proppant transport and fracture width. As the cross-linker increases the viscosity of the fracturing fluid, it increases the fracture width, improves fluid efficiency and proppant transport capability, and otherwise improves fracturing success.

FOAMED FLUIDS

Foamed fluids are created by the addition of nitrogen or carbon dioxide to water-based fracturing fluids so as to create an emulsion of typically 55–85% gas and remainder as liquid. Successful foam generation will allow for good proppant suspension and transport, effective leak-off control from the fracture to the formation, limit the amount of water injected into the formation in the case of water-sensitivity, and bring significant energy for clean-up of the fracturing fluids and kick-off of the well following stimulation.

VISCOELASTIC SURFACTANT GELS

Viscoelastic surfactant gels are fracturing fluids that use nonpolymeric (nonguar or other polymer), low molecular weight, viscoelastic surfactants. The viscoelastic surfactants at threshold concentrations (critical micelle concentration) create large rod-shaped micelle molecules which act to viscosify the formulation for effective proppant transport and creation of fracture width. Once broken, however, the surfactants disperse with an effect of zero residue, higher retained permeability, and minimal damage to the formation. When this type of fluid contacts hydrocarbons, its viscosity decreases substantially, leading to residue-free proppant packs, and thus promotes an efficient hydrocarbon recovery (d'Huteau et al., 2011). Viscoelastic surfactant gels typically display very low friction pressures and excellent proppant transport even at relatively low viscosities on the order of 50–100 cp.

OIL-BASED FLUIDS

Oil-based fluids are often used for water-sensitive formations that may be damaged by water-based fluids. They contain additives for cross-linking and break. Usually they are more expensive than water-based fluids, have more safety requirements, and a higher environmental impact in the event of spill or accident. Their use in unconventional reservoir development has been quite limited.

FRACTURE CONDUCTIVITY

Fracture conductivity is a measure of the property of a particular propped fracture to convey the produced fluids of the well, and is measured in terms of proppant permeability and average propped fracture width (md-ft). The dimensionless fracture conductivity is defined as the ratio of the product of the fracture permeability and fracture width over the product of the formation permeability and fracture half-length. The permeability of the propping agent, sand or ceramics, is typically in hundreds of Darcy without stress applied, but in the subsurface formation, it is reduced greatly because of the stress, proppant embedment, crushing, and other factors such as multiphase fluid flow (gas, oil, or condensate, water).

PAD

The pad consists of a proppant-free portion of the treatment pumped prior to the beginning of proppant addition to the fluid. It is used to initiate and propagate the fracture, generate fracture width, control fluid loss, and otherwise allow for subsequent successful injection of propping agent into the formation. It must be correctly sized based upon reservoir geometry and leak-off considerations typically determined in prefracture injection testing.

PREPAD

A prepad may be pumped prior to the pad in order to cool the tubulars and formation, breakdown and condition the reservoir, and when combined with certain diagnostic tests, allow for data gathering including state-of-stress, leak-off, reservoir deliverability, and pressure.

SLURRY

Slurry is a mixture of fracturing fluid and proppants in a given proportion, it is also called proppant laden fluid.

BLENDER

A blender is a proportioning machine which mixes the water, proppants, and some of the additives used in the fracturing process. As such, it is the key piece of equipment present on all fracturing treatments, and must function properly in order to have a successful treatment.

FLUSH

Following the pumping of the main fracturing treatment, the flush displaces the slurry to typically the top-most perforation in the wellbore, thereby leaving the tubulars clean of any proppant once the treatment is shutdown. It is the last activity in the fracturing pump schedule.

FRACTURE CLOSURE PRESSURE

Fracture closure pressure is the minimum fluid pressure required to initiate the opening of a fracture. Closure pressure is equal to the minimum in situ stress on a layer-basis because the pressure required to open a fracture is the same as the pressure required to overcome the stress in the rock perpendicular to the fracture. When multiple layers are opened by a fracturing treatment, closure pressure will represent the macro impact of all the combined minimum in situ stresses by layer.

NET PRESSURE

Net pressure is the difference between the pressure in the fracture at the entry point into the formation and the closure stress. Classical fracturing models use net pressure to determine fracture width in a linearly elastic fracture model.

PERFORATION

Perforating is the process of creating holes in the casing after the casing has been cemented. The most common perforation technique uses shaped charges. The explosive detonation of these charges creates short tunnels through the casing and cement sheath, and thereby provides hydraulic communications between the wellbore and the reservoir. Perforating for fracturing is a key design criteria for successful treatment placement, including such parameters as perforation diameter, penetration depth, and orientation (phasing) around the wellbore.

19.4 MISCELLANEOUS TERMINOLOGY

PROBABILITY DENSITY FUNCTION

A probability density function (PDF) is a mathematical function that describes the probability of each member of a discrete set or a continuous range of outcomes or possible values of a variable. Common examples of probability density functions are lognormal, normal (Gaussian), exponential, and pareto.

CUMULATIVE PROBABILITY DENSITY FUNCTION

A cumulative probability density function (CDF) is a mathematical function that describes the probability that a variable has a value less than or equal to a specified value. It is the integral (i.e., cumulative) of the PDF, and has values from 0.0 to 1.0.

EXPECTED VALUE

Another term for the expected value is the mathematical expectation. It is mathematically defined as the integral of a random variable with its probability measure. These are theoretical terms. Their experimental counterparts are the mean or average value that is calculated as the probability-weighted sum of sample values.

EXPECTED MONETARY VALUE OR EMV

EMV is the expected value calculated in terms of money, and it is defined as the sum of the products of the probability and monetary value for each possible outcome. It is a measure of reward/benefit. It may also refer to as expected monetary return.

RISK

Risk has several meanings, and is often confused with uncertainty. It is typically used to refer to the consequences of possible outcomes or events, usually undesirable events. Mathematically, it is defined as the product of probability of occurrence of an event and the consequence of that event (Ma and La Pointe, 2011). In developing unconventional resources, one of the biggest risks is the crude oil price (COP) and natural gas price (NGP). Because developing tight formations requires sophisticated technology, the profit margin tends to be lower than developing a conventional reservoir, and thus the projects are more sensitive to the COP and NGP.

OPTIMIZATION

Optimization is a process of finding the optimal solution that minimizes a defined objective function. Typically the objective function is defined, either as a linear or non-linear combination of several input variables and their associated weightings. Thus, the optimal solution typically accounts for all the input variables and honors them to a certain degree, but not 100% unless all the input variables are correlated perfectly. For example, production optimization can be based on maximizing the production for certain period of time giving the input variables including reservoir quality, and completion quality and effort, but economically the optimal solution may not be the highest production, but the highest NPV. Optimization may be also time-dependent; the optimal solution for the highest NPV for 1-year hydrocarbon production may not be the optimal NPV solution for 5 or 10-year production.

VALUE OF INFORMATION

Value of information refers to the reduction of uncertainty as more information becomes available. Developing an unconventional resource has uncertainty and requires evaluation of the play and its uncertainty. Acquisition of different sources of data enables a better reservoir characterization and reducing the uncertainty as a result of using more data. The reduction of uncertainty can narrow the range of possible outcomes and aid in improving decision-making.

SUPPORT EFFECT OR SCALE EFFECT

Support effect is a term from geostatistics to describe the effects of different volumes/sizes in the data. It can also termed scale effect, but scale effect has a broader connotation. In resource evaluation, a core plug may be of size less than a few cubic millimeters while a grid cell in a three-dimensional reservoir model may be in the order of 25 m by 25 m by 1 m. Because variance is related to sample size, proper mathematical treatment of variance for data obtained from different size samples or for use in estimating larger or smaller rock volumes is important. Proper mathematical representation of the support effect is very important in unconventional resource evaluation, as samples or data obtained from core, logs, and seismic data are used for integrated reservoir characterization and resource evaluation.

VARIOGRAM

A variogram, sometimes referred to as a semivariogram, describes the degree of difference of a parameter as a function of their relative distance and direction. Mathematically, a variogram value is the spatial variance as a function of distance and direction between two random variables. Thus, a small variogram value implies a similarity between them and vice versa.

KRIGING

Kriging is an interpolation method from geostatistics. Kriging is a best linear unbiased estimator. Kriging interpolation has a smoothing effect as the kriging estimator reduces the variance in the data. There are various forms of kriging algorithms.

MONTE CARLO SAMPLING OR SIMULATION

Monte Carlo sampling or simulation refers to making probability-weighted random draws from a population of possible outcomes. The probability of randomly selecting a value is proportional to the probability of the value or outcome. Monte Carlo simulation involves making a sufficient number of random draws from the population in order to characterize the population to a specified degree of confidence. It is a very general procedure that can be used to simulate a population when there are many input variables and relations between the variables may be complex. The primary difficulty with Monte Carlo simulation is that a large number of output values may need to be simulated in order to adequately characterize the uncertainty of a complex system, and it may not be computationally feasible to use these values as input to a more complex mathematical model such as a dynamic reservoir simulator.

STOCHASTIC SIMULATION AND SEQUENTIAL GAUSSIAN SIMULATION

Stochastic simulation is a mathematical process that simulates a phenomenon using stochastic theory, instead of using an analytical or deterministic function. Sequential Gaussian simulation is a form of stochastic simulation in which the phenomenon is simulated sequentially while the population of values is assumed to conform to a Gaussian probability distribution. If the phenomenon being simulated is not Gaussian, a normal score transform, which transforms the data into a Gaussian distribution, should be applied before the simulation.

REFERENCES

d'Huteau, E., et al., 2011. Open-channel fracturing—a fast track to production. Oilfield Review 23 (3), 4–17.
Holditch, S.A., 1979. Criteria for Propping Agent Selection. Norton Co, Dallas, Texas.
Jarvie, D.M., 1991. Total organic carbon (TOC) analysis. In: Merrill, R.K. (Ed.), AAPG Treatise of Petroleum Geology, 113–118.
Law, B.E., Spencer, C.W., 1989. Geology of Tight Gas Reservoirs in Pinedale Anticline Area, Wyoming, and Multiwell Experiment Site, Colorado. US Geologic Survey Bulletin 1886.
Ma, Y.Z., La Point, 2011. Uncertainty Analysis and Reservoir Modeling. AAPG Memoir 96.
Montgomery, C., 2013. Fracturing fluids. In: Bunger, A.P., McLennan, J., Jeffrey, R. (Eds.), Effective and Sustainable Hydraulic Fracturing. INTECH, p. 3–24.
Passey, Q.R., et al., June 8–10, 2010. From Oil-Prone Source Rock to gas-Producing Shale Reservoir—Geologic and Petrophysical Characterization of Unconventional Shale-Gas Reservoirs. Paper SPE 131350 Presented at the CPS/SPE International Oil and Gas Conference and Exhibition, Beijing, China.
Sondergeld, et al., February 23–25, 2010. Petrophysical Considerations in Evaluating and Producing Shale Gas Resources. SPE 131768, Presented at the 2010 SPE Unconventional Gas Conference, Pittsburgh, PA, USA.

Wang, F.P., Gale, J.F., 2009. Screening criteria for shale-gas systems. Gulf Coast Association of Geological Society Transactions 59, 779–793.

Waters, G.A., Lewis, R.E., Bentley, D.C., October 30–November 2, 2011. The Effect of Mechanical Properties Anisotropy in the Generation of Hydraulic Fractures in Organic Shale. Paper SPE 146776 Presented at the SPE ATCE, Denver, Colorado, USA.

Zoback, M.D., 2007. Reservoir Geomechanics, sixth ed. Cambridge University Press.

Index

Note: Page numbers followed by "f" and "t" indicates figures and tables respectively.

A

Abnormal pressures, 408f–409f, 409
Acoustic measurements, 201–202
Additives, 520
Advanced diagnostics
  formation evaluation, 377
  hydraulic fracture modeling, 377
  performance analysis, 377
Anadarko Basin, 451
ANN. *See* Artificial neural network (ANN)
Annulus, 518
Artificial neural network (ANN), 417–418, 418f
Asphaltene chemistry, kerogen structure through, 108–114
  molecular diffusion, 109–110
  polycyclic aromatic hydrocarbons (PAH), 108–111
  single pulse excitation (SPE), 110, 111f
  small-angle X-ray scattering (SAXS), 113, 113f
  Yen–Mullins model, 109
Athabasca field, 496f–497f, 497–498

B

Basin and petroleum system modeling (BPSM), 98–99
  example of, 99–101, 101f
  SARA modeling, 99, 100f
B/FIB/SEM approach. *See* Broad/Focused Ion Beam milling and Scanning Electron Microscopy (B/FIB/SEM) approach
BHFP. *See* Bottom hole flowing pressure (BHFP)
Biot's theory, 206
Blender, 522
Bonner treatment, 252–253, 253f
Bottom hole flowing pressure (BHFP), 489–490
Bowers method, 388
BPSM. *See* Basin and petroleum system modeling (BPSM)
Breakers, 520
Broad/Focused Ion Beam milling and Scanning Electron Microscopy (B/FIB/SEM) approach, 129
Brunauer–Emmett–Teller (BET) model, 140
Bubble-point pressure, 182
b-value, 248–249, 270–273, 272f

C

Calibration, core sampling
  optimized sampling, 339–340
  PCA, 340
  sample selection strategy, 340, 341f
  texture and composition, 338–339
Capillary pressure, water saturation and, 438, 439f
Carbonate-rich rocks, 18
Casing, 518
Catagenesis, 72
Cementing, 518
Central Limit Theorem, 349–350
Clayey, 161–164, 163f
Claystone, 160
Coalbed methane (CBM)
  basin evolution and gas generation, 477–480, 478f–480f
  coalification, 476
  development challenges
    dewatering schedule, 489–490
    permeability enhancement operations, 487–489, 489t
    site/interval selection, 486–487, 487t, 488f
    well completion, 485–486, 485t
  dual porosity-type reservoirs, 476
  OGIP, 57–58, 58f, 58t
  overview, 475–477
  Qinshui Basin, 477
  recovery factors (RF), 62, 62f
  reservoir characterization, 479f, 480–485
    gas saturation, 484
    permeability, 481–484, 484f
    sorption time, 485
    thickness and gas content, 481, 481t–482t, 483f
  TRR, 64, 65f, 65t
Coiled tubing (CT), 467–468
Compensated linear-vector dipole (CLVD), 264–266
Completion process, 25–28, 517
Completion quality, 5–6, 13, 18–19, 19f, 21, 32–33
Continuum models, 135
Cookie-cutter approach, 393
Core sampling
  calibration
    optimized sampling, 339–340
    PCA, 340
    sample selection strategy, 340, 341f
    texture and composition, 338–339
  Central Limit Theorem, 349–350
  core data and standardized normal distribution, 351, 351f
  cumulative distribution function, 350, 350f
  gas-filled porosity, predictions, 344, 345f–346f
  Gaussian distributions, 349–350
  GFP, 344
  global regression model prediction, 347, 348f
  heterogeneity, 336
  linear regression, 335
  log data
    cluster analysis, 338

527

Core sampling (*Continued*)
    Gaussian distributions, 336
      principal component analysis, 336–338
      unconventional gas field, 336–338, 337f
    Mahalanobis Distance, 342
    predicted *vs.* measured effective porosity, 347, 349f, 352f
    rock classification scheme, 351, 352f
    well with two cores, 342, 343f
Coulomb frictional law, 293–294
Cross-linked gels, 521
Cross-linked polymer gels, 521
Cumulative probability density function (CDF), 523
Curse of (high) dimensionality (COD), 155

D

DCA. *See* Digital core analysis (DCA)
2D displacement discontinuity method (2D DDM), 301–302
DEPT. *See* Distortionless enhancement by polarization transfer (DEPT)
Diffuse reflectance infrared Fourier transform spectroscopy (DRIFTS), 80–81
Digital core analysis (DCA), 128–134, 128f–129f
    imaging porous samples, 129
    macroscopic properties, determination of, 132–134, 133f
    modeling pore-scale physicochemical processes, 131–132
    reconstruction, 130–131
Direct wireline, 81–82
Discrete fracture network (DFN), 17
    HFN simulation, 317–318
      base case, 319–321, 320t
      Monte Carlo approach, 318
      natural fracture, 321–325
      stochastic generation, 319
      UFM–UPM workflow, 318, 318f
      uncertainty analysis workflow, 318, 319f
    preexisting natural fractures
      length, 311–313, 312f–313f
      orientation, 309–311, 309f–311f
      spacing, 313–314, 314f–315f
Displacement discontinuity method (DDM), 297
Distortionless enhancement by polarization transfer (DEPT), 103
Double couple (DC), 264–266
Downhole monitoring, 246, 278–279
DRIFTS. *See* Diffuse reflectance infrared Fourier transform spectroscopy (DRIFTS)
Drilling, 22–23, 23f, 24t–25t
D-value, 249, 270–273, 272f
Dynamic modeling, 445–446

E

Eaton method, 388
Elemental capture spectroscopy, 87
Elemental concentration logs, 88
Equations-of-state (EOS), 192–194, 194f
Estimated ultimate recovery (EUR), 55
Expected monetary value (EMV), 523
Expected value, 523
Expulsion, 91–93, 92f, 93t

F

FCI. *See* Fracture complexity index (FCI)
Flowback, 34
Fluid saturations, 437–438, 438f–439f
Fluid selection, 216–217, 217t
Flush, 522
Foamed fluids, 521
Focal mechanisms, 249, 281, 281f
Four-dimensional (4D) signal, 262–263
FPHM. *See* Fracture pressure history match (FPHM)
Frac stage, 258
Fracture azimuth, 251–252, 251f
Fracture closure pressure, 522–523
Fracture complexity index (FCI), 266–269, 269f
Fracture conductivity, 522
Fracture diagnosis, fracture geometry constraints, 364–366
    microseismic observation, 365–366
    minifrac analysis, 365
    Nolte–Smith analysis, 365
    step rate tests, 365
Fracture encounter, 256, 256f
Fracture geometry, complexity, 26–27, 27f
Fracture length, 269, 270f
Fracture permeability, 17
Fracture pressure history match (FPHM), 226–227, 229f
Fracturing fluid, proppant, 27–28

G

Gamma rays (GR), 15, 153–154
Gas-in-place and production, adsorbed gas to, 90–91, 91f
Gas-saturation, 484
Geochemical parameters, transient flow effect on, 96–97, 97f–98f
Geomechanics, 263–264, 265f
Geophones/accelerometers, 246
Globalization, 35–40
    global unconventional resources, 35–36
    national security issues and environmental concerns, 36–37, 38f
    resource estimates, uncertainty and risk in, 38
    water stress, 38–40, 39t
Global unconventional resources, 35–36
GR. *See* Gamma ray (GR)
Granite Wash tight gas reservoir, 449–450, 450f
    basin evolution, 449–450, 450f–452f
    horizontal wells, 463–466, 464f–466f

hydraulic fracturing, 467–468, 467f, 469f
multilateral well, 469–470
production history, 461–463, 463f
reservoir architecture and properties, 456–458, 459f–460f
resources and fluid properties, 458–461, 461f–462f
source rock evaluation, 453f, 454
stratigraphy and depositional facies, 455–456, 457t, 458f
trap and seal, 454–455, 455f

H

Heterogeneity, 336
Hierarchical clustering analysis (HCA), 164
Hole effect, 419
Hooke's law, 201
Hydraulic fracture (HF), 518
    classical fracture propagation models, 362, 364f
    design optimization, 375–376, 376f
    fracture diagnosis and fracture geometry constraints, 364–366
        microseismic observation, 365–366
        minifrac analysis, 365
        Nolte–Smith analysis, 365
        step rate tests, 365
    half-length, 518–519
    height and width, 519
    integrated hydraulic fracture design workflow, 375–379
        hydraulic fracture design optimization, 375–376, 376f
        refracture optimization. *See* Refracture optimization
    linear elastic fracture mechanics, 361–362, 363f
    natural fracture. *See also* Natural fractures (NF)
        blocks with shrinkage cracks, 294, 295f
        Coulomb frictional law, 293–294
        crossing-arresting data, 297, 298f
        DDM, 297
        mechanical interaction, 293–294
        MineHF2D, 296–297
        OpenT crossing model, 296
        Renshaw and Pollard crossing criterion, 294
        role, 293
        single-well microseismic event locations, 294–296, 295f
        treatment parameters, 293
        two-dimensional (2D) elasticity equation, 296–297
    overview, 361–366
    pump schedule design, 362–364
    well performance analysis (WPA), 366–375, 367f
        mechanisms affecting, 367–369. *See* Well performance analysis (WPA); mechanisms affecting
Hydraulic fracture network (HFN), 302–303, 304f, 317–318
    base case, 319–321, 320t
    Monte Carlo approach, 318
    natural fracture
        angle, 324–325, 325f–326f
        length, 321–322, 322f–323f
        spacing, 323–324, 323f–325f
    stochastic generation, 319
    UFM–UPM workflow, 318, 318f
    uncertainty analysis workflow, 318, 319f
Hydraulic fracturing treatment (HFT), 5–6, 27–28, 32–33
    completion quality (CQ), 215, 219–220
    economic and operational considerations, 238–239, 238f
        operational and logistic analyses, 239
    fracture fluid and proppant selections, 216–219
        fluid selection, 216–217, 217t
        proppant selection, 217–219, 218f, 219t, 220f
    optimizing fracture design and completion strategies, 219–227
        adequate fracture model, 223–226, 225f–226f
        estimating fracture properties, 226–227, 228f–230f
        mechanical earth model, calibrating, 220–223, 221f, 222t, 223f
    production modeling, 231–238
        analytical *versus* numerical models, 231–234, 231t, 232f–234f
        managing uncertainty, 236–238
        model applications, 238
        predictive model, consolidating, 234–236, 235f–237f
    reservoir quality (RQ), 215, 219–220
Hydraulic stress gradients, 18
Hydrocarbon
    gas retention efficiency, 97–98
    generation, 91–93, 92f, 93t

I

Inclined heterolithic strata (IHS), 506
Indirect wireline, 81
Induced fracture network, 304–317
Inflow performance relationship (IPR), 231–232
Infrared (IR) spectroscopy, 104–106, 105f
Inorganic geochemical logs, 86–88, 87f
    elemental capture spectroscopy, 87
    elemental concentration logs, 88
Integrated hydraulic fracture design workflow, 375–379
    hydraulic fracture design optimization, 375–376, 376f
    refracture optimization
        candidate selection, 377
        evaluation and calibration, 378–379
        refrac design, 377–378
        refrac execution, 378
International Union of Pure and Applied Chemistry (IUPAC), 128–129, 129f, 139, 139f
In-treatment well monitoring, 262, 262f–263f, 278, 279f
Isolation, 258–260, 258f–259f, 259t, 260f
IUPAC. *See* International Union of Pure and Applied Chemistry (IUPAC)

K

Kerogen analyses, 101–108
 content, 6–7, 9, 13
 elemental analysis, 101–103
 infrared spectroscopy, 104–106, 105f
 nuclear magnetic resonance spectroscopy, 103, 104f
 other methods, 108
 types and preparation, 115–116
 X-ray absorption near-edge structure spectroscopy (XANES), 106–108, 107f
KGD-type fracture geometry, 362, 384
Klinkenberg effect, 439
Kriging interpolation method, 525

L

Langmuir isotherm model, 16
Laser-induced pyrolysis (LIPS), 80
Lattice Boltzmann (LB) models, 131
Leakoff, 34
Lennard–Jones (LJ) particles, 176
Lennard–Jones potential parameters, 183–184
Linear gels, 521
Linear polymer gels, 521
Linear regression, 335
LIPS. *See* Laser-induced pyrolysis (LIPS)
Lithofacies, 412–419
 classification
  cutoff methods, problem of, 413–415, 414f–415f
  determining proportions, 417–418, 418f
  stacking patterns and depositional interpretation, 419, 420f
  wireline logs, mixture decomposition of, 415–417, 416f
 clayey/siliceous and carbonate, 161–164, 163f
 conventional formation evaluation, classification in, 155–157, 156f
 multilevel clustering of, 164, 165f
 scope, 154–155
 shale reservoirs, 151–152, 152f, 160–161, 162f
 three-dimensional modeling of, 419–422, 421f
 tight carbonate reservoirs, 157–160, 158f
 wireline logs, 413, 415–417
  responses, 153–154, 153t, 154f
Log data, core sampling
 cluster analysis, 338
 Gaussian distributions, 336
 principal component analysis, 336–338
 unconventional gas field, 336–338, 337f
Logging while drilling (LWD), 23
Log–log plot, 370, 371f
Lower point bars (LPBs), 506

M

Mass balance, 97–98
Maximum horizontal stress, 222
Measurement while drilling (MWD), 23
Mechanical earth model (MEM), 25
 asset development, geomechanical properties impact, 399–401, 399f, 400t
 Bakken pore pressure example, 388
 calibrating, 220–223, 221f, 222t, 223f
 data, 389–390, 389f
 defined, 199–207
 geomechanical properties/reservoir quality and completion strategy, 401–403, 401f–402f
 hydraulic fracture and production modeling, 390–393, 390f, 393f–394f
 mechanical properties, 200–203
 model validation and calibration, 207
 overpressure system, 388
 pore pressure, 203–204, 204f
 rock strength, 203
 scope of, 389
 stresses
  minimum and maximum horizontal stress, 205–206
  stress direction, 206–207
  vertical stress, 205
 well completions, geomechanical properties impact, 393–398, 395f–398f, 399t
MEM. *See* Mechanical earth model (MEM)
Microseismic event parameters, 250, 250t
Microseismicity, 244
Microseismic monitoring, 32–33
 concepts and background
  applications, 245–246, 245f
  microseismicity, 244
 goal, 243
 history, 273–275
 parameters
  b-value, 248–249
  D-value, 249
  focal mechanisms, 249
  moment magnitude, 248
  signal to noise ratio (SNR), 248
  S/P ratio, 249
  velocity, 247–248
 processing, 246–247, 247f, 275–278
 statistical analysis, 266–273
  b-value and D-value, 270–273, 272f
  depth contribution, 266, 268f
  fracture complexity index (FCI), 266–269, 269f
  fracture length and well spacing, 269, 270f
  moment magnitude *versus* distance, 266, 267f
  S/P ratio, 270

unconventional resource development, 250–273
  different completions, 260–262, 261f
  different fracture fluid, 260, 260f–261f
  fracture azimuth, 251–252, 251f
  fracture encounter, 256, 256f
  geomechanics, 263–264, 265f
  in-treatment well monitoring, 262, 262f–263f
  isolation and overlapping, 258–260, 258f–259f, 259t, 260f
  microseismic event parameters, 250, 250t
  natural fractures, 252–253, 252f–254f
  permanent monitoring, 262–263, 264f
  real time processing and analysis, 254–256, 255f
  refracturing and diversion, 256–257, 257f
  source mechanism, 264–266, 265f
Migration fractionation, 76
MineHF2D, 296–297
Mineralogical composition, 13–15, 14f
Minifrac analysis, 365
Minimum horizontal stress, 222
Molecular diffusion, 109–110
Moment magnitude (M_w), 248
  distance vs., 266, 267f
Monte Carlo sampling/simulation, 525

N

Natural fractures (NF), 252–253, 252f–254f
  HFN
    angle, 324–325, 325f–326f
    length, 321–322, 322f–323f
    spacing, 323–324, 323f–325f
  induced fracture network, 307t
    base case pumping schedule, 306, 306t
    base case reservoir zone properties, 304, 305t
    horizontal stress profiles and perforated interval, 304, 306f
    UFM complex fracture model, 304
  multiple sets, 315–317
Near-wellbore pressure calculations, 391
Negative correlation, 30, 31f
Net pressure, 523
Neutron porosity (NPHI), 157–160
Newtonian fluids, 216
Nolte–Smith analysis, 365
Nonideal gas (NIG) effect, 136
NonNewtonian fluids, 216
Normalized root-mean-square (NRMS), 498–499
Nuclear magnetic resonance spectroscopy, 103, 104f, 430

O

Object-based modeling (OBM), 419–422, 421f
Offset well monitoring, 278
OGIP assessment. See Original gas-in-place (OGIP) assessment

Oil-based fluids, 521–522
Open Hole multistage fracturing system (OHMS), 467–468
OpenT crossing model, 296
Optimization process, 524
Organic geochemical logs, ancillary tools
  organic petrography, 85–86
  TOC vs. S2 plots, 84, 84f
  Van Krevelen diagrams, 83, 83f
Organic lithofacies, 161–164
Organic petrography, 85–86
Organic-rich shales, 15, 76
Original gas-in-place (OGIP) assessment, 57–62
  CBM, 57–58, 58f, 58t
  shale gas, 59–62, 60t, 61f
  tight gas, 59, 59t, 60f
Overlapping, 258–260, 258f–259f, 259t, 260f
Overpressure system, 388

P

Pad, 522
PAH. See Polycyclic aromatic hydrocarbons (PAH)
Parameters, microseismic monitoring
  b-value, 248–249
  D-value, 249
  focal mechanisms, 249
  moment magnitude, 248
  signal to noise ratio (SNR), 248
  S/P ratio, 249
  velocity, 247–248
PCA. See Principal component analysis (PCA)
Peng-Robinson equations-of-state (PR-EOS), 192–193
Pennsylvanian-Permian coals, 481, 481t–482t
Perforating process, 523
Permanent monitoring, 262–263, 264f, 279
Permeability, 16–17, 439–440, 440f, 443–445, 444f
  multiplier, 185–188
Petroleum system analysis
  hydrocarbon generation, 9
  processes, 9
  resource prospecting and ranking, 10–12
  shale gas and oil reservoirs, 9
Petrophysical interpretations, 441
Petrophysical measurements, 391
Petrophysical models, 219–220
PKN model, 362, 383–384
PNN. See Probabilistic neural network (PNN)
Polycyclic aromatic hydrocarbons (PAH), 108–111
Pore-network model (PNM), 132
Pore pressure, 19, 203–204, 204f
Pore proximity
  equations-of-state, modifications to, 192–194, 194f
  fluid properties, pore confinement effects on, 175–183

Pore proximity (*Continued*)
　　alkane critical properties, 175–177, 176f
　　multicomponent mixtures, phase behavior of, 178–183, 178t–179t, 180f–183f
　　nanopores, multicomponent fluid transport in, 183–188, 184t
　　produced fluids, compositional variations in. *See* Produced fluids, compositional variations in
　　producers, impact to, 194–195
　　well drainage areas and productivity, implications on, 188–192
　　numerical simulation model, description of, 190–192, 191f–192f
Pore-scale characterization
　　DCA, gas shale by, 128–134, 128f–129f
　　　imaging porous samples, 129
　　　macroscopic properties, determination of, 132–134, 133f
　　　modeling pore-scale physicochemical processes, 131–132
　　　reconstruction, 130–131
　　predicted gas permeability, aggregated effect on, 143–146, 144f–145f
　　shale pores and pore-network models, gas flow behaviors in, 134–137
　　　apparent gas permeability, behaviors of, 136–137, 137f–138f
　　　gas flow regimes, 134–136, 136t
　　surface adsorption/desorption, 137–143
　　　basics, 138–139, 139f
　　　gas–solid adsorption models, 139–140
　　　heterogeneous multilayer gas adsorption and free gas flow, 140–143, 141f–142f, 142t, 143f
Porosity, 435–437, 436f
Preexisting natural fractures
　　branching/turning of fracture, 292
　　complex fracture network
　　　2D DDM, 301–302
　　　fracture geometry and fluid pressure, 302, 303f
　　　fracture width, 300
　　　global volume balance condition, 301
　　　hydraulic fracture network, 302–303, 304f
　　　mass conservation equation, 299
　　　Newton–Raphson method, 301
　　　OpenT crossing model, 301
　　　P3D model, 299
　　　propagation paths, 302, 303f
　　　rheological behavior, 299–300
　　　stress shadow effect, 301–302, 302f
　　　UFM model, 299
　　complex parallel fractures and offsets, 289, 290f
　　crossing with an offset, 292, 292f
　　DFN
　　　length, 311–313, 312f–313f
　　　orientation, 309–311, 309f–311f
　　　spacing, 313–314, 314f–315f
　　direct crossing, 291
　　fluid penetration, 290–291
　　fracture simulation, 289
　　friction coefficient and fluid viscosity, 307–308, 308f
　　HF and NF
　　　blocks with shrinkage cracks, 294, 295f
　　　Coulomb frictional law, 293–294
　　　crossing-arresting data, 297, 298f
　　　DDM, 297
　　　mechanical interaction, 293–294
　　　MineHF2D, 296–297
　　　multiple sets, 315–317, 316f–317f
　　　OpenT crossing model, 296
　　　Renshaw and Pollard crossing criterion, 294
　　　role, 293
　　　single-well microseismic event locations, 294–296, 295f
　　　treatment parameters, 293
　　　two-dimensional (2D) elasticity equation, 296–297
　　HF arrested by NF, 291
　　induced fracture network, 304–317
　　intersecting natural fractures, 292
　　microseismic measurements, 289
　　shear slip along NF, 292
Preliminary diagnostics, 377
Prepad, 522
Pressure transient analysis (PTA), 232–233
Principal component analysis (PCA), 155, 160, 336–338, 339f, 416–417, 416f, 421–422
Probabilistic neural network (PNN), 503–510, 505f, 506t, 507f–510f
Probability density function (PDF), 523
Produced fluids, compositional variations in
　　black oil case study, 188, 188f–190f
　　synthetic oil case study, 185–188, 186f–187f
Production modeling, 231–238
　　analytical *versus* numerical models, 231–234, 231t, 232f–234f
　　managing uncertainty, 236–238
　　model applications, 238
　　predictive model, consolidating, 234–236, 235f–237f
Proppants, 520
　　distribution, 28
　　selection, 217–219, 218f, 219t, 220f
Pseudo-3-D analytical simulator, 224–225, 225f, 299
Pulsed neutron–spectral γ-ray methods, 82
Pump schedule design, 362–364
PVT model, 175, 177, 179–180, 185, 192, 195

R

Rate transient analysis (RTA), 232–233, 391
Real time processing/analysis, 254–256, 255f

Recovery factors (RF)
   coal bed methane, 62, 62f
   shale gas, 63, 64f
   tight gas, 63, 63f
Red Fork Granite Wash, 458
Refrac candidates, 377, 378f
Refrac design, 377–378
Refrac execution, 378
Refracture optimization, 376–379, 376f
   candidate selection, 377
      advanced diagnostics, 377
      preliminary diagnostics, 377
      refrac candidates, 377, 378f
   evaluation and calibration, 378–379
   refrac design, 377–378
   refrac execution, 378
Refracturing
   diversion, 256–257, 257f
   treatment, 32f, 33
Reservoir properties
   fluid saturations, 437–438, 438f–439f
   permeability, 439–440, 440f
   petrophysical interpretations, 441
   porosity, 435–437, 436f
   three-dimensional modeling, 441–446
      dynamic modeling, 445–446
      static models, constructing, 442–445. *See* Static models, constructing
Reservoir quality (RQ), 5–6, 410
   completion effort, 5–6, 33
   completion quality, 5–6, 13, 33
   evaluation of, 6
   kerogen content, 7, 9
   production, 29, 31–32, 32f
Reservoir-related terminology
   adsorption, 514
   diagenesis/catagenesis and metagenesis, 515
   heterogeneity, 515
   kerogen, 514
   pyrolysis, 515
   reservoir quality, 515
   shale, 513
   source rock, 514
   thermal maturation and maturity, 514–515
   total organic carbon (TOC), 514
   unconventional resources, 513
   vitrinite reflectance, 514
Reservoir rocks, 152
Retention, 91–93, 92f, 93t
RF. *See* Recovery factors (RF)
Risk, 523–524
Rock-Eval pyrolysis, 76–77, 77f

Rock mechanics-related terminology
   brittleness index, 516
   completion quality, 515
   overburden stress/maximum and minimum horizontal stresses, 516
   Poisson's ratio, 516
   pore pressure, 517
   in situ stress, 516
   strain, 516
   stress, 516
   tensile strength, 517
   Young's modulus, 517

S

SARA modeling, 99, 100f
Scale effect, 524
Seismic surveys, 12
Sequential Gaussian simulation, 525
Sequential indicator simulation (SIS), 419–421
Shale gas
   DCA, 128–134, 128f–129f
      imaging porous samples, 129
      macroscopic properties, determination of, 132–134, 133f
      modeling pore-scale physicochemical processes, 131–132
      reconstruction, 130–131
   OGIP, 59–62, 60t, 61f
   pores and pore-network models, gas flow behaviors in, 134–137
      apparent gas permeability, behaviors of, 136–137, 137f–138f
      gas flow regimes, 134–136, 136t
      predicted gas permeability, aggregated effect on, 143–146, 144f–145f
   recovery factors (RF), 63, 64f
   reservoirs, 16
   surface adsorption/desorption, 137–143
      basics, 138–139, 139f
      gas–solid adsorption models, 139–140
      heterogeneous multilayer gas adsorption and free gas flow, 140–143, 141f–142f, 142t, 143f
   TRR, 65–67, 67f, 67t
Shale reservoirs, 151–152, 152f
Signal to noise ratio (SNR), 248
Siliceous, 161–164, 163f
Simpson's paradox, 5–6, 15, 19–20, 29–30, 154–155, 165–166, 358
Single pulse excitation (SPE), 110, 111f
Slickwater treatments, 520–521
Slurry, 522
Small-angle X-ray scattering (SAXS), 113, 113f
Source mechanism, 264–266, 265f, 280–282
Source rock, 407–409

Spontaneous potential (SP) log, 430
S/P ratio, 249, 270
SRV. *See* Stimulated reservoir volume (SRV)
Stable carbon isotope rollover, 93–96, 94f
Static elastic properties, 390
Static models, constructing
    modeling porosity, 442–443, 443f
    water saturation and permeability, 443–445, 444f
Statistical analysis, microseismic monitoring, 266–273
    b-value and D-value, 270–273, 272f
    depth contribution, 266, 268f
    fracture complexity index (FCI), 266–269, 269f
    fracture length and well spacing, 269, 270f
    moment magnitude *versus* distance, 266, 267f
    S/P ratio, 270
Steam-assisted gravity drainage (SAGD)
    Athabasca field, 496f–497f, 497–498
    mapping steam, 500–503, 502f–503f
    reservoir characterization
        probabilistic neural network (PNN), 503–510, 505f, 506t, 507f–510f
    seismic and well data, 498–500, 499f–500f
    time-lapse seismic surveying, 495–497
Step rate tests, 365
Stimulated reservoir volume (SRV), 18
Stimulated rock volume (SRV), 245
Stimulation process, 25–28, 517
Stochastic simulation, 525
Stratified reservoirs, transient linear flow in, 370–374, 372f–374f
    transient permeability average, 371–374, 372f–374f
    uncertainty analysis, 374–375, 375f
Streamline simulation, 445
Stresses, mechanical earth model (MEM)
    minimum and maximum horizontal stress, 205–206
    stress direction, 206–207
    vertical stress, 205
Stress shadow effect, 301–302, 302f
Support effect, 524
Surface adsorption/desorption, 137–143
    basics, 138–139, 139f
    gas–solid adsorption models, 139–140
    heterogeneous multilayer gas adsorption and free gas flow, 140–143, 141f–142f, 142t, 143f
Surface monitoring, 246, 279

T

Tangential momentum accommodation coefficient (TMAC), 136–137
Tarim basin, 38–40
Technically recoverable resources (TRR)
    coal bed methane, 64, 65f, 65t
    recovery factors (RF)
        coal bed methane, 62, 62f
        shale gas, 63, 64f
        tight gas, 63, 63f
    shale gas, 65–67, 67f, 67t
    tight gas, 64, 66f, 66t
Three-dimensional modeling, 441–446
    dynamic modeling, 445–446
    lithofacies, 419–422, 421f
    static models, constructing, 442–445. *See* Static models, constructing
Three Forks-2 bench (TF2), 389–390
Tight carbonate reservoirs, 157–160, 158f
    claystone, 160
    neutron porosity (NPHI), 157–160
    PCA, 160
Tight gas sandstone reservoirs
    background, 405–406
    basin-centered extensive deposits/conventional traps, 406–407, 407f
    drilling/completion and development scenarios, 410–411, 411t–412t
    general properties, 407–410
        abnormal pressures, 408f–409f, 409
        reservoir quality, 410
        source rock, 407–409
        stacking patterns, 409–410
    lithofacies, 412–419
        classification. *See* Lithofacies classification
        three-dimensional modeling of, 419–422, 421f
        wireline logs, 413
    OGIP, 59, 59t, 60f
    overview, 405–411
    petrophysical analysis
        common issues in, 430–431, 431f–434f
        reservoir properties. *See* Reservoir properties
    recovery factors (RF), 63, 63f
    TRR, 64, 66f, 66t
TMAC. *See* Tangential momentum accommodation coefficient (TMAC)
TOC. *See* Total organic carbon (TOC)
Total organic carbon (TOC), 78, 153
    direct wireline, 81–82
    geochemical methods for, 78–81, 79t
    indirect wireline, 81
Tracer, 34
Transformation ratio (TR), 72
Triangular mineralogy plots, 74
Triggered microseismicity, 244
TRR. *See* Technically recoverable resources (TRR)

U

UGRAS. *See* Unconventional gas resource assessment system (UGRAS)
Unconfined compressive strength (UCS), 203
Unconventional fracture model (UFM), 299, 326–327
Unconventional gas resource assessment system (UGRAS)
   basins, types and global distribution of, 55, 55f
   coal bed methane (CBM), 53
   energy information administration (EIA), classification system, 54–55, 54f
   global recoverable unconventional gas resource evaluation, 64–67
      coal bed methane technically recoverable resources, 64, 65f, 65t
      shale gas technically recoverable resources, 65–67, 67f, 67t
      tight gas technically recoverable resources, 64, 66f, 66t
   methodology, 56–57, 57f
   Monte Carlo probabilistic approach, 55–56
   original gas-in-place assessment, 57–62
      CBM, 57–58, 58f, 58t
      shale gas, 59–62, 60t, 61f
      tight gas, 59, 59t, 60f
   petroleum resources management system (PRMS), 54, 54f
   shale gas, 53–54
   technically recoverable resources
      coal bed methane recovery factors, 62, 62f
      shale gas recovery factors, 63, 64f
      tight gas recovery factors, 63, 63f
   tight gas, 53
Unconventional reservoirs
   completion applications for, 211
   drilling applications for
      deviation and azimuth, 210–211
      wellbore stability. *See* Wellbore stability
Unconventional resources, 250–273
   asphaltene chemistry, kerogen structure through, 108–114
      molecular diffusion, 109–110
      polycyclic aromatic hydrocarbons (PAH), 108–111
      single pulse excitation (SPE), 110, 111f
      small-angle X-ray scattering (SAXS), 113, 113f
      Yen–Mullins model, 109
   basin and petroleum system modeling (BPSM), 98–99
      example of, 99–101, 101f
      SARA modeling, 99, 100f
   characteristics, 4–5
   completion and stimulation, 24t–25t, 25–28
      fracture geometry and complexity, 26–27, 27f
      fracturing fluid and proppant, 27–28
      fracturing stages, 28, 28f
   completion quality, 5–6
   conventional resources *vs.*, 73
   different completions, 260–262, 261f
   different fracture fluid, 260, 260f–261f
   drilling, 22–23, 23f, 24t–25t
   evaluation, 12–21
      completion quality, 18–19, 19f
      hydrocarbon saturation and types, 16, 17f
      integrated evaluation, 19–21, 21t, 22f
      mineralogical composition, 13–15, 14f
      permeability, 16–17
      pores, 15–16, 16f
      pressure, 19
   exploration and appraisal, 7–12
      petroleum system analysis and modeling, 7–10, 10f
      resource prospecting and ranking, 10–12, 11t, 12f
   fracture azimuth, 251–252, 251f
   fracture encounter, 256, 256f
   gas-in-place and production, adsorbed gas to, 90–91, 91f
   geochemical parameters, transient flow effect on, 96–97, 97f–98f
   geomechanics, 263–264, 265f
   globalization, 35–40
      global unconventional resources, 35–36
      national security issues and environmental concerns, 36–37, 38f
      resource estimates, uncertainty and risk in, 38
      water stress, 38–40, 39t
   hydrocarbon generation/expulsion and retention, 91–93, 92f, 93t
   inorganic geochemical logs, 86–88, 87f
      elemental capture spectroscopy, 87
      elemental concentration logs, 88
   in-treatment well monitoring, 262, 262f–263f
   isolation and overlapping, 258–260, 258f–259f, 259t, 260f
   mass balance and hydrocarbon gas retention efficiency, 97–98
   methods and parameters in, 7, 8t
   microseismic event parameters, 250, 250t
   natural fractures, 252–253, 252f–254f
   organic geochemical and petrophysical characterization, 76–82
      Rock-Eval pyrolysis, 76–77, 77f
      TOC. *See* Total organic carbon (TOC)
   organic geochemical logs and ancillary tools
      organic petrography, 85–86
      TOC *vs.* S2 plots, 84, 84f
      Van Krevelen diagrams, 83, 83f
   organic matter, subsurface evolution of, 71–73, 72f
   permanent monitoring, 262–263, 264f
   production, 28–35
      artificial lift, 33
      drivers, calibrating, 29–32, 30t, 31f–32f
      heavy oil and production from oil sands, 34
      microseismic monitoring, 32–33
      refracturing treatment, 32f, 33

Unconventional resources (*Continued*)
    tracer/leakoff and flowback, 34
    water management, 35
  real time processing and analysis, 254–256, 255f
  refracturing and diversion, 256–257, 257f
  reservoir quality, 5–6
  source mechanism, 264–266, 265f
  stable carbon isotope rollover, 93–96, 94f
  structural elucidation, kerogen analyses for, 101–108
    elemental analysis, 101–103
    infrared spectroscopy, 104–106, 105f
    nuclear magnetic resonance spectroscopy, 103, 104f
    other methods, 108
    X-ray absorption near-edge structure spectroscopy (XANES), 106–108, 107f
  sweet spots, empirical measures of, 73–76, 75f
  unconventional reservoirs, fluid adsorption in, 88–90, 89t
    quantifying adsorption, 89–90
  Young's modulus, 5–6
Unified fracture design, 384–386

V

Value of information, 524
Van Krevelen diagrams, 83, 83f
Variogram, 524
Vclay. *See* Volume of clay (Vclay)
Velocity, 247–248
Vertical proportion profiles (VPPs), 419
Vertical stress, 205
Viscoelastic surfactant gels, 521
Volume of clay (Vclay), 15, 153–154, 154f

W

Water frac treatments, 520–521
Water saturation (Sw), 443–445, 444f
Water stress, 38–40, 39t
Wellbore damage, 209
Wellbore stability, 207–209, 208f
  damage, 209
  depth of failure, 209, 210f
  Kick, 208
  losses and breakdown, 208
Well performance analysis (WPA)
  mechanisms affecting, 367–369
    comparison, 369, 369f
    fracture complexity, 369, 369f
    pressure-dependent fracture conductivity, 368, 368f
    pressure-dependent permeability, 368, 368f
    stratified reservoirs, transient linear flow in, 370–374, 372f–374f
    transient permeability average, 371–374, 372f–374f
    uncertainty analysis, 374–375, 375f
Well spacing, 269, 270f
Winner-take-all approach, 419
Wireline logs, 413, 415–417
  mixture decomposition of, 415–417, 416f
  responses, 153–154, 153t, 154f
WPA. *See* Well performance analysis (WPA)

X

XANES. *See* X-ray absorption near-edge structure spectroscopy (XANES)
X-ray absorption near-edge structure spectroscopy (XANES), 106–108, 107f
X-ray tomography (XRT), 132–133

Y

Yen–Mullins model, 109, 109f
Young's Modulus, 5–6, 18, 20, 200–201
Yule–Simpson's effect, 5–6, 18, 31